T0210183

Never Pure

Never Pure

*Historical Studies of Science
as if It Was Produced
by People with Bodies,
Situated in Time, Space,
Culture, and Society,
and Struggling for Credibility
and Authority*

STEVEN SHAPIN

The Johns Hopkins University Press

Baltimore

© 2010 The Johns Hopkins University Press
All rights reserved. Published 2010
Printed in the United States of America on acid-free paper
9 8 7 6 5 4 3 2 1

The Johns Hopkins University Press
2715 North Charles Street
Baltimore, Maryland 21218-4363
www.press.jhu.edu

Library of Congress Cataloging-in-Publication Data

Shapin, Steven.
    Never pure : historical studies of science as if it was produced by people with bodies,
situated in time, space, culture, and society, and struggling for credibility and authority /
Steven Shapin.
        p.   cm.
    Includes bibliographical references and index.
    ISBN-13: 978-0-8018-9420-6 (hardcover : alk. paper)
    ISBN-10: 0-8018-9420-4 (hardcover : alk. paper)
    ISBN-13: 978-0-8018-9421-3 (pbk. : alk. paper)
    ISBN-10: 0-8018-9421-2 (pbk. : alk. paper)
    1. Science—Social aspects.   2. Science—History.   I. Title.
    Q175.5.S465 2010
    306.4'5–dc22
    2009023834

A catalog record for this book is available from the British Library.

*Special discounts are available for bulk purchases of this book. For more information, please
contact Special Sales at 410-516-6936 or specialsales@press.jhu.edu.*
The Johns Hopkins University Press uses environmentally friendly book materials, including
recycled text paper that is composed of at least 30 percent post-consumer waste, whenever
possible. All of our book papers are acid-free, and our jackets and covers are printed on
paper with recycled content.

CONTENTS

PREFACE

I am a historian of science, but it was only six years ago that I first had the opportunity to work in a history of science department (at Harvard) or to teach a course in an area anywhere near the object of my historical research and writing. Before that (from 1989 to 2003), I was happily, if awkwardly, placed in a sociology department (at the University of California, San Diego). And, before that, I was in the small interdisciplinary Science Studies Unit at Edinburgh University, where I was the nominal historian but where my colleagues—a philosopher and a sociologist—and I were little concerned with who spoke for which academic discipline. When I arrived to take up that position—my first after getting my doctorate (in the newly renamed University of Pennsylvania department of the history and sociology of science)[1]— Edinburgh already had a historian of science (in the history department), who politely warned me off his territory and who was mollified only when I suggested that I label my undergraduate survey course "The Social History of Science." This, he agreed, would sufficiently differentiate what I was doing from the real thing.

The result of this institutional trajectory, over a period of almost forty years, is a body of writing somewhat more heterogeneous than is typical for historians of science and even for humanist and social scientific academics in general. What I've done is various in terms of subject matter: I have moved about from the study of phrenology and "social aspects" of science in industrializing Britain,[2] to the Scientific Revolution of the seventeenth century,[3] to more recent foci on institutional and moral patterns of the scientific life in late modern America,[4] to the history and sociology of dietetics, food, and eating.[5] I have published in scholarly venues frequented by historians, sociologists, anthropologists, geographers, philosophers, and scholars concerned with policy—but rarely by members of all of these disciplines at the same time. Some of my writing is highly focused on particular empirical materials; some is programmatic (what is usually called "methodological" and "theoretical"); most is empirically specific and programmatic at the same time. I have always thought of history as a reflective practice, and I considered it quite normal that historians should tell their rich and detailed stories about the past and also reflect on how they tell those stories and why they tell the stories they do. (From time to time, I have been told that this sort of thing should be left to some other tribe of academics, usually called "philosophers" or "theorists," but I've never fully understood why that is said.) I have written quite a bit of "high journalism" too—though that work is not represented here—and some of my academic writings have been attempts to reach a larger audience, if not precisely "popular," then "the general educated reader" once better catered for by humanist academics and publishers.[6]

For these reasons, I had thought for a while, and more recently had it suggested to me, that there would be some point in making a selection of articles and contributed chapters and putting them together in book form. On the one hand, readers who knew part of my work often did not know that I had written in other areas or in other idioms; on the other hand, such a collection would be an opportunity to make more visible the sensibilities that connected genres of work often seen as having little to do with each other. What might interest readers also interested me, and I welcomed the opportunity, not just to take a backwards look, but to try to point out what intellectual strands ran through this apparent jumble. The first chapter is an attempt to survey changing sensibilities in the history and social studies of science over the past few decades, and it is both a reflection on what I have done—as represented by the contents of this book—and on what has changed over the past several decades in academic historical and sociological engagements with science. Each scholar is a unique product of his or her times, but we are all products of our time. It is never right for historians to think of themselves alone as outside of history.

The rest of the chapters are lightly edited versions of previously published materials. No significant substantive changes have been made, but the opportunity has been taken to make minor stylistic alterations, to correct typographical errors and expository clumsinesses in originals that were sometimes my own fault and sometimes that of aggressive editing, and to add occasional references to work that appeared after the original pieces.

I thank Jacqueline Wehmueller of the Johns Hopkins University Press for seeing some merit in the idea for this book and for her patience during the time it was being prepared; my Harvard colleague Charles E. Rosenberg for suggesting the project; Chris Phillips for working with me in the time-consuming job of getting these papers in suitable form; and two readers for the Press who were gratifyingly positive about the project.

Chapter 2 was first published in *Perspectives on Science* 3 (1995): 255–275; chapter 3 in *The One Culture? A Conversation about Science,* ed. Jay A. Labinger and Harry Collins (Chicago: University of Chicago Press, 2001), pp. 99–115. Chapter 4 has not previously appeared in English: a somewhat different version was published as "Vorurteilsfreie Wissenschaft und Gute Gesellschaft: Zur Geschichte eines Vorurteil," *Transit: Europäische Revue* 16 (Winter 1998–1999): 51–63 (translated by Bodo Schulze). Chapter 5 originally appeared in Isis 79 (1988): 373–404; chapter 6 in *Social Studies of Science* 14 (1984): 481–520; chapter 7 in *Science in Context* 4 (1991): 191–218; chapter 8 in *History of Science* 29 (1991): 279–327; chapter 9 in *Robert Hooke: New Studies,* ed. Michael Hunter and Simon Schaffer (Woodbridge, Suffolk: The Boydell Press, 1989), pp. 253–285; chapter 10 in *The Science–Industry Nexus: History, Policy, Implications,* Nobel Symposium 123, ed. Karl Grandin, Nina Wormbs, and Sven Widmalm (Canton, MA: Science History Publications, 2004), pp. 337–363; chapter 11 in *Science Incarnate: Historical Embodiments of Natural Knowledge,* ed. Christopher Lawrence and Steven Shapin (Chicago: University of Chicago Press, 1998), pp. 21–50; chapter 12 in *Right Living: An Anglo-American Tradition of Self-Help Medicine and Hygiene,* ed. Charles E. Rosenberg (Baltimore: Johns Hopkins University Press, 2003), pp. 21–58; chapter 13 in *Bulletin of the History of Medicine* 77 (2003): 263–297; chapter 14 in *Social Studies of Science* 31 (2001): 731–769; chapter 15 in *The British Journal for the History of Science* 33 (2000): 131–154; and chapter 16 in *The Handbook of Science and Technology Studies,* 3rd ed., ed. Edward Hackett, Olga Amsterdamska, Michael Lynch, and Judy Wajcman (Cambridge, MA: MIT Press, 2007), pp. 433–448.

I am grateful to all the original publishers for permission to use these materials here.

Never Pure

CHAPTER I

Lowering the Tone
in the History of Science

A Noble Calling

Some years ago, a friend summed up what it was I had supposedly accomplished over almost forty years of writing in the history of science. He said that I had *lowered the tone*. Lowering the tone, I've come to think, is a positive achievement in our field—but it's an achievement that belongs to the collective and not to any one individual. So my task is to figure out what my friend might have meant when he said I'd lowered the tone, then to share the credit by saying something about how tone-lowering has been a historically situated response to the times in which we live and changes in our objects of study, then, finally and very briefly, to say why this lowering of tone has some intellectual virtue associated with it.

It was several years ago that I was brought face to face with what it meant to lower the tone. I was reading *The New Yorker,* and there was a story in it by Woody Allen.[1] Posing as a historian, he said he'd come across a hitherto unknown text—*Friedrich Nietzsche's Diet Book.* "Who would have thought," Allen asked, that such a book existed? "Is there a relationship between a healthy regimen and creative genius? . . . The powerful will always lunch on rich foods, well seasoned with heavy sauces, while the weak peck away at wheat germ and tofu, convinced that their suffering will earn them a reward in an afterlife where grilled lamb chops are all the rage. But if the afterlife is, as I assert, an eternal recurrence of this life, then the meek must dine in perpetuity on low carbs."

I don't know what I was most miffed at. Was it that a profound intellectual argument about feeding and thinking had been reduced to a *shtick*? Lesley Chamberlain some years ago wrote a nice piece in *The Times Literary Supplement* about Nietzsche's views on Liebig's Beef Extract[2]—he liked it a lot—and I'd published an essay years before on the dietetics of pure reason, including brief remarks on the eating habits of Isaac Newton, René Descartes, Immanuel Kant, and Ludwig Wittgenstein. I'd even helped myself to Chamberlain's quotation of Nietzsche: "No more greasy, stodgy, beer-washed idealistic Christian German food for me! I shall curl up with gut pain, vomit if you don't give me Italian vegetables."[3] Turin supplied food for the philosopher's body and soul: "I never had any idea," Nietzsche wrote, "what either meat or vegetables or any of these Italian dishes *can* be like . . . Today, for example, the most tender *ossobuchi*—God knows how one says it in German!—the meat on the bones, where the glorious marrow is. Also *broccoli* cooked unbelievably well, and, to start with, the most tender macaroni."[4] And I'd meant the piece to be taken seriously. (Sort of a no-chicken non-joke.) Or was I miffed because I didn't get a chance to make the joke myself, since *The New Yorker* pays rather better than academic publishers? No, the problem is not the opposition between the humor of inversion and intellectual respectability: Woody Allen's story about the Nietzsche diet book versus proper academic seriousness. The very idea of the Nietzsche diet book may be profound, but it *is* funny.

It's funny in pretty much the same way that Monty Python's *Life of Brian* is very funny. The thing about the Sermon on the Mount was poor acoustics. What did he say? asked a Hebrew standing at the bottom of the Mount: "Blessed are the cheesemakers? What's so special about the cheesemakers?" The response was one of the earliest instances of New Testament hermeneutics: "Well, obviously it's not meant to be taken literally; it refers to any manufacturers of dairy products."[5] The Pythons' British joke was to imagine that Jesus was making a Yiddish joke. Now *that's* lowering the tone. Take something very high and juxtapose it to something very low—the sacred and the propane. That's either funny or treasonable, laughable or hangable. The German sociologist Ralf Dahrendorf once wrote that the role of the intellectual historically resembled that of the court jester: the condition of speaking Truth to Power was, if not "smiling when you said that," then at least not being in a position to be taken very seriously.[6] Power, as we now well understand from recent events, has a very dim view of Inconvenient Truths. And so at the core of religion—or at least a religion of belief as opposed to one of law—there is mystery—truths so incredible that you have to make an enormous effort to believe them, things so impossible that you *cannot* laugh at them. Life after death; the Trinity; the Virgin Birth. I suppose that's why Judaism never caught on: no sense of humor.

And while we're on the subject of the sacred, we might as well talk about science. It is a commonplace, or it certainly was a commonplace, in the late nineteenth and early twentieth centuries to say that science was the new religion, that it had supplanted Christianity or, at least, that it had succeeded to the cultural authority that religion once enjoyed. And so in the mid-twentieth century there appeared influential books with such titles as *Science Is a Sacred Cow* and *The New Brahmins*.[7] But the way for these was paved by the aggressive polemics of Victorian Scientific Naturalists and those appropriating Scientific Naturalism in defense of the secular control of education. For John William Draper and Andrew Dickson White, the warfare between science and religion, or at least between science and "dogmatic theology," was one between "the expansive force of the human intellect" and the obscurantism of restrictive religion. And it was one that religion was taken to be inevitably and rapidly losing: "The time approaches when men must take their choice between quiescent, immobile faith and ever-advancing Science." Faith belongs to the past; Science to the bright future: "Science, which is incessantly scattering its material blessings in the pathway of life, elevating the lot of man in this world, and unifying the human race."[8] Not just material benefits but moral: from warlike despotism to peaceable human freedom.

One need not say—and most advocates of science in fact did *not* say— that science was the New Religion. You could hardly say such a thing if you wanted clearly to juxtapose Reason and Faith. What one *could* say was that cultural authority had passed from religious to secular institutions, and some said that much of the moral authority of the priest had similarly passed to the scientist. That was the opinion of the great follower of Auguste Comte, and the founder of our discipline, Harvard's George Sarton. Comte's, and Sarton's, moral vision was one in which science drives history forward, in which science represents humankind's highest achievement, and in which science ultimately frees humankind from its historical shackles. Secular science is shot through with moral significance. The Naturalistic Fallacy noted that you could not logically move from Is to Ought, from description to prescription, but nevertheless the so-called triumph of science over religion signaled for many commentators a shift in the locus of *moral* authority.

As science was humanity's highest and noblest achievement, so the history of science was a celebration of what had been and remained best in human culture. Celebrating science was celebrating the small number of people who had made authentic and lasting discoveries. There were many drones in science but few heroes, and the heroes were the ones who count. The history of science, Sarton said, is "largely the history of a few individuals."[9] This is not like, say, political history, where collective action is more widely considered to be the name of the game. The historian's job—or at least one major job—was to praise these few "famous men"—praising famous men is biblical[10]—and even to establish who deserved lasting credit and fame by sifting and evaluating a range of contributions "to establish the relative truth and relative novelty of scientific ideas."[11]

It was *right* for the historian to be a hagiographer: "above all," Sarton wrote, "we must celebrate heroism whenever we come across it. The heroic scientist adds to the grandeur and beauty of every man's existence."[12] Although Sarton at times observed that scientists had the full range of human vices as well as virtues, at other times he insisted that great scientists represented fallible human nature at its highest stage of development: "truth itself is a goal comparable with sanctity . . . The disinterested and fearless search of truth is the noblest human vocation."[13] Science is "the very anchor of our philosophy, of our morality, of our faith," and it was our proper calling as historians to make that foundational role visible to the wider culture.[14]

Given all this, the possibilities of tone-lowering in writing about the history of science were rich and various, probably uniquely so. It is far easier to lower the tone in the history of science than it is in any other area of historical inquiry. Indeed, the thrust of much social and cultural history in recent years has been explicitly to *raise* the tone of what had once been deemed low: one thinks of the history of lay knowledge, of the history of medicine "from the patient's point of view," of military history in the manner of John Keegan and Richard Holmes, and, notably, of the history of the body, of women, and of whatever counts as "the other." We accept, largely as a matter of course, that the historian's job is not to celebrate—it was more than seventy-five years ago that the American medievalist Charles Homer Haskins said that no historian's task was "to distribute medals for modernity"[15]—and so what the anthropologists call "charitable interpretation" is *just what we do*. But it's understandable if we assume, at the same time, that the rich need no charity. In fact, that what they need is a good boxing round the ears. Hence the popularity, even in forms of biography far removed from science, of equating historical virtuosity with exposé.

There are, however, problems with the view that science replaced religion, and I will return to those at the end. Wherever and whenever it was maintained that science had rightly assumed the cultural and moral authority of religion, the forms of heresy became clear. What might you say about science that would—from this point of view—count as heretical? I can think of many things, but here is a selective list:

- You could say that science happens within, not outside of, historical time, that it has a deep historicity, and that whatever transcendence it possesses is itself a historical accomplishment.
- You could say that science similarly belongs to place, that it bears the marks of the places where it is produced and through which it is transmitted, and that whatever appearance of placelessness it possesses is itself a spatially grounded phenomenon.
- You could say that science is not one, indivisible, and unified, but that *the sciences* are many, diverse, and disunified.
- You could juxtapose ideas of Method and those of genius, the former itself a tone-lowering gesture, as Richard Yeo has taught us,[16] or you could go further and say that there is *no* single, coherent, and effective Scientific Method that does the work that genius was once supposed to do, even that there are no supposedly special cognitive capacities found in science that are not found in other technical practices or in the routines of everyday life.
- You could say that scientists are morally and constitutionally diverse specimens of humankind, that extraordinarily reliable knowledge has been produced by morally and cognitively ordinary people, and, further, that the ordinariness of *individual* scientists was not effectively repaired by any special virtues said to attach to their *communal* way of life.
- You could say that Truth (in any precise philosophical sense) is not a product of science, or that it is not a unique product. Or you could say that the historian is not properly concerned with Truth but with credibility, with whatever it is that *counts as Truth* in a range of historical settings.
- You could say that science is not pure thought but that it is *practice,* that the hand is as important as the head, or even that the head follows the hand.[17]
- You could say that the making and warranting of scientific knowledge are *performances,* that those producing scientific knowledge can and do use a full range of cultural resources to produce these performances, and that these include displaying the marks of integrity and entitlement: expertise, to be sure, but also the signs of dedication and selflessness. The very idea of disembodied knowledge thus becomes a bodily performance, and Newton's, as well as Nietzsche's, diet assumes pertinence, food for thought.

We should understand, of course, that all of these heresies, all these lowerings of tone, are not mere possibilities; they amount to a short-list of the leading edges of change in the historical understanding of science over the past several decades, to some extent in philosophical engagements with science, and more strongly in sociological studies of science. They've become—at least most of them and to some extent—so accepted as just-the-way-we-do-things, and just-what-it-is-to-do-history-of-science, that it's sometimes hard to appreciate how much has changed and how quickly, and it's sometimes hard to recognize the heresy—unless and until weird outbreaks like the so-called Science Wars dramatically remind us. I will say little more here about all of these heresies, and I will draw out some detail for just a few. Then I'll try to say something about *how it is* we've come to write about science as we do, and, finally, why there's a modest sort of virtue in all this lowering of the tone.

The first thing to get out of the way is any idea that these heresies were devised by radical sociologists, or indeed by anyone necessarily concerned to *achieve* a tone-lowering effect, denigration or a reduced estimation of the value of scientific knowledge. So let's start with the historicity of science, remembering that it was the claims of High Criticism about the historicity of Scripture that had such an explosive impact on nineteenth-century intellectual life. (The late twentieth-century alleged "wars" between science and sociology had their parallel in late nineteenth-century wars *within* theology.)[18]

Thomas Kuhn said that the Eureka moment for him came when he looked out the window of his Harvard rooms and realized that Aristotelian physics was as wrong as wrong could be, but that it *worked,* that it was coherent. Past science had its historical integrity, and the task of the historian was not to celebrate its contribution to the future but to describe and interpret its historical situatedness. It was an insight he almost certainly owed not to any sociologist of science—there were few such academics around then—but to the philosopher-historian Alexandre Koyré.[19] Koyré was both enormously exciting—imagine: science as an authentically historical phenomenon—but the radicalism of his work was partly masked by his later recruitment as the Hammer of the Marxists. And so an authentically, and radically, historical sensibility about scientific thought was a marked feature of the new "internalism" associated with the work of such great post–World War II historians of science as A. Rupert and Marie Boas Hall, I. Bernard Cohen, Richard S. Westfall.[20] The standards by which historians should assess past scientific work were not those of the present but those of the pertinent past.

The most innovative, and provocative, work of the 1960s that normalized science as a historical object emerged from historians of scientific ideas at Leeds University or associated with them, including J. R. Ravetz, J. E. Mc-Guire, P. M. Rattansi, Charles Schmitt, and Charles Webster. This was heady stuff, for example, interpreting early modern science, in its historical specificity, as a rich brew of nature study, millenarian Christian religion, mysticism, neo-Platonic philosophy, alchemy, and social Utopianism.[21]

That is one source from which Kuhn's work sprang, and a consequence of saying that science was a cluster of paradigmatic practices was a matter-of-fact acceptance of its disunity. If God is one, unchanging, unitary, and universal, then so is God's Truth, and so is science insofar as it is figured as the New Religion. For many historians of Sarton's generation that was indeed the case. Even as Kuhn's unity-shattering notion of multiple scientific paradigms appeared in 1962 in the *International Encyclopedia of Unified Science,* its Vienna Circle editors evidently neither noticed nor cared about the subversive effect Kuhn's views had on the "unity of science."[22] (The obsession with scientific unity that had its heyday in the early to mid-twentieth century has a bit of the Owl of Minerva feel about it: a systematic search for the grounds of unity was accompanied by a diminished concern among *scientists* about Science as a Whole and their general acceptance of the facts of its accelerating specialization and differentiation.) Historians of science have not, for the most part, had much to say about scientific unity, though the New Historicist impulse to assess past science according to past cultural concerns has had a radically destabilizing effect on notions of that unity over time. It has been left to philosophers—previously much invested in theories of the conceptual or methodological unity of science—to write books identifying the facts and implications of scientific disunity. (One thinks of work by John Dupré, Jerry Fodor, Nancy Cartwright, and Alexander Rosenberg, as well as Peter Galison's co-edited historical collection.)[23]

And now, while historians of science are institutionally content to be located in departments of the history of science, and to be published in history of science journals, those of us who teach introductory survey courses usually have to tell our students that "science" is not a self-evident historical category, that early modern "natural philosophy" was a different thing than "mathematics," that the "Scientific Method" is, and always has been, subject to diverse construals, and even that what counted (and counts) as "the mechanical philosophy" or "the experimental philosophy" varied enormously.[24]

For many years, perhaps even since the 1950s, historians of science stopped writing books called "the history of science," not because they had any systematic argument to make about disunity but because they no longer felt at home with the major narratives that had once given their subject matter its integrity and that had supposedly propelled science forward through history. That situation has recently showed signs of being remedied—possibly in response to publishers' pressure to produce works of greater scope, span, and saleability—but the synthetic responses produced so far display an admirable edginess. They are typically collections of case studies, fascinating in their serial individuality, but either bracketing the question of the integrity of science or turning topic into resource by injecting curiosity about ideas of scientific unity as a historical product:[25] so to speak, "There is no such thing as science and this is a history of it."

From a pertinent point of view, this general silence about the overall identity of our subject matter is one of the crowning achievements of our field. We feel we're *right* to identify the historical specificity and heterogeneity of whatever might count as science; we feel, even if we rarely celebrate it, that this specificity is a sign that we've "made progress," but, under another description, that same sensitivity to specificity and silence about "science" is a disciplinary dirty secret. As we get, we like to think, better and better at doing the history of science, we know less and less about what makes it science and not some other differently designated form of culture.[26] The lowering of tone here is probably more apparent to those who are not members of our tribe than those who are: integrity and value go together. It is widely said that if truth has many faces, not one is worthy of respect.

The tone has also been lowered by our burgeoning fascination with the embodiment of science, with its "personae," and with its performative aspects. Isaac Newton wrote in the General Scholium to the *Principia* that it was only by way of "allegory" that we say that God sees, speaks, laughs, loves, hates, and desires, and anyone who took such divine capacities literally was an idolater.[27] At about the same time, Newton wrote that the Marquis de l'Hôpital wanted to know much the same about Newton: "what color is his hair? Does he eat & drink & sleep? Is he like other men?"[28] Newton's greatest modern biographer knew exactly what l'Hôpital was asking: "He has become for me," Richard Westfall wrote, "wholly other . . . a man not finally reducible to the criteria by which we comprehend our fellow beings."[29]

Now we have an appetite for knowing everything we can about scientists' way of life—we can call it *habitus* if we're feeling in need of a product upgrade. And as we find out about, and write about, such things, we're conscious that this too is going against the historical grain. Nineteenth-century sensibilities, in part reflecting a drift from ideas of genius to ideas of Method, deflected attention away from interest in *who the scientist was*. In 1845, the Scottish politician and man of letters Henry Brougham wrote that "when the studies of a philosopher"—and in this context Brougham included the natural philosopher—"and especially of a mathematician, have been described, his discoveries recorded, and his writings considered, his history has been written. There is little else to say of such a man: his private life is generally uninteresting and unvaried."[30] In the same spirit, T. H. Huxley later wrote about an apocryphal Babylonian philosopher: "Happily Zadig is in the position of a great many other philosophers. What he was like when he was in the flesh, indeed whether he existed at all, are matters of no great consequence. What we care about in a light is that it shows the way, not whether it is lamp or candle, tallow or wax."[31] Claude Bernard insisted on the irrelevance of the individual to the practice of science: "Art is I, Science is We."[32] The supposed elimination of what Thorstein Veblen called the "personal equation" from science also eliminated any *substantive* reason to tell stories about who scientists were.[33] Hagiography, of course, could carry on, but without any significant association to the production or warranting of knowledge, without any epistemological bite.

If our recent fascination with, as the British political historian Sir Lewis Namier put it, "who the guys were" counts as a lowering of the tone,[34] we have our justifications for it. First, attention to scientists' *bodies* is a feature of our increasing interest in scientific practice, in itself a tone-lowering move against the background of a contemplative conception of science as a transcendently intellectual enterprise, generating knowledge disembodied in its outcome and in its mode of production. We now want to know about the sharpness of astronomers' vision, the dexterity of experimentalists' hands, the acuity of chemists' olfactory sense. Attention to tacit knowledge has made us curious about touch: *Fingerspitzengefühl,* even, and especially, if that too is a lowering of an idealist and rationalist tone.[35]

We have wanted to know about the embodied practices of securing and maintaining credibility. How did who you were figure in assessing the worth of what you said, even if that interest also was a lowering of the tone set by some sociologists' mid-twentieth-century insistence on science as no respecter of persons? We have wanted, for similar reasons, to know about scientists' social standing, their manner of living, whether they lived in private or public spaces and how they moved about in the course of the day. *Who was Charles Darwin?* And how did his life at Down House figure in stipulations about the integrity, worth, and consequences of his theory? What do we learn about Darwin when we read that his "guts were noisy and smelly" or have quoted to us his own anxiety that excitement "brings on such dreadful flatulence that in fact I can go no where"?[36] We—and perhaps more among nonacademic than academic writers on science—have wanted to know about scientists' sexuality, or lack thereof. And—here I have to claim a share of the blame—we have been interested in their *diet*. The answer to l'Hôpital's question—did he eat?—would help him judge whether Newton's knowledge was divine or mortal. Why else has the story about Newton's chicken—that is, forgetting whether or not he had eaten it—continued in circulation for more than three hundred years?[37] Did our subjects, as Rebecca Herzig has written, "suffer for science,"[38] and, if so, what did their display of suffering signify for the status and worth of scientific knowledge and for the nature of the scientist's vocation? In another cultural setting, concern with embodied persons could be understood as quite the opposite of tone-*lowering:* nineteenth- and early twentieth-century hagiographical traditions ran parallel with insistence on scientific impersonality. But, for us, pretty much all of this has just been part of writing about science as a human endeavor through and through. It's the late modern normal.

So there are many ways in which the tone has been lowered in writing the history of science. But it would be claiming too much if modern historians took sole credit for that. The tone had *already* been lowered for them, and the academic history of science has been more in this respect a response to cultural and social changes than an inventor of new attitudes. First, there is some sense—only some, but significant enough—in which science was being transformed from a sacred into a secular enterprise from around the middle of the nineteenth century. Even while acquiring enormous social authority, including some of the authority that had been exercised by religious institutions, many scientists insisted that they were not, if they ever were, "priests of nature," and that no moral consequences flowed from the investigation of natural phenomena. Max Weber may have gone a bit far when he described the cultural world of 1918 as being "disenchanted" or when he claimed that only certain "big children" then still believed that science contained any lessons about how one ought to live one's life.[39] But insofar as it was accepted that one could not logically move from an "is" to an "ought," from science to morals, scientists sought to divest themselves of moral authority. After all, it was one thing to study God's Book of Nature, quite another to document the pretty designs accidentally produced by atoms purposelessly bashing into one another. The former had the capacity to give moral uplift to those who studied nature, the latter had none. Secularization and the acceptance of the Naturalistic Fallacy were tone-lowering processes.

Second, the end of the nineteenth century and the early twentieth century witnessed a range of philosophical movements, some of them embraced by scientists, that were either skeptical of the notion of scientific Truth, or of certain absolutist conceptions of Truth, or, more generally, of whether science should be making metaphysical claims, about, for example, "correspondence" and "ultimate realities." These movements go by various names, and in other contexts it would be important to distinguish between them: phenomenalism, operationalism, positivism, conventionalism, and, above all, pragmatism. But equally important is what they have in common: each aimed to sever the links that bound early modern natural philosophy to religion by way of metaphysics and notions of God's Truth. Just as the Scientific Naturalism of the late nineteenth century *lowered the case* of "nature," so all of these characterizations of the quality and character of scientific knowledge *lowered the case* of "truth." And some, indeed, quite explicitly identified the metaphysical tendencies of religious discourse as an intellectual pathology, to be cured by deflationary conceptions of the status of proper scientific knowledge.

By 1960, C. P. Snow was surely speaking for most scientists when he bumptiously stipulated that "by *truth*, I don't intend anything complicated . . . I am using the word as a scientist uses it. We all know that the philosophical examination of the concept of empirical truth gets us into some curious complexities, but most scientists really don't care."[40] If the Science Wars of the 1990s were supposedly about hostile sociologists' attacks on the idea of scientific Truth, it must be as jarring as it is pertinent to note the unpopularity, or just the irrelevance, of notions of Truth *among scientists themselves*.[41] If metaphysical foundations and a stable idea of Truth elevated the tone, then setting such things aside must be said to have lowered the tone. But this too has to be ascribed to changes taking place *within science itself*.

Third, the hagiographical tradition in the history of science celebrated scientists' genius and character, even if the homage was rendered problematic by simultaneous insistence on the impersonality of science. Yet by early in the twentieth century, scientists themselves were repeatedly stipulating that they ought to be regarded as human, if not all too human. Many of them wanted it clearly understood that they had the full range of human foibles, that they were not to be looked upon as paragons, and, when Robert Merton argued in the early 1940s that scientists were motivationally much the same as anybody else, he was actually falling in with sentiments repeatedly expressed within the scientific community.[42] The contexts in which these sentiments were expressed, and the reasons for expressing them, were very various, but they included the professionalization and routinization of science as a remunerated job, a job that was increasingly done not in ivory but in industry. And they included the wish—which acquired salience in the period from World War I to Hiroshima—to stipulate that scientists were not people to be feared, that whatever they were, they were not *worse* than the common run of humanity. One was to understand that neither poison gas nor the atomic bomb was produced by bad people, badly motivated. So if you had an academic reason to do so, you could be curious about who scientists were, freed from either a moral or an intellectual Gold Standard. The moral ordinariness of scientists, and historians' documentation of who they were as moral actors, might be taken as a lowering of tone, but that too was a sensibility arising outside of academic history.[43]

Finally, the very institutional *success* of science over the past century, and especially since World War II, effected wide-ranging changes in how the enterprise was viewed and evaluated. Writing in the late 1930s and the early 1940s, the scientific enterprise seemed—to Merton and to many others—vulnerable, frail, and delicate. If it was the lamp of civilization, its flame was weak, easily extinguished by the blasts of ideology, intolerance, and illiberalism. Threatened by fascism on the right, by communism on the left, and at home supposedly by industrial secrecy, command and control, science—as David Hollinger and others have shown—appeared in need of *protection and celebration*.[44] And its protection involved a proper description of the precise conditions in which it could thrive: autonomy from social forces, above all, but also an appreciation of its essential rationality and of its unique status among other forms of human endeavor.

But what success meant, especially in the West, and most especially in America, was that science became so closely enfolded in the institutions that produced wealth and projected power that accepted accounts of the nature of science and the conditions of its thriving lost their salience and cultural grip.[45] Did science need to be protected from illiberal forces, or had it become one of those forces? What did universalism mean when science became a powerful weapon in state conflict, hot and cold? Conditions of insecurity had been replaced by emerging conditions of complacency, and that freed up historians and sociologists of science to tell all sorts of naturalistic stories about science, just because naturalism seemed no threat. What were the boundaries of science, separating it from other forms of human endeavor, when it had become so bound up with the institutions of business, politics, and war that the very notion of "external context" began to seem something between quaint and bizarre? We could continue to talk about science as a distinct form of culture, uneasily related to disturbing "external" or "contextual" forces only on the condition that we ignored the circumstance that, through much of the twentieth century, most science was done for or in industry, for or in state facilities. Science, in fact, has become so blended with a range of civic, economic, and military projects that we can only appreciate its importance by being puzzled about its identity.[46] And so we are. About the coherent identity of science, we are very much as W. B. Yeats wrote of religion: "The ceremony of innocence is drowned;/The best lack all conviction, while the worst/Are full of passionate intensity."[47]

Historians of science were slow to come to terms with the civic success of science, and have done so only in part. Should this civic success count as a lowering of tone? Only if one identifies science with solitude and the contemplative life. Most citizens in our society, we should understand, regard this civic success wholly or largely as a positive accomplishment. However, if this success has amounted to a lowering of the tone, then that too happened outside of academic history, and historians have responded to it, slowly and indirectly. If I seem to downplay the culpability of academic historians in lowering the tone, I also downplay the credit we may take for initiating these sensibilities. We have reacted to those sensibilities, and we have done so in ways that make what we do *history*—not accusation or apology. That's just to say that we are historians and what we do when we do history is to try to tell it as it really was in the past. That is our institutionalized intention, and we're pretty good at recognizing when someone is trying to tell it like it was as opposed to distributing medals (or punishments) for modernity. But the terms and categories in which we can tell it like it was come from *us*, and from the culture we inhabit. And that was the sense of E. H. Carr's now half-century-old dictum that you "should study the historian before you begin to study the facts."[48] The stories historians tell owe as much to the currents running through their culture as they do to those they seek to tell about. That's not a regrettable circumstance, from which we could extricate ourselves if we tried hard enough or if we had the right rational method. It's our predicament.

So, if nobility is too strong a term for lowering the tone, what is admirable, even virtuous, about this tone-lowering in how we now tend to write about the history of science? Here, Weber's lecture on "Science as a Vocation" gives us a model of what this might mean. As historians of science, we're committed to telling rich, detailed, and, we hope, accurate stories about science without believing that it is cognitively or methodologically or socially unique, without believing that it is integral and unified, without believing that it has a special set of values not possessed by other forms of culture, without believing that it is divinely inspired, without believing that it is produced only by geniuses, without believing that it is the only progressive force in history, or that its practitioners do not eat chicken.

It would be easier to maintain that commitment—to richness, to detail, to accuracy—if we felt that we were doing God's work, but it's more admirable, I think, if we feel that same commitment, that same sense of vocation, when we know that we are not on a divine mission. That we are telling stories—rich, detailed, and, we hope, accurate—about a tone-lowered, heterogeneous, historically situated, embodied, and thoroughly human set of practices. That is, when we are doing what now counts as the history of science.

PART I

Methods and Maxims

Among historical specialties, the history of science has been more disposed than most to reflect upon its methods and objects. It is not hard to understand why, since science has traditionally appeared as uniquely resistant to the procedures historians now use to situate other objects of study in time and context. Is science, indeed, a historical object like any other, or does it stand outside of history? Does science require methodological "special-casing"—notions of divine intervention, inspiration, or genius—or does the usual range of human cognitive capacities and basic modes of interaction suffice for its historical understanding? Over recent decades historians of science have become more comfortable with at least an implicit embrace of the naturalism described in chapter 1 and commended in the programmatic chapters in this Part. But historical naturalism is not the end of inquiry. Resistance to naturalism in understanding science is itself a compelling historical object. What are the historical circumstances that have seemed to push scientific knowledge outside of history and that have made the history of science appear radically different from other sorts of history? Moreover, the struggle for naturalism in the study of science has been only locally won, and, as historians now appreciate, the methods and maxims of historical and sociological naturalism have provoked outrage from a few scientists, and more self-appointed Defenders of Science, who can only think of naturalism as that very unnaturalistic practice, denigration. Those eruptions have been irritating and uncomfortable for some historians and sociologists, but, more important, they are interesting indicators of the state of our culture and of the paradoxical sacredness of our secular.

CHAPTER 2

Cordelia's Love

Credibility and the
Social Studies of Science

When King Lear decides to take early retirement, he announces his intention to divide up the kingdom among his three daughters, each to get a share proportioned to the genuine love she has borne him. Each is asked to testify to her love. For Goneril and Regan that presents no problem, and both use the oily art of rhetoric to good effect. Cordelia, however, trusts the authenticity of her love and says nothing more than the simple truth. For Lear this will not do. Truth is her dower but credibility has she none.

Cordelia, we should understand, is a modernist methodologist. The credibility and the validity of a proposition ought to be one and the same. Truth shines by its own lights. And those claims that need lubrication by the oily art of rhetoric thereby give warning that they are not true. In this sentiment, Cordelia can be celebrated as a neglected forerunner of such plain-speaking English anti-rhetoricians of the seventeenth century as Francis Bacon and Robert Boyle. Use of the arts of persuasion handicapped rather than assisted the perception of truth: it was, Boyle said, like painting "the eye-glasses of a telescope."[1] Bacon urged that all persuasive "ornaments of speech, similitudes, treasury of eloquence, and such like emptiness" be "utterly dismissed."[2] The "truth of knowing and the truth of being are one, differing no more than the direct beam and the beam reflected."[3]

Yet if Cordelia embodies the modernist ideal, Lear represents obdurate reality. Lear makes a mistake such as we are all liable to make. He does not see truth shining by its own lights, and he confuses the pure glow of truth for the artificial brilliance lent by the arts of persuasion. The recognition of truth should be simple. The truth of knowing and the truth of being perhaps *ought* to be the same, but in practice we can never be quite sure that they are. Cordelia loves her father, but she does not evidently understand him as the imperfect being he is. By contrast, the late modern sensibility understands Lear—"human, all too human"—but it is Cordelia who puzzles us. How could anyone not only believe that truth is its own sufficient recommendation, but also consequentially act on that belief?

The changing place of credibility in the understanding of science tracks our move away from Cordelia's innocence. Once upon a time, so the story goes, students of science too believed that truth was its own recommendation, or, if not that, something very like it. If one wanted to know, and one rarely did, why it was that true propositions were credible, one was referred back to their truth, to the evidence for them, or to those methodical procedures the unambiguous following of which testified to the truth of the product. Alternatively, if one wanted to know, and one usually did, why false claims achieved credibility, one pointed to an assortment of contingent circumstances that caused people to hold dear what was in fact worthless. That is to say, once upon a time, pronouncements of validity were considered adequate responses to questions about credibility. And, indeed, it would be a very narrow and pedantic view of the matter to refuse to recognize that, for most students of science and, so far as we know, for most laypeople, they still count as such.

Credibility and the "Big Picture"

It was David Bloor who made the disjunction between validity and credibility into a maxim of method in the social studies of science, and so it has become for those few specialist scholars working in this idiom.[4] Within this practice, it is both appropriate and interesting to ask why it is maintained, by a late modern scientist or by a layperson, that there are such things as neutrinos, or that the pathological signs of Alzheimer's disease include neurofibrillary tangles. The answer to such questions might well include, for example, routine deference to authoritative sources of expert knowledge, just as it might be if one asked why seventeenth-century philosophers and laity maintained the historical reality of Christ's miracles.[5] All claims have to win credibility, and credibility is the outcome of contingent social and cultural practice.

Accordingly, again for those persuaded by this argument, the study of the grounds and means of credibility vastly expanded—from the explanation of false claims to the explanation of all knowledge-claims, whether deemed true or false. The study of credibility then became simply coextensive with the study of knowledge, including scientific knowledge. In sociological terms of art, an individual's *belief* (or an individual's claim) was contrasted to collectively held *knowledge*. The individual's belief did not become collective—and so part of knowledge—until and unless it had won credibility. No credibility, no knowledge.

Credibility has indeed been increasingly identified as a fundamental topic for the social studies of science. And the condition for its emergence is just the (partial and local) decline in authority of the grand old narratives that exempted scientific truth from the need to win credibility. The study of credibility—for those persuaded of Bloor's general point—has expanded to fill the space vacated by the defeat of the grand narratives. So it might be more proper to say that, insofar as we are concerned with scientific knowledge, credibility should not be referred to as a "fundamental" or "central" topic—from a pertinent point of view it is the *only* topic.

The social studies of science has seemingly been going through a period of infatuation with topics of this sort, topics that are not special areas of empirical inquiry—the sort of thing you might do if you studied a discipline or a period of time or a specific set of contextual social relations—or special "factors" that bear on scientific knowledge—sources of patronage, the use of instruments, considerations of the use of science in supporting social hierarchy. Rather, what has seemed to fascinate many of us in recent years might more properly be seen as the presuppositions or necessary preconditions of *any possible body of knowledge*. So, for example, we have the recently popular study of the "spaces of science" or the now-trendy study of the "embodied" nature of science—I plead guilty on both counts.[6] Neither space nor bodies should, strictly speaking, be regarded as "factors": no space, no science; no bodies, no science. And so too with credibility. Science, like finance, is a credit-economy: these are activities in which, if you subtract credibility, there is just no product left—neither a currency nor a body of scientific knowledge. Skepticism in science is like a run on the currency.[7]

We are urged these days—especially in the history of science, but also in sociology—to rise up above our particularism and to retrieve "the big picture,"[8] but the picture framed by the unqualified study of credibility is just too big. It leaves nothing out. For a focus on credibility to do any particular work, some distinctions have to be made. First, some points of methodological principle can be set out. Then, the problem has to be characterized in some useful way, and this I want to do by alluding to recent work in the field that can serve as exemplars of the kinds of things we might attend to and how we might come to grips with the problem of credibility in particular instances. Finally, despite the caveats about a picture that is too big, one can trace some recurrent patterns that might help us recognize classes of credibility-predicaments and the tactics of credibility-management that are pertinent to those classes.

Maxims of Method for the Study of Credibility

Three points of methodological principle: first, if we say that scientific propositions always have to win credibility, then that makes them like the claims of ordinary life, and like those of other specialized practices that have the task of establishing whether claims are the case or not. This principle means that we can make use of many of the resources and procedures that feature in academic inquiries about other cultural practices. Take, for example, the law. Legal proceedings are all about the establishment or erosion of credibility. In law courts utterances are systematically monitored for their veracity and, as Augustine Brannigan and Michael Lynch have noted, legal proceedings are framed as explicit inquiries into the veracity of what is said.[9] What is understood to be at issue is not a notion of philosophic Truth but of something like "truthfulness," an adequate assurance about the case on which a verdict and a penalty are sufficiently warranted—or, as they said in the seventeenth century, a "moral certainty." If that kind of certainty, and that quality of truthfulness, are adequate for understanding how the law works, then perhaps they are relevant to understanding credibility in science, as in fact they are for such versions of science as seventeenth-century English experimental philosophy.[10] Lynch has suggested the use of the term *truthing* to describe the mundane processes by which credibility is established in the law and similar practices.[11] *Truthing*, with that resonance, points to the processes of securing credibility without the neon glow induced by verificationist, confirmationist, or similar versions of Scientific Truth.[12]

Although there are formal writings treating the credibility of witnesses, in the main, law-court assessments of credibility derive from inferential practices that flourish in everyday life—including inferences from the standing of the witness and from postural, gestural, and linguistic manner. Again, as Brannigan and Lynch note, "there is no jurisprudence describing how such inferences are to be made."[13] The procedures for establishing truthfulness are inchoate; they are not formalized; and, perhaps, they are not formalizable. Knowing how to recognize truthfulness is knowing your way around a culture. And, as Mary Douglas repeatedly argued, the procedures by which a culture distributes credibility, like those by which it perceives risk, are so bound up with its moral life that one can give an adequate account of the culture by describing its techniques of credibility- and risk-management: "Credibility depends so much on the consensus of a moral community that it is hardly an exaggeration to say that a given community lays on for itself the sum of the physical conditions which it experiences."[14] There is no set of criteria distinct from a particular culture that uniquely determines what will be believed within it.[15]

The second point follows from this. In principle, there is no limit to the considerations that might be relevant to securing credibility and, therefore, no limit to the considerations to which the analyst of science might give attention: the plausibility of the claim; the known reliability of the procedures used to produce the phenomenon or claim; the directness and multiplicity of testimony; the accessibility and replicability of the phenomenon; the ability to impute bias to the claimants or to assess risks being taken in making the claim; the personal reputation of the claimants or the reputation of the platform from which they speak; knowledge of the friends and allies of claimants, including their personal reputation and power; calculations of the likely consequences of withholding assent; claimants' class, sex, age, race, religion, or nationality and the characteristics associated with these; claimants' expertise, including the means by which that expertise becomes known; the demeanor of claimants and the manner in which claims are delivered; minute aspects of the life-histories of those assessing claims and their knowledge of the life-histories of those making them.[16] Again, in principle, there is no reason why an inquiry into the grounds of scientific credibility might not find itself concerned with the investment portfolios of individual scientists (did Martin "Cold Fusion" Fleischmann own stock in a palladium mine?) or what they eat in the morning (does a medical researcher warning against the risks of dietary cholesterol turn out to eat a "full English breakfast" every day?).

Any aspect of the scene in which credibility is accomplished may prove to be relevant, and the relevance of *nothing* can be ruled out in advance of empirical inquiry—from which the third point of methodological principle follows: there should be no such thing as a *theory* of how credibility is achieved, at least in the sense of one of those grand theories offering an adequate formula for how it is done regardless of setting and the nature of the case at hand. In any particular case, the resources and tactics relevant to the achievement of credibility are likely to be very diverse, and a different array of resources and tactics is likely to bear on different types of case. For that reason alone, we ought to be suspicious of simple and global credibility-stories of whatever sort.[17] Finally, the description or explanation of credibility has got to specify the credibility *of what* and *for whom:* credibility-predicaments vary in interesting ways according to the nature of the claim and according to the relationship between who claims and who is meant to believe.

Metonymy, Induction, and Risk

One way of coming to grips with the scope of the problem we have is to recognize credibility as embedded within what one might call—with some license—a *metonymic* (or "standing-for") relationship. At the most basic level, that relationship is evident when I say "I do all the cooking in my household" and expect you to accept that claim as a fact about me. So all testimony about states of affairs stands in a metonymic relationship to those states of affairs, and the condition of your knowing about these things—otherwise unavailable to you—is your accepting the legitimacy of that relationship. Accordingly, for all the knowledge you have of those states of affairs that you have not yourself experienced, you are dependent on some practical resolution of the problem of credibility.

The same is the case when the claims in question have the character of *inferences* from one state of affairs to another. For example, when in 1648 Blaise Pascal sent his brother-in-law up the Puy-de-Dôme carrying a barometer, that climb of some three thousand feet produced a drop in the level of mercury of three inches. In order for this state of affairs to *stand for* the general phenomenon of the air's weight—which is what it was supposed to stand for—at least three, and very probably many more, metonymic relationships needed to be credited, apart from the credibility of testimony about the event itself. First, the behavior of mercury-in-glass had to be accepted as standing for the weight of the atmosphere above that part of central France; second, that what obtained in this vertical region of space might be extrapolated to stand for the kind of thing that would happen were one to go higher than the Puy-de-Dôme (or lower than the Earth's surface); third, that what happened to the barometer then and there would happen, *ceteris paribus,* to other competently designed barometers at other times and places (and to this one on some other occasion). Likewise, when Robert Boyle put a barometer in the air-pump and then exhausted the air, the behavior of the contained mercury was meant to stand for what would happen were one to walk a barometer up to the top of the atmosphere. Without these metonymic relationships being credited, there would be no philosophic *point* to what was done. In practice, the natural philosopher does not care what happened to *this* mercury in *this* piece of glass apparatus on *this* day and at *this* place, except as these outcomes support inferences to the relatively nonlocal and nonspecific. The local and the specific are not the *point* of these experiments; the philosopher cares, for example, about the *atmosphere* or about *the mechanical nature of the universe*. But in order for specific findings to be *about* the atmosphere or *about* the universe, the credibility of these standing-for relationships has to be accepted. Elements in these relationships are not logically connected, and the metonymic connections between them are defeasible in principle.[18]

Similarly, in Trevor Pinch's work on modern solar neutrino experiments, the detection of surplus Ar^{37} atoms in a vat of dry-cleaning fluid was meant to *stand for* neutrinos emerging from the Sun, while, in Bruno Latour's account, the result of Louis Pasteur's controlled field trials on sheep at Pouilly-le-Fort was meant to *stand for* what would happen to vaccine-protected natural populations of sheep spontaneously exposed to anthrax bacilli.[19] In each case, that to which scientists naturally have, or have worked to secure, effective access is intended to stand for that to which they cannot, or do not yet, have access.[20] That metonymic relationship is a way of pointing to the scope of science: scientific claims—only provided they achieve credibility—act as a shorthand for the natural world. Then we forget, or are obliged to ignore, the defeasible metonymic relationship and accept the claims as simply corresponding to the real states of affairs that are their reference and their point. And put still another way, deciding on the adequacy of the relationship between findings and what they stand for is just the problem of induction—impossible to justify in logical principle, routinely solved for all practical purposes a million times a day.

So far, I have treated this standing-for relationship in very abstract terms. Yet, as we know, the credibility of that relationship is often a highly consequential and politicized affair. Consider, for example, the phenomenon of *testing*.[21] Donald MacKenzie's research on nuclear missile guidance systems notes that no intercontinental ballistic missile has ever yet flown on the north-south polar trajectory it would have to take in the event of U.S.-Russian nuclear war, and only one missile seems ever to have been fired tipped with a live nuclear warhead.[22] Accepting that the results of east-west test firings were credible versions of what would happen when live missiles were fired north-south—and there were important technical reservations about accepting that—was enfolded in Cold War military and political realities.

Consider also the vast range of testing activities involved in the modern pharmaceutical industry and in environmental monitoring and protection, together with the political and legal apparatuses that are fed by test results and that in turn prescribe the adequacy, pertinence, and reliability of test procedures. Billions of dollars depend on the credibility of clinical drug trials as standing for the efficacy and safety of drugs when administered to non-test populations, and thousands of lives depend on whether trials of AIDS drugs and vaccines are designed—in Steven Epstein's nice terminology (appropriated from Dr. Alvan Feinstein's clinical medical usage)—in a "fastidious" or in a "pragmatic" mode.[23] The political and economic interests mobilized around the credibility of such tests are massive. And those interests are pertinent at practically every stage of test design and reporting, and of policy inferences from the tests.

Many years ago, in another incarnation, I worked as a jumped-up laboratory technician (in fact, a summer intern) in the U.S. Food and Drug Administration. Our unit was testing a range of chemicals and drugs—tranquilizers, pesticides, cosmetics, and the like—for possible human mutagenicity and carcinogenicity. For test systems, we used a then relatively well-established type of human tissue culture—inspecting the cells, after exposure, for chromosome breakage—and the new "Ames test," then being developed at another government laboratory. This used *Salmonella typhimurium* bacteria unable to grow on histidine-free plating medium whose back-mutation to the ability to survive on unsupplemented medium containing the substance under study would signal the production of point mutations. What happened in this sensitive bacterial test system was taken to signal possible dangers in human exposure. Our group was interested *inter alia* in possible human genetic damage inflicted by tetrahydrocannabinol (the active component in marijuana), LSD—this was, after all, the 1960s—and also caffeine. It seemed to at least some members of our small cytogenetics unit that the evidence implicating caffeine—in roughly the same concentration reaching your gonads after drinking a cup of strong coffee—was persuasive, while that pointing to the cellular risks associated with smoking marijuana and ingesting LSD was dubious. As it transpired, in subsequent divisional discussions work pointing toward the danger of coffee was deemed at most ambiguous and unconvincing, while the evidence establishing the risks of marijuana and LSD was considered scientifically secure.

Two points: the first is that *absolutely everyone* involved in the discussions—at least at the fairly low levels to which I was privy—understood as a *matter of course* that there was a congenial credibility-environment for claims about the risks of marijuana and LSD while economic and political realities would work strenuously against the public credibility of claims about the dangers of coffee. Accordingly, agency deliberations about the credibility of the different test regimes were political through and through. The second point is that at no stage in the formal discussions I witnessed leading to this outcome was anything *unscientific* said. Nor need it have been. For there was sufficient "play" between the test situation and possible *in vivo* effects for relevant skepticism to be expressed about the caffeine metonymic relationship, and, of course, sufficient grounds of confidence in the pertinence of the marijuana and LSD systems. It is proper usage to say that the legitimacy of inductive inference from *in vitro* to *in vivo* was conceded or contested on scientific grounds *and* on political grounds, yet no one was obliged to depart from a recognizably scientific idiom to give politics a grip.[24] I offer this anecdote both as a typical late modern instance of the politicization of credibility-judgments and as a warning against simplistic ways of understanding the relationship between political interests and technical judgments.

Furthermore, it would be a mistake to think of test situations as having only *one* outcome whose credibility is to suffer the vicissitudes. Consider the relationship that seems very widely to obtain between the credibility of a claim, on the one hand, and the significance and scope of the claim, on the other. In Pinch's work, solar neutrino scientists had the option of choosing among a number of claims, all of which might be deemed to "follow from" experimental findings.[25] You could say that you had observed "splodges" on a graph—which is practically undeniable but uninteresting—or that you had observed a certain number of Ar^{37} atoms in a vat of dry-cleaning fluid—somewhat more deniable and interesting—or, finally that you had observed solar neutrinos—very deniable and very interesting. The series ascending from "splodges" to solar neutrinos progresses along axes—as Pinch says—extending from low to high "evidential significance" and low to high "externality."

What do you say you observed? If you say you saw "splodges," the likely credibility will be high, but the likely interest in such a claim will be low. Moving up the axes of externality and evidential context is to take, and be seen to take, a *credibility risk*—critics can pick away at the gap between elements in the metonymic relationship—but also to bid for rich credibility-rewards. Accordingly, what Pinch in effect offers is a framework for describing the moral economy of risk and reward in the relevant community. And, of course, such decisions can also take place in even more intensely politicized arenas. Sheila Jasanoff's recent work on the U.S. Environmental Protection Agency (EPA) documents the EPA's shift down the axis of evidential significance as it came under increasing pressure of political skepticism in the 1970s—from stating, as she says, "substance X . . . is a carcinogen to giving intricate explanations of the process by which it came to that factual conclusion."[26] The EPA could secure widespread credibility on the condition that it made its processes of inductive inference publicly visible. It responded to political and economic forces by reducing its own exposure to credibility-risk.

"Authorized" and "Conversational" Objects

In such cases credibility arises in part from actors' *judgments* of risk and rewards, and from actors' beliefs about the credibility-economy into which claims will enter. However, the conditions of credibility also flow—it might be said—from the nature of the phenomena or concepts themselves, or, more accurately, from the environment in which they are produced and in which they live out their careers. The English social theorist Zygmunt Bauman has noted that the late modern world is thickly populated with entities about which the authority to speak resides solely with very highly specialized, and very highly bounded, communities. Only certified physicists—and indeed physicists of a certain specialty—can pronounce on the existence and characteristics of intermediate vector bosons, and only very highly specialized astronomers can speak credibly about the existence and characteristics of pulsars:

> The matters dealt with by physics or astronomy hardly ever appear within the sight of non-physicists or non-astronomers. The non-experts cannot form opinions about such matters unless aided by—indeed, *instructed*—by the scientists of the field. The objects which sciences like these explore appear only under very special circumstances, to which no one else has access: on the screen of a multimillion-dollar accelerator, in the lens of a gigantic telescope, at the bottom of a thousand-foot-deep shaft. Only the scientists can see them and experiment with them; these objects and events

are, so to speak, a monopolistic possession of the given branch of science (or even of its selected practitioners); the monopoly has been assured by the fact that the objects and events in question would not occur if not for the scientists' own actions and the deployment of resources those scientists command; and thus the objects and events are, by the very nature of their appearance, a property unshared with anybody who is not a member of the profession. Monopoly of ownership has been guaranteed in advance by the nature of scientific practices, without recourse to legislation and law enforcement (which would be necessary were the dealt-with objects and events in principle a part of a wider practice and hence accessible to outsiders). Being the sole owners of the experience which provides the raw material for their study, the scientists are in full control of the way the material is construed, processed, analysed, interpreted, narrated.[27]

And so long as these things are not taken to be, as Thomas Hobbes said, "contrary to any man's right of dominion," specialized scientists have massive control of the conditions of their credibility.[28] These objects and events are enormously consequential for late modern social and political action, for they include such things as the composition of the protein coat of HIV, the chemical combinations occurring between CFCs (chlorofluorocarbons) and ozone, and the nucleotide sequence of the gene controlling muscular dystrophy. Without them—or, more precisely, without the actions and understandings mobilized around their credibility—late modern society would not be recognizable as the thing it is. Yet the conditions of credibility of such things depend to such an extent on a certain intellectual economy that it is tempting to recognize a distinct class of what might be called "authorized objects."

Not all scientific objects have that authorized character and those conditions of credibility. Consider, for example, the concepts and phenomena of much human and social science. The credibility-economy of such notions as "the unconscious," "psychic regression," "having fixations," "being in denial," or "suppressed memory"—just to take some quasi-Freudian locutions—depends, as Graham Richards has effectively argued, on the extent to which they are vernacularized and, therefore, the extent to which they actually *come to constitute* the phenomenal base to which they refer.[29] If people, as it appears many of them now do, make sense of their lives by organizing them with respect to such notions, to that extent, Richards says, human science has regenerated and re-created *human nature*. There was a time when people did not *have* "unconscious reasons" or see things as "phallic symbols." Now they have and now they do. The credibility of such notions is secured on different conditions than those pertaining to subatomic particles and DNA sequences, so that it might make sense to refer to them as "conversational objects." The ability of specialist communities to speak authoritatively about conversationally constituted subjects is circumscribed by the judgments and decisions of the objects of study. Bauman suggests that, for such reasons, the states of affairs predicated by human and social scientific inquiry have a credibility "handicap"—they can't be authoritatively established in the same way as quarks.[30] Yet it is possible that many notions flowing from expert human and social scientific culture enjoy a credibility *advantage* by virtue of being high-toned versions of locutions already present—in some form and to some degree—in lay culture. To make this point about "conversability" is not merely to pick out some features of *interpretative* human science: arguably, *causal* and deterministic conceptions in human science share the same conditions of credibility since lay actors also routinely use causal notions to make out their own and others' behavior.[31]

The Comparative Economies of Credibility

In 1982, Barry Barnes and David Edge wrote that, as yet, "we know very little about the basis of credibility: the importance of the problem is matched only by its complexity and its comparative neglect."[32] And they warned against those "superficial accounts" and "mythologies" that were best regarded not as attempts to describe and explain the current distribution of cognitive authority but to legitimate it. It was in this connection that the grand narratives of reason, reality, and method were so pointedly criticized. In the social studies of science we have come some way since then. We possess an increasing range of empirical studies of how scientific claims win credibility and many of us now appreciate the complexity of the task involved in giving an account of that credibility.

Following early work by Yaron Ezrahi, Barnes and Edge suggested a preliminary categorization of the "major modulators of credibility" obtaining in the relationship between experts and the modern public in democratic societies.[33] These include the relationship between scientific accounts and prevailing social and cultural beliefs; the relationship between scientifically generated technologies and prevailing social values; the forms of accessibility of a specific science to the public; and the extent of expert consensus. Pay attention to this list of modulators and you will, Barnes and Edge argued, capture important considerations shaping the credibility of expert claims among the public.

This still seems to be pretty good advice. Take account of these sorts of things and you will probably have a decent chance of constructing a rich and interesting story about the conditions of credibility between experts and laity. However, these are not the only conditions of credibility in which we might be interested, and I conclude with some brief speculative remarks about what might be called the *vectors* of credibility and the credibility-economies that arguably obtain along different vectors.

Up to now in science studies most attention has been paid—as Barnes and Edge indicate—to the *public* credibility of expert claims. And rightly so, for there is no doubt that this vector has the greatest practical significance. Yet credibility has other vectors, and the credibility-economy that obtains between experts and laity may not obtain elsewhere. So, for example, some recent science studies work—including my own *Social History of Truth*—has been mainly (though not exclusively) concerned with the economy of credibility *internal* to scientific practices, while other work—notably Theodore Porter's—has focused on the economy of credibility *between* scientific and technical groups in modern differentiated societies.[34] Obviously, the state of credibility-management in one vector bears on credibility-economies in others and this is why, for example, Ezrahi and others are right to note the pertinence of expert consensus and dissensus to lay judgments. But suppose one makes a speculative stab at some distinctions. Within such small interdependent groups as the "core-sets" of specialized scientific practices, the economy of credibility is likely to flow along channels of familiarity.[35] The practitioners involved are likely to know each other very well and to need each other's findings in order to produce their own. Here, the immediate fate of one's claims is in the hands of familiar others, and the pragmatic as well as the moral consequences of distrust and skepticism are likely to be high. In such social settings the analyst should take care not to explain the achievement and maintenance of mutual credibility too aggressively. In a world characterized by familiarity—whether in lay or expert society— taking each other's claims at face value is *normal,* and it is distrust, skepticism, and the demand for explicit warrants for belief that need specially to be justified and accounted for. It is, indeed, hard to conceive how small groups of familiar others could long maintain their cohesiveness were the situation otherwise.

At the apparent opposite pole is the credibility-economy obtaining between expert groups and laity. Here, as such social theorists as Niklas Luhmann and Anthony Giddens have argued, the resources of familiarity for addressing problems of credibility are absent or impoverished.[36] We look instead for formal warrants of credibility—institutional affiliation or standing, the observance of explicitly framed methical procedures, the display of expert consensus, and the like. Accessibility can cut both ways in such an economy. On the one hand, where we have independent access to the "facts of the matter," we may be able to use that knowledge to gauge the claims of experts. On the other hand, the representation of expert knowledge as far beyond lay accessibility can serve as a recommendation for its truth.

Then we come to the economy of credibility obtaining *between* expert groups in modern differentiated societies. This is the world so richly described by Theodore Porter. It is an economy in which, if shared belief is to be secured and maintained, it must travel great distances—in both physical and cultural spaces. This economy substantially lacks the resources of familiarity while it possesses an array of inducements to distrust and skepticism. Here a major recommendation for belief is the public display of the *discipline* to which claimants and their claims are subject. If I *can* impute bias and interest, why else should I believe you except because of a convincing display of the disciplining of bias and interest to which you have been subjected? It is in this connection that the language of quantification and of method has its consequential task in the making of credibility. "Quantification," Porter writes, "is well suited for communication that goes beyond the boundaries of locality and community."[37] I am sure this is quite right—indeed, I have tried to develop parallel arguments about the language of disinterestedness in both seventeenth-century experimental philosophy and early nineteenth-century cerebral anatomy—while reference to "communication beyond the boundaries of community" puzzles.[38] Wherever one has shared knowledge, there, I would suggest, one has a form of "community" where distrust—unlimited in principle—has in fact reached a conventional limit. Why accept the language of quantification as an adequate solution to problems of distrust, since in principle further distrust is always possible?[39]

The world of "community" is not, perhaps, that readily circumscribed. There is no compelling reason to accept that the social theorists' description of the economy of credibility between experts and laity is adequate. If we ask ourselves how it is that we came to believe a whole range of expert scientific and technical claims—probably the majority of those whose truth we now accept—we will discover that we were told these things by familiar others, often in a setting of face-to-face interaction: by people whose names we know (or knew) and whose characteristics we know (or knew)—teachers, professors, physicians, nurses, plumbers, mechanics, colleagues. And it is well to remember that experts commonly stand in something like the position of lay members with respect to the claims of *different* expert groups. The resources of familiarity are not so easily dispensed with, even in the late modern world, and even with respect to the credibility of esoteric scientific claims.

CHAPTER 3

How to Be Antiscientific

I am not a commissioned officer in the Science Wars. If anything, I am something between a common soldier and an interested witness to the current hostilities. I was trained in genetics, but for many years I have been a historian and sociologist of science, writing mostly about the development of science in the seventeenth century.[1] I have suffered some minor shrapnel wounds from wildly aimed shells, but, in the main, the Defenders of Science have had bigger game to stalk and have left me to get on with my work and to reflect from a somewhat disengaged perspective on what has been going on.

The immediate occasion for the Science Wars seems to be a series of claims *about* science made by some sociologists, cultural historians, and fuzzy-minded philosophers. (In my ordinary work, distinctions between these categories—and subdivisions within them—count as crucial, but in this piece for general readers I mainly lump them together.) As a matter of convenience, I refer to propositions *about* science as "metascience," and, because it is very important to be clear about what is at issue, I list here just a few of the more contentious and provocative metascientific claims:

1. There is no such thing as the Scientific Method.
2. Modern science lives only in the day and for the day; it resembles much more a stock-market speculation than a search for truth about nature.
3. New knowledge is not science until it is made social.
4. An independent reality in the ordinary physical sense can neither be ascribed to the phenomena nor to the agencies of observation.
5. The conceptual basis of physics is a free invention of the human mind.
6. Scientists do not find order in nature, they put it there.
7. Science does not deserve the reputation it has so widely gained . . . of being wholly objective.

8. The picture of the scientist as a man with an open mind, someone who weighs the evidence for and against, is a lot of baloney.
9. Modern physics is based on some intrinsic acts of faith.
10. The scientific community is tolerant of unsubstantiated just-so stories.
11. At any historical moment, what pass as acceptable scientific explanations have both social determinants and social functions.

For many readers, even listing such statements is unnecessary: they will already be thoroughly familiar with sentiments like these associated with the writings of sociologists of science and academic fellow-travelers, as they will be equally familiar with the outraged reactions to them expressed by a few natural scientists, convinced that such claims are motivated mainly or solely by hostility to science, or that they proceed from ignorance of science, or both. Science and rationality are said to be besieged by Barbarians at the Gate, and, unless such assertions are exposed for the rubbish they are, the institution of science, and its justified standing in modern culture, will be at risk. It is therefore incumbent on leading scientists themselves to speak out, to say what the real nature of science is, and to take a stand against the ignorance and the malevolence expressed in these claims.[2]

Nevertheless, I have to tell you—in the spirit of our troubled culture—that you have just become a victim of yet another hoax. None of these claims about the nature of science that I have just quoted, or minimally paraphrased, does in fact come from a sociologist, or a cultural studies academic, or a feminist, or a Marxist theoretician. Each is taken from the meta-scientific pronouncements of distinguished twentieth-century scientists, some Nobel Prize winners. (See the end of this chapter for a list of the sources.) Their authors include immunologist Peter Medawar, biochemists Erwin Chargaff and Gunther Stent, entomologist E. O. Wilson, mathematician turned scientific administrator Warren Weaver, physicists Niels Bohr, Brian Petley, and Albert Einstein, and evolutionary geneticist Richard C. Lewontin. This is not a mere party trick—a device to turn the tables or to play intellectual ping pong—though it would seem so if I left it at that. The point I want to make here is substantial, interesting, and potentially constructive: practically all the claims about the nature of science that have occasioned such violent reaction on the part of some recent Defenders of Science have been intermittently but repeatedly expressed by scientists themselves: by many scientists in many disciplines, over many years, and in many contexts.[3]

Accordingly, we can be clear about one thing: it cannot be the claims themselves that are at issue, or the claims themselves that must proceed from ignorance or hostility. Rather, it is *who has made* such claims, and what motives can be attributed—plausibly, if often inaccurately and unfairly—to the *kinds of people* making the claims. So one of the very few, and very minor, modifications I have made in several of the quotations above is the substitution of the third-person "they" or "scientists" or "physicists" for the original "we." We are now, it seems, on the familiar terrain of everyday life: members of a family are permitted to say things about family affairs that outsiders are not permitted to say. It is not just a matter of truth or accuracy; it is a matter of decorum. Certain kinds of description will be heard as unwarranted criticism if they come from those thought to lack the moral or intellectual rights to make them.

Since what scientific family members often do when they make metascientific statements is to *prescribe* how members *ought* to behave—criticizing or praising—there is a tendency to assume that outsiders must be about the same business, though without equivalent entitlements. It is sometimes hard for scientists to understand how the description and interpretation of science could be anything other than coded prescription or evaluation: telling scientists what to do, or sorting out good from bad science, or saying that science as a whole is good or bad. It is hard to recognize, that is, what a naturalistic intention would be like in talking about science, since this is not a luxury readily available to members of the scientific family. Scientists have naturalistic intentions with respect to their objects of study but rarely with respect to the practices for studying those objects. So, for example, some sociologists do indeed insist that scientific representations are "social constructions." And when some scientists read this they assume—wrongly in most cases and in my view—that these sociologists have tacitly prefaced the phrase with the evaluative word *only,* or *merely,* or *just:* science is *only* a social construction. To say that science is socially constructed is then taken as a way of detracting from the value of scientific propositions, denying that they are reliably about the natural world.[4] Scientists do that all the time: that is, they "deconstruct" particular scientific claims in their fields by identifying them as *mere* wish-fulfillment, *mere* fashion, *mere* social construction. But they do so to *do* science, to sort out truth from falsity about the bits of the natural world with which they are concerned. They rarely do so with what might be called a disciplinary intention of just describing and interpreting the nature of science. That is one major reason why we seem to be misunderstanding each other so badly. There are important differences in recognized disciplinary intentions, in seeing their different possibilities and purposes and values. We do not always adequately recognize these differences, and we ought to.

That is one lesson to take away from this little hoax. But it is neither the most interesting nor the most fundamental. The more fundamental observation is just that metascientific statements by scientists vary enormously. I have picked out some that resonate with descriptions offered by sociologists, but, of course, there are many that do not. When scientists say metascientific things, they commonly conflict with each other as well as conflicting, occasionally, with what sociologists say.

Indeed, some scientists' pronouncements on the nature of science stipulate that science is a realist enterprise; others stipulate that it is not. Science, these others say, is a phenomenological, instrumental, pragmatic, or conventional practice. Max Planck, for example, identified the endemic tendency to "postulate the existence of a *real world*" in the metaphysical sense, as "constitut[ing] the irrational element which exact science can never shake off, and the proud name, 'Exact Science,' must not be permitted to cause anybody to under-estimate the significance of this element of irrationality."[5] J. Robert Oppenheimer supposed that laypeople were irritated by scientists' *un*willingness to use words like *real* or *ultimate:* the use of such notions would be a form of metaphysics, and science, Oppenheimer insisted, was a "non-metaphysical activity."[6] These positions are hard to square with such nervously defiant declarations as Steven Weinberg's: "For me as a physicist the laws of nature are real in the same sense (whatever that is) as the rocks on the ground."[7] As it happens, physicists disagree on such things.

Moreover, some scientists—when they say that science is a realist enterprise—mean to pick out a special philosophical position by which theoretical entities are understood to map onto real existents in the world; others seem to be alluding to the sort of robust everyday realism that unites a range of sciences with the practices of everyday life, as when I might say in ordinary conversation, "Look at the cat sitting on the mat," directing someone's attention *over there* and not toward my mouth or my brain. The realism advocated (or rejected) in scientists' metascientific pronouncements is only very occasionally specified in such ways. Some scientists say that science aims at, or arrives at, one universal Truth, others say that the truths of sciences are plural, or that science is just "what works" and that Truth, or even correspondence with the world, is none of their concern—just "what is the case" or "what seems to be the case to the best of our current efforts and beliefs." Some say that science is Coming to an End—about to be completed—but we should understand that this imminent completion has been promised practically as long as there has been science. Other scientists pour scorn on any such idea: science, they say, is an open-ended problem-solving enterprise, where problems are generated by our own current solutions and will continue to be, time without end.[8]

Some scientists' metascientific pronouncements say that there is no such thing as a special, formalized, and universally applicable Scientific Method; others insist with equal vigor that there is. The latter, however, vary greatly when it comes to saying what that method is. Some scientists like Francis Bacon, some prefer René Descartes; some go for inductivism, some go for deductivism; some for hypothetico-deductivism, some for hypothetico-inductivism. Some say—with T. H. Huxley, Max Planck, Albert Einstein, and many others—that scientific thinking is a form of common sense and ordinary inference. "The whole of science," according to Einstein, "is nothing more than a refinement of everyday thinking."[9] Others, like the biologist Lewis Wolpert, vehemently repudiate the commonsense nature of science and suggest that any such idea stems from ignorance or hostility.[10] Few—either for or against the commonsense nature of science—display much curiosity about what common sense is or entertain the possibility that it too might be heterogeneous and protean.

You name it, it's been identified as the Scientific Method, or at least as the method of some practice anointed as the Queen of the Sciences, the most authentically scientific of sciences—usually, but not invariably, some particular version of modern physics. Collect textbook statements about the Scientific Method and see for yourself. Or ask your scientist-friends, one by one, to write down on a piece of paper (no collaborating! no peeking at a philosophy of science textbook!) what they take to be either the Scientific Method or even the formal method thought to be at work in their own practice or discipline. Some of your friends will have heard of Karl Popper, or of Thomas Kuhn, or of Paul Feyerabend and will have their preferences among these—though probably not many of them. (Why should they?) In which case, ask them to write down on another piece of paper what they take to be the position about Scientific Method recommended by their favorite philosopher. (You may find little correspondence with sociologists' or philosophers' professional sense of what Popperianism or Kuhnianism is, and, in any case, sociologists and philosophers also vary in their estimation of what Popper and Kuhn were really saying.)[11]

You might also consider the cultural sources of our current repertoires for talking about Scientific Method. Few chemists, biologists, or physicists will have taken courses on Scientific Method (at least in Anglophone settings), but many psychologists or sociologists will have experienced almost total immersion in such material— modeled on what is taken to be formal natural scientific method. Perhaps no small part of the enormous success of the natural sciences might be ascribed to the relative *weakness* of formal methodological discipline. It is at least a thought worth thinking. This was, for instance, the opinion of the physicist Percy Bridgman: "It seems to me that there is a great deal of ballyhoo about scientific method. I venture to think that the people who talk about it most are the people who do least about it. Scientific method is what working scientists do, not what other people or even they themselves may say about it. No working scientist, when he plans an experiment in the laboratory, asks himself whether he is being properly scientific, nor is he interested in whatever method he may be using as *method* . . . The working scientist is always too much concerned with getting down to brass tacks to be willing to spend his time on generalities . . . Scientific method is something talked about by people standing on the outside and wondering how the scientist manages to do it."[12]

When we consider the *conceptual* identity of science, the situation is much the same. Is science conceptually united? To those scientists who consider that it is, a preferred idiom is a unifying materialist reductionism, though scientists of a mathematical or structural turn of mind reject both materialism and reductionism, while some biologists continue intermittently to ponder whether there is a unique biological mode of thinking and unique biological levels of analysis. Just as E. O. Wilson announced a new—or rather a revived—plan for the reductionist unification of the sciences, natural and human, other scientists rebelled against reductionism, against the claim that "the whole is the sum of its parts," or against its local manifestations in molecular biology, or they say that what had once been a search for understanding has now turned into a reductionist and shallow quest for explanations. Materialistic reductionism is just a sign that a Scientific Age of Iron has followed an intellectual Golden Age.[13]

The conceptual unification of all the sciences on a hard and rigorous base of materialist reductionism is an old aspiration, but it has never commanded (and does not now command) the assent of all scientists. In a whole range of natural sciences—though biology is perhaps the most pertinent case—reductionist unification is rejected, sometimes very violently, and in other parts of science reductionist unification just doesn't figure. It may be somebody's dream, but it's hardly anybody's work.

Recall that I started by picking out claims about the nature of science that

I invited you to associate with ignorant or hostile nonscientists. Then I told you these statements were in fact made by scientists. Taking the argument a step further, I then acknowledged that metascientific statements by scientists were very various—on all subjects, and on all levels—and that many of these conflicted with sentiments in the quoted set, and with each other.

From this circumstance one could draw a number of conclusions. The first would be that a certain set of these statements—say the first set—is hopelessly in error and that their opposites are correct. I don't want to say that. If I did, it would be as much as saying that Medawar, Planck, and Einstein didn't know what they were talking about, nor do the sociologists whose claims resemble theirs most closely. In all honesty, however, I have to admit that when I plow through the range of individual scientists' metascientific statements I often find more internal variability than makes me professionally comfortable. I might even be accused of the sin of quoting isolated remarks out of context, and maybe I have. No one should tendentiously quote out of context, though perhaps quoting Peter Medawar out of context on the Scientific Method is a less serious offense than (I take a randomly chosen example) quoting Steven Shapin out of context on the role of trust in seventeenth-century English science: Medawar's proper business is less damaged by such misleadingly selective quotation than mine. It is bad to quote out of context, or to quote misleadingly. It is bad for sociologists to do when writing about science or metascience, and it is bad for scientists to do when writing about the sociology of science. No, I want to say that the quoted set contains quite a lot of truth—with some qualifications that I am shortly going to make.

The second conclusion would be that all metascientific statements by practicing scientists are best ignored. For this view—at the risk of introducing a Cretan paradox—I can cite prominent scientists' pronouncements, too. It was, after all, Einstein who famously said that we should take little heed of scientists' formal reflections on what they do; we should instead "fix [our] attention on their deeds": "It has often been said, and certainly not without justification, that the man of science is a poor philosopher."[14] So, if we follow Einstein and charitably allow the self-contradiction to pass, what one would be tempted to say is something like this: "Plants photosynthesize; plant biochemists are experts in knowing how plants photosynthesize; reflective and informed students of science are experts in knowing how plant biochemists know how plants photosynthesize."[15] As Aesop put it, the centipede does marvelously well in coordinating the movements of its hundred legs, less well in giving an account of how it does so. No skin off the centipede's back, and no skin off the scientist's back, if it happens that she's not very good at the systematic reflective understanding of her work. That's not her job. And the point, of course, of Aesop's fable is that the centipede pushed to reflective understanding winds up in an uncoordinated heap. Kuhn just follows Aesop in this regard.

That's not really the conclusion I want to press either, though it does have something to recommend it. I see no necessary reason why certain scientists—perhaps not very many, given the pressures on their time and their other interests—shouldn't be just as good at metascience as professional metascientists, nor any necessary reason why professionals in metascience should ignore the pronouncements of amateurs. Nor do professional metascientists—sociologists, historians, and philosophers—globally *have to* concede that practicing scientists "know the science better or best" or "know more science" than they themselves do, though it is very prudent to respect scientists' particular expertise and to make sure, when one is writing about the object of that expertise, to "get it right." They should take great care not to say something about photosynthesis or about the techniques for knowing about photosynthesis that is demonstrably wrong, as judged by the consensus of expert practitioners in that area.

The reason that sociologists, historians, and philosophers do not globally have to concede that "scientists know better about science" is that knowledge about contemporary plant biochemistry, for instance, is not the same thing as "knowledge about science." There are many sciences at time present, and there have been many more sciences, and many versions of plant science, in past times, and who is to say that the historian or sociologist who knows something substantial about these many sciences knows "less science" than the contemporary plant biochemist who, pronouncing on the nature of science, knows less or even nothing at all?

I see no reason to turn the tables and celebrate as a fact that I know "more science" than my friend who is a plant biochemist. As it happens, I know almost nothing about photosynthesis beyond what I was taught in college courses in plant physiology and cell biology, and I would be morally wrong and intellectually careless if I pronounced on how matters stand in that part of present-day science. On the other hand, I have the right to feel slightly miffed if I am lectured about how matters stood in seventeenth-century pneumatic chemistry by practicing scientists who are even more incompetent in that part of science and its history than I am in contemporary plant biochemistry.

Almost needless to say, it's vital that you get your facts right in the subject you're writing about. That obligation is absolute and it's general: it applies to sociologists and historians writing about the aspects of science in which they are interested, and it applies to scientists writing about the sociology and history of science. At the same time, one would hope that normal human and professional frailties would be recognized and that we would pause a nanosecond before ascribing to each other the basest possible motives and the most egregious degrees of incompetence. There is indeed some shoddy work in sociology and cultural studies, and some natural scientists persuasively say in public there is shoddy work in their parts of science. *There is no excuse for shoddiness wherever it is found.* But we should at the same time cut each other a little bit of slack. To err is human, but it is as likely that we err in appreciating each other's intentions as it is that major blunders have been committed or that disciplinary hostility is at work. Before pointing fingers in the press or on public platforms, we might try conversations in a café or a pub. The likely result would be lower blood pressure and a less toxic public culture.

Finally, as I suggested, scientists' metascientific statements often function in the specific context of *doing* science, of criticizing or applauding certain scientific claims or programs or disciplines. That is to say, they may not be pure expressions of institutional intentions to describe and interpret science but tools in saying what *ought* to be believed or done within science as a whole or within a particular discipline or subdiscipline. Viewed in that way, such statements not only *can* be taken seriously by students of science, they *must* be taken seriously, but *in a different way*—as part of the *topic* that the sociologist or historian means to describe and interpret.

The major conclusions I want to come to concern both the variability of scientists' metascientific statements and the nature of their relationship to what might loosely be called "science itself." Here I'd like to say—and again I can call on the additional authority of Einstein and Planck to say it—that the relationships between metascientific claims and the range of concrete scientific beliefs and practices are always going to be intensely problematic. "In the temple of science," Einstein said, "are many mansions."[16] It is a modernist legacy, inherited from the methodological Public Relations Officers of the seventeenth century, that science is one, and, accordingly, that its "essence" can be captured by any one coherent and systematic metascientific statement, methodological or conceptual.[17] But, while the vision of scientific unification remains compelling to some, no plan for unification, and no account of the essence of science, carries conviction for more than a fraction of scientists. And that is one of my points.

So what happens if we follow the sentiments of many scientists (and incidentally that of increasing numbers of philosophers) that the sciences are many and diverse and that no coherent and systematic talk about a distinctive essence of science can make sense of the diversity or the concreteness of practices and beliefs? One thing that may happen is that we take a different view of the variability of metascientific statements, taken, that is, as statements about the distinctive nature of something called "science." We may want to say that different kinds of metascientific statements may pick out aspects of different kinds, or stages, or circumstances of the practices we happen to call scientific. Or different metascientific statements may contingently belong to the practices they purport to be about: as ideals, or norms, or strategic gestures signaling possible or desirable alliances. They may be true, or accurate, about science, but not *globally* true about science, just because no coherent and systematic statement could be globally true or accurate about science and could at the same time distinguish science from other forms of culture. Why ever should we expect that metascientific statements of any sort could hold for particle physics (which kind?) *and* for seismology *and* for the study of the reproductive physiology of marine worms? Some metascientific statements *might be* true about a range of scientific practices localized in time, place, and cultural context, but that is for us to find out, not to assume.

Something else follows from the recognition of diversity for current concern with antiscience. Because scientists' metascientific statements are diverse, and because it is possible that each picks out some real local features of some sciences, when considered from a certain point of view, the relationship between metascience and science is certainly problematic and at most contingent. For that reason alone, one can be allowed to dispute metascientific narratives *of any kind* without being understood to oppose science. If science is really as distinct from philosophy as some Defenders of Science insist it is, then it is puzzling in the extreme why they should be so upset when their favorite *philosophy* is criticized.[18] Natural science justifiably possesses enormous cultural authority; philosophy of science possesses rather little. Some tactical mistake is surely being committed when the Defense of Science appears as a celebration of a particular philosophy, still more when it celebrates versions that have been tried and long abandoned as inadequate by philosophers themselves.

How to be antiscientific, then? I can now tell you some ways in which you *cannot* be coherently and effectively antiscientific. You cannot be against science because you dislike its supposedly unique, unifying, and universally effective Method. You cannot be against science because it is essentially materialistic or essentially reductionist. You cannot be against science because it is essentially "instrumental rationality" or, indeed, because it contains irrationality. You cannot be against science because it is a realist enterprise or because it is a phenomenological enterprise. You cannot be against it because it violates common sense or because it is a form of common sense. Nor can you be against it because it is essentially hegemonic, or essentially bourgeois, or essentially masculinist. And, of course, it should go without saying that you cannot be coherently *for* science for any of these reasons either.

A thought experiment, then a qualification, and finally some remarks on a sense in which one *can* be antiscientific in real, substantial, and constructive ways. First, the thought experiment. I, and some of my colleagues in the history and sociology of science, are methodological relativists. That is to say, I maintain, on the basis of empirical and theoretical work, that the standards by which different groups of practitioners assess knowledge-claims are relative to context and that the appropriate methods to use in studying science should take that relativity into account. So far as the Scientific Method goes, like Peter Medawar and many other scientists, I am a skeptic. Further, this work leads me to believe that the natural world is probably extremely complex and that different cultures can and do stably and coherently classify and construe it in very different ways, according to their purposes and in light of the cultural legacies they bring to their engagements with the natural world. This position has been identified as antiscientific—motivated by ignorance and hostility—and, it is said, that people having such small faith in science should follow its logical conclusions: they should jump in front of cars or consult witch-doctors rather than neurologists when their heads ache.

It is a silly and misguided argument, but nonetheless an interesting one to consider. I do not jump in front of cars and I do consult physicians when I feel a need to do so. What does this prove? Not that I am insincere in my methodological relativism, or that I have contradicted myself, but that my genuine confidence in a range of modern scientific and technical practices and claims proceeds from different sources than my belief in some set of methodological metascientific stories. My confidence in science is very great: that is just to say that I am a typical member of the overall overeducated culture, a culture in which confidence in science is a mark of normalcy and which produces that confidence as we become and continue to be normal members of it.

I have been to the same sorts of schools as Alan Sokal, Steven Weinberg, Paul Gross, and Norman Levitt; we share other important cultural legacies and sensibilities; we probably vote the same way and like the same sorts of movies, though that's just a guess. Apart from our different academic disciplines, our institutional environment is much of a muchness; and if we met each other at a party with our name tags off, there's a decent chance we'd hit it off pretty well. But, for all that, my professional confidence in a range of metascientific global stories about the Scientific Method, and its warrant for scientific effectiveness, is very low. So *this* is what is proved by my preference for physicians over witch-doctors, for astronomers over astrologers: the grounds of my confidence in science have very little to do with metascientific stories, of any kind. And, arguably, the same situation obtains over a broad range of educated, and perhaps of not-so-educated, people.

Now the qualification: in my academic work I have made, and I continue to make, claims about science that have an apparently global character, though to be honest I've become a bit more circumspect about making them as time has gone on. And I want to defend the character, pertinence, and legitimacy of these claims. So, for example, I've been known to say that the social dimension of science is constitutive and that trust is a necessary condition for the making and maintenance of scientific knowledge. These *are* metascientific statements, and they *are* meant to apply to all scientific practices that I know of. So am I not hoist on my own petard? I don't think so. The reason is that when I say such things about science I am theorizing about the conditions for having knowledge of any kind. I am, so to speak, doing cognitive science without a license. What I am *not* doing is picking out a unique essence of science, meant to hold good for invertebrate zoology and for seismology and for particle physics (all kinds) and not to hold good for phrenology or for accountancy or for the empirical and theoretical projects of everyday life. I may be right or wrong in the domain of theorizing-about-knowledge-of-any-kind, but I am not theorizing about a unique scientific essence. And that is the matter at issue.

Again the question: how to be antiscientific? As I said, being against the essence of science and being against one or other metascientific story uniquely about science are not very good ways of being antiscientific, nor do I find that my skepticism about the Scientific Method frees me in any way and to any degree from belief in the existence of electrons or in DNA as the chemical basis of heredity. Those who are against the methodological or conceptual essence of science are against nothing very much in particular. And those who might be genuinely hostile to what they take to be the essence of science are probably just as ineffective as they are misguided. Who reads this stuff anyway? In order to corrupt the youth of Athens, you first have to get this stuff in their hands, then you have to get them to read it, and understand it, and care about it; then you have to persuade them—against the background of everything else they've been told—that you're right. Not such an easy business, really, as any teacher in my line of work knows.

But being against something in particular about science is both possible and legitimate. How to be against something in particular about science *should one wish to*? Here again it is good to listen to what some scientists themselves have to say. And if we listen to scientists (other than those who took the lead in the Science Wars), what we can hear is not a global defense of science, nor, of course, a global criticism of science. Rather, we can hear local criticisms of certain tendencies *within* science, or within parts of it— criticisms that are often substantial and vehemently expressed.

Some scientists are now violently critical of what *they* take to be the shallowness of reductionist programs, the tyrannizing and stultifying effects of bureaucratization in science, the dedicated following of scientific fashions and the attendant loss of the Big Picture and of imagination, the hegemony of Big Science at the expense of Little, the incompetence of the peer review system, the commercialization of science and the attendant ethical and intellectual erosion, and many other ills *they* diagnose in the contemporary Body Scientific. Some of these internal criticisms happen to look to professional metascience and even to the history of science for aids in understanding how current arrangements came to be and as tools in making things better; many do not.

It is not difficult to find these public internal criticisms: recent issues of biological periodicals are full of them, and memoirs and reflections by eminent scientists—including those by E. O. Wilson, Erwin Chargaff, Gunther Stent, and Richard Lewontin—are another rich seam of such criticisms. The striking thing, given the ultimate vacuity of the Science Wars, is just how little professional metascientists have concerned themselves with these *internal* contests, and, indeed, how little sociologists and historians have even noticed them as topics. That is almost certainly a Bad Thing: if being against science is, as I am suggesting, being against nothing very much in particular, being against the current peer review system, or against the hegemony of Big Science, or against the way in which clinical trials are constituted and funded, is being against something substantial and important. Is it sociologists' and historians' role to take sides in such debates? I don't think so (though I know of some sociologists who disagree). But these debates do offer a venue in which we can have interesting and substantial conversations with our scientist-colleagues. It would be mutually beneficial to have these conversations.

Finally, we need to remember that professional metascientists, like professional scientists, are also citizens. We are equal members—many of us—of institutions of higher education, and we all pay our share in the state support of scientific research. So far as the first type of citizenship is concerned, no one, I think, should rule it out of order, or identify it as *lèse majesté*, to take one side or another in university discussions over, for example, how much science should be taught in the curriculum or how scientific subjects should be taught. If one wants to say (as I do *not*) that there's too much science in the required curriculum, or if wants to say (as I *do*) that the philosophical, historical, or sociological dimensions should have a place in the science curriculum, then one should be free to do so. And, should one want to make such arguments, one should not have to face accusations of being antiscientific. Similarly, as citizens paying the bill for much scientific research, one should be free to say if one wants—on an informed basis—that the Superconducting Supercollider cost too much relative to its advertised benefits, or that too much money is going to a cure for AIDS and too little for an AIDS vaccine, or that governments have got their priorities wrong as between AIDS research and diarrhea research, or that some science supported by the public purse is trivial or intellectually unimaginative, or that certain links between publicly funded science and the commercial world are becoming worrying. And one should be able to say such things—again, if one wants—without being denounced as antiscientific. Some scientists say such things on a professional basis, and some citizens may want to say such things as responsible members of democratic societies. They must be free to do so, not intimidated into deferential silence.

The fear is that, if we carry on in our present courses, the ultimate and consequential casualties of the Science Wars will not be the job security of sociologists of science, but free, open, and informed public debate about the health of modern science. And the health of science ultimately depends on that debate.

Here are the sources for the notorious metascientific claims at the beginning of this essay:

1. Many sources, including Peter B. Medawar (immunologist), *The Art of the Soluble* (London: Methuen, 1957), p. 132; James B. Conant (chemist), *Science and Common Sense* (New Haven, CT: Yale University Press, 1951), p. 45; Lewis Wolpert

(biologist), *A Passion for Science* (Oxford: Oxford University Press, 1988; compiled with Allison Richards), p. 3; Richard Lewontin, "Billions and Billions of Demons," *New York Review of Books* 44, no. 1 (9 January 1997): 28–32, on p. 29: "The case for the scientific method should itself be 'scientific' and not merely rhetorical."

2. Erwin Chargaff (biochemist), *Heraclitean Fire: Sketches from a Life before Nature* (New York: Rockefeller University Press, 1978), p. 138.

3. Edward O. Wilson (entomologist, sociobiologist), *Naturalist* (New York: Warner Books, 1995), p. 210.

4. Niels Bohr, quoted in Abraham Pais, *Niels Bohr's Times, in Physics, Philosophy, and Polity* (Oxford: Clarendon Press, 1991), p. 314.

5. Albert Einstein (physicist), *Out of My Later Years* (New York: Philosophical Library, 1950), p. 96; also idem, *Ideas and Opinions* (New York: Crown Publishers, 1954), p. 355. I have here slightly paraphrased Einstein's original statement that the bases of physics cannot be inductively secured from experience, but "can only be attained by free invention." Geometrical axioms—the bases of the deductive structure of physics—are, Einstein, said, "free creations of the human mind" (*Ideas and Opinions*, p. 234).

6. Jacob Bronowski (mathematician), "Science is Human," in *The Humanist Frame,* ed. Julian Huxley (New York: Harper and Brothers, 1961), pp. 83–94, on p. 88. I have here altered the first-person "we" to the third-person "scientists."

7. Warren Weaver (mathematician and scientific administrator), "Science and People," in *The New Scientist: Essays on the Methods and Values of Modern Science,* ed. Paul C. Obler and Herman A. Estrin (Garden City, NY: Anchor, 1962), pp. 95–111, on p. 104.

8. Gunther Stent (biochemist), interviewed in Lewis Wolpert and Allison Richards, *A Passion for Science* (Oxford: Oxford University Press, 1988), p. 116.

9. Brian Petley (physicist), *The Fundamental Physical Constants* (Bristol: Adam Hilger, 1985), p. 2: "Modern physics is based on some intrinsic acts of faith, many of which are embodied in the fundamental constants."

10. Richard Lewontin (evolutionary geneticist), "Billions and Billions of Demons," p. 31: "[The public] take the side of science *in spite* of the patent absurdity of its constructs, *in spite* of its failure to fulfill many of its extravagant promises of health and life, *in spite* of the tolerance of the scientific community for unsubstantiated just-so stories, because we have a prior commitment, a commitment to materialism."

11. Richard Lewontin, Steven Rose (neurobiologist), and Leon J. Kamin (psychologist), *Not in Our Genes: Biology, Ideology, and Human Nature* (New York: Pantheon, 1984), p. 33; see also "The internalist, positivist tradition of the autonomy of scientific knowledge is itself part of the general objectification of social relations that accompanied the transition from feudal to modern capitalist societies" (p. 33). It would not be easy to find such a sweepingly didactic statement expressed by present-day historians or sociologists of science!

CHAPTER 4

Science and Prejudice
in Historical Perspective

There has always been, and there probably always will be, an intimate connection between how people recognize good knowledge and how they conceive a good society. The relationship is not one of mere analogy; it is constitutive. The making of reliable, objective, and robust knowledge—that is, for us, *science*—is accomplished by communities of expert human beings—scientists—and it has widely been supposed unlikely that an unjust community of knowers can produce anything but distorted knowledge: not science but ideology or dogma or error. Verity and virtue march in lock-step through history, as do error and evil. The Republic of Science is modeled on the City of God.

And that is perhaps why the communities of authentic scholars and scientists have intermittently been held up to the wider society as models of communal virtue. If, it has been said, matters were ordered in the wider society as they are in the Republic of Science, the result would not only be a more knowledgeable society but a more just society. Observe and imitate. We find it hard to imagine—we may even find it inconceivable—that the knowledge fueling social hatred could emerge from a properly organized and properly regulated intellectual community. How could a genuine Republic of Science produce such pathologies as racialist biology or the imputation of so-called "Aryan" or "Jewish" physics?

The late twentieth and early twenty-first centuries inherit from the European historical past several ways of talking about the good society that uniquely produces good knowledge. By far the most influential of these ways of specifying the relations between knowledge and virtue comes down to us from the Scientific Revolution of the seventeenth century and the Enlightenment of the eighteenth.[1] The story goes like this: human beings are intellectually imperfect and limited; they are subject to tidal currents of passion and interest; these currents flow against their rational faculties and hinder or distort the operation of rationality. Instead of reckoning rightly or seeing what is authentically there to be seen, passion- and interest-influenced human beings tend to think and see not what *is* but what they *wish* to be the case.

Francis Bacon enumerated the "Idols" that corrupt human judgment and perception. We are prone to mistake because each of us has his or her individual prejudices or enthusiasms (the "Idols of the Cave"), because we use the common language and misapply words to things (the "Idols of the Market-Place"), because some of us are drawn to the dogmas of systematic philosophies (the "Idols of the Theatre"), and, lastly, because we are all emotional and interested people who tend to accept the evidence of our own senses when the information delivered by those senses is likely to be unreliable, colored by emotion and interest (the "Idols of the Tribe").[2] We are human, all-too-human; we are fallen and fallible; and such is our fate. The order of human thought will never reflect the order of reality unless we can purposefully join a reflective awareness of our imperfect nature to our own cultural innovations—innovations that have the capacity to discipline, to manage, and to mitigate the corrupting effects of the distortions to which human beings are prone. How can we, if it is not possible to eliminate prejudice and bias, nevertheless make their workings as harmless as possible?

In seventeenth-century Europe these innovations took several forms. Most were methodological. If only the fallible human mind was directed and disciplined by rational method, then our knowledge would be well founded. We would be able to penetrate to the hidden causal structure of things, or, if not that far, we would at least recognize the limits of what human beings could know with certainty. The rational methods offered up by early modern philosophers varied enormously—the English tended to embrace Baconian induction and probabilism, the French, Cartesian deduction and logical certainty—but method was meant to mend all. Yet method could do little by itself: it had to be encouraged, implemented, and enforced by intellectual communities, whose members willingly accepted and enforced its disciplines. The rational methods that were supposed to deliver correct knowledge had to be embedded within communities of virtue. Rational method required, so to speak, its ventriloquists—those collectively organized bodies that could speak with authority in its favor and whose virtues could commend it.[3]

These virtues were both intellectual and moral. Even to enumerate the norms, obedience to which was supposed to guarantee proper knowledge of the natural world, is to list what was, and is, widely counted as proper behavior in the social world.[4] So seventeenth-century scientific and philosophical reformers commended an openness and freedom of mind. The genuine philosopher was a person of integrity and the Republic of Science was a community of integrity. René Descartes, for example, announced his personal openness and freedom by claiming that he had set aside everything he had been taught at the Jesuit college of La Flèche, and Robert Boyle later celebrated his openness and freedom by saying that he hadn't read Descartes.[5] For Boyle, as for Bacon, commitment to existing systems of thought compromised the authenticity of individual judgment—whether that was individual sensory experience of the world or individually conducted trains of rational reflection—and such commitments just had to be shrugged off. Traditional knowers were weighed down by accretions of custom and convention whose acceptance caused both the unfreedom of knowers and the distortion of their knowledge. The modern knower was supposed to be an unencumbered self whose freedom allowed his knowledge to mirror nature.[6] There was nobility in that view, for, as modernist rhetoric put the question, who would rather be a slave to Aristotle than to submit his judgment alone to the tribunes of reason or reality?

And who would ignobly accept another's word when they could—by experiment or observation—submit themself directly to nature's testimony? So it was not just Aristotle and ancient authority that had to be set aside; routine reliance upon the testimony of other contemporary observers was likewise rejected. If you really want to secure truth about the natural world, you must not only forget tradition and ignore authority; you must also be skeptical of what others say about the world and you must rely only upon what you yourself can see and show. As John Locke said, "we may as rationally hope to see with other men's eyes, as to know by other men's understandings. So much as we ourselves consider and comprehend of truth and reason, so much we possess of real and true knowledge . . . In the sciences, every one has so much as he really knows and comprehends. What he believes only, and takes on trust, are but shreds." To know about the natural world was to know *by yourself*.[7]

Within that reformed and virtuous community of knowers, all were to be accounted equal. The Republic of Science was no traditional school, where some taught and the rest submitted. Instead, several of the new scientific societies founded from the middle of the seventeenth century insisted upon the equal capacity of all men to make contributions to the stock of knowledge: all men, whatever their wealth, their nation, or their social station. If the wider society was hierarchical, and riven by distinctions of religion, rank, and nation, there was every reason for the Republic of Science to reject these distinctions as prejudices. The Thirty Years' War had recently written the lesson of such prejudices in rivers of European blood, and Gottfried Leibniz was not alone in trying to build a new pan-European harmony on the model of the reformed scientific academy.[8] It was considered that the Republic of Science simply could not afford to be carved up into cultural nation-states with divided confessional allegiances in the same way as political Europe. There was no place for intolerant and divisive patriotism or religious bigotry in a community whose products counted as universal knowledge about a shared and mutually accessible reality. In seventeenth- and eighteenth-century Europe, as in ancient Greece, philosophers might describe themselves not as citizens of Athens or Rome but as "citizens of the world." If European nations and faiths battled each other, still, it was recurrently said, "the sciences were never at war."[9]

So Edward Gibbon wrote that "It is the duty of the patriot to prefer and promote the exclusive interest and glory of his native country but a philosopher must be permitted to enlarge his views and to consider Europe as one great Republic whose various inhabitants have attained almost to the same level of politeness and cultivation." Just as the object of scientific and philosophical inquiry was the whole world, so the authentic intellectual was said to be a stranger nowhere in the world, subject to no merely national or confessional constraints: "Die Gedanken sind frei." Montesquieu wrote that "I prefer my family to myself, my country to my family, but the human race to my country," and he was much admired by the Scot David Hume for so saying. The man of science or letters—free of national or confessional prejudice—was the only authentic European.[10] Freedom from intellectual prejudice; freedom from divisive parochialisms; freedom from custom and tradition; freedom from arbitrary or conventional authority; freedom from distinctions among the ranks and sorts of human beings—this was a vision of philosophical virtue handed down from the seventeenth-century Scientific Revolution to the eighteenth-century Enlightenment and from the Enlightenment to liberal and pluralistic traditions of nineteenth- and twentieth-century culture and social thought.

In eighteenth-century France this vision could be politically radical. Just because the Republic of Science was free of prejudice, arbitrary authority, and distinction of rank, its advocates could advertise it as a standing criticism of Old Régime society. Seven years before the storming of the Bastille, a radical French journalist celebrated the egalitarianism of the ideal scientific community, which "can know neither despots, nor aristocrats, nor electors . . . To admit a despot, aristocrats, or electors who by edicts set a seal upon the products of geniuses is to violate the nature of things and the liberty of the human mind. It is an affront to public opinion which alone has the right to crown genius."[11] And a year after the Bastille was taken, Condorcet's *éloge* of the American Benjamin Franklin announced that "Forever free amidst all manners of servitude, the sciences transmit to their practitioners some of their essence of independence or either fly from countries ruled by arbitrary power or gently prepare the revolution that will eventually destroy it."[12] Some English natural philosophers hoped for their own (kinder, gentler) revolution, and likewise pointed to the Republic of Science as a model of social equality and justice. The chemist Joseph Priestley wrote that "Any man has as good a power of distinguishing truth from falsity as his neighbours"; "This rapid progress of knowledge will, I doubt not, be the means under God of extirpating all error and prejudice, and of putting an end to all undue and usurped authority in the business of religion as well as of science . . . The English hierarchy (if there be anything unsound in its constitution) has . . . reason to tremble even at an air pump or an electrical machine."[13]

The liberal vision of social equality and fairness modeled on the scientific community continued in vigor in the twentieth century. The founding father of the sociology of science—the influential American sociologist Robert K. Merton—famously described the unique and effective "norms of science"— those values held dear by the scientific community and enforced upon its members. Just on the condition that the scientific community embraces these values and punishes those who betray them, that community will be able to fulfill its institutional function of extending certified knowledge.[14] First articulated in the late 1930s and early 1940s, Merton's norms nonetheless drew upon sentiments about the scientific community and scientific practice expressed from the seventeenth century. Members of that community should be, and Merton said they were, open-minded, skeptical, disinterested, and universalistic. They obeyed no irrelevant distinctions of rank, race, religion, or sex, evaluating all contributions to scientific knowledge on intellectual merit alone. Science, indeed, was the great meritocracy of modern times. Submission to these norms guaranteed the production of authentic science, while at the same time holding up to Western society a mirror of what it ought to be.[15] Merton meant readers to draw lessons from epistemic virtue to social justice. The situation in Nazi Germany, he wrote, is an object lesson in how political interference with scientific norms yields such corrupt products as "Aryan physics." That scientized corruption then goes on to fan the flames of social hatred. When external social prejudice intrudes itself upon scientific judgment, the result is at most only the appearance of objective science but no longer its reality.

Knowledge free of prejudice—coming to experience without any prior judgment or expectation—and knowledge that is objective because of that freedom; guaranteed by a community that had found a formula for keeping prejudice at bay, a community whose unique virtue could provide a concrete model for social virtue. It is a noble vision and it has been called the Enlightenment Vision. Yet that Vision is still with us. And those who have dissented from it have done so in the cause of some of the greatest crimes committed in the present century. For these and other reasons, one criticizes the Enlightenment Vision with caution. Yet exempting that Vision from dispassionate and disinterested scrutiny would be to betray it, just as it would be to accept its historical accuracy merely on the grounds of faith, because it counted as an authoritative statement about the nature of science and the communities that have produced it. Freedom from arbitrary and destructive prejudice is so appealing and important as a social goal that one justification for criticizing the Enlightenment Vision is the hope that such freedoms might be *better* secured by other means.

There are two major faults with the Enlightenment Vision of the relations between science, prejudice, and social virtue. The first may be called historical and the second moral and political. Descartes was wrong: it is not humanly possible to build a house of knowledge on foundations wholly new, wholly to reject traditionally received knowledge; and his own medical and physiological writings—owing so much to Galenical thought—prove him wrong out of his own mouth. Locke was wrong too: to reject taking knowledge on trust is not to have a purer form of knowledge but to have no knowledge at all. How else could Robert Boyle know the shape and size of icebergs off the coast of Newfoundland—and he *did* know such things— except by trusting trustworthy sources? How else could the community of seventeenth-century English astronomers know the apparent motion of the comet of 1665 across the skies except through trusting the observations of astronomers in Poland, Germany, Italy, Spain, and even at a rustic institution in Massachusetts called Harvard?[16] That reliance on trust and testimony might be—indeed, ought to be—thought innocuous. Making scientific knowledge is a communal matter, and scientists—whatever their expressed skepticism or their faith in method—have the same task as the rest of us, deciding whom to trust. And there is no unambiguous rational formula for making that decision.

Trust is central to social order, but the attribution of trustworthiness is not equally distributed among all human beings. Early modern scientists often reflected about whom to trust, but their solutions broadly followed the contours of social power. The word of gentlemen might be trusted—the honor-code stipulated that it must be trusted—while the testimony of the vulgar, the unlettered, and women might be held suspect. So when the un-educated Dutch draper and microscopist Antoni van Leeuwenhoek reported seeing hosts of small animals in a drop of pond-water, the gentlemen of the Royal Society required that his skill and probity be vouched for—not by equivalent experts, for there were none as skilled as him, but by the ministers and lawyers of Delft.[17]

But there are other respects in which the Enlightenment Vision bears historical scrutiny, and these might be thought less innocuous, and more central to concern with social prejudice. So, when, for example, members of the seventeenth- or eighteenth-century Republic of Science announced their openness to everyone, the everyone they had in mind was definitely *not* all human beings. Their academies, their salons, and their deliberations included few non-Europeans and few women. The domestic membership of the Royal Society of London in the seventeenth century encompassed few Roman Catholics, almost no Quakers, no Jews, and no women. The Paris Academy of Sciences was more catholic—in both senses of the word—but, after the Revocation of the Edict of Nantes in 1685, French academic life just became impossible for Huguenot scholars.[18]

The exclusions of women and Jews at least are pretty much what you would expect of seventeenth-century English literate culture, largely matters of course. Like the vast majority of the illiterate poor, they had no effective access to the institutions of higher education, and so one did not *have to* exclude them by statute, just because they were effectively excluded by lack of relevant expertise. So one historian has wittily written that the Royal Society of London was open to "everybody" in just the same way that the Ritz Hotel is open to "everybody."[19] The warning "Let no one ignorant of mathematics enter here"—inscribed on the gateway to Plato's Academy—is socially innocuous just on the condition that everyone has equal access to acquiring mathematical expertise. But there was no such equal access. How sure are we that there now is? Intellectual communities not open to all are prone to produce knowledge unattractive to those who are excluded. So the condition for a biology supporting the inferiority of women or Jews or Slavs is a biological community purely composed of males or Gentiles or non-Slavs, or, at least, a community whose structure of authority is dominated by them. The independent causal consideration here is not a science of hatred but the system of exclusions that permit such a science ever to develop.[20]

Both the cosmopolitanism of the Republic of Science and the universalism of its knowledge were key articles of its advertised social justice. But the cosmopolitanism needs qualification and the universalism needs to be treated with caution. Until the late eighteenth century confessional allegiances were almost certainly more important criteria to be rejected by the virtuous Republic of Science than were national origins, just because centralized nation-states were relatively weaker than they later became, and because science in the seventeenth and eighteenth centuries could contribute far less to state power than it later came to do.[21] If "the sciences were never at war" in the early modern period, by the time of Los Alamos and Peenemünde no such global statement was any longer sensible.

Latin was the universal language of the Republic of Science through much of the seventeenth century—the intellectual Euro-currency of its time—but its universality was very like that of mathematical and scientific expertise. When it was said that "everyone spoke it," the sense, of course, was "everyone that mattered."[22] Moreover, the very claims to universal knowledge and universal method that give the air of nobility to the Enlightenment vision have their darker sides. By the early eighteenth century, as Isaiah Berlin noted in one of his more lapidary essays, such critical voices as that of the Neapolitan Giambattista Vico were bridling at the illegitimate vaulting ambition—and the cultural intolerance—contained in the conviction that to every question there was only one true answer, "true universally, eternally and immutably."[23]

For these and many other reasons, the historical case for the existence of knowledge without prejudice is not good. Knowledge free of prejudice has not been obtained in historical practice, and, it is probably impossible to obtain in principle. The Republic of Science seems rather to reflect the most widely distributed prejudices of its time and of its citizens. And, insofar as these are so widely distributed, they may appear to its citizens as no prejudices at all, though hindsight (if not academic history) reserves the right to judge otherwise. For the historian to *understand* the general taken-for-grantedness of early modern exclusions of Jews, women, and the vulgar is not—of course, it cannot be—to approve them for us.

For all that, the rejection of prejudice is neither empty of content nor devoid of consequence. Properly understood, such rejection is not absolute but relative, not global but partial. Rhetoric rejecting authority and testimony translates as the rejection of certain *kinds* of authority and certain *types* of testimony; rhetoric commending the openness of intellectual forums means openness to relevant others; the participation of "all people" decodes as everyone possessing the credentials deemed necessary for competent participation. And, because the rejection of prejudice cannot be absolute, we have no universal rational formula for which prejudices to reject and which we cannot reject. Absent such formulas, both intellectuals and others have to do the best they can.

The absolutist version of the Enlightenment Vision is also at the root of its moral and political faults. While there is nobility in the vision, there is also the possibility of hubris and inhumanity. In its absolutist form it charges the individual with terrifying responsibility: it leaves us all alone, mistrustful of our emotions and instincts, mistrustful of our community's customs, and mistrustful our ancestors' wisdom. In the name of the liberty of everyman, in its purer forms it enjoins everyman to be skeptical of every other man.[24] Edmund Burke's reflective defense of reasoned prejudice picked out just this point. French Jacobin philosophers and politicians have, Burke wrote, "no respect for the wisdom of others; but they pay it off by a very full measure of confidence in their own." The English were more sensible—this was Burke's prejudice: "We are afraid to put men to live and trade each on his private stock of reason; because we suspect that this stock in each man is small, and that the individuals would do better to avail themselves of the general bank and capital of nations, and of ages. Many of our men of speculation, instead of exploding general prejudices, employ their sagacity to discover the latent wisdom which prevails in them. If they find what they seek, and they seldom fail, they think it more wise to continue the prejudice, with the reason involved, than to cast away the coat of prejudice, and to leave nothing but the naked reason; because prejudice, with its reason, has a motive to give action to that reason, and an affection which will give it permanence."[25]

If there is much nobility in the Enlightenment Vision, there is much humanity in recognizing its limits. We depend for our knowledge not just on our individual reason and individual experience but on our ancestors and on each other. If we expect to know together, then we must expect to live together, in all our diversity. Just as people have hated in the name of their religion, their sex, their nation, and their race, so they have expressed intolerance and committed injustices in the name of the one universal reason that secures the one universal truth about the world.

So the moral of the story told by a historian of science is at once simple and endlessly complex. Knowledge without prejudice is not possible and neither is social life. Prejudice can be selectively managed and disciplined but it cannot be eliminated. We have to pick out those prejudices that we find intolerable and oppose them as vigorously as we can with whatever resources we can. But we are going to have to do so without a rational master formula derived from the history of the Republic of Science.[26]

PART II

Places and Practices

There are two notions that have been closely attached to the cultural value placed on scientific knowledge. One is placelessness. Of all forms of culture, science has been thought least marked by the places in which it is made and evaluated. The universal validity of scientific knowledge has been taken as testimony to the irrelevance of the particular physical and social sites in which it happened to be produced. The place of science is at once everywhere and nowhere. Assuming or insisting upon placelessness is a move in practical epistemology—a way of stipulating intellectual *value*. But suppose we take a close look at the culturally demarcated physical sites from which scientific knowledge has historically emerged—houses, courts, laboratories, workshops, cities (and parts of cities), and so forth. Who was present, and who absent, in a range of science-making scenes? What sorts of people, behaving in what sorts of ways, with what sorts of purposes and carrying with them what sorts of cultural baggage? How was scientific knowledge physically and culturally transferred from one place to another, and how did it circulate within specific places? And what follows from the answers to these questions for historical appreciations of the nature and authority of different types of knowledge? A similar distinction-making role is played by understandings that science is *discovered* rather than *made*. We have always known that making scientific knowledge involves things like the construction and plying of instruments, the mobilization and evaluation of testimony, the making of textual and visual representations, and the persuasive techniques used in winning of credibility for claims made. What we have not always appreciated is the historical interest of such things, that all these things, and many more aspects of mundane practice, constitute the human and material stuff out of which scientific knowledge is produced, and that they themselves work to secure the idea of scientific knowledge as discovered rather than invented.

CHAPTER 5

The House of Experiment in Seventeenth-century England

That which is not able to be performed in a private house will much
less be brought to pass in a commonwealth and kingdom.

WILLIAM HARRISON, *The Description of England* (1587)

My subject is the place of experiment. I want to know where experimental
science was done. In what physical and social settings? Who was in attendance at the scenes in which experimental knowledge was produced and
evaluated? How were they arrayed in physical and social space? What were
the conditions of access to these places, and how were transactions across
their thresholds managed?

The historical materials with which I am going to deal are of special interest. Seventeenth-century England witnessed the rise and institutionalization of a program devoted to systematic experimentation, accompanied by
a literature explicitly describing and defending practical aspects of that program. Nevertheless, the historiography informing this essay may prove of
more general interest. Historians of science and ideas have not, in the main,
been much concerned with the siting of knowledge production.[1] This essay
offers reasons for systematically studying the venues of knowledge. I want
to display the network of connections between the physical and social settings of inquiry and the position of its products on the map of knowledge. I
show how the siting of knowledge-making practices contributed toward a
practical solution of intellectual problems. The physical and the symbolic
siting of experimental work was a way of bounding and disciplining the
community of practitioners; it was a way of policing experimental discourse;
and it was a way of publicly warranting that the knowledge produced in
such places was reliable and authentic. That is to say, the place of experiment counted as a partial answer to the fundamental question, Why should
one give one's assent to experimental knowledge-claims?

I start by introducing some connections between empiricist processes of
knowledge-making and the spatial distribution of participants, pointing to

the ineradicable problem of trust that is generated when some people have direct sensory access to a phenomenon and others do not. I then mobilize some information about where experimental work was in fact performed in mid- to late seventeenth-century England, focusing upon sites associated with the work of the early Royal Society and two of its leading fellows, Robert Boyle and Robert Hooke. The question of access to these sites is then considered: who could go in and how was the regulation of entry implicated in the evaluations of experimental knowledge? The public display of the moral basis of experimental practices depended upon the form of social relations obtaining within these sites as much as it did upon who was allowed within. Indeed, these considerations were closely related, and I discuss how the condition of gentlemen and the deportment expected of them in certain places bore upon experimental social relations and, in particular, upon the problems attending the assessment of experimental testimony. The essay concludes by analyzing how the stages of experimental knowledge-making mapped onto physical and symbolic patterns of movements within the rooms of a house, particularly the circulation between private and public spaces.

On the Threshold of Experiment

The domestic threshold marks the boundary between private and public space. Few distinctions in social life are more fundamental than that between private and public.[2] The same applies to the social activities we use to make and evaluate knowledge: on either side of the threshold the conditions of our knowledge are different. While we stand outside, we cannot see what goes on within, nor can we have any knowledge of internal affairs but what is related to us by those with rights of access or by testimony still more indirect. What we cannot see, we must take on trust, or trust being withheld, continue to suspect. Social life as a whole and the social procedures used to make knowledge are spatially organized.[3] The threshold is a social marker: it is put in place and maintained by social decision and convention. Yet once in place it is a constraint upon social relations. The threshold acts as a constraint upon the distribution of knowledge, its content, quality, conditions of possession, and justification, even as it forms a resource for stipulating that the knowledge in question really is the thing it is said to be.

Within empiricist conceptions of knowledge the ultimate warrant for a claim to knowledge is an act of witnessing. The simplest knowledge-producing scene one can imagine in an empiricist scheme would not, strictly speaking, be a social scene at all. It would consist of an individual, perceived as free and competent, confronting natural reality outside the social system. Although such a scene might plausibly be identified as the paradigm case of knowledge production in seventeenth-century empiricist writings, it was not, in fact, recommended. Three sorts of problems were recognized to attend the privacy of solitary individual observation.[4] First, the transformation of mere belief into proper knowledge was considered to consist of the transit from the perceptions of cognitions of the individual to the culture of the collective. Empiricist writers therefore looked for the means by which such a successful transit might be managed. The second problem was connected with the view that the perceptions of postlapsarian people were corrupt and were subject to biases deriving from interest. Although these factors could not be eliminated, their consequences might be mitigated by ensuring that both witnessing and the consideration of knowledge-claims took place in a certain kind of social setting. Third, there were often contingent practical problems attending the circumstances of observation, which meant that social relations of some sort had to be established for the phenomena in question to be dealt with. Certain observations, particularly in natural history sciences but also in experimental science, could, for instance, be made only by geographically privileged persons. In such cases, there was no practical way by which a witnessing public could be brought to the phenomena or the phenomena brought to the public. Testimony was therefore crucial: the reception of testimony constituted a rudimentary social scene, and the evaluation of testimony might occur in an elaborately constructed social scene.

English empiricists did not think that testimony could be dispensed with, but they worked strenuously to manage and discipline it. Most empiricist writers recognized that the bulk of knowledge would have to be derived from what one was told by those who had witnessed the thing in question, or by those who had been told by those who had been told, and so on. If, however, trust was to be a basis for reliable knowledge, the practical question emerged: whom was one to trust? John Locke, among others, advised practitioners to factor the creditworthiness of the source by the credibility of the matter claimed by that source.[5] One might accept the report of an implausible phenomenon from a creditworthy source and reject plausible claims from sources lacking that creditworthiness. Credibility as an attribute ascribed to people was not, therefore, independent of views of what the world was like. One might calibrate persons' credibility by what it was they claimed to have witnessed, just as one might use their accepted credibility to gauge what existed in the world.[6] Nevertheless, credibility has other sources: certain *kinds* of people were independently known to be more trustworthy sources than others. Roughly speaking, the distribution of credibility followed the contours of English society, and that it did was so evident that scarcely any commentator felt obliged to specify the ground of this creditworthiness. In such a setting one simply knew what sorts of people were credible, just as one simply knew whose reports might be suspect.[7] Indeed, in certain instances Robert Boyle recommended that one ought to credit the testimony of things rather than the testimony of certain types of persons. Discussing one of his hydrostatical experiments of the 1660s, Boyle argued that "the pressure of the water in our . . . experiment having manifest effects upon inanimate bodies, which are not capable of giving us partial informations, will have much more weight with unprejudiced persons, than the suspicious, and sometimes disagreeing accounts of ignorant divers, whom prejudicate opinions may much sway, and whose very sensations, as those of other vulgar men, may be influenced by predispositions, and so many other circumstances, that they may easily give occasion to mistakes."[8] When in 1667 the Royal Society wished to experiment on the transfusion of animal blood into a human being, they hit upon an ingenious solution to the problem of testimony posed by such an experiment. The subject, Arthur Coga, was indigent and possibly mad (so it was expedient to use him), but he was also a Cambridge graduate (so his testimony of how he felt on receipt of sheep's blood might be credited).[9]

Experimental Sites

One of the considerations that recommended the program of systematic artificial experimentation launched in the middle of the seventeenth century by Boyle and his associates was that experimental phenomena could be arranged and produced at specific times and places. Such phenomena were disciplined, and disciplined witnessing might be mobilized around them. What sorts of places were available for this program? What conditions and opportunities did they provide? Put simply, the task resolves into the search for the actual sites of seventeenth-century English experiment. Where and what was the laboratory?

Two preliminary cautions are necessary. The first is a warning against linguistic anachronism. The word *laboratory* (or *elaboratory*) was not in common English usage at the middle of the seventeenth century. For example, despite his extensive description of ideal experimental sites, I cannot find the word used by Francis Bacon, in *The New Atlantis* or elsewhere. As Owen Hannaway has shown, there is some evidence of medieval Latin usage (*laboratorium*), but the word did not acquire anything of its modern sense until the late sixteenth century. It seems that the word was transmitted into English usage in the late sixteenth century, carrying with it alchemical and chemical resonances.[10] Among scores of English uses I have registered through the 1680s, I have not encountered one in which the space pointed to was one without a furnace, used as a non-portable source of heat for chemical or pharmaceutical operations. The word did become increasingly common during the course of the seventeenth century, although even by the early eighteenth century it was not used routinely to refer to just any place dedicated to experimental investigation.[11] On the founding of the Royal Society, there were a number of plans for purpose-built experimental sites, none of which materialized, even though the new Oxford Ashmolean Museum (1683) did contain a chemical laboratory in its basement.[12] By the end of the century there still did not exist any purpose-designed and purpose-built structure dedicated to those non-heat-dependent sciences (such as pneumatics and hydrostatics) that were paradigmatic of the experimental program. The new experimental science was carried on in existing spaces, used just as they were or modified for the purpose.

Second, the status of spaces designated as laboratories and of experimental venues generally in seventeenth-century England was intensely contested. Were they private or public, and what status ought they to have? In the rhetoric of English experimental philosophers, what was wrong with existing forms of practice was their privacy. Neither the individual philosopher in his study nor the solitary alchemist in his "dark and smokey" laboratory was a fit actor in a proper setting to produce objective knowledge.[13] In contrast, spaces appropriate to the new experimental program were to be public and easy of access. This was the condition for the production of reliable knowledge within.[14] In stipulating that experiment was to take place in public spaces, experimental philosophers were describing the nature of the physical and social setting in which genuine knowledge might be made.

The performance and the consideration of experimental work in mid- to late seventeenth-century England took place in a variety of venues. These sites ranged from the apothecary's and instrument-maker's shop, to the coffeehouse, the royal palace, the rooms of college fellows, and associated collegiate and university structures. But by far the most significant venues were the private residences of gentlemen or, at any rate, sites where places of scientific work were coextensive with places of residence, whether owned or rented. The overwhelming majority of experimental trials, displays, and discussions that we know about occurred within private residences. Instances could be enumerated *ad libitum:* the laboratory equipped for Francis Mercury van Helmont at Anne Conway's Ragley House in Warwickshire; the role of Towneley House in Lancashire in the career of English pneumatics; Clodius's laboratory in the kitchen of his father-in-law Samuel Hartlib's house in Charing Cross; Kenelm Digby's house and laboratory in Covent Garden after the Restoration; the Hartlibian laboratory worked by Thomas Henshaw and Thomas Vaughan in their rooms at Kensington; William Petty's lodgings at Buckley Hall in Oxford, where the Experimental Philosophy Club originated in 1649; Thomas Willis's house, Beam Hall, where the club met during the early 1660s.[15]

In the following sections I display the conditions and opportunities presented by the siting of experiment in the private house. In particular, I point to the role of conditions regulating access to such venues and to conventions governing social relations within them. I argue that these conditions and conventions counted toward practical solutions of the questions of how one produced experimental knowledge, how one evaluated experimental claims, and how one mobilized and made visible the morally adequate grounds for assenting to such claims. To this end I concentrate on three of the most important sites in the career of experiment in mid- to late seventeenth-century England: the various residences and laboratories of Robert Boyle, the meeting places of the Royal Society of London, and the quarters occupied by Robert Hooke.

ROBERT BOYLE

Boyle had laboratories at each of the three major residences he successively inhabited during his mature life. From about 1645 to about 1655, he was mainly in residence at the manor house of Stalbridge in Dorset, an estate acquired by his father, the first Earl of Cork, in 1636 and inherited by his youngest son on the earl's death in 1643. By early 1647, Boyle was organizing a chemical laboratory at Stalbridge, perhaps with the advice of the Hartlibian circle, whose London laboratories he frequently visited.[16] Late in 1655 or early in 1656, he removed to Oxford, where his sister Katherine, Lady Ranelagh, had searched out rooms for him in the house of the apothecary John Crosse, Deep Hall, in the High Street. He was apparently able to use Crosse's chemical facilities, and his own rooms contained a pneumatic laboratory, where, assisted by Hooke, the first version of the air-pump was constructed in 1658–1659.[17] During his Oxford period, Boyle also had access to a retreat at Stanton St. John, a village several miles to the northeast, where he made meteorological observations but probably did not have a laboratory of any kind.

Boyle was away from Oxford for extended periods, staying sometimes at a house in Chelsea, sometimes with Katherine in London, and sometimes with another sister, Mary Rich, Countess of Warwick, at Leese (or Leighs) Priory in Essex.[18] In 1666, he had Henry Oldenburg look over possible lodgings in Newington, north of London, but there is no evidence he ever occupied these. And he periodically stayed at Beaconsfield in Buckinghamshire, possibly at the home of the poet Edmund Waller. But Oxford remained his primary residence and experimental workplace until he moved into quarters with Katherine at her house in Pall Mall in 1668. This was a house (actually two houses knocked into one) assigned to Lady Ranelagh by the Earl of Warwick in 1664. It stood on the south side of Pall Mall, probably on the site now occupied by the Royal Automobile Club. Although luxury building in this area was proceeding apace in the Restoration, at the time Boyle moved in Pall Mall still retained a rather quiet and semi-rural atmosphere. During the 1670s, Boyle's neighbors included Oldenburg, Dr. Thomas Sydenham, and Nell Gwyn.

Boyle's laboratory in Katherine's house was probably either in the basement or attached to the back, and there is some evidence to suggest that one could obtain access to the laboratory from the street without passing through the rest of the house.[19] The unmarried Boyle seems to have dined regularly with his sister, who was a major social and cultural figure in her own right, living "on the publickest scene," and who entertained his guests at the family table.[20] He remained there until his death in 1691, which closely followed Katherine's.

THE ROYAL SOCIETY

After its founding in 1660, the Royal Society held weekly meetings in Gresham College in Bishopsgate Street, originally in the rooms of the professor of geometry, afterward in rooms specifically set aside for its use. The Great Fire of London in September 1666 made Gresham College unavailable, and temporary hospitality was extended by Henry Howard, later sixth Duke of Norfolk, at Arundel House, his residence in the Strand. The society met there for seven years, from 1667 to 1674, until Gresham became available again. Gresham continued to be its home until 1710, when the society for the first time became the owner of its premises, purchasing the former home of a physician, Crane Court in Fleet Street.[21]

During the 1660s and 1670s, the society was continually searching for alternative accommodation and making plans, all of which proved abortive, for purpose-built quarters of its own. In the event, for the first half century of its existence the public business of the Royal Society was transacted largely within private residences. Arundel House was unambiguously such a place, and Gresham College, built in the late sixteenth century as the residence of the great merchant banker Sir Thomas Gresham and transformed into a place of public instruction in 1598, had by the 1660s changed its character. When the Royal Society met there, it was a place where some professors lived and taught; where other sinecurist professors lived and did not teach; and where still others, who were not professors, lived in quarters hired out to them. According to its modern historian, Gresham College had by the mid-1670s "declined from a seat of learning into a lodging house." The significance of Arundel House in seventeenth-century English culture and social life cannot be overestimated. Until his death in 1646, it was the residence of Thomas Howard, second Earl of Arundel, who (despite the Catholicism he abandoned in 1616) as Earl Marshal was the head of the English nobility and the "custodian of honor." Arundel was one of the greatest collectors and patron of the arts of his age, and the house that contained his collections was made into a visible symbol of how a cultivated English gentleman ought to live. Indeed, his patronage of the educationalist Henry Peacham resulted in the production of an influential vade mecum for the guidance of English gentlemen. Arundel and his circle set themselves the task of modeling and exemplifying the code of English gentility, drawing liberally upon Italianate patterns. His grandson Henry Howard continued the great Arundel's proclivities, and it was through the encouragement of his friend John Evelyn that the society was offered space in the gallery of Arundel House and, ultimately, became one of the beneficiaries of the celebrated Arundel Collection of books, manuscripts, and objets d'art.[22]

On the founding of the Royal Society Robert Hooke was still serving Boyle as his technical assistant, lodging with Boyle in Oxford and when in London staying at least occasionally with Lady Ranelagh. When in November 1662 he was appointed by the society to the position of curator of experiments, Boyle was thanked "for dispensing with him for their use." By the next year Hooke was made a fellow (with charges waived) and was being paid by the society to lodge in Gresham College four days a week.[23] In 1664, he was elected professor of geometry at Gresham, with its associated lodgings, and there he remained, even during the society's absence at Arundel House, until his death in 1703. His quarters apparently opened behind the college "reading hall" and contained an extensive pneumatical, mechanical, and optical workshop, supplemented in 1674 by a small astronomical observatory constructed in a turret over his lodgings.[24]

The conditions in which Hooke lived and worked were markedly different from those of his patron Robert Boyle. Margaret 'Espinasse has vividly described his personal life at Gresham, where he "lived like a rather Bohemian scientific fellow of a college." His niece Grace was sharing his quarters from 1672 (when she was eleven years old) and was evidently sharing his bed sometime afterward. Hooke was also having sexual relations with his housekeeper Nell Young and, on her departure, with her successors. To what extent Hooke's domestic circumstances were known to his associates among the fellowship is unclear, though it is possible that there was some connection between those circumstances and the relative privacy of his rooms. It was Hooke who visited his high-minded patron Boyle; Boyle almost never visited Hooke. Hooke's relations with his various technicians were, in a different way, also very intimate. He took several of them into his lodgings, where they were treated in a manner between sons and apprentices (three of them becoming fellows of the Royal Society and one succeeding him as curator). Although his rooms were rarely frequented by gentleman-fellows on other than scientific and technical matters, and although his table was not a major venue for their discourses, Hooke lived on a public stage. He circulated through the taverns and the coffeehouses of the City of London and was a fixture at the tables of others. Hooke's place of residence, probably the most important site for experimental trials of Restoration England, was in practice a private place, while he himself lived an intensely public life. It is questionable indeed whether Hooke's quarters constituted a "home" in seventeenth-century gentlemanly usage. It was a place fit for Hooke to live and work; it was not a place fit for the reception and entertainment of gentlemen.[25]

Access

The threshold of the experimental laboratory was constructed out of stone and social convention. Conditions of access to the experimental laboratory flowed from decisions about what kind of place it was. In the middle of the century, those decisions had not yet been made and institutionalized. Meanwhile, there were a variety of stipulations about the functional and social status of spaces given over to experiment, and a variety of sentiments about access to them.

To the young Robert Boyle the threshold of his Stalbridge laboratory constituted the boundary between sacred and secular space. He told his sister Katherine that "*Vulcan* has so transported and bewitched me, that as the delights I taste in it make me fancy my laboratory a kind of *Elysium*, so as if the threshold of it possessed the quality the poets ascribe to *Lethe*, their fictions made men taste of before their entrance into those seats of bliss, I there forget my standish [inkstand] and my books, and almost all things."[26] The experimenter was to consider himself "honor'd with the Priesthood of so noble a Temple" as the "Commonwealth of Nature." And it was therefore fit that laboratory work be performed, like divine service, on Sundays. (In mature life Boyle entered his Pall Mall laboratory directly after his morning devotions, although he had apparently given up the practice of experimenting on the Sabbath.)[27] In the 1640s, he told his Hartlibian friends of his purposeful "retreat to this solitude" and of "my confinement to this melancholy solitude" in Dorset. But it was said to be a wished-for and a virtuous solitude, and Boyle complained bitterly of interruptions from visitors and their trivial discourses.[28]

Transactions across the experimental threshold had to be carefully managed. Solitude appeared both as a mundanely practical consideration and as a symbolic condition for the experimentalist to claim authenticity. Models of space in which solitude was legitimate and out of which valued knowledge emerged did exist: these included the monastic cell and the hermit's hut. The hermit's hut expressed and enabled individual confrontation with the divine; the solitude of the laboratory likewise defined the circumstances in which the new "priest of nature" might produce knowledge as certain and as morally valuable as that of the religious isolate. Here was a model of space perceived to be insulated from distraction, temptation, distortion, and convention.[29] Yet experimentalists like Boyle and his Royal Society colleagues in the 1660s were engaged in a vigorous attack on the privacy of existing forms of intellectual practice. The legitimacy of experimental knowledge, it was argued, depended upon a public presence at some crucial stage or stages of knowledge making. If experimental knowledge did indeed have to occupy private space during part of its career, then its realization as authentic knowledge involved its transit to and through a public space.

This transit was particularly difficult for a man in Boyle's position to accomplish and make visible as legitimate. He presented himself as an intensely private man, one who cared little for the distractions and rewards of ordinary social life. This presentation of self was successful. Bishop Gilbert Burnet, who preached Boyle's funeral sermon, described him as a paragon: "He neglected his person, despised the world and lived abstracted from all pleasures, designs, and interests."[30] At the same time, Boyle effectively secured the character of a man to whom justified access was freely available. He was entitled by birth and by wealth (even as diminished by the Irish wars) to a public life, and, indeed, there were forces that acted to ensure that he did live in the public realm. He advertised the public status of experimental work, and, from his first publication, condemned unwarranted secrecy and intellectual unsociability.[31] Yet he chose much solitude, was seen to do so, and was drawn only fitfully into the company of fellow Christian virtuosi, extended exposure to which drove him once more to solitude. In constructing his life and making it morally legitimate, Boyle was endeavoring to define the nature of a space in which experimental work might be practically situated and in which experimental knowledge would be seen as authentic. Such a space did not then clearly exist. The conditions of access to it and the form of social relations within it had to be determined and justified. This space had necessarily to be carved out of and rearranged from existing domains of accepted public and private activity and existing stipulations about the proper uses of spaces.

Many contemporary commentators remarked upon the ease of access to Boyle's laboratory. John Aubrey wrote about Boyle's "noble laboratory" at Lady Ranelagh's house as a major object of intellectual pilgrimage: "When foreigners come to hither, 'tis one of their curiosities to make him a Visit." This was the laboratory that was said to be "constantly open to the Curious, whom he permitted to see most of his Processes." In 1668, Lorenzo Magalotti, emissary of the Florentine experimentalists, traveled especially to Oxford to see Boyle and boasted that he was rewarded with "about ten hours" of his discourse, "spread over two occasions." John Evelyn noted that Boyle "had so universal an esteeme in Foraine parts; that not any Stranger of note or quality; Learn'd or Curious coming into England, but us'd to Visite him." He "was seldome without company" in the afternoons, after his laboratory work was finished.[32]

But the strain of maintaining quarters "constantly open to the curious" told upon him and was seen to do so. As an overwrought young man he besought "deare Philosophy" to "come quickly & releive Your Distresst Client" of the "vaine Company" that forms a "perfect Tryall of my Patience." Experimental philosophy might rescue him "from some strange, hasty, Anchoritish Vow"; it could save him from his natural "Hermit's Aversenesse to Society."[33] When, during the plague, members of the Royal Society descended upon him in Oxford, he bolted for the solitude of his village retreat at Stanton St. John, complaining of "ye great Concourse of strangers," while assuring Oldenburg that "I am not here soe neere a Hermite" but that some visitors were still welcome. Even as John Evelyn praised Boyle's accessibility, he recorded that the crowding "was sometimes so incomodious that he now and then repair'd to a private Lodging in another quarter [of London], and at other times" to Leese or elsewhere in the country "among his noble relations."[34]

Toward the end of his life Boyle took drastic and highly visible steps to restrict access to his drawing room and laboratory. It was reported that when he was at work trying experiments in the Pall Mall laboratory and did not wish to be interrupted, he caused a sign to be posted on his door: "Mr. Boyle cannot be spoken with to-day." In his last years and in declining health, he issued a special public advertisement "to those of his friends and Acquaintance, that are wont to do him the honour & favour of visiting him," to the effect that he desired "to be excus'd from receiving visits" except at stated times, (unless upon occasions very extraordinary)."[35] Bishop Burnet said that Boyle "felt his easiness of access" made "great wasts on his time," but "thought his obligation to strangers was more than bare civility."[36]

That obligation was a powerful constraint. The forces that acted to keep Boyle's door ajar were social forces. Boyle was a gentleman as well as an experimental philosopher. Indeed, as a young man he had reflected systematically upon the code of the gentleman and his own position in that code.

The place where Boyle worked was also the residence of the son of the first Earl of Cork. It was a point of honor that the private residence of the son of a gentleman should be open to the legitimate visits of other gentlemen. Seventeenth-century handbooks on the code of gentility stressed this openness of access, one such text noting that "Hospitalitie" was "one of the apparentest Signalls of *Gentrie*." Modern historians confirm the equation between easy access and gentlemanly standing: "generous hospitality was the hallmark of a gentleman"; "so long as the habit of open hospitality persisted, privacy was unobtainable, and indeed unheard of." And as the young Boyle himself confided in his *Commonplace Book*, a "Noble Descent" gives "the Gentleman a Free Admittance into many Companys, whence Inferior Persons (tho never so Deserving) are . . . excluded."[37] Other gentlemen knew who was a gentleman, they knew the code regulating access to his residence, and they knew that Boyle was obligated to operate under this code. But they did not know, nor could they, what an experimental scientist was, nor what might be the nature of a different code governing admittance to his laboratory. In the event, as Marie Boas wrote, they might plausibly come to the conclusion that Boyle "was only a virtuoso, amusing himself with science, [that] he could be interrupted at any time . . . There was always a swarm of idle gentlemen and ladies who wanted to see amusing and curious experiments."[38] When, however, Boyle wished to shut his door to these distractions, he was able to draw upon widely understood moral patterns that enabled others to recognize what he was doing and why it might be legitimate. The occasional privacy of laboratory work could be assimilated to the morally warrantable solitude characteristic of the religious isolate.[39]

Rights of Passage

What were the formal conditions of entry to experimental spaces? We do have some information concerning the policy of the early Royal Society, particularly regarding access of English philosophers to foreign venues. It was evidently common for the society's council to give "intelligent persons, whether Fellows of the Society or not, what are styled 'Letters Recommendatory.'" These documents, in Latin, requested "that all persons in authority abroad would kindly receive the bearer, who was desirous of cultivating science, and show him any attention in their power."[40] Similarly, in 1663, the society drafted a statute regulating access to its own meetings. As soon as the president took the chair, "those persons that [were] not of the society, [were to] withdraw."

There was, however, an exemption for certain classes of persons to remain if they chose, that is, "for any of his majesty's subjects . . . having the title and place of a baron, or any of his majesty's privy council . . . or for any foreigner of eminent repute, with the allowance of the president." Other persons might be permitted to stay with the explicit consent of the president and fellows in attendance. Barons and high-ranking aristocrats could become fellows on application, without the display of philosophical credentials.[41]

Too much should not be made of such fragmentary evidence of formal conditions granting or withholding rights of entry to experimental sites. It is noteworthy how sparse such evidence is, even for a legally incorporated body like the Royal Society. In the main, the management of access to experimental spaces, even those of constituted organizations, was effected more informally. For example, there was the letter of introduction to an experimentalist, a number of which survive. In 1685, one visitor carried with him a letter of introduction, from someone presumably known to Boyle, which identified the bearer as "ambitious to be known to you, whose just character of merit is above his quality . . . being the eldest son of [a diplomat] and brother-in-law to the king of Denmark's envoy."[42] In this instance and in others like it, it was not stated that the proposed entrant to Boyle's society possessed any particular technical competencies, nor even that he was "one of the curious," merely that he was a gentleman of quality and merit, as vouched for by the correspondent. In other cases, "curiosity" was explicitly stipulated as a sufficient criterion for entry.

Generally speaking, it appears that access to most experimental venues (and especially those located in private residences) was achieved in a highly informal manner, through the tacit system of recognitions, rights, and expectations that operated in the wider society of gentlemen. With respect to Boyle's laboratories and drawing rooms, it seems that entry was attained if one of three conditions could be met: if the applicant was (1) known to Boyle by sight and of a standing that would ordinarily give rights of access; (2) known to Boyle by legitimate reputation; (3) known to Boyle neither by sight nor by reputation, but arriving with (or with an introduction from) someone who satisfied condition (1) or (2). These criteria can be expressed more concisely: access to experimental spaces was managed by calling upon the same sorts of conventions that regulated entry to gentlemen's houses, and the relevant rooms within them, in general.[43] These criteria were not codified and written down because they did not need to be. They would be known and worked with by every gentleman. Indeed, they would almost certainly be known and worked with by those who were not gentlemen, shaping their understanding of the grounds for the denial of entry. Standing gave access. Boyle was perhaps unusual among English gentlemen in reflecting explicitly upon this largely tacit knowledge: "A man of meane Extraction (tho never so advantag'd by Greate meritts) is seldome admitted to the Privacy & the secrets of greate ones promiscuously & scarce dares pretend to it, for feare of being censur'd saucy, or an Intruder."[44]

I have alluded to some formal criteria governing entry to the rooms of the Royal Society. For all the significance of such considerations, informal criteria operated there as well, just as they did in the case of Boyle's laboratories, to manage passage across the threshold. These almost certainly encompassed not only the informal criteria mentioned in connection with Boyle, for whom standing gave access, but also other sorts of tacit criteria. When Lorenzo Magalotti visited England in 1668 (after the Royal Society's removal to Arundel House), his arrival at the society's weekly meeting was apparently expected and special experiments had been made ready to be shown. But Magalotti had second thoughts about the advisability of attending and the terms on which he thought entry was offered. "I understood," he wrote to an Italian prelate, "that one is not permitted to go in simply as a curious passer by, [and] I would not agree to take my place there as a scholar, for one thing because I am not one . . . Thus, therefore, I got as far as the door and then went away, and if they do not want to permit me to go and be a mere spectator without being obliged to give opinions like all the others, I shall certainly be without the desire to do so."[45] There are reasons to doubt the absolute reliability of Magalotti's testimony. Nevertheless, he pointed to a crucially important tacit criterion of entry. Magalotti was, of course, the sort of person, carrying the sort of credentials, who would have had unquestioned access to the Royal Society meeting at Arundel House, or, indeed to Arundel House itself. His claims indicate, however, that the experimental activities that went on within its interior imposed further informal criteria regulating entry. These included the uncodified expectation that, once admitted, one would act *as a participant*. The notion of participation followed from a distinction, customary but not absolute, between *spectating* and *witnessing*. The Royal Society expected those in attendance to validate experimental knowledge as participants, by giving witness to matters of fact, rather than to play the role of passive spectators to the doings of others.[46] But there was a further consideration: those granted entry were tacitly enjoined to employ the conventions of deportment and discourse deemed appropriate to the experimental enterprise, rather than those current in, say, hermetic, metaphysical, or rationalist practice. Those unwilling to observe these conventions could exclude themselves. These are the grounds on which one might rightly say that a philosopher like Thomas Hobbes was, in fact, excluded from the precincts of the Royal Society, even though there is no evidence that he sought entry and was turned away. Margaret 'Espinasse was therefore quite correct in saying that the society was "open to all classes rather in the same way as the law-courts and the Ritz," and Quentin Skinner was also right to characterize it as "like a gentlemen's club," even if he unnecessarily contrasted that status with the society's ostensible role as "the conscious centre of all genuinely scientific endeavour."[47]

Relations in Public

If we are able to recognize what kind of place we are in, we find we already possess implicit knowledge of how it is customary to behave there. But in the middle of the seventeenth century the experimental laboratory and the places of experimental discourse did not have standard designations, nor did people who found themselves within them have any tacit knowledge of the behavioral norms obtaining there. On the one hand, publicists of the experimental program offered detailed guidance on the social relations deemed appropriate to experimental places; on the other, there is virtual silence about some of the most basic features of these places. The situation is about what one would expect if new patterns of behavior in one domain were being put together out of patterns current in others.

In 1663, the Royal Society was visited by two Frenchmen, Samuel Sorbière, physician and informal emissary of the Montmor Academy, and the young Lyonnaise scholar Balthasar de Monconys. Both subsequently published fairly detailed accounts of the society's procedures. Sorbière recorded that the meeting room at Gresham was some sort of "Amphitheatre," possibly the college reading hall or an adaptation of a living room of Gresham's sixteenth-century cloistered house to make it suitable for public lecturing. The president sat at the center of a head table, with the secretary at his side and chairs for distinguished visitors. The ordinary fellows sat themselves on plain wooden benches arranged in tiers, "as they think fit, and without any Ceremony."[48] An account dating from around 1707, toward the end of the society's stay at Gresham, gives a description of three rooms in which it conducted its affairs but omits any detail of the internal arrangements or of social relations within them.[49]

When Magalotti visited the Royal Society at Arundel House in February 1668, he described the assembly room off the gallery, "in the middle of which is a large round table surrounded by two rows of seats, and nearer to it by a circle of plush stools for strangers." On his second visit in April 1669, he recorded that the president sat "on a seat in the middle of the table of the assembly."[50] No visitor, or any other commentator, provided a detailed account of the physical and social arrangements attending the performance of experiments in the Royal Society. Monconys offered a recitation of experiments done, without describing the circumstances in which they were done. Sorbière mentioned only that there was brief discussion of "the Experiments proposed by the Secretary." Magalotti recorded that he saw experiments performed, demonstrated by "a certain Mr Hooke." These were set up on a table in the corner of the meeting room at Arundel House. When working properly, experiments were transferred to a table in the middle of the room and displayed, "each by its inventor." Experimental discussion then ensued.[51] By the 1670s, it is evident that experimental "discourse," or formal presentation setting forth and interpreting experiments tried elsewhere, was much more central to the society's affairs than experiments tried and displayed within its precincts.

Sorbière, Monconys, Magalotti, and other observers all stressed the civility of the Royal Society's proceedings. The president, "qui est toujours une personne de condition," was clearly treated with considerable deference, by virtue of his character, his office, and, most important, his function in guaranteeing good order. Patterns provided by procedure in the House of Commons are evident. Fellows addressed their speech to the president, and not to other fellows, just as members of the House of Commons conventionally address the Speaker. So, the convenient fiction was maintained that it was always the matter and not the man that was being addressed. Both Sorbière and Magalotti noted that fellows removed their hats when speaking, as a sign of respect to the president (again following Commons practice). Whoever was speaking was never interrupted, "and Differences of Opinion cause no manner of Resentment, nor as much as a disobliging Way of Speech."[52] An English observer said that the society "lay aside all set Speeches and Eloquent Haranques (as fit to be banisht out of all Civil Assemblies, as a thing found by woful experience, especially in *England,* fatal to Peace and good

Manners)," just as the reading of prepared speeches was (and is) conventionally deprecated in Commons. "Opposite opinions" could be maintained without "obstinacy," but with good temper and the language of civility and moderation."[53]

This decorum was the more remarkable in that it was freely entered into and freely sustained. Sorbière said that "it cannot be discerned that any Authority prevails here"; and Magalotti noted that at "their meetings, no precedence or distinction of place is observed, except by the president and the secretary." As in the seventeenth-century House of Commons, the practice of taking any available seat (with the exception of the president and the secretaries, who, like the Commons Speaker, his clerks, and privy councillors, sat at the head of the room) constituted a visible symbol of the equality in principle of all fellows and of the absence of sects, even if the reality, in both houses, might be otherwise.[54] All visitors found it worth recording that the society's mace, laid on the table before the president when the meetings were convened, was an emblem of the source of order. Again, as in Commons, the mace indicated that the ultimate source was royal. The king gave the society its original mace even as he replaced the Commons mace that had disappeared in the Interregnum. The display of the mace in the Royal Society confirmed that its authority flowed from, and was of the same quality as, that of the king. Nevertheless, Thomas Sprat took exception to any notion that mace ceremonials constituted rituals of authority: "The Royal Society itself is so careful that such ceremonies should be just no more than what are necessary to avoid Confusion."[55] Sprat took the view that the space occupied and defined by the fellowship was truly novel: it was regulated by no traditional set of rituals, customs, or conventions. An anonymous fellow writing in the 1670s agreed: the society's job was "not to whiten the walls of an old house, but to build a new one; or at least, to enlarge the old, & really to mend its faults."[56]

Yet no type of building, no type of society is wholly new. And despite the protestations of early publicists, it is evident that the social relations and patterns of discourse obtaining within the rooms of the Royal Society were rearrangements and revaluations of existing models. Aspects of a parliamentary pattern have already been mentioned. The relationship between the proceedings of the early Royal Society and the Interregnum London coffeehouse merits extended discussion, most particularly in connection with the rules of good order in a mixed assembly. Other elements resonate of the monastery, the workshop, the club, the college, and the army.[57] But the most potent model for the society's social relations was drawn from the type of place in which they actually occurred. The code that is closest to that prescribed for the experimental discourses of the Royal Society was that which operated within the public rooms of a gentleman's private house.

The Experimental Public

What was the experimental public like? How many people, and what sorts of people, composed that public? In order to answer these questions we have to distinguish rhetoric from social reality. When, for example, Sprat referred to the Royal Society's experimental public as being made up of "the concurring Testimonies of threescore or an hundred" and pointed to "many sincere witnesses standing by" experimental performances, he was, it seems, referring to an ideal state. The Royal Society was, of course, the most populated experimental space of Restoration England, but the effective attendance at weekly meetings probably averaged no more than two score, and by the 1670s meetings were being canceled for lack of attendance.[58] More intimate groups assembled as "clubs" of the society, centered particularly upon Hooke and usually meeting at coffeehouses near Gresham College.

In the event, historians have rightly questioned whether the rooms of the Royal Society should properly be regarded as a major experimental site.[59] Most actual experimental research was performed elsewhere, most notably in private residences like Boyle's Oxford and Pall Mall laboratories and in Hooke's quarters. Unsurprisingly, evidence about the population in these places is scarce. Boyle frequently named his experimental witnesses, and in no case does that named number exceed three. We do also have commentators' testimony about the throngs of visitors, but these are probably best regarded as genuine spectators rather than witnesses.[60] I will mention the circumstances of experimental work in Hooke's lodgings later, but his laboratory was certainly more thinly populated than that of his patron. Apart from Hooke himself, the population of Hooke's laboratory seems mainly to have been composed of his various assistants, technicians, and domestics.

I need in this connection to make a distinction between a real and a relevant experimental public, between the population actually present at experimental scenes and those whose attendance was deemed by authors to be germane to the making of knowledge. We have, for example, conclusive evidence of the presence in Boyle's laboratories of technicians and assistants of various sorts. As we might say, their role was vital, since Boyle himself had little if anything to do with the physical manipulation of experimental apparatus, and since at least several of these technicians were far more than mere laborers.[61] Yet their presence was scarcely acknowledged in the scenes over which Boyle presided. Two of them, Hooke and Denis Papin, were named and responsible elements in those scenes, although even here Boyle's account probably understates their contribution. Toward the end of his career, Boyle acknowledged Papin's responsibility for the writing of experimental narratives as well as for the physical conduct of air-pump trials. "I had," he wrote, "cause enough to trust his skill and diligence." But Boyle still insisted on his own ultimate responsibility for the knowledge produced, and the manner in which he did so is instructive: Boyle asked Papin to "set down in writing all the experiments and the phaenomena arising therefrom, as if they had been made and observed by his own skill . . . But I, myself, was always present at the making of the chief experiments, and also at some of those of an inferior sort, to observe whether all things were done according to my mind." Certain interpretations of experiments were indeed left to Papin: "Some few of these inferences owe themselves more to my assistant than to me."[62] Still, Boyle, not Papin, was the author of this text.

For the most part, however, Boyle's host of "laborants," "operators," "assistants," and "chemical servants" were invisible actors. They were not a part of the relevant experimental public. They made the machines work, but they could not make knowledge. Indeed, their greatest visibility (albeit still anonymous) derived from the capacity of their *lack* of skill to sabotage experimental operations. Time after time in Boyle's texts, technicians appear as sources of trouble. They are the unnamed ones responsible for pumps exploding, materials being impure, glasses not being ground correctly, machines lacking the required integrity.[63]

Technicians had skill but lacked the qualifications to make knowledge. This is why they were rarely part of the relevant experimental public, and when they were part of that public, it was because they were only ambiguously functioning in the role of technician. Ultimately, their absence from the relevant experimental public derived from their formal position in scenes presided over by others. Boyle's technicians, including those of mixed status like Hooke and Papin, were paid by him to do jobs of experimental work, just as both were paid to do similar tasks by the gentlemen of the Royal Society. As Boyle noted in connection with his disinclination to become a cleric, those that were paid to do something were open to the charge that this was *why* they did it.[64] A gentleman's word might be relied upon partly because what he said was without consideration of remuneration. Free verbal action, such as giving testimony, was credible by virtue of its freedom. Technicians, as such, lacked that circumstance of credibility. So far as their capacity to give authentic experimental testimony was concerned, they were truly not present in experimental scenes. Technicians were not *there* in roughly the same way, and for roughly the same reasons, that allowed Victorian families to speak in front of the servants. It did not matter that the servants might hear: if they told what they heard to other servants, it did not signify; and if they told it to gentlemen, it might not be credited.

The Condition of Gentlemen

The early Royal Society set itself the task of putting together, justifying, and maintaining a relevant public for experiment. Its publicist Thomas Sprat reflected at length on the social composition of this public and its bearing on the integrity of knowledge-making practices. Historians are now thoroughly familiar with the Royal Society's early insistence that its company was made up of "many eminent men of all Qualities," that it celebrated its social diversity, and that it pointed to the necessary participation in the experimental program of "vulgar hands." Nevertheless, this same society deemed it essential that "the farr greater Number are Gentlemen, free, and unconfin'd." In the view of Sprat and his associates, the condition of gentlemen was the condition for the reliability and objectivity of experimental knowledge.[65]

There were two major reasons for this. First, an undue proportion of merchants in the society might translate into a search for present profit at the expense of luciferous experimentation and even into an insistence upon trade secrecy, both of which would distort the search for knowledge. This is what Joseph Glanvill meant in praising the society for its freedom from "sordid Interests."[66] More important, the form of the social relations of an assembly composed of unfree men, or, worse, a society divided between free and unfree, would corrupt the processes by which experimental knowledge ought to be made and evaluated, and by which that knowledge might be advertised as reliable. Unfree men were those who lacked discretionary control of their own actions. Technicians, for example, belonged to this class—the class of servants—because their scientific labor was paid for. Merchants might be regarded as compromised in that their actions were geared to achieving the end of present profit. One could not be sure that their word corresponded to their state of belief. Put merchants and servants in an assembly with gentlemen and you would achieve certain definite advantages. But there was also a risk in the shape of the knowledge-making social relations that might be released. Inequalities of rank could, in Sprat's view, corrode the basis of free collective judgment on which the experimental program relied.[67]

As Sprat said, the trouble with existing intellectual communities was the master-servant relationship upon which their knowledge-constituting practices were founded, the scheme by which *"Philosophers* have bin always *Masters, & Scholars;* some imposing, & all the other submitting; and not as equal observers without dependence." He judged that "very mischievous . . . consequences" had resulted because "the Seats of Knowledg, have been for the most part heretofore, not *Laboratories,* as they ought to be; but onely *Scholes,* where some have *taught,* and all the rest subscrib'd." So the school-room was a useful resource in modeling a proper experimental space, precisely because it exemplified those conventional social relations deemed grossly inappropriate to the new practice: "The very inequality of the Titles of *Teachers,* and *Scholars,* does very much suppress, and tame mens Spirits; which though it should be proper for Discipline and Education; yet it is by no means consistent with a free Philosophical Consultation. It is undoubtedly true; that scarce any man's mind, is so capable of *thinking strongly,* in the presence of one, whom he *fears* and *reverences;* as he is, when that restraint is taken off."[68]

The solution to the practical problem thus resolved into the description and construction of a social space that was both free and disciplined. Sprat said that the "cure" for the disease afflicting current systems of knowledge "must be no other, than to form an *Assembly* at one time, whose privileges shall be the same; whose gain shall be in common; whose *Members* were not brought up at the feet of each other." Such disinterested free men, freely mobilizing themselves around experimental phenomena and creating the witnessed matter of fact, could form an intellectual polity "upon whose labours, mankind might . . . freely rely."[69] The social space that Sprat was attempting to describe was a composite of a number of existing and past spaces, real and ideal. Still, one model for such a space was, perhaps, more pertinent than any other, precisely because it corresponded to the type of space within which experimental discourses typically occurred. This, again, was the gentleman's private residence and, within it, its public rooms. The conventions regulating discourse in the drawing room were readily available for the construction of the new space and for making visible the social relations appropriate to it. It was the acknowledged freedom of the gentleman's action, the honor accorded to his word, the moral discipline he imposed upon himself, and the presumed moral equality of the company of gentlemen that guaranteed the reliability of experimental knowledge. In other words, gentlemen in, genuine knowledge out.

Gentlemen were bound to credit the word of their fellows or, at least, to refrain from publicly discrediting it.[70] These expectations and obligations were grounded in the face-to-face relations obtaining in concrete spaces. The obligation to tell the truth, like the consequences of questioning that one was being told the truth, were intensified when one looked the other "in the face," and particularly when it was done in the public rooms of the other's house. The disastrous effects of violating this code were visible to the Royal Society in the quarrel between Gilles Roberval and Henri-Louis Habert de Montmor in the latter's Parisian town house. As Ismael Boulliau told the story to Christiaan Huygens, Roberval "has done a very stupid thing in the house of M. de Montmor who is as you know a man of honor and position; he was so uncivil as to say to him in his own house . . . that he had more wit than he, and that he was less only in worldly goods . . . Monsieur de Montmor, who is very circumspect, said to him that he could and should behave more civilly than to quarrel with him and treat him with contempt in his own house." Roberval never returned to the Montmor Academy, and the group never recovered. The Parisians tried to learn a lesson: as this dispute was over doctrine, they resolved to move "towards the study of nature and inventions," in which civility could be more easily maintained since the price of dissenting publicly from a gentleman's testimony on matters of fact would dissuade others from the contest.[71]

The code relating to face-to-face interactions in the house could be, and was, extended to the social relations of experimental knowledge production generally. It was rare indeed for any gentleman's testimony on a matter of experimental fact to be gainsaid. In the early 1670s, Henry More disputed Boyle's report of a hydrostatical matter of fact. The manner of Boyle's response is telling: "Though [More] was too civil to give me, *in terminus*, the lye; yet he did indeed deny the matter of fact to be true. Which I cannot easily think, the experiment having been tried both before our whole society, and very critically, by its royal founder, his majesty himself."[72] Boyle appealed to the honor of a *company* of gentlemen, and, ultimately, to the greatest gentleman of all. In 1667, Oldenburg specifically cautioned fellows not to deny within the society's rooms experimental testimony deriving from foreign philosophers. Oldenburg took an offending fellow aside afterward and asked him "how he would resent it, if he should communicate upon his own knowledge an unusual experiment to [those foreign experimenters], and they brand it in public with the mark of falsehood: that such expressions in so public a place, and in so mixed an assembly, would certainly prove very destructive to all philosophical commerce."[73]

The same relationship of trust that was enjoined to govern experimental discourse in the drawing room was constitutive of transactions between public and private rooms of the experimental house. I noted at the beginning the central problem posed in empiricist practice by the indispensable role of testimony and trust. The Royal Society was evidently quite aware that the population of direct witnesses to experimental trials in the laboratory was limited by practical considerations if by nothing else. Nevertheless, the trajectory of a successful candidate for the status of matter of fact necessarily transited the public spaces in which it was validated. The practical solution offered by the society was the acceptance of a division of experimental labor and the protection of a relationship of trust between those within and without the laboratory threshold. Sprat said that there was a natural division of labor among the fellowship: "Those that have the best faculty of *Experimenting,* are commonly most averse from reading Books; and so it is fit, that this *Defect* should be supply'd by others pains." Those that actually performed experimental trials, and those that accompanied them as direct witnesses, were necessarily few in number, but they acted as representatives of all the rest. One could, and ought to, trust them in the way one could trust the evidence of one's own senses: "Those, to whom the conduct of the *Experiment* is committed . . . do (as it were) carry the eyes, and the imaginations of the whole company into the *Laboratory* with them." Their testimony of what had been done and found out in the laboratory, undoubted because of their condition and quality, formed the basis of the assembly's discursive work, "which is to *judg,* and *resolve* upon the matter of Fact," sometimes accompanied by a showing of the experiment tried in the laboratory, sometimes on the basis of narration alone. Only when there was clear agreement ("the concurring Testimonies") was a matter of fact established. Such procedures were advertised as morally infallible. Glanvill reckoned that "the relations of your Tryals may be received as undoubted Records of certain events, and as securely be depended on, as the Propositions of Euclide." The very transition from private to public space that marked the passage from opinion to knowledge was a remedy for endemic tendencies to "over-hasty" causal conjecturing, to "finishing the *roof,* before the *foundation* has been well laid." Sprat assured his readers that "though the *Experiment* was but the private task of one or two, or some such small number; yet the *conjecturing,* and *debating* on its *consequences,* was still the employment of their full, and solemn Assemblies."[74] An item of experimental knowledge was not finished until it had, literally, come out into society.

Trying It at Home

A house contains many types of functionally differentiated rooms, each with its conditions of access and conventions of appropriate conduct within. Social life within the house involves a circulation from one sort of room to another. The career of experimental knowledge follows the same sort of circulation. So far I have spoken of the making of experimental knowledge in a loose way, scarcely differentiating between its production and its evaluation. I now deal more systematically with the stages of knowledge-making and relate these to the physical and social spaces in which they take place. In mid- to late seventeenth-century England there was a linguistic distinction the force and sense of which seem to have escaped most historians of science. This was the discrimination between "trying" an experiment, "showing" it, and "discoursing" upon it. In the common usage of the main experimental actors of this setting, the distinction between these terms was both routine and rigorous. The trying of an experiment corresponds to research proper, getting the thing to work, possibly attended with uncertainty about what constitutes a working experiment. Showing is the display to others of a working experiment, which is commonly called demonstration.[75] And experimental discourses are the range of expatiatory and interpretative verbal behaviors that either accompany experimental shows or refer to shows or trials done at some other time or place. I want to say that trying was an activity that in practice occurred within relatively private spaces, whereas showing and discoursing were events in relatively public space. The career of experimental knowledge is the circulation between private and public spaces.[76]

We can get a purchase upon this notion by considering a day in the experimental life of Robert Hooke. I have noted that Hooke lived where he worked, in rooms at Gresham College with an adjacent laboratory, rooms that were little visited by fellow experimentalists, English or foreign. He rose and then dined early, usually at home and frequently with his technicians, some of whom lodged with him. Before issuing forth, Hooke worked at home, trying experiments, as his diary records: "tryd experiment of fire," or "tryd experiment of gunpowder." Some of these were preparations for displays at the Royal Society, either next door or, during the Arundel House period, a mile and a half away. It was in the assembly rooms of the society that these experiments were to be shown and discoursed on: "tryd expt of penetration of Liquors . . . shewd it at Arundell house." Experimental discourses could also take place elsewhere. When Hooke left his rooms, he would invariably resort to local coffeehouses or taverns, where he would expect to meet a small number of serious and competent philosophers for experimental discussion. In the evenings he was a fixture at the tables of distinguished fellows of the society, notably at Boyle's, Christopher Wren's, and Lord William Brouncker's houses, where further experimental discourse occurred.[77]

Thomas Kuhn has written about what he sees as a crucial difference between the role of experiment in mid-seventeenth-century England and preceding practices. In the experimental program of Boyle, Hooke, and their associates, Kuhn says, experiments were seldom performed "to demonstrate what was already known . . . Rather they wished to see how nature would behave under previously unobserved, often previously nonexistent, circumstances."[78] Broadly speaking, the point is a legitimate one. However, it applies only to one stage of experimentation and to one site at which experimental activity occurred. Hooke and Boyle might, indeed, have undertaken experimental trials without substantial foreknowledge of their outcome, although they could scarcely have done so without *any* foreknowledge, since they would then have been unable to distinguish between experimental success and failure. An experimental trial could fail; indeed, trials usually did fail, in the sense that an outcome was achieved out of which the desired sense could not be made. So, Hooke's diary records, among many other instances: "Made tryall of Speculum. not good"; "Made tryall upon Speculum it succeeded not"; "at home all day trying the fire expt but could not make it succeed." So far as trials are concerned, a failure might legitimately be attributed to one or more of a number of causes: the experimenter was inept or blundered in some way; the equipment was defective or the materials impure; relevant background circumstances, not specifiable or controllable at the time of trial, were unpropitious; and so on.[79] However, a further possibility was open and, indeed, sometimes considered by experimenters, namely that the theory, hypothesis, or perspective that informed one's sense of what counted as a successful outcome was itself incorrect. In a trial it was therefore always possible that an outcome deemed unsuccessful might come to be regarded as the successful realization of another theory of nature. In this way, the definition of what counted as a well-working experimental trial was, in principle, open-ended. In the views of the relevant actors, nature might perhaps speak unexpected words, and the experimenter would be obliged to listen.

The notion of the experimental trial therefore carried with it a sense of indiscipline: the experimenter might not be fully in control of the scene. The thing might fail. It might fail for lack of technical competence on the part of the experimenter, or it might fail for want of theoretical resources required to display the phenomena as docile.[80] Trials were undisciplined experiments, and these, like undisciplined animals, children, and strangers, might be deemed unfit to be displayed in public. This is why experimental trials were, in fact, almost invariably performed in relatively private spaces (such as Hooke's rooms and Boyle's laboratory) rather than in the public rooms of the Royal Society.

The weekly meetings of the Royal Society required not trials but shows and discourses.[81] It was Hooke's job as curator of experiments to prepare these performances for the society's deliberation, instruction, and entertainment. His notes entitled "Dr. Hook's Method of Making Experiments" stipulate that the curator was to make the trial "with Care and Exactness," then to be "diligent, accurate, and curious" in "shewing to the Assembly of Spectators, such Circumstances and Effects . . . as are material." Even a visitor like Magalotti observed that he who was in charge of the society's experiments "does not come to make them in public before having made them at home."[82] Hooke had specific directions to this effect. For instance, in connection with a set of magnetic experiments, "It was ordered, that Mr. Hooke . . . try by himself a good number of experiments . . . and draw up an account of their success, and to communicate it to the Society, so that they might call for such of them as they should think good to be shewn before them." And in the case of a transfusion trial, Hooke and others were "appointed to be curators of this experiment, first in private by themselves, and then, in case of success, in public before the society."[83] Hooke did labor assiduously "at home," disciplining the trials and, when they had been made docile, bringing them to be shown.

He was a success at his job. His first biographer said that his experiments for the Royal Society were "performed with the least Embarrassment, clearly, and evidently."[84] There was always the risk of "embarrassment" precisely because these were to be not trials but shows, performed not in private but in public. "Embarrassment" was avoided, and the society had a successful meeting, when "the experiments succeeded," that is, when they met the shared expectations attending their outcome (and, presumably, when they offered a certain amount of amusement and entertainment).

But even Hooke did not always succeed. When an experimental show failed, the reasons were more circumscribed than in the case of a trial. With any event labeled as, and intended to be, a show, failure could mean only that the experimenter or the materials under his direction were in some way wanting. Accordingly, the Royal Society was not tolerant of failed shows. Hooke's wrist was smartly slapped when he produced in public the undisciplined phenomena that abounded in private settings: "The operator was ordered to make his compressing engine very staunch; and for that end to try it often by himself, that it might be in good order against the next meeting"; "Mr. Hooke was ordered to try this by himself at home"; "He made an experiment of the force of falling bodies to raise a weight; but was ordered to try it by himself, and then to shew it again in public."[85]

The relations between trials and shows, between activities proper to private and to public spaces, were, however, inherently problematic. The status of what had been produced or witnessed was a matter for judgment. A clear example of this is the case of the so-called anomalous suspension of water. In the early 1660s, there was serious dispute in the Royal Society over the factual status and correct interpretation of this phenomenon. (Water that is well purged of air bubbles will not descend from its initial standing in the Torricellian apparatus when it is placed in an evacuated air-pump. Boyle had pointed to descent as crucial confirmation of his hypothesis of the air's spring.) Huygens had produced the alleged phenomenon in Holland, and Boyle disputed its status as an authentic fact of nature by suggesting that non-descent was due to the leakage of external air into Huygens's pump. Hooke was directed to prepare the experiment for the Royal Society. During the early phases of the career of anomalous suspension in England, the experimental leaders of the Royal Society were of the opinion that no such phenomenon legitimately existed. Any experiment that showed it was considered to have been incompetently performed—the apparatus leaked. Since members of the society had considerable experience of Hooke's bringing them experiments in pumps that were not "sufficiently tight," they readily concluded that Hooke's first productions of anomalous suspension were instances of experimental failure.[86] The experimental phenomena had not been made sufficiently docile. Hooke had indeed tried the experiment at home and had deemed it ready to be shown. The leaders of the society concluded otherwise: Hooke had produced only a trial, a failed show. What Hooke claimed to be knowledge, the society rejected as artifact. They disputed his claim by stipulating that the thing was not proper to be shown in a public place.[87]

When the Royal Society was at Arundel House, its curator Robert Hooke was continually ordered to bring the air-pump to their meetings from its permanent lodgings in Hooke's rooms a mile and a half away at Gresham. In the course of being trundled back and forth, the brittle seals that ensured the machine against leakage were liable to crack, so that the curator's experimental shows sometimes failed. Hooke made a modest proposal. He suggested that, in this one instance and for this circumscribed practical reason, the honorable fellows who wished to satisfy themselves how matters stood should come to him, instead of Hooke and the machine going to them. Hooke "moved that . . . a committee might be appointed to see some experiments made with [the air-pump] at his lodgings."[88]

An ad hoc committee was constituted and the visit to Hooke's rooms was made. In this instance, the normal pattern of movement in seventeenth-century experimental science was reversed: those who wanted to witness experimental knowledge in the making came to where the instruments permanently lived, rather than obliging the instruments to come to where witnesses lived. This inversion of the usual hierarchical ordering of public and private spaces was exceptional in seventeenth-century practice, and, in the event, it was rarely repeated. The showing of experimental phenomena in public spaces to a relevant public of gentleman-witnesses was an obligatory move in that setting for the construction of reliable knowledge. What underwrote assent to knowledge-claims was the word of a gentleman, the conventions regulating access to a gentleman's house, and the social relations within it.

The contrast with more modern patterns is evident. The disjunction between places of residence and places where scientific knowledge is made is now almost absolute. The separation between the laboratory and the house means that a new privacy surrounds the making of knowledge whose status as open and public is often insisted upon. The implications of this disjunction are both obvious and enormously consequential. Public assent to scientific claims is no longer based upon public familiarity with the phenomena or upon public acquaintance with those who make the claims. We now believe scientists not because we know them, and not because of our direct experience of their work. Instead, we believe them because of their visible display of the emblems of recognized expertise and because their claims are vouched for by other experts we do not know. Practices used in the wider society to assess the creditworthiness of individuals are no longer adequate to assess the credibility of scientific claims. We can, it is true, make the occasional trip to places where scientific knowledge is made. However, when we do so, we come as visitors, as guests in a house where nobody lives.

CHAPTER 6

Pump and Circumstance

Robert Boyle's Literary Technology

The production of knowledge and the communication of knowledge are usually regarded as distinct activities. I argue to the contrary: speech about natural reality is a means of generating knowledge about reality, of securing assent to that knowledge, and of bounding domains of certain knowledge from areas of less certain standing. I display the conventional status of specific ways of speaking about nature and natural knowledge, and I examine the historical circumstances in which these ways of speaking were institutionalized. Although I will be dealing with communication within a scientific community, there is a clear connection between this study and the analysis of scientific popularization. The popularization of science is usually understood as the extension of experience from the few to the many. I show here that one of the major resources for generating and validating items of knowledge within the scientific community under study was this same extension of experience from the few to the many: the creation of a scientific public. The etymology of some of our key terms is apposite: if a *community* is a group sharing a common life, *communication* is a means of making things common.

The materials used to address this issue come from episodes of unusual interest to the history, philosophy, and sociology of science. Robert Boyle's experiments in pneumatics in the late 1650s and early 1660s represent a revolutionary moment in the career of scientific knowledge. In his *New Experiments Physico-Mechanical* (1660) and related texts of the early Restoration, Boyle not only produced new knowledge of the behavior of air, he exhibited the proper experimental means by which legitimate knowledge was to be generated and evaluated. And he did so against the background of alternative programs for the production of knowledge, the proponents of which subjected Boyle's recommended methods to explicit criticism. At issue in the controversies over Boyle's air-pump experiments during the 1660s was the question of how claims were to be authenticated as knowledge. What was to count as knowledge, or "science"? How was this to be distinguished from other intellectual categories, such as "belief" and "opinion"? What degree of certainty could be expected of various intellectual enterprises and items of knowledge? And how could the appropriate grades of assurance and certainty be secured?[1]

These were all practical matters. In the setting of early Restoration England there was no one solution to the problem of knowledge that commanded universal assent. The technology of producing knowledge had to be built, exemplified, and defended against attack. The categories of knowledge and their generation that seem to us self-evident and unproblematic were neither self-evident nor unproblematic in the 1660s. The foundations of knowledge were not matters merely for philosophers' reflections; they had to be constructed and the propriety of their foundational status had to be argued. The difficulties that we evidently have in recognizing this work of construction arise from the very success of that work: to a very large extent we live in the conventional world of knowledge-production that Boyle and his colleagues among the experimental philosophers labored to make safe, self-evident, and solid.

Robert Boyle sought to secure universal assent by way of the experimental *matter of fact*. About such facts one could be highly certain; about other items of natural knowledge more circumspection was indicated. Boyle was, therefore, an important actor in the probabilist and fallibilist movement of seventeenth-century England. Before circa 1660, as Ian Hacking and Barbara J. Shapiro have shown, the designations of "knowledge" and "science" were rigidly distinguished from "opinion."[2] Of the former one could expect the absolute certainty of *demonstration,* exemplified by logic and geometry. The goal of physical science had been to attain to this kind of certainty compelling assent. By contrast, the English experimentalists of the mid-seventeenth century increasingly took the view that all that could be expected of physical knowledge was a high degree of *probability,* so breaking down the radical distinction between "knowledge" and "opinion." Physical hypotheses were provisional and revisable; assent to them was not necessary, as it was to mathematical demonstration; and physical science was, to varying degrees, removed from the realm of the demonstrative.[3] The probabilistic conception of physical knowledge was not regarded as a regrettable retreat from more ambitious goals; it was celebrated by its proponents as a wise rejection of failed dogmatism. The quest for necessary and universal assent to physical propositions was seen as improper and impolitic.

If universal assent was not to be expected of explanatory constructs in science, how, then, was proper science to be founded? Boyle and the experimentalists offered the *matter of fact*. The fact was the item of knowledge about which it was legitimate to be "morally certain." A crucial boundary was drawn around the domain of the factual, separating it from those items that might be otherwise and from which absolute and permanent certainty should not be expected. Nature was like a clock: people could be certain of its effects, of the hours shown by its hands, but the mechanism by which these effects were produced, the clock-work, might be various.[4]

It is in the understanding of how matters of fact were produced and how they came to command universal assent that historians have tended to succumb to the temptations of self-evidence.[5] This essay displays the processes by which Boyle constructed experimental matters of fact and thereby produced the conditions in which assent could be mobilized.

The Mechanics of Fact-making

Boyle proposed that matters of fact be generated by a multiplication of the witnessing experience. An experience, even of an experimental performance, that was witnessed by one person alone was not a matter of fact. If that witness could be extended to many, and in principle to all people, then the result could be constituted as a matter of fact. In this way, the matter of fact was at once an epistemological and a social category. The foundational category of the experimental philosophy, and of what counted as properly grounded knowledge generally, was an artifact of communication and of whatever social forms were deemed necessary to sustain and enhance communication. The establishment of matters of fact utilized three technologies: a *material technology* embedded in the construction and operation of the air-pump; a *literary technology* by means of which the phenomena produced by the pump were made known to those who were not direct witnesses; and a *social technology* that laid down the conventions natural philosophers should employ in dealing with each other and considering knowledge-claims.[6] Given the concerns of this essay, I pay most attention to Boyle's literary technology: the expository means by which matters of fact were established and assent mobilized. Yet the impression should not be given that we are dealing with three distinct technologies: each embedded the others. For example, experimental practices employing the material technology of the air-pump crystallized particular forms of social organization; desired forms of social organization were dramatized in the exposition of experimental findings; the literary reporting of air-pump performances provided an experience that was said to be essential to the propagation of the material technology or even to be a valid substitute for direct witness. In treating Boyle's literary technology we are not, therefore, talking about something that is merely a "report" of what was done elsewhere; we are dealing with a most important form of experience and the means for extending and validating experience.

THE MATERIAL TECHNOLOGY OF THE AIR-PUMP

I start by noting the obvious: Boyle's matters of fact were *machine-made*. In his terminology, performances using the air-pump counted as "unobvious" or "elaborate" experiments, contrasted to either the "simple" observation of nature or the "obvious" experiments involved in reflecting upon common artifacts like the gardener's watering-pot.[7] The air-pump (or "pneumatic engine") constructed for Boyle in 1659 (largely by Robert Hooke) was indeed an elaborate bit of scientific machinery (see the figure below).[8]

It consisted of a glass "receiver" of about 30-quarts volume, connected to a brass "cylinder" ("3") within which plied a wooden piston or "sucker" ("4"). The aim was to evacuate the receiver of atmospheric air and so achieve a working vacuum. This was done by manually operating a pair of valves: on the downstroke, valve "S" (the stop-cock) was opened and valve "R" was inserted; the sucker was then moved down by means of a rack-and-pinion device ("5" and "7"). On the upstroke, the stop-cock was closed, the valve "R" removed, and a quantity of air drawn into the cylinder was expelled. This operation was repeated many times until the effort of moving the sucker became too great, at which point a working vacuum was deemed to have been attained. Great care had to be taken to ensure that the pump was sealed against leakage, for example, at the juncture of receiver and cylinder and around the sides of the sucker. Experimental apparatus could be placed into the receiver through an aperture at the top of the receiver ("B-C"), for instance, a barometer or simple Torricellian apparatus. The machine was then ready to produce matters of fact. Boyle used the pump to generate phenomena that he interpreted in terms of "the spring of the air" (its elasticity) and the weight of the air (its pressure).

Boyle's air-pump was, as he said, an "elaborate" device; it was also tem-

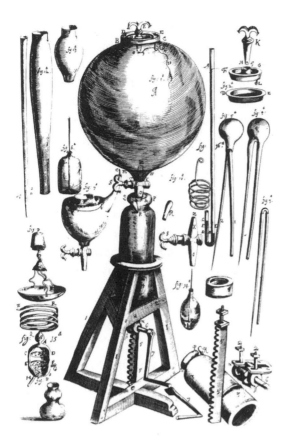

Boyle's Air-Pump of 1660. From Boyle's *New Experiments Physico-Mechanical Touching the Spring of the Air* (1660).

peramental (difficult to operate properly) and very expensive: the air-pump was seventeenth-century "Big Science." To finance its construction on an individual basis it helped mightily to be a son of the Earl of Cork. Other natural philosophers, almost as well supplied with cash, shied away from the cost of having one built, and a major justification for founding scientific societies in the 1660s and afterward was the collective financing of the instruments upon which the experimental philosophy was deemed to depend. Air-pumps were not widely distributed in the 1660s. They were scarce commodities: Boyle's original machine was quickly presented to the Royal Society of London; he had one or two redesigned instruments built for him by 1662, operating mainly in Oxford; Christiaan Huygens had one made in The Hague in 1661; there was one at the Montmor Academy in Paris; there was probably one at Christ's College, Cambridge, by the mid-1660s; and Henry Power may have possessed one in Halifax from 1661. So far as can be found out, these were all the air-pumps that existed in the decade after their invention.[9]

So, air-pump technology posed a problem of access. If knowledge was to be made using this technology, then the numbers of philosophers who could produce it were limited. Indeed, in Restoration England this restriction was one of the chief *recommendations* of "elaborate" experimentation: knowledge could no longer legitimately be generated by alchemical "secretists" and sectarian "enthusiasts" who claimed individual and unmediated inspiration from God. Experimental knowledge was to be tempered by collective labor and disciplined by artificial devices. The very intricacy of machines like the air-pump allowed philosophers, it was said, to discern which cause, among the many possible, might be responsible for observed effects. This was something, in Boyle's view, that the gardener's pot could not do.[10] However, access to the machine had to be opened up if knowledge-claims were not to be regarded as mere individual opinion and if the machine's matters of fact were not to be validated on the bare say-so of an individual's authority. How was this special sort of access to be achieved?

WITNESSING SCIENCE

In Boyle's program the capacity of experiments to yield matters of fact depended not only upon their actual performance but crucially upon the assurance of the relevant community that they had been so performed. He therefore made an important distinction between actual experiments and what are now termed "thought experiments."[11] If knowledge was to be empirically based, as Boyle and other English experimentalists insisted it should, then its experimental foundations had to be attested to by eyewitnesses. Many phenomena, and particularly those alleged by the alchemists, were difficult to credit; in which cases Boyle averred "that they that have seen them can much more reasonably believe them, than they that have not."[12] The problem with eyewitnessing as a criterion for assurance was one of discipline. How did one police the reports of witnesses so as to avoid radical individualism? Was one obliged to credit a report on the testimony of any witness whatever?

Boyle insisted that witnessing was to be a collective enterprise. In natural philosophy, as in criminal law, the reliability of testimony depended upon its multiplicity:

For, though the testimony of a single witness shall not suffice to prove the accused party guilty of murder; yet the testimony of two witnesses, though but of equal credit . . . shall ordinarily suffice to prove a man guilty; because it is thought reasonable to suppose, that, though each testimony single be but probable, yet a concurrence of such probabilities, (which ought in reason to be attributed to the truth of what they jointly tend to prove) may well amount to a moral certainty, i.e. such a certainty, as may warrant the judge to proceed to the sentence of death against the indicted party.[13]

And Thomas Sprat, defending the reliability of the Royal Society's judgments in matters of fact, inquired "whether, seeing in all Countreys, that are govern'd by Laws, they expect no more, than the consent of two, or three witnesses, in matters of life, and estate; they will not think, they are fairly dealt withall, in what concerns their *Knowledg*, if they have the concurring Testimonies of *threescore or an hundred*."[14]

The thrust of the legal analogy should not be missed. It was not just that one was multiplying authority by multiplying witnesses (although this was part of the tactic); it was that right *action* could be taken, and seen to be taken, on the basis of these collective testimonies. The action concerned the positive giving of assent to matters of fact. The multiplication of witnesses was an indication that testimony referred to a real state of affairs in nature. Multiple witnessing was counted as an active, and not just a descriptive, license. Does it not force the conclusion that such and such an action was done (a specific trial), and that subsequent action (offering assent) was warranted?

In experimental practice one way of securing the multiplication of witnesses was to perform experiments in a social space. The "laboratory" was contrasted to the alchemist's closet precisely in that the former was said to be a public and the latter a private space. The early air-pump trials were routinely performed in the Royal Society's ordinary public rooms, the machine being brought there specially for the occasion.[15] In reporting upon his experimental performances Boyle commonly specified that they were "many of them tried in the presence of ingenious men," or that he made them "in the presence of an illustrious assembly of virtuosi (who were spectators of the experiment)."[16] Boyle's collaborator Robert Hooke worked to codify the society's procedures for the standard recording of experiments: the register was "to be sign'd by a certain Number of the Persons present, who have been present, and Witnesses of all the said Proceedings, who, by Sub-scribing their Names, will prove undoubted Testimony."[17] And Sprat described the role of the "Assembly" in "resolv[ing] upon the matter of Fact" by collectively correcting individual idiosyncrasies of observation and judgment.[18] In reporting experiments that were particularly crucial or problematic, Boyle named his witnesses and stipulated their qualifications. The experiment of the original air-pump trials that was "the principal fruit I promised myself from our engine" was conducted in the presence of "those excellent and deservedly famous Mathematic Professors, Dr *Wallis,* Dr *Ward,* and Mr *Wren* . . . whom I name, both as justly counting it an honour to be known to them, and as being glad of such judicious and illustrious witnesses of our experiment." Another important experiment was attested to by Wallis, "who will be allowed to be a very competent judge in these matters." And in his censure of the alchemists Boyle generally warned natural philosophers not "to believe chymical experiments . . . unless he, that delivers that, mentions his doing it upon his own particular knowledge, or upon the relation of some credible person, avowing it upon his own experience." Alchemists were recommended to name the putative author of these experiments "upon whose credit they relate" them.[19] The credibility of witnesses followed the taken-for-granted conventions of that setting for assessing individuals' reliability and trustworthiness: Oxford professors were accounted more reliable witnesses than Oxfordshire peasants. The natural philosopher had no option but to rely for a substantial part of his knowledge on the testimony of witnesses; and, in assessing that testimony, he (no less than judge or jury) had to determine their credibility. This necessarily involved their moral constitution as well as their knowledgeableness, "for the two grand requisites, of a witness [are] the knowledge he has of the things he delivers, and his faithfulness in truly delivering what he knows." In this way, the giving of witness in experimental philosophy transitted the social and moral accounting systems of Restoration England.[20]

Another important way of multiplying witnesses to experimentally produced phenomena was to facilitate their replication. Experimental protocols could be reported in such a way as to enable readers of the reports to perform the experiments for themselves, so ensuring distant but direct witnesses. Boyle elected to publish several of his experimental series in the form of letters to other experimentalists or potential experimentalists. The *New Experiments* of 1660 was written as a letter to his nephew Lord Dungarvan; the various tracts of the *Certain Physiological Essays* of 1661 were written to another nephew Richard Jones; the *History of Colours* of 1664 was originally written to an unspecified friend. The purpose of this form of communication was explicitly to proselytize. The *New Experiments* was published so "that the person I addressed them to might, without mistake, and with as little trouble as possible, be able to repeat such unusual experiments." The *History of Colours* was designed "not barely to relate [the experiments], but . . . to teach a young gentleman to make them."[21] Boyle wished to encourage young gentlemen to "addict" themselves to experimental pursuits and, thereby, to multiply both experimental philosophers and experimental facts.

Replication, however, rarely succeeded, as Boyle himself recognized. When he came to prepare the *Continuation of New Experiments* seven years after the original air-pump trials, Boyle admitted that, despite his care in communicating details of the engine and of his procedures, there had been few successful replications: "in five or six years I could hear but of one or two engines that were brought to be fit to work, and of but one or two new experiments that had been added by the ingenious owners of them."[22]

This situation had not notably changed by the mid-1670s. In the seven or eight years after the *Continuation,* Boyle said that he heard "of very few experiments made, either in the engine I used, or in any other made after the model thereof." By this time a note of despair began to appear in Boyle's statements concerning the replication of his air-pump experiments. He was "more willing to set down divers things with their minute circumstances; because I was of opinion, that probably many of these experiments would be never either re-examined by others, or re-iterated by myself. For though they may be easily read . . . yet he, that shall really go about to repeat them, will find it no easy task."[23]

The Literary Technology of Virtual Witnessing

The third way by which witnesses could be multiplied is far more important than the performance of experiments before direct witnesses or the facilitating of actual replication: it is what I call "virtual witnessing." The technology of virtual witnessing involves the production in a reader's mind of such an image of an experimental scene as obviates the necessity for either its direct witness or its replication. Through virtual witnessing the multiplication of witnesses could be in principle unlimited. It was therefore the most powerful technology for constituting matters of fact. The validation of experiments, and the crediting of their outcomes as matters of fact, necessarily entailed their realization in the laboratory of the mind and the mind's eye. What was required was a technology of trust and assurance that the things had been done and done in the way claimed.

The technology of virtual witnessing was not different in kind to that used to facilitate actual replication. One could deploy the same linguistic resources in order to encourage the physical replication of experiments or to trigger in the reader's mind a naturalistic image of the experimental scene. Of course, actual replication was formally preferred, for this eliminated reliance upon testimony altogether. Yet, because of natural and legitimate suspicion among those who were neither direct witnesses nor replicators, a greater degree of assurance was required to produce assent in virtual witnesses. Boyle's literary technology was crafted to secure this assent.

PROLIXITY AND ICONOGRAPHY

In order to understand how Boyle deployed his literary technology of virtual witnessing we have to reorientate some of our common ideas about the status of the scientific *text*. We usually think of an experimental report as a narration of some prior visual experience: it points to sensory experience that lies behind the text. This is correct. However, we should also appreciate that the text itself constitutes a visual source. We need to see how Boyle's texts were constructed so as to provide a source of virtual witness that was agreed to be reliable. The best way to fasten upon the notion of the text as this kind of source might be to start by looking at some of the pictures that Boyle provided alongside his prose.

The figure above, for example, is an engraving of his original air-pump, appended to the *New Experiments*. Producing these kinds of images was an expensive business in the mid-seventeenth century and natural philosophers used them sparingly. The figure is not a schematized line-drawing but an attempt at detailed naturalistic representation, complete with the pictorial conventions of shadowing and cut-away sections of parts. This is not a picture of the "idea" of an air-pump but of a particular existing air-pump.[24] The same applies to Boyle's pictorial representations of his particular pneumatic experiments: in one, we are shown a mouse lying dead in the receiver; in another, images of experimenters. Boyle devoted great attention to the manufacture of these engravings, sometimes consulting directly with artist and engraver, sometimes by way of Hooke.[25] Their role was to supplement the imaginative witness provided by the words in the text. In the *Continua-*

tion Boyle expanded upon the relationships between the two sorts of exposition. He told his readers that "they who either were versed in such kind of studies or have any peculiar facility of imagining, would well enough conceive my meaning only by words," but others required visual assistance. He apologized for the relative poverty of the images, "being myself absent from the engraver for a good part of the time he was at work, some of the cuts were misplaced, and not graven in the plates."[26]

Visual representations, few as they necessarily were in Boyle's texts, were mimetic devices. By virtue of the density of *circumstantial* detail that could be conveyed through the engraver's laying of lines, the images imitated reality and gave the viewer a vivid impression of the experimental scene. The sort of naturalistic images Boyle favored provided a greater density of circumstantial detail than would have been proffered by more schematic representations. The images served to announce, so to speak, that "this was really done" and that it was done in the way stipulated; they allayed distrust and facilitated virtual witnessing. So understanding the role of pictorial representations offers a way of appreciating what Boyle was trying to achieve with his literary technology.[27]

In the introductory pages of the *New Experiments,* Boyle's first published experimental findings, he directly announced his intention to be "somewhat prolix." His excuses were threefold: first, delivering things "circumstantially" would facilitate replication; second, the density of circumstantial details was justified by the fact that these were "new" experiments, with novel conclusions drawn from them: it was therefore necessary that they be "circumstantially related, to keep the reader from distrusting them"; third, circumstantial reports such as these offered the possibility of virtual witnessing. As Boyle said, "these narratives [are to be] as standing records in our new pneumatics, and [readers] need not reiterate themselves an experiment *to have as distinct an idea of it,* as may suffice them to ground their reflexions and speculations upon."[28] If one wrote an experimental report in the proper way, the reader could take on trust that these things happened. It would be as if readers had been present at the proceedings. They would be recruited as witnesses and be put in a position where they could validate experimental phenomena as matters of fact.[29] Attention to the writing of experimental reports was of at least equal practical importance to doing the experiments themselves.

In the late 1650s, Boyle devoted himself to laying down the rules for the literary technology of the experimental program. Stipulations about how to write proper scientific prose were dispersed throughout his experimental reports of the 1660s, but he also composed a special tract on the subject of "experimental essays." Here Boyle offered extended justifications for his "prolixity": "I have," he understated, "declined that succinct way of writing"; he had sometimes "delivered things, to make them more clear, in such a multitude of words, that I now seem even to myself to have in divers places been guilty of verbosity." Not just his "verbosity" but also Boyle's ornate sentence-structure, with piled-up appositive clauses, was, he said, part of a plan to convey circumstantial details and to give the impression of verisimilitude: "I have knowingly and purposely transgressed the laws of oratory in one particular, namely, in making sometimes my periods [i.e., complete sentences] or parentheses over-long: for when I could not within the compass of a regular period comprise what I thought requisite to be delivered at once, I chose rather to neglect the precepts of rhetoricians, than the mention of those things, which I thought pertinent to my subject, and useful to you, my reader."[30] Elaborate sentences, with circumstantial details encompassed within the confines of one grammatical entity, might mimic that immediacy and simultaneity of experience afforded by pictorial representations.

Boyle was constituting himself as a reliable purveyor of experimental testimony and offering conventions by means of which others could do likewise. The provision of circumstantial details of experimental scenes was a way of assuring readers that real experiments had yielded the findings stipulated. It was also necessary, in Boyle's view, to offer readers circumstantial accounts of *failed* experiments. This performed two functions: first, it allayed anxieties in those neophyte experimentalists whose expectations of success were not immediately fulfilled; second, it assured readers that the relator was not willfully suppressing inconvenient evidence, that the author was in fact being faithful to reality. Complex and circumstantial accounts were to be taken as undistorted mirrors of complex experimental performances, in which a wide range of contingencies might influence outcomes.[31] So, for example, it was not legitimate to hide the fact that air-pumps sometimes did not work properly or that they often leaked: "I think it becomes one, that professeth himself a faithful relator of experiments not to conceal" such unfortunate contingencies.[32] It is, however, vital to keep in mind that the contingencies proffered in Boyle's circumstantial accounts represent a selection of possible contingencies. There was not, nor can there be, any such thing as a report noting all circumstances that might affect an experiment. Circumstantial, or stylized, accounts do not, therefore, exist as pure forms but as publicly acknowledged moves toward or away from the reporting of contingencies.

THE MODESTY OF EXPERIMENTAL NARRATIVE

The ability of reporters to multiply witnesses depended upon readers' acceptance of them as providers of reliable testimony. It was the burden of Boyle's literary technology to assure his readers that he was such a man as should be believed. He therefore had to find the means to make visible in the text the accepted tokens of a man of good faith. One technique has just been discussed: the reporting of experimental failures. A man who recounted unsuccessful experiments was such a man whose objectivity was not distorted by interests. So, the literary display of a certain sort of morality was a technique in making of matters of fact. A man whose narratives could be credited as mirrors of reality was a "modest man"; his reports should make that modesty visible.

Boyle found a number of ways of displaying modesty. One of the most straightforward was the form of the experimental *essay*. The essay (i.e., the piecemeal reporting of experimental trials) was explicitly contrasted to the natural philosophical *system*. Those who wrote entire systems were identified as "confident" individuals, whose ambition extended beyond what was proper or possible. By contrast, those who wrote experimental essays were "sober and modest men," "diligent and judicious" philosophers, who did not "assert more than they can prove." This practice cast the experimental philosopher into the role of intellectual "under-builder," or even that of "a drudge of greater industry than reason." This was, however, a noble character, for it was one that was freely chosen to further "the real advancement of true natural philosophy" rather than personal reputation.[33] The public display of this modesty was an exhibition that concern for individual celebrity did not cloud judgment and distort the integrity of one's reports. In this connection it is crucial to remember who it was that was portraying himself as a mere "under-builder." He was the son of the Earl of Cork, and everyone knew that very well. It was plausible that such modesty could have a noble character, and Boyle's presentation of self as a role model for experimental philosophers was powerful.[34]

Another technique for displaying modesty was Boyle's professedly "naked way of writing." He would eschew a "florid" style; his object was to write "rather in a philosophical than a rhetorical strain." This plain, puritanical, unadorned (yet convoluted) style was identified as *functional*. It served to exhibit, once more, the philosopher's dedication to community service rather than to his personal reputation. Moreover, the "florid" style to be avoided was a hindrance to the clear provision of virtual witness: it was, Boyle said, like painting "the eye-glasses of a telescope."[35]

The most important literary device Boyle employed for demonstrating modesty acted to protect the fundamental epistemological category of the experimental program: the matter of fact. There were to be appropriate moral postures, and appropriate modes of speech, for items on either side of the crucial boundary that separated matters of fact from the locutions used to account for them: theories, hypotheses, speculations, and the like. So, Boyle told his nephew, "in almost every one of the following essays I . . . speak so doubtingly, and use so often, *perhaps, it seems, it is not improbable,* and such other expressions, as argue a diffidence of the truth of the opinions I incline to, and that I should be so shy of laying down principles, and sometimes of so much as venturing at explications." Since knowledge of physical causes was only "probable," this was the correct moral stance and manner of speech, but things were otherwise with matters of fact, and here a confident mode was not only permissible but necessary: "I dare speak confidently and positively of very few things, except of matters of fact."[36]

It was necessary to speak confidently of matters of fact because, as the foundations of proper philosophy, they required protection. And it was proper to speak confidently of matters of fact because they were not of one's own making; they were, in the empiricist model, discovered rather than invented. As Boyle told one of his adversaries, experimental facts can "make their own way" and "such as were very probable, would meet with patrons and defenders."[37] The separation of modes of speech, and the ability of facts to make their own way, was made visible on the printed page. In *New Experiments* Boyle said he intended to leave "a conspicuous interval" between his narratives of experimental findings and his occasional "discourses" upon their interpretation. One might then read the experiments and the "reflexions" separately.[38] Indeed, the construction of Boyle's experimental essays makes manifest the proper balance between the two categories: *New Experiments* consists of a sequential narrative of forty-three pneumatic experiments; *Continuation* of fifty; and the second part of *Continuation* of an even larger number of disconnected experimental observations, only sparingly larded with interpretative locutions.

The confidence with which one ought to speak about matters of fact extended to stipulations about the proper use of authorities. Citations of other writers should be employed to use them not as "judges, but as witnesses," as "certificates to attest matters of fact." If this practice ran the risk of identifying the experimental philosopher as an ill-read philistine, it was, however, necessary: "I could be very well content to be thought to have scarce looked upon any other book than that of nature."[39] The injunction against citing of authorities played a significant role in the mobilization of assent to matters of fact. It was a way of displaying that one was aware of the workings of the Baconian "Idols" and was taking measures to mitigate their corrupting effects on knowledge-claims.[40] A disengagement between experimental narrative and the authority of systematists served to dramatize the author's lack of preconceived expectations and, especially, of theoretical investments in the outcome of experiments. For example, Boyle several times insisted that he was an innocent of the great theoretical systems of the seventeenth century. In order to reinforce the primacy of experimental findings, "I had purposely refrained from acquainting myself thoroughly with the intire system of either the Atomical, or the Cartesian, or any other whether new or received philosophy." And, again, he claimed that he had avoided a systematic acquaintance with the systems of Pierre Gassendi, René Descartes, and even, remarkably, of Francis Bacon, "that I might not be prepossessed with any theory or principles."[41]

Boyle's "naked way of writing," his professions and displays of humility, and his exhibition of theoretical innocence all complemented each other in the establishment and the protection of matters of fact. They served to portray the author as a disinterested observer and his accounts as unclouded and undistorted mirrors of nature. Such an author gave the signs of a man whose testimony was reliable. Hence, his texts could be credited and the number of witnesses to his experimental narratives could be multiplied indefinitely.

Scientific Discourse and the Community

I have said that the matter of fact was a social as well as an intellectual category. And I have argued that Boyle deployed his literary technology so as to make virtual witnessing a practical option for the validation of experimental performances. Here I examine the ways in which Boyle's literary technology dramatized the social relations proper to a community of experimental philosophers. Only by establishing right rules of discourse between individuals could matters of fact be generated and defended, and only by constituting these matters of fact into the agreed foundations of knowledge could a moral community of experimentalists be created and sustained. Matters of fact were to be produced in a public space: a particular space in which experiments were collectively performed and directly witnessed and an abstract space constituted through virtual witnessing. The problem of producing this kind of knowledge was, therefore, the problem of maintaining a certain form of discourse and a certain form of social order. In the following sections I discuss the ways in which Boyle's literary technology worked to create and maintain this social solidarity among experimental philosophers.

THE LINGUISTIC BOUNDARIES OF THE
EXPERIMENTAL COMMUNITY

In the late 1650s and early 1660s, when Boyle was formulating his experimental and literary practices, the English experimental community was still in its infancy. Even with the founding of the Royal Society, the crystallization of an experimental community centered on Gresham College, and the network of correspondence organized by Henry Oldenburg, the experimental program was far from securely institutionalized. Criticisms of the experimental way of producing physical knowledge emanated from English philosophers (notably Thomas Hobbes) and from continental writers committed to rationalist methods and to the practice of physics as a demonstrative discipline. Experimentalists were made into figures of fun on the Restoration stage: Thomas Shadwell's *The Virtuoso* dramatized the absurdity of weighing the air, and scored most of its good jokes by parodying the convoluted language of Sir Nicholas Gimcrack (almost certainly a pastiche of Boyle and Hooke).[42] The practice of experimental philosophy, despite what numerous historians have assumed, was *not* overwhelmingly popular in Restoration England.[43] In order for experimental philosophy to be established as a legitimate activity, several things needed to be done. First, it required *recruits:* experimentalists had to be enlisted as neophytes, and converts from other forms of philosophical practice had to be obtained. Second, the social role of the experimental philosopher and the linguistic practices appropriate to an experimental community needed to be defined and publicized.[44] What was the proper nature of discourse in such a community? What were the linguistic signs of competent membership? And what uses of language could be taken as indications that an individual had transgressed the conventions of the community?

The entry fee to the experimental community was to be the communication of a candidate matter of fact. In *The Sceptical Chymist,* for instance, Boyle extended an olive-branch even to the alchemists. The solid experimental findings produced by some alchemists could be sifted from the dross of their "obscure" speculations. Since the experiments of the alchemists (and of the Aristotelians) frequently "do not evince what they are alleged to prove," the former could be accepted into the experimental philosophy by stripping away the theoretical language with which they happened to be glossed. As Carneades (Boyle's mouthpiece) said, "your hermetic philosophers present us, together with divers substantial and noble experiments, theories, which either like peacocks feathers make a great shew, but are neither solid nor useful; or else like apes, if they have some appearance of being rational, are blemished with some absurdity or other, that, when they are attentively considered, make them appear ridiculous."[45] Those alchemists who wished to be incorporated into a legitimate philosophical community were instructed what linguistic practices could secure their entry. The same principles were laid down with respect to any practitioner: "let his opinions be never so false, his experiments being true, I am not obliged to believe the former, and am left at liberty to benefit myself by the latter."[46] By arguing that there was only a contingent, not a necessary, connection between the language of matters of fact and theoretical language, Boyle was defining the linguistic terms upon which existing communities could join the experimental enterprise. They were liberal terms, which might serve to maximize potential membership.[47]

There were other natural philosophers Boyle despaired to recruit. Hobbes, notably, was the kind of philosopher who, on no account, ought to be admitted, for he denied the value of systematic and elaborate experimentation, the foundational status of the matter of fact, and the distinction between causal and descriptive language. Of Hobbes's *Dialogus physicus,* Boyle asked, "What new experiment or matter of fact Mr *Hobbes* has therein added to enrich the history of nature?" In his criticisms of Boyle's experiments Hobbes "does not, that I remember, deny the truth of any of the matters of fact I have delivered." According to Boyle, both Hobbes and another critic, the Jesuit Franciscus Linus, had not "seen cause to deny any thing that I deliver as experiment."[48] One could not be regarded as a competent member of the experimental community if one failed to communicate experimental matters of fact, or if one did so in a manner that failed to recognize the linguistic boundaries between factual and causal locutions.

LINGUISTIC BOUNDARIES WITHIN
THE EXPERIMENTAL COMMUNITY

Just as linguistic categories were used to manage entry to the experimental community, distinctions between the language of facts and that of theories were deployed to regulate discourse within it. In broad terms, Boyle insisted upon a separation between "physiological" and "metaphysical" languages: experimental discourse was to be confined to the former. One of the central categories of Boyle's "new pneumatics" also happened to be a major preoccupation of the old physics—namely, vacuism versus plenism, and the judgment whether a vacuum was possible in nature. How was it proper to speak of the contents of the receiver of an evacuated air-pump? And how did this speech relate to traditional usages of the term *vacuum*?

A practical problem was posed by the fact that the lexicon of the new philosophy was largely compiled out of the usages of old discursive practices. Old words had to be given new meanings. So it was proper to apply the term *vacuum* to the contents of the exhausted receiver, but it was improper to take this to mean that the space was absolutely devoid of all matter. Such an absolutely void space was the "vacuum" of metaphysical discourse. What Boyle meant by the air-pump's "vacuum" was "not a space, wherein there is no body at all, but such as is either altogether, or almost totally devoid of air."[49] If contemporary plenists maintained that this vacuum might be filled by a subtle form of matter, or "aether," Boyle could reply with a series of experiments that showed that such an aether could not be made "sensible," that is, it had no physical manifestations. And speech of entities that were not amenable to sensible experimentation was not permissible within experimental philosophy.[50]

The separation of "physiological" from "metaphysical" language was most crucial to Boyle's strategy for dealing with causal inquiry in physical science. In keeping with his probabilist conception of knowledge, Boyle wished to bracket off speech about matters of fact, about which one might be certain, from speech of their physical causes, which were at best probable. In terms of Boyle's air-pump program, the most important instance of this bracketing concerned the notion that was the main product of these experiments: the "spring of the air." Boyle said that his "business" was "not to assign the adequate cause of the spring of the air, but only to manifest, that the air hath a spring, and to relate some of its effects." The cause of the air's elasticity *might* be accounted for variously: by Cartesian vortices, or by the real physical existence in the corpuscles of the air of "slender springs" or of a fleecy structure.[51] The job of the experimental philosopher was to speak of experimentally produced matters of fact, not to conjecture further.[52]

Boyle had considerable problems in diffusing this new mode of speech. Plenist critics persisted in understanding Boyle to be using *vacuum* in its metaphysical sense, and Boyle was obliged persistently to reiterate its proper usage.[53] Other writers either refused to conceive of a natural philosophy that bracketed off causal speech, or reckoned that Boyle must be committed to some (illegitimate and unacknowledged) causal account of the spring of the air.[54] So far as the "spring of the air" was concerned, Boyle's stipulation that it had been made experimentally "manifest" and his disinclination to speak of its cause had an interesting effect. By putting the spring on the other side of the boundary from causal locutions, Boyle constituted the spring, for all practical purposes, into a matter of fact. When it came to labeling the epistemological status of the spring, Boyle variously referred to it as an "hypothesis" or even as a "doctrine." However, by talking of the spring as something made manifest through experiment, and by protecting it from the uncertainties that afflicted epistemological items like causal notions, Boyle treated this "hypothesis" in the same way that he treated other matters of fact.[55]

The vital difference between matters of fact and all other epistemological categories was the degree of assent one might expect to them. To an authenticated matter of fact all men will assent. In Boyle's system that was taken for granted because it was through the technologies that multiplied witness that matters of fact were constituted. General assent was what made matters of fact, and general assent was therefore mobilized around matters of fact. With "hypotheses," "theories," "conjectures," and the like, the situation was quite different. These categories threatened that assent that could be crystallized in the institution of the matter of fact. So the linguistic conventions of Boyle's experimental program separated speech appropriate to the two categories as a way of drawing the boundaries between that about which one was to expect certainty and assent and that about which one could expect uncertainty and divisiveness. The idea was not to eliminate dissent or to oblige people to agree to all items in natural philosophy (as it was for Hobbes); rather, it was to *manage* dissent and to keep it within safe bounds. An authenticated matter of fact was treated as a mirror of nature; a theory, by contrast, was clearly man-made and could, therefore, be contested. Boyle's linguistic boundaries acted to segregate what could be disputed from what could not. The management of dispute in experimental philosophy was crucial to protecting the foundations of knowledge.

MANNERS IN DISPUTE

Since natural philosophers were not to be compelled to give assent to all items of knowledge, dispute and controversy was to be expected. How should this be dealt with? The problem of conducting dispute was a matter of intense practical concern in early Restoration science. During the civil wars and Interregnum, the divisiveness of "enthusiasts," sectarians, and hermeticists threatened to bring about radical individualism in philosophy. Nor did the various sects of peripatetic natural philosophers display a public image of a stable and united intellectual community. Unless the new experimental community could exhibit a broadly based consensus and harmony within its own ranks, it was unreasonable to expect it to secure the legitimacy within Restoration culture that its leaders desired. Moreover, that very consensus was vital to the establishment of matters of fact as the foundational category of the new practice.

By the early 1660s, Boyle was in a position to give concrete exemplars of how disputes ought to be conducted. Three critics—Linus, Hobbes, and Henry More—published their responses to his *New Experiments,* and he replied to each one. But even before he had been engaged in dispute, Boyle laid down a set of rules for how controversies were to be handled by the experimental philosopher. For example, in *A Proëmial Essay* (composed 1657), Boyle insisted that disputes should be about findings and not about persons. It was proper to take a hard view of reports that were inaccurate but most improper to attack the character of those who rendered them: "for I love to speak of persons with civility, though of things with freedom." The *ad hominem* style must at all costs be avoided, for the risk was that of making foes out of mere dissenters. This was the key point: potential contributors of matters of fact, however wrong they may be, must be treated as possible converts to the experimental philosophy. If, however, they were bitterly treated, they would be lost to the cause and to the community whose size and consensus validated matters of fact:

And as for the (very much too common) practice of many, who write, as if they thought railing at a man's person, or wrangling about his words, necessary to the confutation of his opinions; besides that I think such a quarrelsome and injurious way of writing does very much misbecome both a philosopher and a Christian, methinks it is as unwise, as it is provoking. For if I civilly endeavour to reason a man out of his opinions, I make myself but one work to do, namely, to convince his understanding; but, if in a bitter or exasperating way I oppose his errors, I increase the difficulties I would surmount, and have as well his affections against me as his judgment: and it is very uneasy to make a proselyte of him, that is not only a dissenter from us, but an enemy to us.[56]

Furthermore, it was impolitic to acknowledge the existence of "sects" in natural philosophy. One way by which one could hope to overcome sectarianism was to decline public recognition that it existed: "it is none of my design," Boyle said, "to engage myself with, or against, any one sect of Naturalists." The experiments would decide the case. The views of these "sects" should be noted only insofar as they were founded upon experiment. Therefore, it was right and politic to be harsh in one's writings against those who did not contribute experimental findings, for they had nothing to offer to the constitution of matters of fact. Finally, the experimental philosopher must show that there was point and purpose to legitimately conducted dispute. He should be prepared publicly to renounce positions that were shown to be erroneous. Flexibility followed from fallibilism. As Boyle wrote, "till a man is sure he is infallible, it is not fit for him to be unalterable."[57]

The conventions for managing dispute were dramatized in the structure of *The Sceptical Chymist.* These fictional conversations (between an Aristotelian, two varieties of hermeticists, and "Carneades" as mouthpiece for Boyle) took the form, not of a Socratic dialogue, but of a *conference.*[58] They were a little piece of theater that exhibited how persuasion, dissensus, and, ultimately, conversion to truth ought to be conducted. Several points about Boyle's theater of persuasion can be briefly made: first, the "symposiasts" are imaginary, not real. This means that opinions can be confuted without exacerbating relations between real philosophers. Even Carneades, although he is manifestly "Boyle's man," is not Boyle himself: Carneades is made actually to quote "our friend Mr *Boyle*" as a device for distancing opinions from individuals. The author is insulated from the text and from the opinions he may actually espouse. Second, truth is not inculcated from Carneades to his interlocutors; rather it is dramatized as emerging through the conversation.[59] Everyone is seen to have a say in the consensus, which is the dénouement.[60] Third, the conversation is, without exception, civil: as Boyle said, "I am not sorry to have this opportunity of giving an example, how to manage even disputes with civility."[61] No symposiast abuses another; no ill temper is displayed; no one leaves the conversation in pique or frustration.[62] Fourth, and most important, the currency of intellectual discourse, and the means by which agreement is reached, is the experimental matter of fact. Here, matters of fact are not treated as the exclusive property of any one philosophical sect. Insofar as the alchemists have produced experimental findings, they have minted the real coins of experimental exchange. Their experiments are welcome, while their "obscure" speculations are not. Insofar as the Aristotelians produce few experiments, and insofar as they refuse to dismantle the "arch"-like "mutual coherence" of their philosophical system into facts and theories, they can make little contribution to the experimental conference.[63] In these ways, the structure and the linguistic conventions of this imaginary conversation make vivid the rules for real conversations proper to experimental philosophy.

Real disputes followed hard upon the imaginary ones of *The Sceptical Chymist,* providing Boyle with valuable opportunities of putting his principles into practice. Linus was the adversary who experimented but who denied the power of the "spring of the air"; More was the adversary whom Boyle wished to be an ally—offering what he regarded as a theologically more appropriate explanation of Boyle's pneumatic findings; and Hobbes was the adversary who denied the value of systematic experiment and the foundational status of the matter of fact. Each carefully crafted response that Boyle produced was labeled as a model for how disputes should be managed by the experimental philosopher.[64]

First, all public disputes had to be justified: the experimental philosopher should be loath to engage in controversy. As Boyle claimed, "I have a natural indisposedness to contention."[65] The justification was not the defense of one's reputation but the protection of what was vital to the collective practice of proper philosophy: the value of systematic experimentation, the matters of fact that experiment produced, the boundaries that separated those facts from less certain epistemological items, and the rules of social life that regulated discourse in the experimental community. Boyle took care to identify the *object* of controversy as interpretations of facts, not the facts themselves. Neither Linus nor Hobbes, he said, denied "any thing that I deliver as experiment . . . so that usually . . . they are fain to fall upon the hypotheses themselves." This was a crucial stipulation, because, if it was accepted, then the arena of disagreement could be so defined as to protect the status of matters of fact. The very phenomenon of public disputation about "hypotheses" could be contrasted to the absence of controversy about that which Boyle "deliver[ed] as experiment."[66]

The importance of protecting experimental practice is evident in the differing tones of Boyle's responses to Linus and to Hobbes. While Linus attacked the spring of the air, the major interpretative resource of Boyle's pneumatics, "he takes no exceptions at the experiments themselves, as we have recorded them." Boyle concluded that this "is no contemptible testimony, that the matters of fact have been rightly delivered." The Jesuit was congratulated for essaying to experiment himself and for his diligence in understanding what Boyle had written.[67] He was a good adversary and was dealt with as a potential convert. With Hobbes the situation was different. This adversary, "not content to fall upon the explications of my experiments, has (by an attempt, for aught I know, unexampeled) endeavoured to disparage unobvious experiments themselves, and to discourage others from making them."[68] Hobbes was a dangerous adversary; there was no possibility of recruiting such a man to the experimental program, and his objections had to be publicly exploded.

For all that, Hobbes, no less than Linus and More, had to be dealt with civilly. Boyle aimed, he said, "to give an example of disputing in print against a provoking, though unprovoked, adversary, without bitterness and incivility." He hoped that his own *Examen* "will not be thought to have less of reason for having the less of passion."[69] Managing a dispute with Hobbes was a hard case, and, if it could be conducted in a decent tone, it would offer a model of the language of controversy appropriate to a moral community of experimental philosophers. Boyle did not have far to look to find examples of improper disputation, in which the language of controversy acted to exacerbate divisions in natural philosophy. From the mid-1650s, Hobbes's natural philosophy and geometry had been attacked by the Oxford professors John Wallis and Seth Ward. Wallis, one of the toughest street-fighters of the new philosophy, had not only shown his adversary's notions to be erroneous, he had punned upon the plebeian origins of Hobbes's name and insinuated improper political affiliations and motivations. Hobbes, who professed himself concerned for maintaining good manners in dispute, showed his foes the sharp side of his tongue: "So go your ways, you *Uncivil Ecclesiastics, Inhuman Divines, Dedoctors of morality, Unasinous Colleagues, Egregious pair of Issachars, most wretched Vindices and Indices Academiarum.*"[70] And again, summing up the value of one of Wallis's criticisms, "all error and railing, that is, stinking wind; such as a jade lets fly, when he is too hard girt upon a full belly."[71]

This is the sort of thing Boyle wished to avoid. It was not merely a matter of Boyle's individual "modest" temperament or what he reckoned was owing to fellow Christian philosophers. What was at issue was the creation and preservation of a calm public space in which natural philosophers could heal their divisions, collectively agree upon the foundations of knowledge, and, thereby, establish their credit in Restoration culture. Such a calm space was vital to achieving these goals. As Boyle reminded his readers in the introduction to his *New Experiments,* published in that "wonderful pacifick year" of the Restoration of the monarchy, "the strange confusions of this unhappy nation, in the midst of which I have made and written these experiments, are apt to disturb that calmness of mind and undistractedness of thoughts, that are wont to be requisite to happy speculations."[72] And Sprat recalled the circumstances of the Oxford group of experimentalists that had spawned the Royal Society: "Their first purpose was no more, then onely the satisfaction of breathing a freer air, and of conversing in quiet one with another, without being ingag'd in the passions, and madness of that dismal Age." He described the difference between "humane affairs," which "may affect us, with a thousand various disquiets," and the experimental study of nature: "*that* gives us room to differ, without animosity; and permits us to raise contrary imaginations upon it, without any danger of a *Civil War.*"[73]

This calm space that experimental philosophy was to inhabit would be created and maintained through the deployment within the moral community of appropriate linguistic practices.[74] An appropriate language had to perform several functions. First, it had to be a resource for managing dissent and conflict in such a way as to make it possible for philosophers to express divergent views while leaving the foundations of knowledge intact, and, in fact, buttressing these foundations. We have seen this in the linguistic separation Boyle wished to make between speech of matters of fact and speech of explanatory items. Second, it had to facilitate reconciliation among existing sects of philosophers, mobilizing that reconciliation so as to reinforce the foundational status of matters of fact. We have seen this in Boyle's distribution of authentic matters of fact among groups with divergent theoretical commitments and in his identification of experimental matters of fact as the medium of exchange in the new practice. Third, such a language had to constitute a vehicle whereby matters of fact could effectively be generated and validated by a community whose size was, in principle, unlimited. And this we have seen in the role played by Boyle's literary technology in multiplying the witnessing experience.

Scientific Knowledge and Exposition: Conclusions

I have shown that three technologies were involved in the production and validation of Boyle's experimental matters of fact: the material, the literary, and the social. Although I have concentrated here upon the literary technology, I have also suggested that the three technologies are not distinct: the working of each depends upon and incorporates the others. I want now briefly to develop that point by showing how each technology contributes to a common strategy for constituting matters of fact.

What makes a fact different from an artifact is that the former is not perceived to be man-made. What people make, people may unmake, but a matter of fact is taken to be the very mirror of nature. To identify the role of human agency in the making of an item of knowledge is to identify the possibility of its being otherwise. To shift agency onto natural reality is to stipulate the grounds for universal assent. Each of the three technologies works to achieve the appearance of matters of fact as *given* items: each functions as an objectifying resource.

Take, for example, the role of the air-pump in the production of matters of fact. Pneumatic facts were machine-made. The product of the pump was not, as it is for the modern scientific machines studied by Bruno Latour, an "inscription": it was a visual experience that had to be transformed into an inscription by a witness.[75] However, the air-pump of the 1660s has this in common with the gamma counter of the present-day neuroendocrinological laboratory: it stands between the perceptual competencies of a human being and natural reality itself. A "bad" observation taken from a machine need not be ascribed to cognitive or moral faults in human beings, nor is a "good" observation their personal product. It is the machine that has generated the finding. A striking instance of this usage arose in the 1660s when Christiaan Huygens offered a matter of fact produced by his pump that appeared to conflict with one of Boyle's central explanatory resources. Boyle did not impugn Huygens's integrity or his perceptual and cognitive competences. Instead, he suggested that the fault lay with the machine: "[I] question not his Ratiocination, but only the staunchness of his pump."[76] The machine constitutes a resource that may be used to factor out human agency in the intellectual product: so to speak, "it is not I who says this: it is the machine that speaks," or "it is not your fault; it is the machine's."

Boyle's social technology constituted an objectifying resource by making the production of knowledge visible as a collective enterprise: "it is not I who says this; it is all of us." As Sprat insisted, collective performance and collective witness served to correct the natural working of the "Idols": the faultiness, the idiosyncrasy, or the bias of any individual's judgment and observational ability. The Royal Society advertised itself as a "union of eyes, and hands"; the space in which it produced its experimental knowledge was stipulated to be a *public space*. It was public in a very precisely defined and very rigorously policed sense: not everyone could come in; not everyone's testimony was of equal worth; not everyone was equally able to influence the official voice of the institution. Nevertheless, what Boyle was proposing, and what the Royal Society was endorsing, was a crucially important *move toward* the public constitution and validation of knowledge. The contrast was, on the one hand, with the private work of the alchemists, and, on the other, with the individual dictates of the systematical philosophers.

In the official formulation of the Royal Society, the production of experimental knowledge commenced with individuals' acts of seeing and believing, and was completed when all individuals voluntarily agreed with one another about what had been seen and ought to be believed. This freedom to speak had to be protected by a special sort of discipline. Radical individualism—each individual setting himself up as the ultimate judge of knowledge—would destroy the conventional basis of knowledge, while the disciplined collective social structure of the experimental language-game would create and sustain that factual basis. The experimentalists were on guard against "dogmatists" and "tyrants" in philosophy, just as they abominated "secretists" who produced their knowledge-claims in private space. No one person was to have the right to lay down what was to count as knowledge. Legitimate knowledge was objective insofar as it was produced by the collective, and agreed to voluntarily by those who comprised the collective. The objectification of knowledge proceeded through displays of the communal basis of generation and evaluation. Human coercion was to have no visible place in the experimental way of life.[77]

It was the function of the literary technology to create that communal way of life, to bound it, and to provide the forms and conventions of social relations within it. The literary technology of virtual witnessing supplemented the public space of the laboratory by extending a valid witnessing experience to all readers of the text. The boundaries stipulated by Boyle's linguistic practices acted to keep that community from fragmenting and served to protect items of knowledge to which one could expect universal assent from items that produced divisiveness. Similarly, Boyle's stipulations concerning proper manners in dispute worked to guarantee that social solidarity that generated assent to matters of fact and to rule out of order those imputations that would undermine the moral integrity of the experimental way of life.

I have attempted to display these linguistic practices in the making, and, within restrictions of space, I have alluded to sources of seventeenth-century opposition to these practices. It is important to understand two things about these ways of expounding scientific knowledge and securing assent: that they are historical constructions and that there have been alternative practices. It is particularly important to understand this because of the problems of givenness and self-evidence that attend the institutionalization of these practices. Just as the three technologies operate to create the illusion that matters of fact are not man-made, so the institutionalized and conventional status of the scientific discourse that Boyle helped to produce makes the illusion that scientists' speech about natural reality is simply a reflection of that reality. In this instance, and in others like it, the historian has two major tasks: to display the man-made nature of scientific knowledge, and to account for the illusion that this knowledge is *not* man-made. It is one of the recommendations of the sociology of knowledge that its practitioners often attempt to accomplish these two tasks in the same exercise.[78]

In the twenty-first century scientific papers are rarely, if ever, written with the depth of circumstantial detail that Boyle's reports contained. Why might this be? The answer to this question leads us to the study of linguistic aspects of scientific institutionalization and differentiation. In discussing the characteristics of a *Denkkollektiv,* Ludwik Fleck noted that such a group cultivates "a certain exclusiveness both formally and in content . . . A thought commune becomes isolated formally, but also absolutely bonded together, through statutory and customary arrangements, sometimes a separate language, or at least special terminology . . . The optimum system of a science, the ultimate organization of its principles, is completely incomprehensible to the novice or, Fleck might have added, to any non-member."[79]

Fleck was suggesting that the linguistic conventions of a body of practitioners constitute an answer to the question "Who may speak?" The language of an institutionalized and specialized scientific group is removed from ordinary speech, and from the speech of scientists belonging to another community, both as a sign and as a vehicle of the group's special and bounded status. Not everyone may speak; the ability to speak entails the mastering of special linguistic competences; and the use of ordinary speech is taken as a sign of non-membership and non-competence. Such a group gives linguistic indications that the generation and validation of its knowledge does not require the mobilizing of belief, trust, and assent outside its own social boundaries. (Yet, when external support or subvention is required, special *occasional* modes of speech may be resorted to, including the various languages of "popularization.")

By contrast, Boyle's circumstantial reporting was a means of involving a wider community and soliciting its participation in the making of factual experimental knowledge. His circumstantial language was a way of bringing readers into the experimental scene, indeed of making the *reader* an actor in that scene. Readers were to be shown not just the products of experiments but their mode of construction and the contingencies affecting their performance, *as if they were present*. Boyle aimed to accomplish this, not by inventing a totally novel language (although it was novel to the natural philosophical community of the time), but, it could be argued, by incorporating aspects of ordinary speech and lay techniques of validating knowledge-claims. The language of early Restoration experimental science was, in this sense, a public language. And the use of this public language was, in Boyle's work, vital to the creation of both the knowledge and the social solidarity of the experimental community. Trust and assent had to be won from a public that might deny trust and assent.

PART III

The Scientific Person

Much traditional thinking about the person of the scientist sits uneasily astride some rather big cultural fault-lines. The so-called hagiographical conception of the history of science—praising famous scientists and displaying their unique intellectual powers and moral virtues—has only recently been put aside by academic historians, even though it remains vigorous in the wider culture. In the hagiographical tradition, great scientists possess *genius;* they are taken to be constitutionally and morally distinct from the common run of humankind; and the job of the historian is frank celebration. Yet, apart from the current preferences of an academic discipline, several culturally pervasive sentiments run in opposition to hagiography. One is the idea that scientific knowledge is produced, evaluated, and even made credible through the operation of *Method*—a formal Scientific Method that is unique, historically stable, effective, and, above all, impersonal. Why should the persons of scientists be worth attending to if Method is all? Second, there has emerged, over the past century or so, a growing insistence—vigorously expressed by many scientists themselves—that scientists are nothing very special in terms of their cognitive abilities or moral makeup. What might be learned if we turn resource into historical topic? If we ask how our past and present cultures think as they do about the scientific person? And if we go on to investigate the cultural materials from which portrayals of the scientific person have been historically assembled—the characters, roles, typifications, and other more or less robust expectations about what different sorts of people are like, what they do and value, how they are marked by different virtues and vices? And, finally, what may be learned if we consider the consequences of various historically situated understandings of the scientific person for views of the nature, quality, and authority of scientific knowledge?

CHAPTER 7

"The Mind Is Its Own Place"

Science and Solitude in Seventeenth-century England

Call this, truth—
Why not pursue it in a fast retreat,
Some one of Learning's many palaces,
After approved example?

ROBERT BROWNING, *Paracelsus*

There is a solitude of space
A solitude of sea
A solitude of death, but these
Society shall be
Compared with that profounder site
That polar privacy
A soul admitted to itself—
Finite Infinity.

EMILY DICKINSON, *To Alice Dickinson,
Mathematician, Teacher, Ringer of Changes*

If we find it difficult to point to the social place of knowledge, it is partly because we inherit the historical legacy of so much testimony that the producers of our most valued knowledge are not *in* society. At the point of securing their knowledge, they are said to be outside the society to which they mundanely belong. And when they are being most authentically intellectual agents, they are said to be most purely alone. The social place of knowledge is nowhere.

Solitude as the setting for the transfer of divine knowledge to human agents is a familiar topos. God prefers to hold His conversation with isolates. The Commandments were imparted to Moses alone on Mount Sinai; the Lord spoke to Paul alone on the road from Jerusalem to Damascus; Mohammed shut himself up in the cave of Hera to receive the word of Allah;

George Fox, the founder of Quakerism, went alone up Pendle Hill in York-shire to hear the Lord tell him in what places He had a great people to be gathered; Joseph Smith uncovered the message of Mormonism alone on a hill in upstate New York. More commonly, the voice of God is not physi-cally accessible to others at all. Mohammed was one of many prophets who heard God's speech within his own head, or within the privacy of dreams, though physical separation from society seems to enhance the probability of divine conversation. Having received divine afflatus and publicly communi-cated the holy word, the prophet demonstrates and confirms his authenticity by removing himself from society: John the Baptist was among very many inspired voices crying in the wilderness. Genuine religious knowledge and genuine piety are not, in this code, to be attained so long as one lives in human society. Holy and civil society stand in contrast.

The truths of artistic creation, no less than those of religion, are similarly said to be most accessible to those who place themselves outside the polity. From the metaphysicals to the romantics, poets have limned their aloneness. Walter Raleigh compared himself to "a Hermite poore in place obscure"; John Keats sojourned "alone and palely loitering," found inspiration "by my soli-tary hearth" and succor for his fears standing "on the shore/Of the wide world . . . alone"; Percy Bysshe Shelley celebrated the uncorrupted young poet Alastor: "He lived, he died, he sung, in solitude"; and William Wordsworth "wandered lonely as a cloud" to experience his transcendental daffodils:

> For oft, when on my couch I lie
> In vacant or in pensive mood,
> They flash upon the inward eye
> Which is the bliss of solitude.

Eighteenth-century British poets developed a conventional poetics of mel-ancholic solitude. From Thomas Gray's *Elegy Written in a Country Church-yard* to James Thomson's *Seasons,* authentic poetic sensibility and creative genius were attached to lonely places. William Cowper placed the virtuous as well as the creative man in nature's solitude, where "traces of Eden" were still to be seen. Thomson found his inspiration in the "deepening dale, or inmost sylvan glade":

> These are the haunts of meditation, these
> The scenes where ancient bards the inspiring breath

Ecstatic felt, and, from this world retired,
Conversed with angels and immortal forms.

Literary historians are wholly familiar with the theme, variously referred to as "literary loneliness" and "gloomy egoism."[1]

We are accustomed, indeed in most modern sensibilities we prefer, to hear of our writers, painters, and composers struggling in garrets and studios, unrecognized and alone. Modern tastes tend to distrust the authenticity and sincerity of art produced in and for a social setting. The difference between solitary ("Rembrandt") and social ("Studio of Rembrandt") painting is counted in hard cash.[2] And no aesthetic, or indeed intellectual, epithet is currently deemed more damning than "fashionable." Samuel Johnson described the garret as "the usual receptacle of the philosopher and poet": it was a place from which he could look at the world from a distance.[3] Moreover, solitude is often said to provide the setting for profound understanding of both self and society. Edward Gibbon reflected on the relationship between his own solitude and his understanding of the historical past. Isolation was the price of insight—as much for historical actors as for the historian. Membership in human society imparted mundane knowledge, but only separation from that society yielded heroic knowledge. His example was the prophet Mohammed: "Conversation enriches the understanding, but solitude is the school of genius."[4] The Quaker William Penn observed that solitude was "a school few care to learn in, though none instructs us better."[5] And a modern psychoanalyst practically equates genius with the capacity to be creatively alone.[6]

Jean-Jacques Rousseau dwelt extensively on solitude as the proper condition for securing self-knowledge. His most painful dissections of self and his most anguished commentaries on the relations between self and society were achieved in isolation: at the Hermitage in the forest of Montmorency and St. Peter's Island in Lake Bienne.[7] Solitude taught Robinson Crusoe the holy basis of human happiness, which living in society had masked.[8] The young Ralph Waldo Emerson reflected on solitude as the fittest condition for self-knowledge and self-cultivation; for him, as for Gibbon, "greatness is the fruit of solitary effort."[9] Henry David Thoreau left Concord for the solitude of Walden Pond in order to understand the nature of the individual and his proper place in society. The more one lives an individually authentic life, the more one is really free and really wise: "the laws of the universe will appear less complex, and solitude will not be solitude, nor poverty poverty, nor weakness weakness."[10] Alexis de Tocqueville analyzed American democracy as institutionalized aloneness. It was a polity that made each man forget his ancestors, and that separated him from his contemporaries: "It throws him back for ever upon himself alone, and threatens in the end to confine him entirely within the solitude of his own heart."[11]

Solitude is identified as a proper setting for obtaining philosophical knowledge and for analyzing the methods by which it is to be won. A fundamental item of seventeenth-century scientific knowledge—the principle of universal gravitation—was said to have been conceived by a philosopher "as he sat alone in a garden."[12] Over a century later, Isaac Newton's solitude was assimilated to the heroic epistemology of the romantic poets. From his rooms in St. John's, Wordsworth looked out upon

> The antechapel where the statue stood
> Of Newton with his prism and silent face,
> The marble index of a mind for ever
> Voyaging through strange seas of thought, alone.

To his modern biographer, Newton was the definitive "solitary scholar."[13] And the most far-reaching methodological insights of the Scientific Revolution were also said to have been secured in solitude. René Descartes prefaced his *Discourse on Method* with a picture of chilling aloneness. The methodological principles for attaining indubitable knowledge were vouchsafed to him during the course of a single day when he "remained the whole day shut up alone in a stove-heated room." His conclusion was that those who wished correct knowledge could not seek it in society, or in the stock of knowledge available in society, but only in themselves: "I could not . . . put my finger on a single person whose opinions seemed preferable to those of others, and I found that I was, so to speak, constrained myself to undertake the direction of my procedure." Descartes decided that his renewal of philosophic method depended on separating himself from society, resolving "to remove myself from all places where any acquaintance were possible, and to retire to a country such as this [Holland] . . . where . . . I can live as solitary and retired as in deserts the most remote."[14] Into the nineteenth century scientists portrayed themselves as alone during crucial phases of their work and used that display to assist the appropriate evaluation of ideas, methods, and roles. Dorinda Outram's study of Georges Cuvier shows how, in the midst of an intensive and compromising political career, he painted a picture of himself, and the evolution of his thought, in solitude.[15]

I want to explore some aspects of discourse concerning the setting of intellectual life and work and, especially, repertoires that have historically placed the philosopher in solitude. These repertoires and their attendant structures of evaluation are very widely distributed in Western culture, and I will briefly trace some patterns of their use from antiquity to the early modern period. I then want to pause at a particular historical conjuncture, and indicate how these repertoires were drawn upon and put to work in making stipulations and rendering evaluations concerning the meaning and worth of specific cultural activity. The setting is seventeenth-century England, and the culture is that of the new experimental and, later, mathematical natural philosophy. This is a particularly pertinent exercise, since discursive features of that context are still evident in modern portrayals of scientific knowledge and its proper setting. Finally, I speculatively summarize some persistent uses of these repertoires in present-day science and the evaluations they import into epistemology.

Two features of this discourse need to be stressed at the outset. First, one is dealing here largely with symbolic locations. When solitude was spoken of as the setting of intellectual life, it rarely meant absolute aloneness. (The sociologist from Mars, unfamiliar with our culture and conventional ways of talking about it, would probably not be convinced that philosophers engaged in less social interaction than others. Nor should an earthly sociologist accept that a culture-producing individual, however sociometrically isolated, is rightly to be regarded as an isolate.) The historical rhetoric of solitude typically signified a series of normatively patterned disengagements from specific institutions or sectors of society. Such disengagements might, and typically did, place the "solitary" not in a vacuum but in a different social institution from the one from which he or she professed separation. Solitude might be achieved, for example, in a sizable community of like-minded fellows—as in a monastery—or within the considerable household of a gentleman's country seat.[16] Moreover, the solitude referred to might be regarded as specifically non-sociometric: there was an aloneness that might be achieved even in company. Second, solitude was often an intensely public pose, intended to express an evaluation of the society from which isolates represented themselves to be disengaged and of the activities that went on in their chosen solitude. It made no sense without that public audience, and its meaning depended utterly on a publicly understood language and stock of evaluations.

What was meant when it was said that knowledge was produced either alone or in company? How did the symbolic placement of knowledge bear upon the perceived worth of that knowledge and those who made it? I will display both the weight of history and the inventiveness of historical actors in specific settings—creatively adapting, rearranging, and redefining the cultural resources they inherit. History prestructures and preselects the resources available to comment on and evaluate practices. But when historical actors put those resources to work they do ingenious *bricolage*.

Classical and Christian Repertoires

Aristotle understood man to be a naturally political animal, "one whose nature is to live with others." Life in society was the condition for the exercise of that activity—even contemplative activity—which made man happy and virtuous.[17] For Aristotle and his followers, the ascription to man in general of a naturally sociable character was basic to a particular understanding of how *philosophers* were situated. While no one could fulfill the ends of their nature as a solitary, the philosopher was, of all people, the best equipped to work outside the polity. Compared to other people, those who dedicated themselves wholly to the search for truth were relatively unencumbered: "The man who is contemplating the truth needs no such thing [as money or power], at least with a view to the exercising of his activity; indeed they are, one may say, even hindrances, at all events to his contemplation";[18] the contemplative life "is wholly independent of external goods."[19] Philosophers' conditions are thus independent in comparison to those of other people. For their activity to be acquitted philosophers need less of the world and its goods than other people. Like people leading other types of life, they require the necessities to support their material existence. But Aristotle conceded that those contemplating truth were set apart from other people by the fact that they may, in practice, achieve their goals alone: "the philosopher, even when by himself, can contemplate truth, and the better the wiser he is; he can perhaps do so better if he has fellow-workers, but still he is the most self-sufficient."[20] As philosophers imitate the gods in their dedication to pure contemplation, so they mirror divine freedom and integrity.[21] Greek thought therefore displayed divergences between, on the one hand, the view that man was naturally sociable and ought to live actively in society and, on the other, the portrayal of the search for truth as a praiseworthy pursuit that might, in practice, be carried out in relative solitude.

Cicero and the Stoics deepened the evaluative distinction between those living public lives, acquitting their obligations to the state, and the "barren and fruitless" private lives of people who wholly gave themselves over to the contemplative search for truth. The unsociability of those entirely dedicated to philosophy was its own condemnation: by "being engaged in their learning and studies, they abandon their friends to be injured by others, whom in justice they ought to have protected and defended." The entirely sequestered contemplative life was culpably egoistical; society must stand before self and the pleasures of privately seeking truth.[22]

Hannah Arendt has argued that early Christian and medieval uses of the tag *vita activa* were fundamentally different from their Greek counterparts. Where the Greeks meant specifically to designate the action required to sustain the political life of the city-state, writers from Augustine onward treated the "active life" as the total pattern of engagements with the things of this world, including "work" and "labor," leaving contemplation "as the only truly free way of life."[23] For Arendt this counted as a massive "misunderstanding," but the translation enabled Greek and Ciceronian-Stoical conceptions to remain visibly current while being artfully adapted to new social and cultural circumstances. Yet Christian conditions also altered the classical framework in which the place of those committed to the search for truth was evaluated. First, the Christian contrast between the city of man and the city of God introduced a basic dichotomy into commentary about how one ought to conduct oneself in this life: the codes that operated in civil society might have a legitimate call upon us, but they might be fundamentally different from the codes of holy conduct. Second, the evolution of dual religious and civil cultural and social structures in European society meant that individuals might choose which structures, and hence which codes, to live under; and, so long as these parallel structures continued in authoritative form, equally legitimate different forms of life might coexist in one society. Those who wished entirely to devote themselves to preparation for heaven might legitimately separate themselves from the city of man and live in a society peripheral or parallel to it. Finally, Christian, unlike classical, society licensed and sustained a set of institutions wholly committed to separation from civil society, and these institutions constituted the major intellectual sites of the early Christian and medieval periods. The effect of these developments was that there now existed a significant practical challenge to the

Greek and Roman stress on the natural sociability of man, and thought about the location of intellectual activity was now implicated in much more polarized conceptions of the nature and value of active and contemplative lives.

The most visible contrast to early and medieval Christian thought stressing man's natural sociability was monasticism and associated forms of divinely sanctioned withdrawal from civil society.[24] From the early Christian period, the desert or wilderness formed a potent symbol of religious authenticity as well as an important site of religious testing. Early saints such as Anthony went to the desert to subject themselves to the same temptations that Christ experienced and, in withdrawing themselves from human society, to forge closer links with God. Through the Middle Ages, the iconography of St. Jerome in his desert cell was a pervasive visual display of the solitude of the holy intellectual.[25]

The significance of these patterns of withdrawal is twofold. First, it was widely considered among both laity and clerics that such separation from society was the most authentic life for a Christian, that it was the ideal to which all should aspire. And those whose separation was most total—the hermits and anchorites— were viewed as carriers and exemplars of the religious values of the Middle Ages. Second, it was understood that such separation and relative solitude were implicated in the attainment of genuine religious knowledge. The monastery, the hermitage, and the reclusorium were the major medieval sites for the production of the highest forms of religious knowledge— which is to say, the highest forms of knowledge recognized by that society. (Friar Roger Bacon was said to have marked his turn away from natural magic and his embrace of wholly divine forms of knowledge by walling himself up in an anchorhold.) These were places of contemplation, and the contemplation of God was the fundamental goal of the solitary life. One walled oneself off from civil society, or one wandered alone in the wilderness, to engage the devil, to suffer temptation, and also, crucially, to put oneself in a position for the most direct communion with God. Only by being dead to the society of men could the living enter into society with God: "Stripped . . . of human contact, dead to the world, the soul would be freest to find its home." For St. Jerome, solitude was a "provisional paradise."[26]

Even so, late medieval and Renaissance understandings and evaluations of the act of withdrawing oneself from civil society diverged radically.[27] The dominant opinion among secular writers, and many clerics, followed the drift of classical, and especially Stoical, thought: man was naturally sociable; privacy, retirement, and the wholly sequestered contemplative life were culpable. However, Christian culture and Christian institutions expressed and sustained patterns of separation. In the endless discussions of the active and contemplative lives through the Renaissance and early modern periods, one particular Aristotelian epigram was endlessly quoted and paraphrased: the person who lives alone, it was said, is either a saint or a savage, a God or a beast.[28] No other sentiment so accurately and economically expressed the contrasting evaluations of solitude that coexisted in early modern European society. The late Renaissance and early modern period thus inherited *two* repertoires for identifying and evaluating the place in society of those committed to the production of intellectual goods. One repertoire held that no person, however fitted and disposed for a contemplative life, should separate himself or herself from civil society. To do so was egoistical, corrupting, unnatural. The other repertoire maintained that those seeking the highest forms of knowledge should (indeed, must) live in relative solitude. Isolation from this world, from the society of man, enabled the closest and most direct engagement with the divine and transcendent source of truth. A major task undertaken by many sixteenth- and seventeenth-century commentators was to rearrange these repertoires and, especially, by redefining each of the components, to arrive at a conception of the happy and virtuous life that might be seen as suitably situated in relation to the extremes of solitude and active engagement in society.[29]

The Scholar versus the Gentleman

Through the Renaissance, and especially in secular circles, a solid consensus developed against the legitimacy of a wholly retired and contemplative life for those of gentle standing. At the same time, the role of the scholar as a relative isolate was increasingly acknowledged as a fact of social life. The monastery, the college, the closet or study, and even the laboratory, observatory, and garden, were places where the scholar could be found at work. By contrast, the active citizen was associated with the court, the exchange, the theater, the gaming house, and the tavern.[30] The result was the elaboration of an important topos that contrasted the "scholar" and the "gentleman" and their respective situations in civil society and that evaluated their patterns of life accordingly.[31]

Recommendations varied about how the scholar ought to live and ought to be placed with respect to society. Nevertheless, through the early seventeenth century there was general understanding of the practical fact of scholarly solitude, as well as widespread appreciation of the meaning and consequences of this isolation. The Jacobean physician Robert Burton, for example, noted that "enforced solitariness" was an endemic feature of the lives of students as well as monks and anchorites; and an Italian courtesy writer[32] whose work was enormously influential in late Tudor and Stuart England agreed that for the ascent to heavenly intellectual benefits "the desartes, al by places and solitarie, are the right ladders. And contrariwise, companies are nought els but hookes and tonges, which withdrawing us by force out of the course of our good thoughts, set us in the way of distruction."[33] Nevertheless, scholars—divine and secular—were considered to have paid a price for solitude. Melancholy was the scholar's "inseparable companion." Lack of exercise "dries the brain and extinguisheth natural heat"; lack of company and distraction encourages mania and wild swings between delight and brooding.[34] The poet Samuel Butler described a melancholy man as "one, that keeps the worst Company in the World, that is, his own."[35]

If solitude was widely regarded as a practical necessity for the scholar, it was considered neither necessary nor legitimate for the gentleman, whose retirement from active public concerns and rejection of his "calling" were typically read as licenses to idleness, trivial pursuits, and debauch. A sixteenth-century humanist asked of the scholarly recluse: "Doe you not thinke that these men may bee called wise by learning, and fooles in respect of the common people?"[36] Montaigne observed that "The most great Clerkes are not the most wisest men." The multitude disdained scholars "as ignorant of the first and common things . . . as incapable of publike charges, as leading an unsociable life, and professing base and abject customes, after the vulgar kind."[37] Sir Henry Wotton, diplomat and provost of Eton during Robert Boyle's pupilage, recognized that "slovenliness is the worst sign of a hard Student" and, endorsing Plato's advice to Xenocrates, recommended that the philosopher occasionally "offer Sacrifice at the Altars of the Graces."[38] Indeed, in early modern Europe the public display of carelessness, unkemptness, distractedness, and social solecism came to count as emblematic of authentically scholarly status.

These patterns and their associated evaluations were both pervasive and persistent. While the thrust of English and continental humanist thought from the sixteenth century was to urge the gentleman to become more intellectually accomplished, practical courtesy texts uniformly traded upon legitimate distinctions between the scholar and the gentleman. The scholar's solitary life, it was said, unfitted him for the public life of a gentleman. "The study" was contrasted to "the world."[39] From the sixteenth through the eighteenth centuries, social commentators manipulated repertoires that drew contrasts between the lives of gentlemen and scholars and that situated those contrasts in the relative engagements with the world that did, and should, characterize each. So far as gentlemanly society was concerned, there was general agreement that the life of unremitting retirement and solitude was grossly inappropriate for the citizen. So far as scholarly society was concerned, there was considerable disagreement about where the life of the mind ought to be located. Should the life of the intellectual be acted out in the same venues inhabited by the active citizen? Or should the contrast between the public citizen and private scholar be accepted? What considerations bore upon such deliberations? How did they affect public appreciations of the value of knowledge and those who produced it?

The Scholar and Public Affairs: Bacon's Reformation

Stefano Guazzo, Richard Brathwait, and a host of courtesy writers of the sixteenth and seventeenth centuries were primarily concerned with the making (or "institution") of a gentleman. They argued, against more traditional patterns, that the gentleman could and should bear the tincture of the scholar, that he was more authentically a gentleman if he acquired the lineaments of learning. Accordingly, such writers recommended that a gentleman's life might legitimately transit the scholar's study, while arguing for a new openness of that study. But what did these efforts at redefinition look like from the point of view of the world of learning and philosophy? Francis Bacon's *Advancement of Learning* (1605) systematically addressed the charges laid by civil authorities at the scholar's door.[40] Learning and the learned were to be acquitted from the denigrations and reservations of princes and politicians. In so doing, Bacon attempted to resituate the world of learning in public space.

Bacon's defense of learning, and the particular argument that princes should actively encourage its bearers and institutions, rebutted five major "disgraces which learning receiveth from politiques."[41] Given the reforms in learning that Bacon proposed, (1) it was not true that learning worked against the military and civic virtues; (2) it was not true that the life of learning was, or conduced to, sloth, sensuality, irreligion, or subversion; (3) it was not true that immersion in scholarly pursuits corrupted manners; (4) most significant for the seventeenth-century career of natural history and natural philosophy, it was not true that a *correctly conceived* philosophy was a solitary activity—the making of knowledge was profoundly social; and (5) precisely because proper philosophy took place in public space, it was not a luxury charged to the public purse; instead it was, and must be regarded as, a public good, contributing importantly to the public welfare.

The "Idols" that afflicted human cognitions and judgments could not be dissolved, but their effects could be mitigated by ensuring that individuals criticized and corrected one another's deliberations. This could be achieved only if individual philosophers were led out of their solitude into social interaction with one another and with civil society.[42] Bacon specifically criticized so-called voluntaries—that is, those who asserted themselves to be their own masters, whether in knowledge or in political action. Because of the workings of the "Idols," individual voluntaries could not produce reliable natural knowledge. The individual human understanding needed to be disciplined by method, namely an instrument of true induction. And that instrument was implemented not by an individual but by a complexly organized, densely interacting collectivity.[43]

Solitude and the New Science

The advertised public character of the experimental science propagated by the early Royal Society is such a common theme among historians that it scarcely bears reiteration. The debt to Bacon so fulsomely acknowledged by the leaders of the society freely specified his resituation of natural philosophy in civic space. And, like Bacon, the Royal Society regarded the social setting of scientific activity as a strong guarantee of the reliability of its intellectual products.[44] Thomas Sprat's official *History of the Royal Society* (1667) commenced with a historical criticism of philosophical privacy:

> They who retire from humane things, and shut themselves up in a narrow compass, keeping company with a very few, and that too in a solemne way, addict themselves, for the most part, to some melancholy contemplations, or to *devotion,* and the thoughts of another world. That therefore which was fittest for the *School-mens* way of life, we will allow them. But what sorry kinds of Philosophy must they needs produce, when it was part of their Religion, to separate themselves, as much as they could, from the converse of mankind?[45]

And even some of the "moderns" failed to appreciate the philosophical costs of a secluded contemplative life. Sprat fastened on the philosophical aloneness recommended by "the excellent *Monsieur des Cartes*" in the *Discourse on Method*. Descartes's retirement to his stove-heated room and his immersion in self-reflection were systematically rejected. This method is perhaps "allowable in matters of Contemplation, and in a Gentleman, whose chief aim was his own delight," but it will not serve as the basis for "practical and universal *Inquiry*." The unsociability of Cartesian method was the basis of its philosophical illegitimacy.[46]

Yet at the same time a quite different repertoire concerning the place of philosophical life was also in circulation. This repertoire was inherited from antiquity and the Middle Ages; it stipulated that solitude was the proper setting for the scholar and philosopher, and it displayed that solitude as a warrant for the value and authenticity of the knowledge issuing forth. Indeed, solitude figured importantly in rhetoric surrounding the new experimental science, and even in practical measures for its institutionalization. In 1648, William Petty laid plans for a "College of Tradesmen," which, despite its strongly practical purposes, was described as a residential institution with celibate "ministers."[47] In the year prior to the formation of the Royal Society, John Evelyn solicited Robert Boyle's support for the partial realization of Solomon's House, pointing to Carthusian cenobitic models for rural retreat and disengagement: "Is this not the same that many noble personages did at the confusion of the empire by the barbarous Goths, when Saint Hierome [Jerome], Eustochius, and others retired from the impertinences of the world?"[48] A year after the society was established, the poet Abraham Cowley, while condemning "the solitary and unactive Contemplation of Nature," still proposed a cloistered "Philosophicall Colledge" set outside the city and housing thirty-five unmarried philosophers.[49] And Sprat's version of the society's prehistory reeked with the rhetoric of retirement.[50]

Until Isaac Newton entered the natural philosophical arena, no English life was more influential in forging the identity of the new scientific practitioner than that of Robert Boyle. Again, historians are familiar with Boyle's character as a paragon of the civic philosopher. Contemporaries and eulogists celebrated the dedication of his scientific work to the public welfare and his easiness of access to the philosophical public.[51] Yet throughout his life Boyle also portrayed himself as a solitary and his philosophical work as taking place in seclusion from the civic world. And this portrayal, while less familiar to historians, was equally influential among Boyle's contemporaries.

Bishop Burnet's funeral sermon applauding Boyle's public life also celebrated his separation from the world and stipulated its holiness: Boyle "withdrew himself early from affairs and courts"; "his mind was . . . entirely disengaged from all the prospects and concerns of this world."[52] Eulogists' displays of Boyle's life as retired and withdrawn were grounded in the philosopher's own self-presentation. As a young man Boyle recorded a conversion experience while walking alone toward a Carthusian monastery in Savoy.[53] During the mid- to late 1640s, Boyle repeatedly made personal notes recording his exasperation at the triviality of normal social intercourse, confessing to a "Hermit's Aversenesse to Society" and pleading to be rescued by "deare Philosophy" from "some strange, hasty, Anchoritish Vow."[54] Through the 1660s, Boyle publicly yearned for solitude and debated his identity as hermit or anchorite.[55] By the 1670s, Boyle was publicly warning the prospective experimentalist about the inconveniences of acquiring a philosophical reputation: "if he should affect a solitude . . . yet he will not escape" the solicitations and visits of the curious.[56]

In *Seraphick Love* (written in 1648 and published in 1659) Boyle counseled a holy renunciation of "unmanly sensualities" and recommended retreat as the proper setting for a pious life—in his case to "my own western hermitage" at Stalbridge.[57] In fact, Boyle remained celibate, and almost certainly a virgin, advertising practitioners' chastity and disengagement as guarantees of a genuinely Christian philosophy. Harold Fisch and J. R. Jacob have discussed Boyle's identification of the Christian natural philosopher as a "priest of nature."[58] And, indeed, Boyle was widely recognized as such by contemporaries.[59] The stipulated solitude of his laboratory was publicly understood as holy.[60] It was the setting from which divinely endowed practitioners might, while alone, enjoy divine conversation.[61] This newly described solitude was identified as an appropriate place in which to make right philosophy and right religion.

Solitude and the Making of Natural Philosophical Knowledge

The retirement of the new natural philosopher was a retreat to a redefined and relegitimized solitude. Moreover, it was an intermittent retreat, temporally linked to active work in a public forum. The cultural place of the new practitioner was artfully reassembled from existing repertoires associating places and the respective virtues that flourished therein. Much the same sort of *bricolage* was involved in situating the knowledge that this new practitioner produced. The career of an item of experimental knowledge similarly traced a trajectory between private and public spaces.

Yet, as Thomas Kuhn and others have pointed out, there were at least two traditions of scientific work current in the seventeenth century, differing in their respective placements in civic space and in the roles they identified for a philosophical public.[62] If the role of a public in the making of empirical knowledge was considered vital, such a role in the constitution of mathematical or logical knowledge was highly problematic.[63] In the more pure forms of such mathematical practice, and in the most extreme versions of what mathematical discourse consisted in, it was unclear what role could or should be played by philosophers assembled at a particular place and time. In the seventeenth century it was customarily understood that mathematical operations could be followed by anyone possessing natural reason who had been adequately informed of a finite number of axioms and procedures.[64] The rightness of mathematical inferences might, therefore, be checked by any competent members in the privacy of their studies; indeed, the presumption would be that this had already been done by a competent mathematical author. If the public had a role, it was only in confirming what could equally have been concluded by each individual working alone. And the judgment collectively rendered could only be one of the rightness or wrongness of rule-following. Establishing mathematical knowledge did not involve the pooling of individual stocks of empirical experience. In this ideal-type practice, the outcome of the conversation of a mathematical public could not modify, but only check, the claim under consideration.

To be sure, in seventeenth-century England the real culture of mathematics rarely corresponded in practice to the ideal type. The early Royal Society always maintained a strong research program in mathematical physics, even during the highly empirical years of strongest Boylean influence, and in the early 1670s Isaac Newton presented it with a radical challenge to the Boylean enterprise in the form of his first optical papers. In this work Newton elaborated a demonstrative practice on an experimental foundation. Once the veracity of the experimental claim was granted, the inferences made from it were meant to follow with the certainty of mathematics. Where Boyle stressed the importance of reiterating large numbers of experiments, Newton argued that his demonstrative enterprise could as well proceed on the basis of one well-judged experiment. In 1676, Newton told Henry Oldenburg, "It is not number of Expts, but weight to be regarded; & where one will do, what need of many?" If any experiments are "demonstrative, they will need no assistants nor leave room for further disputing about what they demonstrate."[65] It is significant that this highly mathematical natural philosophy emerged together with Newton's presentation of self as solitary scholar and with his denial of the importance and legitimacy of the role of a philosophical public as it had been understood in the previous decades.[66]

Newton's contemporaries were as aware as his biographers of his devotion to solitude. At Cambridge he became legendary for his studious retirement, his neglect of appearance and bodily needs, and his disengagement from the ordinary social and intellectual life of college and university.[67] Through the nineteenth century, Newton's solitude was pointed to as the school of his genius; his devotion to scientific truth was manifested in his neglect of the mundane social world. It was a philosophical as well as a social aloneness. He advertised himself as having been no man's pupil—as Richard Westfall says, "an autodidact in mathematics, as he was in natural philosophy."[68] Indeed, Newton generalized the importance of autodidacticism in making an authentic mathematician. Writing a letter of reference for a mathematician applying for a London teaching position, Newton pointed to "ye surest character of a true Mathematicall Genius"—namely, that the candidate had learnt mathematics "by his owne inclination, & by his owne industry without a Teacher."[69] Contemporaries and later commentators both recognized close links between Newton's solitary meditations and his method of discovering scientific truths. William Whewell wrote that "often, lost in meditation, he knew not what he did, and his mind appeared to have quite forgotten its connexion with the body . . . Even with his transcendent powers, to do what he did, was almost irreconcilable with the common conditions of human life." And David Brewster argued against any view that Newton made his discoveries merely by applying known rules of method, such as Bacon's: this would "tend to depose Newton from the high priesthood of nature."[70]

Newton's discoveries having been made in solitude, it became a highly contested matter what the public forum of the Royal Society should, or was intended to, do with them. When the first optical papers of 1672—"my poore & solitary endeavours"—emerged from Newton's "darkened chamber" at Trinity, they met with initial resistance at the Arundel House meetings of the Society.[71] Zev Bechler has shown that the controversy between Newton and Robert Hooke initially stemmed from divergent conceptions of the certainty appropriate to mathematically versus empirically conceived enterprises.[72] To Hooke, publicly disciplined to Boylean probabilism, the mathematical certainty that Newton announced he had secured was illegitimate. To Newton, refusal by critics to grant that certainty could only mean that his "veracity" or mathematical competence was being impugned.[73] While Boyle's and Hooke's public was accustomed to weigh, consider, and modify empirical claims, the public Newton wrote for was instructed to assent to competent mathematical demonstration. Correspondingly, Newton's early attitude toward the existing philosophical public was either ambivalent or hostile. Within a month of communicating his optical papers to the Royal Society, Newton was approving what he then took to be the *private* status of that organization. He welcomed publication in Oldenburg's *Philosophical Transactions* "instead of exposing discourses to a prejudic't & censorious multitude (by wch means many truths have been bafled & lost)."[74] Four years later, he knew better. Encouraging Hooke to limit their future exchanges to "private correspondence," Newton turned upside down the Royal Society's previous justification of the public constitution of scientific knowledge: "What's done before many witnesses is seldome wthout some further concern then that for truth: but what passes between friends in private usually deserves ye name of consultation rather then contest."[75] The public that guaranteed scientific objectivity for Boylean experimentalists became, in Newtonian practice, a continuing potential source of corruption and distortion.

So different forms of seventeenth-century natural philosophical practice were considered to intersect different points of the axis connecting solitude and public life. However, even if attention is confined to empirical-experimental practice, different stages in the career of knowledge-making occur at different sites. I have noted the thrust of seventeenth-century experimentalist rhetoric, which insisted on its location in public space. Yet that rhetoric tended to equate the constitution of knowledge with just one phase of its career and with one particular image of how scientific knowledge was secured. In particular, the rhetoric of "public" science highlighted the stage of showing experiments to witnesses and expatiating upon their interpretation, while drawing a veil over processes that occurred distal to that stage. In more modern terminology, one might say that this rhetoric dwelt on a certain view of the "context of justification" as opposed to the "context of discovery."

In the English seventeenth-century context, one appreciation of legitimate scientific solitude was freely available: scientific discovery—empirical or mathematical—could proceed along the same channels as religious enlightenment. The solitary philosopher, like the religious isolate, might be seen as separated from the corruptions and contaminations of social life. Just as the astronomer must for practical reasons move away from the glare of the city to observe the heavens, so natural philosophers discover truth by removing themselves from the conventions, interests, authoritative beliefs, and distortions of society. And if the object of scholarly gaze was a divine Book of Nature, then the same sorts of understandings that were traditionally available to appreciate the religious recluse could also be drawn upon to establish the natural philosopher's identity as "nature's priest."[76] Both Boyle and Newton were so identified, and their solitude was so appreciated. Yet neither religious nor natural philosophical exercises were traditionally thought to be sufficiently constituted through individual acts of belief or witness. Individual belief had to be turned into knowledge; witness had to be shared. In both cases, private contexts of discovery had to be linked with public contexts of justification.

A public had to be shown what had been won in private. But this public display of private events was rarely what it seemed. The experimental displays and "shows" that characterized the public meetings of the seventeenth-century Royal Society were not simple reiterations of events that took place in the experimentalist's laboratory. Rather, they were *demonstrations* of ideal experiments, made ready to be displayed in public through endless private work devoted to making their phenomena docile, amplifying their read-outs, and routinizing their performances.[77] The seventeenth-century public for experiment neither requested, nor was it offered, displays of nature's recalcitrance, the ambiguity of experimental judgment, and the uncertainty of experimental integrity. What it witnessed, validated, and discoursed upon were demonstrations and displays—experiments specially prepared and adapted for public consumption. These characteristics of private and public science are, of course, fully general. David Gooding's work on Michael Faraday splendidly documents the trajectory traced from the "experiments" in the Royal Institution's basement laboratory to the "demonstrations" in the lecture theater, and H. M. Collins has recently analyzed the "public experiments" laid on to convince a general audience of the safety of flasks for nuclear fuel and the effectiveness of anti-fire additives to aviation kerosene.[78] Collins notes that demonstrations work "because of the smoothness of performance, distancing the audience from the untidy craft of the scientist—

caging Nature's caprices in thick walls of faultless display. Whereas most experiments are directly witnessed only by their perpetrators . . . a demonstration is made for direct witnessing by strangers."[79] The context of discovery is thus pushed into solitude, because processes of persuasion cannot tolerate a full release of what happens there into public space.

Such tendencies are to be found in a wide range of cultural performances. The segregation of culture-making away from public space is an important means of securing the integrity and stability of cultural goods. If we have no appreciation of how things are made, we are very unlikely to be able to take them apart. In fact, we may even come to believe that the things are not made at all, but that they are divinely gifted or are part of the natural order. Erving Goffman documented these processes in his study of the "regions" in which "self" is socially constituted.[80] "Front regions" are those in which social actors expressively accentuate to others certain features of their activities; "back regions," by contrast, are those containing suppressed facts, those that might discredit the impression encouraged in the front region. All performances have their back regions, and these may correspond to divisions in physical space. For example, it is "backstage" that the means to create theatrical illusion are located. But the illusion itself is generated "frontstage," and any accidental intrusion of backstage reality into the audience's view has a powerful capacity to destroy the illusory effect. Impression management is not, however, the exclusive concern of professional actors, though it could be said that "professionals" are people who have achieved peculiarly effective means of controlling access to their back regions. People who labor to produce goods and services for others have a practical requirement to divide front from back regions and to regulate entry to the latter. Back region work needs to be conducted in relative solitude because public knowledge of it erodes characteristics of the product or service that carries its value. The *Rheingold* is tin foil; the gas station staff are not as caring about your car as is portrayed; the restaurant kitchen is not clean; scientific knowledge is not generated by following clear and universal methodological principles. If no man is a hero to his valet, it is because the valet has access to the inside leg. The protection of back regions is, therefore, a practical necessity in the making of all sorts of cultural goods. Whatever cultural resources are contingently available to warrant solitude while those goods are in the making is available to protect those regions. In making seventeenth-century science, religious understandings of solitude were available for the task.

The Place of Knowledge

The uses of solitude in seventeenth-century English scientific discourse were strongly directed toward goals specific to a range of social and cultural contexts. Contemporary actors creatively adapted, redefined, and reassembled the repertoires for talking about the place of knowledge and action that they had inherited from antiquity. For all that ingenious *bricolage*, certain understandings and stipulations about the place of knowledge were scarcely changed from their Greek and Roman origins and, indeed, remain fundamentally unchanged today. Within the Western heritage of thinking about people, society, and knowledge, there are certain patterns that persist and that have altered little over the broad sweep of history.

In our culture we do not have to listen hard to hear the hermit's voice. Everywhere, there are voices claiming to speak from solitude, reporting on the solitary state, commenting on social life, as it were, from the other side. The hermit's voice is consequential. It enjoys an audience widely distributed among groups on the margins of many social institutions. Yet, as Mary Douglas noted, the hermit's voice is fatally compromised: "The more that the hermit thinks it worthwhile speaking out and seeking for his voice to be heard abroad, the more he is edging onto the social map and becoming part of the throng to which he preaches."[81] So the voices that emerge from "solitude," which define its characteristics and virtues, make sense only within some public setting, commenting on and evaluating features of a communal life. The hermit's voice speaks to an audience. To be audible it must use a public language and address itself to public concerns.

So far as the constitution of knowledge is concerned, the hermit's voice has spoken with remarkable consistency in our culture. The most valuable forms of knowledge, it has said, are not attached to place at all. If knowledge is to be regarded as universal and transcendent, then by definition it is not knowledge belonging to any particular social place. Society is conceived of as a set of forces and structures that work *against* the constitution of proper knowledge. Its conventions, customs, structures of authority, interests, and exigencies all act to compromise the integrity of knowledge. In our culture, the rhetoric of solitude is powerfully supported by individualistic views of society and empiricist theories of knowledge. If the model of genuine knowledge-making places an isolated individual in direct contact with reality, then all that society can do for proper knowledge is get out of the way. The greatest knowledge-makers are most wholly alone, and in relation to the individual genius, society is, as Jonathan Swift suggested, but a confederacy of dunces.[82] (The structures collectively used by knowledge-makers to secure and assess their products are not typically regarded as "society" at all: they are simply arrangements found maximally to assist, and minimally to interfere with, the workings of individual minds. "Society" is what happens outside the life of the mind.)[83] To be anywhere in particular in society is to be poorly positioned to make global knowledge. What is believed in society may be contrasted to the transcendentally true. Local knowledge need not even be dignified by the term: it can instead be called "lore," "custom," or "skill." But, since the Greeks, we have become used to equating the idea of knowledge with the idea of transcendence. The very idea of mathematics and science encompasses a global domain. It is knowledge that is taken to apply everywhere and at every time.[84] Any attempt, therefore, to show the situatedness of knowledge is likely to be regarded as denigration: what we thought was genuine knowledge was, in fact, just local lore.[85]

Where, then, are the producers of this global and transcendent knowledge to be found? Again, from the Greeks to modern times, the answer has been notably uniform. Unlike the makers of handicrafts, the makers of universal knowledge are not tied to the workshop; unlike the hunter or fisherman, they are not dependent on the movement of their prey; unlike the professional, they need not await the solicitation of clients or patients; and unlike the athlete, their work is not conditional on the state of their bodies. Philosophers express ultimate freedom from the world's particular ties and demands. They are at home everywhere and nowhere. Their disengagement from the social world is a symbolic voucher of their integrity. Diogenes the Cynic, like many other ancient philosophers, was a man exiled from his own country. On being reproached for that banishment, Diogenes replied: "You wretched man, that is what made me a philosopher." When asked what country he belonged to, he gave the classic philosopher's response that he was "A Citizen of the World." When Anaxagoras was criticized for having no affection for his country, he said "Be silent . . . for I have the greatest affection for my country," pointing up to heaven. Crates of Thebes announced that "'Tis not one town, nor one poor single house,/That is my country; but in every land/Each city and each dwelling seems to me,/A place for my reception ready made."[86] The Stoic Epictetus distinguished the philosopher from other men on the grounds of his superior integrity: "The ignorant man's position and character is this: he never looks to himself for benefit or harm, but to the world outside him. The philosopher's position and character is that he always looks to himself for benefit and harm."[87] Democritus's solitude in his sepulcher and Pythagoras's in his cave publicly testified to their independent disengagements from particular worldly concerns and constraints.

The wise man and the philosopher are said to live anywhere at all and nowhere in particular. This is the ultimate basis of their wisdom and their integrity. Not only are philosophers free of place; through the exercise of pure intellect they constitute their own place. Their minds elaborate a world wholly independent of their corporeal situation. Philosophers, like any person of absolute integrity, need no company to distract, amuse, or instruct them. Hannah Arendt identified with the Greek tradition when she wrote that the "philosopher can always rely upon his own thoughts to keep him company":

> The philosopher, even if he decides with Plato to leave the "cave" of human affairs, does not have to hide from himself; on the contrary, under the sky of ideas he not only finds the true essences of everything that is, but also himself, in the dialogue between "me and myself" in which Plato saw the essence of thought. To be in solitude means to be with one's self, and thinking, therefore, though it may be the most solitary of all activities, is never altogether without a partner and without company.[88]

Scipio Africanus's claim that he was never "less alone than when he was alone" was echoed through the early modern period.[89] Montaigne underlined the importance of mental solitude compared to mere separation from society: "It is not enough, for a man to have sequestred himselfe from the concourse of people: it is not sufficient to shift place, a man must also sever himselfe from the popular conditions, that are in us." Removing one's self from social intercourse was but an aid toward achieving the true solitude that was the cleansing of self from dependence on the social and physical worlds. And that solitude, once achieved, *might as well be enjoyed in company.* "We have a mind moving and turning in it selfe; it may keep it selfe companie; it hath wherewith to offend and defend, wherewith to receive, and wherewith to give . . . Vertue is contented with it selfe, without discipline, without words, and without effects."[90] In the seventeenth century John Milton expressed the Puritan impulse to move eschatological geography into mental space: "The mind is its own place, and in itself/Can make a Heav'n of Hell, a Hell of Heav'n."[91]

Within the space defined as being everywhere and nowhere we find philosophers, the transcendent knowledge they make, and the ultimate source of transcendent knowledge. For God is also defined as being everywhere and nowhere. And only God, it is said, enjoys the unity and integrity of absolute solitude. In the seventeenth century, Sir Thomas Browne identified the limits of human aloneness: "There is no such thing as solitude, nor any thing that can be said to be alone and by itself, but God, Who is His own circle, and can subsist by Himself; all others . . . cannot subsist without the concourse of God, and the society of that hand which doth uphold their natures. In brief, there can be nothing truly alone and by it self, which is not truly one; and such is only God."[92] The solitary philosopher is therefore only a man imitating God.

CHAPTER 8

"A Scholar and a Gentleman"

The Problematic Identity of the Scientific
Practitioner in Seventeenth-century England

In a small parish church in Fladbury, Worcestershire, there is a funerary plaque commemorating a local Anglican minister who died in 1786, though the plaque itself may date from some years later. The inscription describes him as "a scholar, a Christian, and a gentleman." Some years ago, I found the plaque noteworthy enough to photograph, and I have been trying ever since to articulate why I found it so interesting. "A scholar, a Christian, and a gentleman"—possibly I found it remarkable because of the stranger's natural curiosity: I have not the good fortune to be any one of those things, let alone all three. Perhaps because, though the phrase is and has been for many years a commonplace, I had not at the time noticed its use in materials relating to research I had been doing on seventeenth-century English scientific culture and its social setting. I thought there might be some significance in that observation.

So far as I now can tell, the commonplace that applauded an individual as "a scholar, a Christian, and a gentleman," and that implied that the attributes of one enhanced or supported the others, developed in English usage sometime between the middle of the eighteenth and the middle of the nineteenth centuries. For reasons I later note, I did not feel that the "gentleman and Christian" conjunction was as interestingly problematic as that juxtaposing the "gentleman and scholar," and I set about some unsystematic investigation of the latter usage. Standard reference sources record Robert Burns using it in his Kilmarnock poems of 1786: "His lockèd, letter'd, braw brass collar,/Shew'd him the gentleman an' scholar"—though the individual concerned was not a person but one of *The Twa Dogs*—as, more influentially, did William Wordsworth, referring to Cambridge college fellows, in *The Prelude* of 1805: "Brothers all/In honour, as in one company,/Scholars and gentlemen."[1] My own reading has turned up a few further early usages not generally known, mostly dating from the late eighteenth century.[2]

The formula possibly developed through the nineteenth century in association with Newmanite educational philosophy and the muscular Christian and classical culture of the great Victorian public schools.[3] Going back in time through the eighteenth century, references to "the gentleman and the scholar" seem to become more sparse, and, by the seventeenth century, they appear very rarely, if at all. An alert and provoked friend drew my attention to a late seventeenth-century description of the poet Hugh Crompton (fl. 1657) as "born a Gentleman and bred up a Scholar."[4] On closer inspection, however, the message is that Crompton was born into a comfortable landed family, but that his father fell on evil times, and the son was obliged to earn his living by his wits—as a scholar. A *contrast* is indicated between the two roles rather than (the more modern usage) an approving gesture toward two compatible and mutually supporting dimensions of an individual. I have persuaded myself that the usage with which we are familiar did not exist as a commonplace in the late sixteenth or seventeenth century. Nevertheless, I neither wish nor need to argue a null hypothesis. While I am reasonably confident in my impression about the temporal distribution of the usage, the main thrust of my argument proceeds through an assessment of publicly voiced sixteenth- and seventeenth-century English *attitudes* toward the attributes of the gentleman and the scholar.[5]

There are two general reasons why I thought it worthwhile to offer this preliminary archaeology of "the gentleman and the scholar." First, although historians now intermittently make gestures toward informal theories of cultural change as *bricolage,* I want to offer a detailed exploration of the ways in which very old and pervasive cultural topics are recombined and revalued in processes of social change. I refer to stable constellations of such attributions and evaluations as *repertoires.*[6] And I argue that the use of these repertoires in social action worked further to entrench their stability. I note that specific early modern descriptions and evaluations of the characters of scholars and gentlemen have ancient sources, and I devote some attention to the sixteenth-century humanist background to seventeenth-century initiatives that sought, by reforming a body of culture, to respecify the identity of both gentlemen and scholars. I display the restricted materials with which cultural innovators work and the recombination of which constitute the cultural dimensions of social change.

Second, I want to contribute to a rather well-established body of sociological theory that deals with the relationship between culture and institutional change. Historians and sociologists of science are familiar, for example, with Robert Merton's views on the institutionalization of a reformed natural philosophy in seventeenth-century England. Merton here followed Max Weber in arguing that institutionalization proceeded by attaching yet-to-be-legitimated social practices to major reservoirs of legitimacy in the local culture. The reformed philosophy, in Merton's materials, had to display its standing vis-à-vis predominantly religious values. If it could do that, then it might become institutionalized, and, ultimately, a value in its own right. This general view of institutionalization is undoubtedly correct, useful, and underexploited by cultural and social historians. By extension, it amounts to a program of research into legitimation and social order, an inquiry into how culture is constituted by the stock of available justifications for social practices, and a framework for assessing the constitution of culture as local repertoires for managing social stasis and prompting social change.[7]

More recently, Quentin Skinner attempted to merge Weberian views of legitimation and social action with the categories of analytic philosophy. Situated repertoires (my puffed-up version of what Skinner calls "evaluative-descriptive terms") define the domain of legitimacy in any culture. If actions and roles are to be legitimated, it must be *in some contextually specific way.* The range of application of these repertoires at any time defines, as it were, a state of rest, and the cultural innovator must, in giving them new *reference,* set them in motion. Skinner plausibly speculates that the institutionalized uses of these repertoires constitute a constraint upon social action. Whatever the innovator may intend, he "cannot hope to stretch the application of the existing principles indefinitely; correspondingly, he can only hope to legitimate a restricted range of actions."[8]

Freely conceived this way, the Weber–Merton–Skinner theory needs only extension and minor modification. It is, of course, unlikely that any society can be characterized by a single coherent value system, or, indeed, that the articulation of such values in situated legitimations is straightforward and uncontested: there are likely to be many possible paths from values to justifications for courses of action. Accordingly, there can be no such thing as a crucial or irresistible legitimation—one that is *bound to work* and to which there is no effective riposte—and idealist approaches to the relations between culture and social action are likely to look less credible as historical inquiries become more detailed.

My specific interest concerns the cultural resources available for legitimating the new scientific practices of seventeenth-century England, the use of relevant repertoires in justifying (and condemning) these new practitioners, and the practical consequences of those cultural maneuvers. In the particular case at hand, there is little question that Puritan strands of religion provided important sources of legitimacy for many groups in seventeenth-century English society, even, especially, for those groups that offered significant support and approval to the new cultural forms of empirical and experimental natural philosophy. There were, however, other sources of legitimacy in contemporary English society, and among the most consequential of these for the new practices was the pervasive and quite traditional culture that specified who *gentlemen* were, how it was proper for them to live and behave among their fellows, and how the gentlemanly role stood with respect to various forms of cultural practice and their practitioners.

I argue that a major (and, perhaps, the major) problem for the proponents of new scientific practices was the exhibition of their suitability for gentlemen. And insofar as legitimacy was to be sought through a display of the new practices as genteel, then a further, more negative, task was also implicated: in that setting it had to be shown that the new knowledge was not just another form of traditional learning and that its practitioners were not just traditional scholars. In seventeenth-century England what was understood of gentlemen and what was understood of scholars were in conflict. The story that Merton told ended in rapid triumph; this story, by contrast, does not really end at all. I will show that the reformed scientific practices secured little legitimacy within seventeenth- or eighteenth-century gentlemanly society. It is far more likely that these new scientific practices ultimately gained what legitimacy they did partly by processes of intercalation into existing roles for the learned and partly by the gradual development of new sources of support and approval outside of genteel society.

The essay has three parts: in the first, I argue (against some apparent evidence to the contrary) that both the ascribed attributes and the concrete circumstances of the gentleman and the scholar set them in opposition; in the second, I show that the publicists of the new practices attempted to develop a scholarly way of life suitable for gentlemen, in the course of which they aimed to respecify what it was to be a gentleman and what it was to be a scholar; finally, I analyze the obstacles confronted in that attempted respecification, and I ask what the reformed scientific enterprise looked like from the point of view of genteel society. To this end, I turn to sources not commonly used by historians of science and philosophy— the so-called "courtesy" literature, which detailed codes of practical conduct and manners for gentlemen.[9]

The Scholar and the Gentleman

There is no reason to deny much evidence from sixteenth- and seventeenth-century England that seems to suggest an environment favorably disposed to the conjunction of the gentleman and the scholar. Although classical resources were mainly mobilized to condemn the excessively solitary and sequestered nature of the scholar's and philosopher's lives, Tudor and Stuart commentators occasionally pointed to Greek and Roman patterns that identified similar bases for gentle and learned lives. *The Institucion of a Gentleman* (1555) went so far as to assert that the very idea of nobility originated in a learned class.[10] Some neo-Stoicists argued that the defining characteristics of the gentleman were his *integrity* and *independence*. The gentleman wanted and depended upon nothing external to himself—*nil admirari*. The philosopher too needed less of the world's goods, its applause or sustenance. Indeed, he provided his own company, established and conversed with his own society, within the privacy of his mind. So, an early seventeenth-century English translator of Epictetus noted that "It is a true marke of vulgar basenesse, for a man to expect neither good nor harme from himselfe, but all from externall events. Contrariwise, the true note of a Philosopher, is to repose all his expectation, upon himselfe alone."[11]

More pervasively, sixteenth-century English *humanist* writers (for example, Thomas More, Thomas Elyot, Lawrence Humphrey, Roger Ascham) strenuously argued that learning was appropriate for those of gentle standing, that gentlemen required learning to acquit their duties to king and country, and that gentlemen might legitimately be recognized through their cultural accomplishments.[12] Both Erasmian and Italian humanist models acquired an avid following in England. Arguments in favor of genteel learning appeared in the practical courtesy literature as well as in texts of more elevated tone. The translator of Stefano Guazzo's *Civile Conversation* made the point most forcibly: "You will be but ungentle gentlemen, if you be no Schollers . . . Therefore (gentlemen) never deny your selves to be Schollers, never be ashamed to shewe your learnyng, confesse it, professe it, imbrace it, honor it; for it is it which honoureth you, it is only it which maketh you men, it is onely it which maketh you Gentlemen."[13]

Castiglione's *Book of the Courtier* urged the case for regarding "letters," besides virtue, as "the true and principal adornment of the mind."[14] In England, as in Urbino, "a noble mind" expressed itself in "The courtier's, soldier's, scholar's, eye, tongue, sword."[15] An Interregnum courtesy text went so far as to claim that "knowledge . . . must be the principal accomplishment of the Gentleman."[16] James Cleland's influential *The Instruction of a Young Noble-man* argued a utilitarian case for learning. It fitted a man to be his prince's valued counselor, to take active part in public life, to be an effective military leader: "A learned Courtier is capable of his Maiesties profound discourses at al times." Learning also assisted genteel voluptuousness. A well-read and well-spoken gentleman "can court the ladies with discretion and intertaine them in wise and honest conference"; an ignorant courtier "maketh the chamber maids laugh at his discourses."[17] *The Rich Cabinet* noted the power of learning in maintaining the station of the well-born and in advancing that of the humbly born: by its aid "men of base parentage have come to place of high preheminency."[18] English Puritan courtesy writers, otherwise suspicious of "prideful" learning, argued a case for learning as an antidote to genteel indolence: "*Learning* hee holds not only an additament, but ornament to *Gentry*. No compliment gives more accomplishment. He intends more the tillage of his minde, than his grounde."[19]

In 1622, Henry Peacham's *The Complete Gentleman* came as close as any early seventeenth-century English text to mobilizing a coherent vision of what the "gentleman-scholar" might look like. Since learning "is an essential part of nobility . . . it followeth that who is nobly born and a scholar withal deserveth double honor." Learning polished "inbred rudeness"; helped one to live with "honor" and "dignity"; moderated the brutish passions. Largely through the example of Peacham and the Arundel circle that provided his patterns, "virtuosity" became an increasingly common vehicle for the self-fashioning of the English gentleman in the seventeenth century.[20]

Sentiments approving of learning were accompanied by significant shifts in the exposure of the gentry and aristocracy to the institutions of higher education—the universities and Inns of Court. The Tudor Revolution in government provoked widespread recognition among the gentry that their neglect of learning might result in the loss of effective political power to ambitious and humbly born clerks.[21] For these and other reasons, the period from ca. 1580 to ca. 1640 saw very substantially increased demand from the gentry for the intellectual and moral goods purveyed at Oxford and Cambridge. The growth in university enrollments as a whole, and the rising proportion of matriculants from the gentry and aristocracy, have been well documented.[22] If, at the beginning of the sixteenth century it was true that the universities "were little more than seminaries for the education of the clergy of the Established Church," by the early seventeenth century, according to Mark H. Curtis, "the academic haunts of the medieval clergy had become a normal resort for the sons and heirs of the English gentry and nobility."[23] A university education became the norm for the great crown servants of the Elizabethan era—the Cecils, the Bacons, and their like.[24] And while Hugh Kearney reckons that the clerical character of the universities was not wholly lost but only diluted during the sixteenth and seventeenth centuries, there is little doubt that the English gentry did make an increasing mark upon institutions of higher education and that, as Lawrence Stone puts it, they were, in significant measure, "converted" by the humanists' arguments.[25] As a gentleman wrote to the Master of St. John's, Cambridge, in 1614, upon sending his son up to university, "my greatest care hitherto hath bene, and still is, to breed my sonne a scholar."[26]

Moreover, there is evidence that the universities responded to the gentlemanly influx by changing the goods they had to offer their students. Modern subjects and a curriculum less orientated to the needs of future clerics were gradually introduced. As early as 1550, Hugh Latimer exaggerated that "There be none now but great men's sons in the colleges, and their fathers look not to have them preachers."[27] In the middle of Elizabeth's reign, Gabriel Harvey observed that Cambridge scholars were becoming "active rather than contemplative philosophers" and that they were reading such practical and worldly works as Castiglione's *Courtier* and Guazzo's *Civile Conversation*. Later on, Peacham's conception of the virtuoso became modish, and the landed aristocrat in early seventeenth-century Oxford might take advantage of lessons in riding, dancing, and vaulting.[28] In the 1620s, John Earle described "A meer yong Gentleman in the university" as one whose "father sent him thither, because he heard there were the best dauncing and fencing schooles; from those he has his education."[29] And in the 1650s, no less a person than the Oxford professor of astronomy defended the universities from sectarian attack by *denying* that the gentry expected them to teach their sons materially useful knowledge: "the desire of their friends is . . . that their reason, and fancy, and carriage, be improved by lighter Institutions and Exercises, that they may become Rationall and Gracefull speakers, and be of an acceptable behaviour."[30] The argument that sixteenth-century humanists are said to have "won" concerned the definition of "nobility" or "gentility." Nobility was now said to consist not only in birth but also in "virtue." And virtue was not only a matter of piety and morality but also, as Stone noted, "the mastery of certain technical proficiencies," among which were now counted a degree of "book-learning."[31]

That is one story: during the course of the sixteenth and early seventeenth centuries learning triumphed over ignorance, as braying hordes of rising gentry tramped through the scholar's study and emerged on the other side canny, smooth, and virtuous. Another, more negative, story is more difficult to summarize and make coherent. First, the real changes that occurred in Tudor and Stuart conceptions of the gentleman were matters of degree, and the extent to which they turned traditional patterns upside down looks to have been overstated. Second, much of what has been rightly noted of "learning" and the universities in relation to the gentry does *not* in fact simply translate into the constitution of a "gentleman-scholar" as a recognized role in sixteenth- and seventeenth-century English culture. And most important, through the seventeenth century, and into later periods, that culture continued to be informed by a pervasive sensibility about what scholars and the scholarly life were like—a sensibility that persistently worked against approval of commitment to deep or systematic learning in gentlemanly society.

Portrayals of the English gentleman as ignorant and proud of it were hardly new in the sixteenth century: what was new was the notion that things might be otherwise.[32] Nor did those depictions disappear with the supposed humanist "victory." Both the rustic squire and the urban gallant of Restoration comedy testify to the limited extent of humanist success: the one was defiantly unlettered and the other used what learning he had in the cause of dissipation. For all that has been written about the exposure of the English gentry and aristocracy to the world of learning, the impact of scholarly goals and values upon gentlemanly polite culture was circumscribed. The tide that increasingly swept the gentry into the universities had, in any case, run out by the Restoration. Despite complaints that the aristocracy were pushing poor scholars out of the colleges, the evidence suggests that the proportion of gentlemen and above never rose much over 50 percent, and after 1660 the lesser gentry in particular again regarded higher education with a jaundiced eye, reluctant to rub shoulders with their social inferiors. According to Anthony à Wood, they thought "an University too low a breeding."[33]

Moreover, it would be wrong to infer from the "astonishing explosion of higher education in England" that what the gentry wanted from the university was that for which the university had traditionally been known. In 1561, his son about to go up to university, William Cecil said, "I mean not to have him . . . scholarly learned but civilly trained." The universities were merely becoming places where it was more and more possible for "civil training," practical learning, and polite accomplishments to be acquired, if that was what was wanted.[34] Nevertheless, through the seventeenth century, the universities continued vigorously to fulfill their traditional functions of training clerics and sustaining scholarship. Among other things, this meant that universities were sites where several cultures came briefly into contact before—in the main—going their separate ways. The gentry's elder sons passed through the university on their way to succession and/or crown service; the poor secured training as clerics, or, if exceptionally able and lucky, ultimately rose through the ranks to join their betters; and the fellowship and professoriate, augmented by the most able students wishing to become "gown-men," maintained institutional and cultural continuity.[35] The external social landscape was inscribed within the colleges. "Scholars" and "fellows" on the college establishment were generally of plebeian origin while aristocratic "commoners" and "gentlemen-commoners" paid their way and used their inferiors as "battelers" or "servitors."[36] The contours of social standing roughly followed the boundaries between those who were professionally committed to a scholarly life and those who were not.

A Scholar and a Pedant

In sixteenth- and seventeenth-century usage, the term *scholar* might refer to individuals fulfilling several loosely related but different roles. Most generally, the "scholar" could be anyone who received his instruction from an acknowledged "master." Any child or young person undergoing a course of education (privately or publicly) might be counted a "scholar." The term could (as an evaluative discrimination) designate someone in the process of being educated who was especially able, or (in the administrative sense mentioned above) who received emoluments from the institution by virtue of his abilities. It might more specifically refer to someone whose learning was considered to be especially extensive or deep, whatever his means of support; and, finally, it might indicate someone professionally engaged in the pursuit or transmission of knowledge, particularly in institutions of higher education.[37] The problems set by contemporary culture for the role of the "gentleman-scholar" pertain largely to the last two usages.

Frictions between sixteenth- and seventeenth-century conceptions of the attributes of the gentleman and those of the scholar were the actively sustained and occasionally contested legacies of Antiquity. They were inherited from Greek and Roman repertoires concerning the form of life that was desirable for citizens and that which was expedient or even necessary for those giving their lives over to the search for truth. Aristotle argued for the natural sociability of man and for the responsibility all citizens bore to the polis, while recognizing that philosophers might effectively pursue their quest separated from society. Cicero sharpened the evaluative contrast between the active and contemplative lives, constituting one of the most important medieval and early modern resources approving civic commitment and condemning the wholly withdrawn contemplative life. And, while monasticism provided the sequestered intellectual with sites, roles, and justifications, Christian patterns deepened the divide between the perceived attributes of the clerk and the citizen or knight.[38]

Through the late Middle Ages and early modern period, a view of the scholar crystallized in secular aristocratic culture that pervasively identified his character and way of life as incompatible with those of the gentleman.[39] While genteel society did on occasion express great respect for scholarly goods (and in the case of sacred scholarship was formally bound to do so), the attributes of a scholar and those of a gentleman were increasingly seen in opposition. Aspects of that opposition derived from social geography. Early medieval society was marked by a deep social gulf separating the world of the learned (or even the lettered) from that of the gentleman or aristocrat. Those who bore and transmitted literate knowledge were a small minority of the population, and they were almost always clerics. Toward the end of the fourteenth century, the monastery and the university lost their monopoly as centers of literate culture, and, while literacy and a degree of learning now became more common courtly attributes, the professional intellectual was even more separated from political life and mundane society. Scholars—both sacred and secular—generally lived in worlds removed from those of the court or the exchange; their identities, as well as the meaning and worth of their intellectual goods, were associated with their contemplative sequestration. The late fifteenth and early sixteenth centuries saw the strong development in some Italian, central, and northern European states of court patronage and a courtly setting for philosophical and mathematical scholars. Important studies by Bruce Moran, Robert Westman, and, especially, Mario Biagioli have shown how significant client-patron relations at court became for some Renaissance careers in the sciences.[40] Nonetheless, patterns revealed by the study of a small number of scientific clients do not equate with an overall appreciation of the scholar by polite and courtly society. While some professional scholarly lives did indeed depend crucially upon court patronage—Galileo's is now one of the best studied—professional scholars' claims to gentility and civic utility were scarcely widely acknowledged or endorsed in the writings of European courtiers about the relationship between learning and the noble life. Even at the height of the humanist movement in the sixteenth century, professional intellectuals were still routinely associated with sites separated from secular society.[41]

The scholar was also distinguished from the gentleman by facts of economic life. Clerks were largely recruited from the common people, and remained poor in relation to simple gentlemen and aristocrats. While the upper echelons of the clergy might accumulate vast wealth, both the clerical scholar and the university fellow or professor of the Middle Ages or early modern period were noted for their relatively humble origins, for their poverty, and for their routinely expressed disdain of worldly goods.[42] In the early sixteenth century, an English diplomat observed that learned men were "beggars" and that he would rather see his son hanged than learned. A century later, a courtesy text noted that "A scholer is an enemy of fortune." James Cleland lamented that "manie growing in yeares professe nothing more then scoffing at learning & the professors therof, in calling them al *clerks* or *pedants*. If they perceive anie Noble man better disposed to learning then themselues, presentlie after a scorning manner they wil baptize him with the name of Philosopher." And the Jacobean writer Robert Burton complained that the scholar's life was esteemed base and mean, "and no whit beseeming the calling of a Gentleman."[43] While the opposition between the scholar and the gentleman referred to social and economic facts, it was expressed in a series of cultural attributions that spoke of scholars' temperamental, moral, and social characteristics. Those attributions formed an important normative resource in medieval and early modern Europe, allowing the simultaneous judgment of virtue and cultural value, of appropriate conduct and appropriate knowledge.

While the definition and justification of gentle standing was highly contested throughout early modern Europe, there was little disagreement concerning the overall mode of conduct to which the gentleman was enjoined.[44] First, he was to live in society and to take part in its affairs; he was not to live for himself but for others; he was to express his valor in his selflessness, whether on the field of battle or in civic settings. Second, he was in all things to seek to please his fellows, adapting his behavior to his circumstances. In early modern parlance, the preference was for an "active" rather than a "contemplative" life for those of gentle standing, *negotium* rather than *otium*, and the terms used to designate measured and contextually appropriate behavior were *decorum* and *courtesy*. The gentleman was to avoid excess in all things; to exercise self-control; to avoid opinionated and verbose behavior or the visible effort to press his own views on his fellows. If he wished to dominate, he must do so through studied effortlessness and agreeableness. The quest for fame must seem like its nonchalant denial. Everything he did must express magnanimity. His conduct must work to uphold social harmony and the presumed equality of gentle society; nothing he did should divide, disrupt, or disturb. Conduct must be fitted to social circumstance, and the art of doing so ought to appear utterly artless. Affectation was to be avoided at all costs. Castiglione's influential version identified decorum as tacit knowledge: "I think it difficult to give any rule in this . . . Whoever has to engage in conversation with others must let himself be guided by his own judgment and must perceive the differences between one man and another, and change his style and method from day to day, according to the nature of the person with whom he undertakes to converse."[45] And Guazzo's humanism aimed at "civile conversation," "an honest, commendable and vertuous kinde of living in the world." Decorum was at once means and end: "Wee must yeelde humbly too our Superiour, perswade gently with our inferior, and agree quietly with our equall. And by that meanes there shall never bee any falling out."[46]

The identification of gentlemanly decorum and active engagement in the world provided a sharp contrast to attributions of scholars' temperament and mode of life. Aristotle asked why it was "that all men who have become outstanding in philosophy . . . are melancholic," and concurrence that they *were* morose and withdrawn resonated through medieval and early modern culture. While early Christian thinking condemned the melancholic temperament and its "disordered wanderings of the imagination," a more approving attitude emerged from the late Middle Ages. Melancholia and attachment to solitude became for many intellectuals a recognized mode of self-presentation, "a way of feeling and being" that identified the intellectual and that vouched for the authenticity of his cultural goods.[47] In this respect, a withdrawn and depressive persona marked out the intellectual in cultural space and constituted an understood token of the value of his knowledge, while contrasting his character and circumstances to those of the actively engaged public citizen. Throughout early modern Europe the portrayal of the scholar and philosopher as solitary melancholics was an institution in social commentary, medicine, poetry, and painting.[48]

The scholar's melancholic temperament and solitary situation were widely mobilized as explanations of the most damaging attribution about his character. Throughout early modern Europe professional scholars were characterized as *pedants*. The characters of the pedant and the gentleman were set in radical opposition. Indeed, in the courtesy literature and practical social philosophies of early modern Europe the disposition and deportment of the pedant appeared as a practical inversion of the gentlemanly type. The root meaning of *pedant* was simply someone who taught, who acted as a pedagogue. This neutral usage was still found in late sixteenth-century England,[49] but, as time went on, to say that someone was a pedant was invariably to identify a damaged and unsatisfactory character.

What were the characteristics of the pedant considered to be in sixteenth- and seventeenth-century England? He was someone whose manner was disputatious, litigious, affected, and hectoring; he lectured rather than conversed. He was selfish, aiming not to please but to dominate. He was, therefore, deficient in the key gentlemanly virtue of magnanimity. He was both temperamentally unbalanced and the source of discord in company; his gravity and melancholia annoyed and disturbed; his disputatiousness subverted social harmony and order. He was obsessively concerned with matters of little interest to polite society; his objects of study were frequently trivial and ignoble. He was ill-mannered, rude, and uncivil; living separated from polite society, he could not master its mores and was a figure of fun and ridicule when obliged to enter that world; he might be mad. He was lacking in the knightly virtue of valor; his blind reliance upon ancient authority over prudence was an expression of timidity; he who was a slave to Aristotle was not a free man. Withdrawn study worked against the acquisition of that sense of emulation and responsibility that made men do brave deeds. His objects of study were trivial and ignoble, and, hence, unsuitable for polite conversation. He was a bore. Since his knowledge was based upon books and authority, rather than upon worldly experience, what he knew was impractical, of no use to civil society. For that reason, gentlemanly society might judge the pedant, however learned, to be lacking in sense, a fool, a bookish idiot.

From the sixteenth to the eighteenth centuries, condemnation of scholarly pedantry and incivility was pervasive. Montaigne's defense of *appropriate* learning, for example, proceeded by way of an attack on the pedant. The "opinions and ways" of present-day schoolmen and philosophers make them "ridiculous"; they fill their memory but leave their understanding empty:

> Just as birds sometimes go in quest of grain, and carry in their beak without tasting it to give a beakful to their little ones, so our pedants go pillaging knowledge in books and lodge it only on the end of their lips, in order merely to disgorge it and scatter it to the winds . . . It passes from hand to hand for the sole purpose of making a show of it, talking to others and telling stories about it . . . I know a man who, when I ask him what he knows, asks me for a book in order to point it out to me, and wouldn't tell me that he has an itchy backside unless he goes immediately and studies in his lexicon what is itchy and what is a backside.[50]

Guazzo poked fun at the learned recluse—"Doe you not think that these men may bee called wise by learning, and fooles in respect of the common people"—and precisely similar sentiments about the scholar-pedant were expressed by Montaigne, Burton, Lodowick Bryskett, and many courtesy writers.[51] Roger Ascham observed that "all Mathematicall heades, which be onely and wholy bent to those sciences," were "unfit to live with others" and "unapt to serve in the world"; "as they sharpen mens wittes over much, so they change mens maners over sore."[52] Francis Bacon said that to "spend too much time in studies, is sloth; to use them too much for ornament, is affectation; to make judgment wholly by their rules, is the humour of a scholar."[53] Montaigne agreed "that the pursuit of learning makes men's hearts soft and effeminate more than it makes them strong and warlike."[54] Bryskett noted that "we see oftentimes men of great learning in sundry professions, to be nevertheless rude and ignorant in things that concerne their carriage and behaviour."[55] And John Evelyn, defending the view that knowledge and action *ought* to go together, granted that it was both true and culpable that philosophers and "our learned *book-worms* come forth of their Cells with so ill a grace into *Company.*" A life spent wholly "in *Theory* and *Fancy, Extasie* and *Abstractions*" was "fitter for Bedlam . . . then for sober men."[56]

In the Restoration Samuel Butler's satires on the abuse of learning linked "the peccant Humors of Learnd Men" to the circumstances in which the scholarly life was led:

> For whether 'tis their want of Conversation,
> Inclines them to al Sorts of Affectation:
> Their Sedentary Life, and Melancholy,
> The Everlasting Nursery of Folly;
> Their Poring upon Black and White too subtly
> Has turnd the Insides of their Brains to Motly . . .
> Their Constant over-straining of the minde
> Distort[s] the Braine, as Horses break their wind.[57]

And Burton acknowledged that the polite world considered solitary scholars to be quite mad: "How many poor scholars have lost their wits, or become dizzards, neglecting all worldly affairs and their own health, wealth, being & well being, to gain knowledge! for which, after all their pains, in the world's esteem they are accounted ridiculous and silly fools, idiots, asses, and (as oft they are) rejected, contemned, derided, doting, and mad! . . . Go to Bedlam and ask."[58] In 1632, John Milton, then a student at Christ's, Cambridge, censured Scholastic pedantry and acknowledged that scholarly solitude was considered to render the learned unfit for polite society: "Many people . . . complain that the learned are often hard to please, rude, poorly trained in manners, and lacking in agreeable speech . . . I admit that a man who is for the most part alone and absorbed in his studies . . . is less suited for the finer conventions of social behavior."[59]

In 1651, the provost of Eton, Sir Henry Wotton, repeated ancient Greek advice to the philosopher "sometimes to offer Sacrifice at the Altars of the Graces," thinking "Knowledg to be imperfect without Behaviour." It was true that deficiencies in manners were the cause of "disrespect to many Scholers [from] Observers of Ceremonies." Indeed, "slovenliness is the worst sign of a hard student, and civility the best exercise of the remiss." Even so, in Wotton's view, the gentleman and the scholar had their separate spheres of appropriate conduct, and it would be wrong to impose on either an alien code: "Somewhat of the Gentleman gives a tincture to a Scholer, too much stains him."[60] And in 1660 a Christian courtesy text pointed to the dilemma faced by university tutors: in "*making* the *Scholar*" one is thought to be "*spoiling* the *Gentleman*."[61]

The Reform of Learning and the End of Pedantry

While the identification of the scholar with the pedant was quite general in polite Tudor and Stuart society, particular *forms* of scholarly life and endeavor were marked out for special odium. Literally trivial pursuits—grammar, rhetoric, and logic, and, by imputation, the whole form and substance of Scholasticism—were increasingly condemned as nothing but pedantry. In the view of those advocating a reform of learning, what was wrong with Scholasticism was that it proceeded from, and fostered, a form of life that was in no way suitable for a civic gentleman.

Lord Herbert of Cherbury recommended an ordinary course of university study for elder sons, but disapproved of learning the subtleties of logic, "which, as it is usually practised, enables them for little more than to be excellent wranglers, which art, though it may be tolerable in a mercenary lawyer, I can by no means commend in a sober and well-governed gentleman." Oratory was, of course, an important study, yet "I can by no means yet commend an affected eloquence, there being nothing so pedantical . . . than to use overmuch the common forms prescribed in schools." The ritual forms of Scholastic *disputation* publicly dramatized the argumentative and combative character of its practitioners.[62] Milton criticized the Scholastic curriculum for making students both litigious and useless: "And these are the fruits of mispending our prime youth at the Schools and Universities as we do, either in learning meer words or such things chiefly, as were better unlearnt." Scholastic "ragged Notions and Babblements" were the cause of the current "hatred and contempt of Learning." What Milton called "a compleat and generous Education" was one that fit a man "to perform justly, skilfully and magnanimously all the offices both private and publick of Peace and War."[63]

The imputation of pedantry and incivility to the traditional scholarly form of life was an important resource in the arguments of those who wished to replace Scholasticism with a new type of knowledge and a new type of knower. Ramists and Paracelsians are not, of course, widely identified with genteel circles, but their condemnation of Scholastic disputatiousness and impracticality associated existing scholarly patterns with acknowledged disqualifications to gentility.[64] John Webster's analysis of Scholastic logic-chopping was a catalogue of the recognized vices of pedantry: "a civil war of words, a verbal contest, a combat of cunning craftiness, violence and altercation, wherein all verbal force, by impudence, insolence, opposition, contradiction, derision, diversion, trifling, jeering, humming, hissing, braw[l]ing, quarreling, scolding, scandalizing, and the like, are equally allowed, and accounted just."[65] Much sectarian criticism of the universities for their neglect of practical studies (chemistry, agriculture) during the civil war decades is better understood as the mobilization of widespread humanist sentiments in favor of a civically virtuous and useful life for the gentleman than as advocacy of artisan takeover.

Criticism of Scholastic incivility, uselessness, and pedantry was pervasive among seventeenth-century "modern" natural philosophers. Indeed, identification of Scholasticism as unsuitable for the practice or fashioning of a gentleman was a powerful argument in favor of new knowledge and new intellectual institutions. In 1669, Gottfried Leibniz urged the Duke of Württemberg that a new university be established in an *urban* setting: "This may well seem a peculiar view to him who believes in the vulgar maxim that Universities should be situated in quiet solitary places, so that young men should not be disturbed at their studies . . . Yet, to tell the truth, nothing has so much contributed to the spirit of Pedantry . . . than just this rule." Whenever this "monkish erudition" made its "appearance in the world or in conversation," it constituted little more than an "object of ridicule."[66] René Descartes's *Discourse on Method* recommended self-knowledge and experience over school-philosophy in terms instantly recognizable from sixteenth- and early seventeenth-century humanist and courtesy literature. Thomas Hobbes systematically excoriated Scholastic universities as places of pedantry, litigiousness, and, ultimately, of social subversion. School "Aristotely" bred disputatious men, who were in turn the fulminators of civil strife: "The Universities have been to this nation, as the wooden horse was to the Trojans." Yet "men may be brought to a love of obedience by preachers and gentlemen that imbibe good principles in their youth at the Universities . . . We shall never have a lasting peace, till the Universities themselves be . . . reformed."[67] Schoolmen's discourse was nothing but "canting"; their blind reliance upon authority "is a signe of folly, and generally scorned by the name of Pedantry."[68]

English arguments in favor of a reformed natural philosophy similarly proceeded from the humanist case and its analysis of the malaises of the existing scholarly life. Advocates of the new experimental science in seventeenth-century England were simultaneously arguing for a new type of knowledge, a new type of knower, and a new mode of relationship between the knower and what was to be known. If the gentleman was to be refashioned through learning, then the world of learning had also to be reformed to fit it for the task. Such a new form of learning would produce a scholar who suffered from none of the vices of the pedant—a virtuous scholar whose pursuit of knowledge was a civic act, bringing civic benefits. It would, that is, fashion a gentleman-scholar, fit to live in, and to benefit, the polis.[69]

No formulation of this humanist case for new knowledge was as locally potent as Bacon's *Advancement of Learning.* A major theme in Bacon's commendation of proper learning for a gentleman was the evidence of Antiquity when, he said, uncorrupted knowledge was the proud possession of kings and warriors. Genuine knowledge did not "soften men's minds," make them "more unapt for the honour and exercise of arms," or "mar and pervert men's dispositions for matter of government and policy." Instead, *properly constituted* knowledge nurtured the noble virtues of magnanimity and valor, and fitted men for active and valuable roles in civic life. While it "hath been ordinary with politic men to extenuate and disable learned men by the names of *Pedantes,*" yet ancient states flourished best when they were in the hands of truly learned persons.[70] In fact, Bacon's argument here was a respecification of a centerpiece of English humanist writing from Thomas Elyot's *Governour* (1531) onward. According to Elyot, Alexander the Great "often tymes sayd that he was equally as moche bounden to Aristotle as to his father kyng Philip, for of his father he receyued lyfe, but of Aristotle he receyued the waye to lyue nobly." Who condemned Epaminondas "for that he was excellently lerned and a great philosopher"? Who reproved the Emperor Hadrian "for that he was so exquisitely lerned . . . in all sciences liberall"? There was no lack of ancient patterns for "howe much excellent lernynge commendeth, and nat dispraiseth, nobilitie."[71] What was novel about Bacon's argument was only that it required a radical reform of learning to accomplish the ends that Elyot and other English humanists reckoned could be achieved largely by existing resources.

The alleged "distempers" of learning were, in Bacon's view unjustly, attached to learning itself, constituting real obstacles to the humanist program for the simultaneous reform of culture and gentry. It was said that learning, as well as making men soft, made them morose, ill-mannered, tumultuous, litigious, and civically worthless. These distempers were, according to Bacon, the real flaws of a secluded, private, abstract, authoritarian, overly systematic, and disputatious practice. Once these distempers were remedied by a new and purified form of intellectual practice, it would be seen "without all controversy that learning doth make the minds of men gentle, generous, maniable, and pliant to government." Nor was it true that "any disgrace to learning can proceed from the manners of learned men; not inherent to them as they are learned." True, learned men "contend sometimes too far to bring things to perfection," but this was because "the times they read of are commonly better than the times they live in."[72] And while Bacon strove to resuscitate the civic reputations of ancient *Pedanti,* he was, of course, perfectly able to point the finger at the "errors and vanities," the "peccant humours," of present-day scholars: the "fantastical," "contentious," and "delicate" forms of Scholastic practice. This was what "traduced" learning as a whole; this was what was responsible for the tension between the very idea of the scholar and the very idea of a gentleman.[73]

The Scholar, the Gentleman, and the Experimental Philosopher

The establishment of the Royal Society of London was advertised as the practical realization of the humanist-Baconian vision. Here was a real community of gentleman-scholars, practicing a form of inquiry that was suitable for the noble and that ennobled those who practiced it. Here at last was an antidote to the distempers of existing intellectual forms that had resulted in genteel contempt for learning and the learned.[74] The Royal Society announced itself to be at once a philosophical community and "a great assembly of Gentlemen"; it actively enlisted the membership and support of the gentry and aristocracy, and it proffered reasons why the new experimental learning was uniquely suitable for English gentlemen.[75] The objects of the new practice elevated those who participated. The study of nature was not ignoble because it was the decoding of God's Book of Nature; the application of natural knowledge to the improvement of trades was not ignoble because England's imperial destiny depended upon its commerce and its military technology.[76]

The civic legitimacy of the new experimental philosophy was to be secured by showing in just what ways it was *not* the old learning and its practitioners were *not* traditional scholars:

> The common Accusations against *Learning* are such as these; That it inclines men to be unsetled, and *contentious;* That it takes up more of their time,

than men of business ought to be bestow; That it makes them *Romantic*, and subject to frame more perfect images of things, than the things themselves will bear; That it renders them overweening, unchangeable, and obstinat; That thereby men become averse from a practical cours, and unable to bear the difficulties of action; That it emploies them about things, which are no where in use in the world; and, That it draws them to neglect and contemn their own present times, by doting on the past. But now I will maintain, that in every one of these dangers *Experimental Knowledge* is less to be suspected than any other; that in most of them (if not all) it is absolutely innocent; nay, That it contains the best remedies for the distempers which some other sorts of Learning are thought to bring with them.[77]

The old learning was by merit raised to its bad eminence. According to John Aubrey, the whole of learning in the sixteenth and early seventeenth centuries was "Pædentry, i.e. criticall learning"; scholarly style was "pedantique, stuff't with Latin, & Greeke sentences, like their Clothes"; it "was but a sad, & slavish labour."[78] Joseph Glanvill was unrestrained in diagnosing the old scholarship as pedantry: it was a "*disputing physiology*"; it taught men to "cant endlessly," to "dress up *Ignorance* in words of *bulk* and *sound,* which shall stop the mouth of *enquiry,* and make *learned fools* seem *Oracles* among the *populace.*" So long as school-philosophy defined what it was to be learned, how could scholarship attract "Men of *generous Spirit,*" how could learning seem "otherwise then *contemptible,* in the esteem of the more *enfranchised* and *sprightly* tempers"? Its dissolution by the new experimental practice was nothing less than "striking at the root of *Pedantry.*"[79]

The traditional scholar was solitary, morose, and melancholic. So far as gentlemanly society was concerned, "If they can bring such Inquirers under the scornfull Titles of *Philosophers,* or *Schollars,* or *Virtuosi,* it is enough: They presently conclude them, to be men of another World, onely fit companions for the shadow, and their own melancholy whimsies." The experimental philosopher was, however, of a quite different humor: sociable, pliant, and polite. Indeed, these virtues were the condition for the production of reliable knowledge: "*The Natural Philosopher is to begin, where the Moral ends.* It is requisite, that he who goes about such an undertaking, should first know himself, should be well-practis'd in all the modest, humble, friendly Vertues: should be willing to be taught, and to give way to the Judgement of others."[80] I have argued elsewhere that a gentlemanly code regulating manners in dispute was *locally* effective within the community of those willing to subscribe to the rules of the new philosophical game. The peaceable commonwealth of experimental natural philosophers liberally helped itself to codes of conduct preexisting in genteel society.[81] The school-philosopher was, from this perspective, rightly despised as opinionated, magisterial, and litigious. His dogmatism was a token of his baseness and unfitness for genteel society: "To be *confident in Opinions* is *ill manners* and *immodesty* . . . It betrayes a *poverty* and *narrowness* of spirit, in the Dogmatical assertors. There are a set of Pedants that are born to slavery." And while experimentalists formed a community of free agents, submitting to no man's authority, they used that integrity and freedom to secure cultural and civic order: "The more generous spirit preserves the liberty of his judgement, and will not pen it up in an *Opinionative Dungeon* . . . When as the Pedant can hear nothing but in favour of the conceits he is amorous of; and cannot *see,* but out of the grates of his *prison;* the determinations of the Nobler mind, are but *temporary,* and he holds them, but till better evidence repeal his former apprehensions."[82] Good manners made good knowledge.

The experimental community was a "*Society* of persons of *Quality* and *Honour,*" incorporated to serve the civic interest. Its celebrants trusted "that the *eminence* of your *condition,* and the *gallantry* of your *Principles,* which are *worthy* those that own them, will invite Gentlemen to the *useful* and *enobling* study of *Nature,* and make *Philosophy* fashionable." Glanvill announced that the Royal Society had already "redeemed the *credit* of *Philosophy*" and made it suitable for gentlemanly participation: "I hope to see it accounted a piece of none of the *meanest breeding* to be acquainted with the *Laws* of *Nature* and the *Universe.* And doubtless there is nothing wherein men of *birth* and *fortune* would better consult their *treble interest* of PLEASURE, ESTATE, AND HONOUR, then by such *generous researches.*"[83]

Moreover, publicists of the new experimental philosophy had secured a precious rhetorical prize—the person of the Honourable Robert Boyle. For all that has been written in recent years about Boyle, his emblematic significance for the legitimacy of the experimental enterprise has yet to be fully appreciated. Boyle was at once the most noble and the most intellectually central of the early Royal Society's active fellows. The son of the Earl of Cork, the possessor of a major fortune, and a man of widely known personal probity and Christian piety, Boyle was not just a learned man but a man of learning, identified with the pursuit and dissemination of knowledge. His participation was seen as giving the experimental community its particular legitimacy as resolution of the endemic problems attending the relations between the polite and the learned worlds. From early in his life, Boyle wrestled with the great topics in that problematic. Was the happy and virtuous life to be lived in society or in seclusion? How could one, while engaged in the search for truth, acquit one's responsibility to one's fellow men? What form of culture was it that allowed one to pursue truth virtuously and that promoted virtue in its practitioners? In a series of publications from the early 1660s, Boyle announced that the experimental philosophy was just that practice: it was a form of learning that was not only suitable for the gentleman, but that was necessary for the gentleman to fulfill his divinely ordained role in civil society.[84] The "Christian Virtuoso" was *more* rather than less authentically a gentleman for his philosophical pursuits. It was experimental natural philosophy that elevated him both in its objects and in its social practice; it was this study of nature that made him magnanimous, disciplined, modest, generous, peaceable, and civically useful. It was Christianity, and whatever cultural forms most conduced to Christian belief, that sustained the originally secular Renaissance commitment to an active life for a gentleman. The new scientific practices buttressed Christianity, and these new practices were for that reason handmaidens to the civic life.[85]

In these ways, Boyle was offered, and offered himself, as a pattern of the English Christian gentleman that was being canvassed by such religiously orientated English courtesy texts as Richard Braithwait's *The English Gentleman* (1630), Richard Allestree's *Gentleman's Calling* (1660), Clement Ellis's *The Gentile Sinner* (1660), William Ramesey's *The Gentlemans Companion* (1669), and such moral tracts as Sir Thomas Browne's *Religio medici* (1643) and *Christian Morals* (comp. 1650) and Sir George Mackenzie's *Religio stoici* (1665), *Moral Essay* (1665), *Moral Paradox* (1667), and *Moral Gallantry* (1667).[86] While still a young man, Boyle was pervasively pointed to as the exemplar of what it was to be a Christian gentleman and a Christian philosopher. Glanvill suggested that in other ages he would have been regarded as a "deified Mortal." His writings combined "the *gentilest smoothness,* the most *generous* knowledge, and the *sweetest* Modesty" with "the most *devout, affectionate* Sense of *God* and of *Religion.*"[87] At his funeral his life was painted as the "Character of a Christian Philosopher." He was "the purest, the wisest, and the noblest" of men. His life was "a pattern of living"; "nature seemed entirely sanctified in him"; he never "offended any one person, in his whole life, by any part of his deportment"; he "was exactly civil, rather to ceremony"; he "had nothing of the moroseness, to which philosophers think they have some right."[88] In specific terms, the

Honourable Robert Boyle embodied and instantiated the Christian experimental natural philosopher; in general terms, his life and work were mobilized as concrete representations of the new gentleman-scholar.

Pedantry and the Idea of the Gentleman, 1660–1760

The early Royal Society sought nothing less than the effective respecification of the ideas of knowledge and gentility. Repertoires descriptively and normatively opposing the idea of the gentleman and the idea of the scholar were to be dissolved and recombined in a new culture and a new role of the gentleman-scholar. Local claims for the potency of that respecification were vehement. According to publicists like Sprat and Glanvill, the mere existence of the Royal Society as a thriving community of gentlemanly scholars proved the success of that respecification. Henry Oldenburg used his network of correspondence to circulate the pattern throughout the world of letters. Both the polite and the learned worlds were said to be flocking to the new standard. In the late 1660s, a Christian courtesy writer argued against the deep study of logic and rhetoric for the gentleman—they resulted in "Contention and Ostentation"—but much preferred "*Experimental Philosophy* . . . especially the *Spagyrical,* and *Cartesian;* Experience being that chief thing indeed, that perfects our Studies."[89] Gilbert Burnet's *Thoughts on Education,* written for a Scots aristocrat, recommended "the chieffe experiments that are of late made; and this is the best apparatus for philosophy." If the young man "be of a composed mind and moderate spirit," he may also "look discreetly into chymistry," as this "may oblige him to love home, and seek a retired life."[90] In the late seventeenth-century "Battle of the Books" between "ancients" and "moderns," the great defender of modern learning William Wotton announced the end of pedantry and the dissolution of the distinction between scholars and gentlemen. The ancients charged that the cause "of the great Decay of Modern Learning is *Pedantry,*" but this was a sign that their proper target was not learning "as it now stands, but as it was Fifty or Sixty Years ago." The difference was made by "the new Philosophy," which had, Wotton claimed, succeeded in introducing a great "Correspondence between Men of Learning and Men of Business." That "*Pedantry* which was formerly almost universal, is now in a great Measure dis-used."[91]

The British natural philosophical community and its publicists developed the celebration of Boylean nobility and piety into a centerpiece of natural theological rhetoric and cultural legitimation. Gilbert Burnet's eulogy of Boyle as gentleman-scholar remained in print and influential well into the nineteenth century. In America, Cotton Mather's *Christian Philosopher* (1721) dispersed the Boylean template to a new world. Among courtesy texts *The Gentlemans Companion* recommended a few well-chosen books for the Christian gentleman's library, notably including "The Honourable Mr. *Boyl* his *Stile of the Scripture*" and "the Great *Boyl* his *Experimental Natural Philosophy,* and the rest of his Works."[92] And Daniel Defoe's *The Compleat English Gentleman* strongly recommended "experimentall as well as naturall phylosophy" as "the most agreeable as well as profitable study in the world." He claimed that rural gentleman-virtuosi were becoming among the most admired men in their counties. As an apocryphal minister said of the local virtuoso, "I never met so with so much wit . . . and so much polite learning in any gentleman of his age in my life. We were talking phylosophy the other day with him; why he has treasur'd up a mass of experiments of the nicest nature that I ever met with; he has the Phylosophic Transaccions allmost by heart." And, while Defoe condemned the continuing "voluntary and affected stupidity and ignorance" of the English gentry, he pointed to an influential pattern of nobility combined with disinterested dedication to science: "the great and truly honorable Mr. Boyl, who was not a gentleman onely, not a man of birth and blood as to antiquity onely, but in degree also, being of noble blood and one of the families that has the most enobl'd branches of any in England and Ireland."[93] A late eighteenth-century editor reminded readers that Boyle "was himself 'The Christian Virtuoso' which he has described."[94] And in 1796 the editor of Izaak Walton's *Lives* wrote of Boyle: "To the accomplishments of a scholar and a gentleman, he added the most exalted piety, the purest sanctity of manners."[95]

Despite the grand ambitions of the early Royal Society and this fragmentary evidence apparently pointing toward success, it would be pointless to ask whether the experimental community—however loosely defined—achieved the wholesale overthrow of traditional attitudes to the gentleman and the scholar, or even to what counted as legitimate and valuable *knowledge*. Of course, no such revolution occurred. Nonetheless, patterns of British attitudes toward knowledge and gentility in the "long eighteenth century" were enormously consequential, bearing upon the legitimacy of the new knowledge and the institutionalization of the new knower's role. These matters need to be considered mainly from the point of view of gentlemanly polite society, for it was overwhelmingly that society through the eighteenth century that had the gift of support and legitimacy in its hands. Although the social distribution of literate culture underwent substantial change from the Restoration to the French Revolution, within polite society ideas of gentility in general and attitudes toward the relationship between the gentleman and the scholar in particular changed, if at all, with glacial slowness.[96] When marked changes in attitudes did develop in the latter part of the eighteenth century, they did so outside the boundaries of traditional gentlemanly society. Meanwhile, within polite society traditional attributions about learning and men of learning tended to persist. The same Defoe text that castigated the gentry for their sloth and ignorance and that advertised Boyle as a paragon of learned noble virtue also made clear the recommended limits of genteel learning: "I distinguish between a learned man and a man of learning as I distinguish between a schollar and a gentleman."[97]

In fact, late seventeenth- and eighteenth-century social and cultural commentary, tracts on education, and practical courtesy literature continued to characterize pedantry as a major disqualification for membership in genteel society. Scholarly pedantry was pervasively treated, and few texts dealing with the subject regarded pedantry as a thing of the past. Negotiating one's way between wisdom and prudence, on the one side, and pedantry, on the other, was an immensely important practical problem in establishing standing within the society of gentlemen. Lingard's *Advice to a Young Gentleman* warned the new graduate: "When you come into Company, be not forward to show your *Proficiency*, nor impose your *Academical Discourses*, nor glitter affectedly in *Terms of Art*"; *The Rules of Civility* sternly cautioned against speaking in company "in a language that the rest do not understand"; *Rules of Good Deportment* abominated the "Spirit of Contradiction" and counseled that "Obstinacy is contrary to the Laws of Civility": "It is not civil in all Occasions to use Syllogisms, and to deliver our thoughts in Mood and Figure; such Philosophical Cant suits the School, yet not common Converse."[98] Samuel Butler said that the pedant violated decorum. He does not converse but rather lectures: "He speaks in a different Dialect from other Men, and much affects forced Expressions, forgetting that *hard Words,* as well as *evil ones, corrupt good Manners.*"[99] The scholar's learning was not genuine knowledge at all:

> The artificialst Fooles
> Have not been changd i' th' Cradle but the Schooles:
> Where Error, Pædentry, and Affectation
> Run them, behind Hand, with their Education.[100]

In 1673, Obadiah Walker, master of University College, Oxford, published an influential tract on education for young gentlemen that pointed to pedantry as a continuing problem for polite society. Well-born youth should acquire a "sufficient perfection" in their studies, "not so much as is required for a *Professor*, but so much, as is necessary or requisite for a *Gentleman*." "Do not," Walker warned, "in *ordinary company* treat of matters too subtil and curious, nor too vile and mean":

> *Pedantry* is a vice in all Professions, it self no Profession. For a School-Master is not therefore a *Pedant*; but he onely who importunately, impertinently, and with great formality, shews his learning in scraps of *Latin* and *Greek*; or troubles himself with knowledge of little use or value; or values himself above his deserts, because of something he knows (as he conceives) more then ordinary; or despiseth others not skilled in his impertinences; or censures all Authors and persons confidently without reason. And whoever doeth thus, be he Divine, Lawyer, Statesman, Doctor, or Professor, he is a Pedant.[101]

Even writers, like William Temple, concerned vigorously to defend the worth of traditional forms of scholarship, acknowledged that genteel contempt for learning had been largely brought about by the deplorable behavior of many scholars: "The last maim given to learning, has been by the scorn of pedantry, which the shallow, the superficial, and the sufficient [over-confident] among scholars first drew upon themselves, and very justly, by pretending to more than they had . . . by broaching it in all places, at all times, upon all occasions, and by living so much among themselves, or in their closets and cells, as to make them unfit for all other business, and ridiculous in all other conversations."[102]

The Earl of Shaftesbury developed the criticism of pedantry and systematic learning into one of the foundations of his good-humored and common-sensical civil polity. The pedant's canting and hectoring causes "a breach of the harmony of public conversation"; it "puts others into silence, and robs them of their privilege of turn." All knowledge worthy of the name was self-knowledge, to be gained by "soliloquy" and the interrogation of self, and not by turning a philosophic gaze onto the world of nature and its supposed underlying processes. That way, the objects of knowledge stood in a direct relationship to the civic uses of knowledge: "A good poet and an honest historian may afford learning enough for a gentleman . . . The truth is, as notions now stand in the world with respect to morals, honesty is like to gain little by philosophy, or deep speculations of any kind. In the main, 'tis best to stick to common sense and go no farther . . . The mere amusements of gentlemen are found more improving than the profound researches of pedants."[103]

Such fashionable eighteenth-century periodicals as *The Tatler* and *The Spectator* dilated upon scholarly incivility and pedantry, nowhere betraying any sense that pedantry had been either eliminated or eroded by the new philosophy. Joseph Addison reckoned that "A Man who has been brought up among Books, and is able to talk of nothing else, is a very indifferent Companion, and what we call a Pedant," generalizing the term to anyone who cannot think "out of his Profession, and particular way of Life."[104] Such a man could give no pleasure in company. Richard Steele pointed to scholarly lecturing, carping, and quibbling as the bane of civil conversation. Scholars insisted upon teaching and instructing others, while the good order of gentlemanly conversation depended utterly upon participants' presumed equality: "it is the greatest and justest skill in a man of superior understanding, to know how to be on a level with his companions." The greatest enemies of "good company," and those who most violated "the laws of equality (which is the life of it), are the clown, the wit, and the pedant."[105] To Dr. Johnson pedantry was "the unseasonable ostentation of learning." The solitary meditation of scholars was responsible for their social solecisms and clumsiness: "To trifle agreeably, is a secret which schools cannot impart; that gay negligence and vivacious levity, which charm down resistance wherever they appear, are never attainable by him who having spent his first years among the dust of libraries, enters late into the gay world with an unpliant attention and established habits." In Johnson's view, a necessary condition for the pursuit of knowledge was "retirement," and retired settings in turn bred up scholars unsuited for polite society.[106]

Chesterfield's *Letters to His Son* represents eighteenth-century English polite society's view of learning and the learned at the most practical level. By all means, "hoard up, while you can, a great stock of knowledge" and do it in your youth so that you can in later life draw upon it "to maintain you." However, do not equate valuable knowledge with that found in books or academic researches: "Let the great book of the world be your serious study." When you have laid up your stock of knowledge, do not display it promiscuously or lecture others: "From the moment you are dressed and go out, pocket all your knowledge with your watch, and never pull it out in company unless desired . . . Company is a republic too jealous of its liberties to suffer a dictator even for a quarter of an hour." The society of men of learning is not what is meant by "good company"; "they cannot have the easy manners and *tournure* of the world, as they do not live in it." The pedant does not communicate his knowledge, "but he inflicts it upon you"; "such manners as these . . . shock and revolt." Scholars' abstract and generally useless learning was never to be identified with genuine knowledgeability. Gentlemanly society required prudence and experience, not bookishness or philosophy. Indeed, the civic irrelevance of scholars' knowledge derived from the solitary circumstances in which it was produced. This was why genteel society considered that it might take scholars' boorishness and incivility as reliable gauges of their uselessness. The man of learning "has no knowledge of the world, no manners, no address . . . His theories are good, but unfortunately are all impracticable. Why? Because he has only read and not conversed . . . His actions are all ungraceful; so that, with all his merit and knowledge, I would rather converse six hours with the most frivolous tittle-tattle woman, who knew something of the world, than with him."[107]

Pedantry and the "New Science"

It might be supposed that gentlemanly criticisms of pedantry distinguished between traditional scholarly enterprises and the new scientific practices. After all, Bacon and the early Royal Society had sought to eliminate the old pedantry and to reform learning so as to make it appropriate for gentlemen. Yet there is little indication in the literature produced by and for polite society that the new science was in any way exempted from the pervasive condemnation of pedantry and the scholar's way of life. Of all writers on the practical education of gentlemen in the late seventeenth century, John Locke had the most reason to recommend the new philosophy, both natural and moral. Fellow of the Royal Society, friend and collaborator of Boyle, Locke was intimately familiar with the experimental program and its justification. Yet there was little difference between Locke and the courtesy writers when it came to the diagnosis of scholarly distempers. "Prudence and good Breeding" are necessary; most of what passes as "Learning" is not. Only "a very foolish Fellow" would not "value a virtuous or a wise Man infinitely before a great Scholar." Learning can as easily mar as improve a character. "*Learning* must be had, but in the second Place, as subservient only to greater Qualities." Locke specifically warned well-born youth against acquiring "the just Furniture of a Pedant . . . than which there is nothing less becoming a Gentleman."[108] In fact, there *were* reasons for giving the natural sciences some place in the gentleman's education—insofar as they could offer solidity and *certainty*. Yet, no matter how favorably disposed he was to Boylean corpuscularianism and probabilism, Locke did not see that *any* system of natural philosophy could offer uncontrovertibility: "Tho' the World be full of Systems of [Natural Philosophy], yet I cannot say, I know any one which can be taught a young Man as a Science wherein he may be sure to find Truth and Certainty, which is what all Sciences give an Expectation of." There was no actual *harm* in acquiring a passing familiarity with the systems of Descartes, Boyle, and Newton; indeed, "There are very many Things in [them] that are convenient and necessary to be known to a Gentleman." But neither natural philosophy nor any other reformed study of nature was to have more than a peripheral place in the fashioning of a gentleman. The success of Locke's program can be gauged from the mature views of the young man he was then tutoring, the future third Earl of Shaftesbury.[109]

The courtesy literature and related tracts clearly failed to exempt the new science from its continuing criticism of pedantry and the learned. Moreover, very many polite cultural productions specifically identified Royal Society learning as, at best, just another sort of learned nonsense and, at worst, as a particularly virulent and offensive type of pedantry, *more* unsuited to gentlemanly participation than traditional practices. It is not my purpose here to review the literature on Restoration and Augustan scientific "satire" so well treated by R. H. Syfret and Marjorie Hope Nicolson.[110] Rather, I want briefly to show in what ways rejection of the new practice produced by and for genteel society mobilized cultural repertoires contrasting the gentleman and the scholar that had been current at least since the sixteenth and early seventeenth centuries. In this connection, the identification of the new science as trivial, silly, or just *funny* was arguably more devastating than its portrayal as *dangerous*. Defenders of the new learning acknowledged this fact: "Nothing wounds more effectually than a Jest; and when Men once become ridiculous, their Labours will be slighted, and they will find few Imitators."[111] Amused or exasperated condescension was the language traditionally used by polite society to reject learning and the learned: the language of threat and danger in relation to cultural forms was more customarily employed *within* clerical circles. The humanistic respecification of knowledge and the knower instigated by Bacon and exemplified by the early Royal Society had not only failed to institutionalize the new gentleman-scholar, it had failed to bring about a significant change in the repertoires that polite society used to keep learning and the learned at arm's length. What was said of the new learning and the new learned through the eighteenth century was *mutatis mutandis* much the same as had been said of empty scholarship and "dizzard" scholars in the sixteenth century.

The themes are familiar. It was said, first, that the objects of the new practice were trivial or ignoble, unworthy of the attention of gentlemen, irrelevant to their concerns, and inappropriate to polite conversation; second, that the new knowledge was abstract and useless, making no contribution to the polity, and perpetrating fraud when useful outcomes where claimed for it; third, that the new practice was in no way less disputatious or litigious than the old Scholasticism, and that the new practitioners were as given to pomposity, ostentation, canting, hectoring, and lecturing; and, finally, that the new knowledge was no better, or even worse, than the old in its effects on practitioners' manners, civility, and fitness for polite society.

Samuel Butler's satires were among the earliest to attach the charge of pedantic triviality to the Royal Society's empirical and experimental practice. From 1663, *Hudibras* and other mock-heroic poems specifically targeted the new natural philosophy as one of the more prominent contemporary pompous frauds and "abuses of learning." Butler's *Pædants* tendered a stock early Restoration joke about the vacuousness of Boylean science:

> A Pædagogue, that mounted in his Schoole
> Is but a Kinde of Master of Misrule,
> Is Puft up with his own conceipt, and Swels
> With Pride and vanity and Nothing else,
> Like Bladders in the Late Pneumatique Engine,
> Blown up with nothing but their owne Extension.[112]

The Elephant in the Moon identified the triviality of the objects of the Royal Society's inquiry (flies, mice); the grandiose, pompous, and deluded claims made for them (elephants); and the dishonorable conduct of virtuosi in attempting to suppress the truth.[113] Butler's *Satyr upon the Royal Society* was one of several Restoration japes that showed how little the claimed utility of the new science was credited in polite society:

> These were their learned Speculations
> And all their constant Occupations;

To measure *Wind,* and weigh the *Air,*
And turn a *Circle* to a *Square.*[114]

Shadwell's *The Virtuoso* mercilessly lampooned the *ignobility* of the phenomena so obsessing Sir Nicholas Gimcrack. The natural theological rhetoric that saw divine wisdom in the architecture of a louse and heard sermons in gall-stones clearly won little applause in fashionable circles. In his nieces' opinion Gimcrack was a bathetic fool: "A sot that has spent two thousand pounds in microscopes to find out the nature of eels in vinegar, mites in cheese, and the blue of plums which he has subtly found to be living creatures. One who has broken his brains about the nature of maggots, who has studied these twenty years to find out the several sorts of spiders, and never cares for understanding mankind." The foppish pedant Sir Formal Trifle agreed: "no man upon the face of the earth is so well seen in the nature of ants, flies, humble-bees, earwigs, millepedes, hog's lice, maggots, mites in a cheese, tadpoles, worms, newts, spiders, and all the noble products of the sun by equivocal generation."[115]

The satirist William King ridiculed the Royal Society's virtuosi and philosophers for mucking around with "the very dregs of Nature." Boyle's swilling about in human urine and feces to extract the phosphorus and Leeuwenhoek's investigations into the globular structure of mouth-slime elicited a polite retch-reflex as well as a smirk. The air-pump was never funnier: "I had several discourse[s] with Mr. Muddifond, about 'an old cat and a young kitling in an air pump, and how the cat died after 16 pumps, but the kitling survived 500 pumps.' Upon which, he fell into a learned discourse, of the lives of cats; and at last agreed upon this distinction, That it ought not to be said that *cats,* but that *kitlings,* have nine lives."[116] Royal Society apologists from John Wilkins to Robert Hooke and Robert Boyle urged the nobility as well as the sublimity of inquiry into nature's smallest and most wretched creatures: "The creation of a glorious angel did not cost [God] more than that of a despicable fly."[117] Polite society remained largely unpersuaded that the flies and lice were anything but despicable.

The Tatler judged the "immensity" of nature a more noble philosophical object than its "minuteness." Just because "the world abounds in the noblest fields of speculation," it was the "mark of a little genius to be wholly conversant among insects, reptiles, animalcules, and those trifling rarities that furnish out the apartment of a virtuoso." Men whose "heads are so oddly turned this way" tend to be "utter strangers to the common occurrences of life":

> I would not have a scholar wholly unacquainted with these secrets and curiosities of nature; but certainly the mind of man, that is capable of so much higher contemplations, should not be altogether fixed upon such mean and disproportioned objects. Observations of this kind are apt to alienate us too much from the knowledge of the world, and to make us serious upon trifles . . . In short, studies of this nature should be the diversions, relaxations, and amusements; not the care, business, and concern of life.[118]

Natural philosophy in its modern virtuosic practice did not "so much tend to open and enlarge the mind" as to "fix it upon trifles": "This in England is in a great measure owing to the worthy elections that are so frequently made in our Royal Society. They seem to be in a confederacy against men of polite genius."[119] Feminist satirists endorsed their polite brothers' strictures. *An Essay in Defense of the Female Sex* characterized the "virtuoso" as one who had "abandoned the Acquaintance and Society of Man for that of Insects, Worms, Grubbs, Maggots, [and] Fleas."[120]

Shaftesbury systematized polite opinion on the matter. There was nothing about learning as such that was at all culpable for gentlemanly pursuit:

> But when we push this virtuoso character a little further and lead our polished gentleman into more nice researches, when from the view of mankind and their affairs, our speculative genius and minute examiner of Nature's works proceeds with equal or perhaps superior zeal in the contemplation of the insect life, the conveniencies, habitations, and economy of a race of shell-fish; when he has erected a cabinet in due form, and made it the real pattern of his mind, replete with the same trash and trumpery of correspondent empty notions and chimerical conceits, he then indeed becomes the subject of sufficient raillery, and is made the jest of common conversations.[121]

Systematic learning or philosophy had nothing to teach the gentleman; it had no lessons for civil society. As Joseph Levine has rightly noted, many late seventeenth- and early eighteenth-century critics "found the whole business of examining nature irrelevant, *however it was conducted.*"[122] There was little if any relevant contrast drawn by polite critics between an antiquarian pedant and a Newton. What was important in civil society was knowing what the right thing was and then doing it. What had *any* brand of "philosophy" to contribute to that end? The proper study of mankind was man. If scholars and divines wanted the moral order underwritten indirectly, for example, via the study of God's design manifest in nature, gentlemanly society preferred to take its ethics neat. Whether the atomists or the plenists were in the right simply did not *matter* to practical life: "My mind, I am satisfied, will proceed either way alike, for it is concerned on neither side. 'Philosopher, let me hear concerning what is of some moment to me. Let me hear concerning life what the right notion is, and what I am to stand to upon occasion.'"[123]

Syfret followed Walter Houghton in discerning in Restoration criticisms of the new science a significant distinction between "Baconians," pursuing natural knowledge as it tended toward use, and "the virtuosi," cultivating knowledge for its own sake.[124] Such a distinction may indeed capture something about practitioners' varying goals, but little was made of it in contemporary censure and satire. In the view of polite critics, it amounted to a distinction between those who never thought to turn knowledge to civic account and those who fraudulently claimed to have done so. The utilitarian claims of the "Baconians" were not credited; the aims of virtuosic *curiosi* were ridiculed in their own terms; and the practice of both was roughly equated with the civically useless sequestered and abstracted modes that polite society since the Renaissance had overwhelmingly rejected. Shadwell's Gimcrack said that he was concerned with the "speculative part" of knowledge: "I care not for the practic. I seldom bring anything to use, 'tis not my way."[125] And when Gimcrack's house was besieged by proto-Luddite ribbon-weavers, convinced that the virtuoso had invented an engine-loom putting them all out of work, the virtuoso protested his innocence: "I never invented anything of use in my life"; "We virtuosos never find out anything of use, 'tis not our way."[126]

William Temple spoke not just for "ancient" scholars but for the generality of polite society when he asked, "What has been produced for the use, benefit, or pleasure of mankind, by all the airy speculations of those who have passed for the great advancers of knowledge and learning these last fifty years[?]"[127] Temple's client Jonathan Swift captured the technological ambition of the Royal Society in his "Grand Academy of Projectors" in

Lagado, "where the professors contrive new rules and methods of agriculture and building," and where, in consequence, "the whole country lies miserably waste, the houses in ruins and the people without food or clothes."[128] *The Tatler*'s Gimcrack, believing on philosophical grounds that "there was no such thing in nature as a weed," sacked his gardener, reducing his previously well-ordered grounds to a tangle of weeds. He ruined his fortune, spending lavishly on cabinets stuffed with "rat's testicles, and Whales pizzle," and dying destitute.[129] William Temple and William King both found the Royal Society's utilitarian pretensions so obviously bogus that they had only to be enumerated for their fraudulence to be evident: rejuvenation by transfusion, a universal language, the art of flying.[130]

If the perpetration of fraud was one specific distemper of the new learning, its unmannerly *ostentation* was a more general vice disqualifying practitioners from polite society. If, as Dr. Johnson said, pedantry was the display of knowledge out of place, then the new learning was no less culpable than the old. Samuel Butler saw little distinction in this respect between the naturalist-virtuoso and the antiquary-grammarian: "[The virtuoso] is a Haberdasher of small Arts and Sciences . . . He differs from a Pedant, as *Things* do from *Words;* for he uses the same Affectation in his Operations and Experiments, as the other does in Language."[131] *The Tatler* cautioned that "It is the duty of all who make philosophy the entertainment of their lives, to turn their thoughts to practical schemes for the good of society, and not pass away their time in fruitless searches, which tend rather to the ostentation of knowledge than the service of life."[132] Temple recommended ancient learning over modern because the former was decorously modest while "ours leads us to presumption, and vain ostentation of the little we have learned."[133] King identified ostentation rather than utility or curiosity as the motive force of the new philosophical cant. "Gentleman" asks "Virtuoso" about the *use* of narratives about fossil bones, and Virtuoso replies: "You mistake the Design: it was never intended to advance Natural knowledge; for who is the wiser for knowing that the bones of a dead fish have been dug up, or where? No, the true use of the story is to amuse the ignorant; for, if they talk of things that are out of the way, we presently make an harangue about 'the Mandibulum of a *Pastinaca Marina* found fossile' in 'Maryland;' and then they 'are silenced at an instant.'"[134]

Pomposity is ostentation without a hint of a smile, and the piously po-faced earnestness of some notable expressions of the new natural philosophy attracted the ridicule of those polite critics who found it all too tedious.

The Virtuoso cruelly spoofed Boyle's profession that he had "read a Geneva Bible" by the "light of a rotten leg of pork." (Shadwell exaggerated quite unfairly: it was *veal* and the text was the *Philosophical Transactions*.)[135] The bonhommic wit Sir Samuel Hearty had to apologize for giving offense to Gimcrack: "you know I am an airy, brisk, merry fellow, and facetious, and his grave philosophical humour did not agree with mine. Besides, he does not value wit at all . . . he's an enemy to wit as all virtuosos are."[136] Samuel Butler's satire of Boyle's *Occasional Reflections* picked out the attempted ennobling of the louse as pompous twaddle. Jonathan Swift followed with a spoof meditation on divine design as manifested in a broomstick that was too deft to raise a laugh from Lady Berkeley, and there is even evidence that the libidinous Rochester circle had fun at pious Boyle's expense.[137] Shaftesbury urged "raillery" and good humor as a test of veracity, "For without wit and humour, reason can hardly have its proof": "Grimace and tone are mighty helps to imposture. And many a formal piece of sophistry holds proof under a severe brow, which would not pass under an easy one. 'Twas the saying of an ancient sage 'that humour was the only test of gravity; and gravity of humour. For a subject which would not bear raillery was suspicious; and a jest which would not bear a serious examination was certainly false wit.'"[138] By this criterion, the pomposity of the new science gave adequate grounds for polite society to suspect its truthfulness.

Moreover, it was said that the practitioners of the new learning were just as quarrelsome and disputatious as their predecessors. Butler's character of "A Philosopher" tarred the new natural philosophers with the same brush they had used for the Schoolmen:

> When his Profession was in Credit in the World, and Money was to be gotten by it, it divided itself into Multitudes of Sects, that maintained themselves and their Opinions by fierce and hot Contests with one Another; but since the Trade decayed, and would not turn to Account, they all fell of themselves, and now the World is so unconcerned in their Controversies, that three Reformado Sects joined in one, like *Epicuro-Gassendo-Charletoniana*, will not serve to make one Pedant.[139]

Swift's Laputans were not only "very bad reasoners" but "vehemently given to opposition, unless when they happen to be of the right opinion, which is seldom the case": "I have indeed observed the same disposition among most of the mathematicians I have known in Europe."[140] Genteel society in the late seventeenth and early eighteenth centuries was well apprised of the acrimonious disputes involving such major modern English natural philosophers, naturalists, and mathematicians as Robert Boyle, Thomas Hobbes, Martin Lister, Robert Hooke, John Flamsteed, and, spectacularly, Isaac Newton.[141] Irenic professions by leading figures of the early Royal Society were, for these and other reasons, not widely credited. The period from ca. 1695 to ca. 1710 was marked by particularly serious personal bickering in the Royal Society. Accusations of disingenuousness were broadcast; challenges were made; duels were threatened. For polite society the naturalist and antiquarian William Woodward became the pattern of a litigious and quarrelsome philosopher as well as a pedant.[142]

What gentleman would wish to spend an evening in such company? Certainly, polite opinion was united in the view that practitioners of the new science were every bit as boorishly ill-mannered as the general run of scholars had always been. Again, in fashionable condemnation of the virtuoso and the naturalist, Dr. Woodward came to stand for the type. Courtesy to guests was strongly enjoined in gentlemanly society, and accounts of the doctor's epic rudeness to curious visitors became collectors' items of their kind. A German visitor was intentionally kept waiting for hours in the parlor: "This is the discourteous little ceremony that the affected and pedantic mountebank makes a habit of going through with all strangers who wait on him." The English naturalist Samuel Dale was smartly reprimanded by Woodward for speaking to a friend while their host was holding forth on a fossil: "'it was not manners to speak while another was speaking.' [Dale] thought he had never met such 'an ungenteel Banquet after a seemingly courteous invitation to a Rara show as at that Gentleman's Lodgings.'"[143] Swift squelched the Laputan philosophers for their total lack of decorum: "in the common actions and behaviour of life I have not seen a more clumsy, awkward, and unhandy people."[144]

An early eighteenth-century courtesy text specifically identified the new natural philosophy as just another form of pedantry:

I would not have you upon all Occasions discourse in Syllogism, nor deliver your thoughts in Mood and Figure: such Philosophical Cant suits better with a Pedant, than a Gentleman; and may pass in the School, but not in the Parlour. Neither press upon Company a *Vacuum,* nor Mr. *Boyle's Pondus*

Atmosphæræ, a civil Conversation may be managed handsomly in either *Hypothesis;* and I conceive Discourse prospered no less in the Days of good old *Materia Prima*, than in the Reign of *des Cartes*'s Third Element.[145]

Jolly gentlemen need not trouble their heads.

Dr. Johnson's *Rambler* related a story about the social solecisms of a scientific scholar bred up in solitude and quite unable to make agreeable conversation in polite company. He had nothing suitable to say to the ladies, and found himself standing apart. Approached by a clergyman who asked the scholar "some questions about the present state of natural knowledge" and who affected interest in "the Newtonian philosophy," the scholar was drawn into pedantic babble grossly violating "the laws of conversation." The bored company turned its chatter to "the uselessness of universities, the folly of book-learning, and the awkwardness of scholars," and the philosopher made haste to leave, though not before spilling his tea, staining a lady's petticoat, and scalding the lap-dog.[146] If, as gentlemanly circles had agreed since the Renaissance, the test of proper knowledge was its ability to sustain and contribute to civil society, then incivility could be taken as an absolutely reliable sign that the new scientific knowledge was not a suitable object of polite participation or support. Bad manners meant bad knowledge.

The attempted respecification of conceptions of learning and the learned by practitioners of the new scientific practices was a substantial failure. The new culture did not precipitate a new role of gentleman-scholar nor was polite society ever persuaded that a reformed systematic natural knowledge was a necessary gentlemanly accomplishment. Of course, during the eighteenth century *certain forms* of knowledge were persistently, and at times effectively, recommended to gentle society, and these notably included antiquarianism, civil and natural history, and even mathematics.[147] Yet, while conceptions of appropriate knowledge changed, English polite culture continued to be marked by practical opposition between the ideal of the gentleman and the imputed characteristics of the scholar. Knowledge must be useful to a gentleman, but those whole role and identity were constituted by the serious pursuit of that knowledge were still regarded as having a spoiled identity. For the English upper classes (not quite the same thing as the body of gentlemen), the danger of learning was (and is) the risk of appearing "too clever by half," and thus, by conventionally understood innuendo, to be neither the gentleman, the Englishman, nor the Christian. And nothing argued or achieved by the practitioners of the new science from the time of Bacon to the time of Newton made much difference to that assessment.

The seventeenth-century failure of respecification was consequential for the institutionalization of science in England. It meant that the ambitions of Bacon and the early Royal Society for the flow of the new learning through the channels of polite society were not significantly realized. Yet failure to legitimize the new practice through one set of institutions, roles, and values did not mean that it could not be done through other vehicles. I offer some brief speculations about the identity of two such alternative frameworks available during the period from the seventeenth to the nineteenth centuries.

First, from the middle of the eighteenth century, England developed venues and vehicles for the support of scientific culture that were outside of traditional polite society and that partly developed in explicit opposition to polite values. From early in the eighteenth century, nonconformist circles, particularly in the Midlands and North, supported dissenting academies deeply committed to "modern" education, including the natural sciences. Later in the century, provincial scientific societies emerged strongly associated with the mercantile and manufacturing classes. We now have a generation of research in the social history of science that establishes significant links between approval of and participation in scientific culture and the emergence of the "new men" of industrializing Britain.[148] Many "new men" took seriously the scientific culture that the old elites had ridiculed and continued to reject. And texts such as Joseph Priestley's *Essay on a Course of Education for Civil and Active Life* (1765) sketched a new pattern of gentility answering to the circumstances and goals of mercantile and manufacturing dissenters. Natural knowledge and the mechanical arts were not, in this view, funny; they were fundamental to the formation of the new model gentleman and the new moral order. Elsewhere in industrializing circles, gentility was not co-opted but condemned, in favor of a seriously impolite utilitarian culture. Insofar as the classes that bore the new culture prospered, thus far British society threw up significant alternative vehicles for the support and legitimization of science. By the 1830s, a new utilitarian culture articulated political demands for the professionalization of the scientific role and the state subvention of science—not because it fostered or was compatible with gentility but because it was materially useful to society. What was now asked for and achieved was state support, not for underwriting the role of the Christian philosopher and gentleman-scholar, but for that of the scientific *expert.* By the turn of the twentieth century, it was no longer required that scientific learning have any effect at all upon individual or communal virtue.[149] Polite society, whenever it formed an opinion of professional men of science, could think what it liked. And if the Brideshead *jeunesse dorée* despised scientists as "northern chemists," nevertheless pater's taxes ultimately went to pay scientific salaries.[150]

The second set of vehicles for the support and legitimization of the new science is wholly traditional: the traditional scholar's role, the traditional institutions that sustained him, and traditional gentlemanly repertoires specifying the identity of the scholar and the value of his work. If the scholar was never to be a gentleman, still he might be tolerated, sustained, and, in certain important respects, valued. He might be a holy man or a solitary genius.[151] Precisely because he was not as other men, did not live with them, did not do as they did, and did not value what they valued, what he came to know might have a peculiar standing and worth. Indeed, the public display of the scholar's solitude and incivility could be taken as his intellectual *bona fides*. If Robert Boyle was offered up as the pattern of a gentleman-scholar, in late seventeenth- and early eighteenth-century English culture, Isaac Newton provided an extremely potent, and highly traditional, presentation of the man of science as otherworldy holy recluse.[152]

It was a presentation of scholarly self that was readily understood by polite society. It was attached to familiar institutions for its legitimacy and subvention. It drew upon familiar cultural repertoires. The value set upon the scholar's work was the price polite society paid for the goods he was considered in principle capable of producing: truth, the preservation and transmission of tradition, the legitimations needed by civil society, sacred, and, latterly, secular power. It was a presentation that identified the scholar, philosopher, or intellectual as a special sort of person. Seen from the point of view of civil society, the scholar was at once commendable, contemptible, and strange. Whoever he was, he was not *one of us*.

I know little about the developing reference and resonance of the "gentleman-scholar" usage as it emerged in the late eighteenth century. However, it would seem that the commonplace was rarely applied to practitioners of technically useful learning. Indeed, Cardinal Newman's influential version of "a gentleman's knowledge" was defined precisely by its non-utilitarian character, and, while the Edinburgh Whigs peddled useful knowledge to the lower and middling classes, Thomas Carlyle spoke for much gentlemanly society in associating mechanic knowledge with mechanic minds.[153] One reference for the "gentleman-scholar" of the late eighteenth and early nineteenth centuries was, in all probability, someone legitimating claims to social standing by displaying morally textured antiquarian, civil or natural historical knowledge of *locale*—the "squarson"-naturalist or antiquary exemplified by White of Selborne.[154] Perhaps the commonplace also developed through the nineteenth century in association with the ancient universities. If so, then it needs to be stressed that changes in access and reforms of the curriculum signified far-reaching shifts in institutional conceptions of both gentility and the scholarly life.

In modern educated middle-class company, both "the scholar" and "the gentleman" have for decades taken mandatory air-quotes. The gentleman's traditional civic exercise of moral authority and the scholar's disengaged moral proprietorship of systematic knowledge have both been supplanted by the same social person—the *expert*—leaving both the "scholar" and the "gentleman" as almost empty linguistic shells, roles well on their way to losing their institutions, their legitimacy, and, ultimately, their members.[155] For this reason, I suspect that the commonplace has for some time been wide open to ironic appropriation, increasingly so as one approaches time present. When I first started asking around about "the scholar and the gentleman," one friend speculated that these days the formula was more commonly heard in lager advertisements and in the public bar as a sarcastic pleasantry than anywhere else or in any other sense. He was probably right. In *The Scholar's Arms* every man now goes through the door marked "Gents."

CHAPTER 9

Who Was Robert Hooke?

The easy answer to the question of Robert Hooke's identity is also an intractably difficult answer: Hooke was an experimental philosopher, or, as we might now be tempted to say, a scientist. The answer is easy in that it commands instant recognition from present-day audiences. We know what it is to say someone is a scientist; we have plenty of examples around us in case we want to check the characteristics of a scientist or to show someone unfamiliar with our usage what a scientist is. The images of the scientist's identity are on display as models for anyone who wants to become one; occupants of the role can defend their behavior by saying that it is scientific, just as they can condemn other behavior by claiming that it violates expectations of a scientist's proper conduct.

Yet this easy answer to Hooke's identity begins to look indefensibly glib if only we consider matters from his point of view and from the situation in which he found himself. Mid- to late seventeenth-century English society recognized what it was to be a gentleman, a professor, a physician, an architect, an operator, a mechanic, an instrument-maker. It did not, however, readily comprehend the role of experimental philosopher, nor were resources routinely available to explain or justify behavior by referring to what was normal and proper for a person performing this role.

The historian can call Hooke an experimental philosopher, even a scientist: the anachronism is unfortunate but not necessarily vicious. However, Hooke's contemporaries did not call him an experimental philosopher. To John Aubrey, Hooke was known either by his official capacities (one of the "surveyors" of the city; curator of experiments to the Royal Society; professor at Gresham College), by his relationship to others ("assistant" to Thomas Willis and to Robert Boyle), or according to his practical skills and his routine deployment of them ("He is certainly the greatest mechanick this day in the world").[1] Nor did Hooke systematically refer to himself as a philosopher. He did not call himself an experimental philosopher, but he occupied social and cultural terrain staked out by those who did so identify themselves. Where was the role of the experimental philosopher located on the social and cultural map of mid- to late seventeenth-century England? How did one go about establishing one's entitlement to the position? What obstacles might be confronted in the course of this establishment? Finally, what connections were there between the terms of occupancy of the role and the terms in which experimental knowledge was made and justified? Hooke is an apt study for these questions precisely because his entitlement to the ascribed attributes of the experimental philosopher was problematic. I want to examine aspects of Hooke's problematic identity as a way of understanding his particular place in the moral economy of the contemporary scientific community. I also want to use this material to illuminate certain fundamental features of that moral economy, especially the relationship between the ascribed characteristics of individuals variously situated on the social map and their capacity to make scientific knowledge.

A Day in a Life

One way of establishing Hooke's identity—who he was understood to be in his society—is to follow him around through the course of a day's work, to trace the quotidian patterns of his movements through his physical, social, and cultural environments. What work did he do and where did he do it? What social relationships were transitted and constructed in the course of doing his work, and what was the moral texture of those relationships? What were the connections between Hooke's quotidian movements and the economy of knowledge in which he was a key actor? We cannot, of course, re-create a day in the life of Hooke in its entirety: *cinema verité* is not a seventeenth-century technology and the flies on his walls are long since dead. However, we can pick out certain recurrent features of his movements that bear importantly upon who Hooke was understood to be in his mid- to late seventeenth-century environments.

From the mid-1660s, the basic quotidian structures of Hooke's life were set in place. After a fitful night's sleep, or sometimes no sleep at all, he rose at varying hours, but usually early, in his rooms at Gresham College. He dined there, sometimes alone, more commonly with his various resident technicians. Alone or assisted by his technicians, he set to work, at his own mechanical contrivances, at architectural models and drafts, at experiments required by the Royal Society, or at discourses for their benefit. Leaving his rooms in the afternoon, Hooke then met friends and philosophical or mechanical colleagues at one or another local coffeehouse or tavern. There he discussed his work and learned of the work of others. Moving about London, he visited the booksellers of St. Paul's churchyard; the laboratory of Robert Boyle in Pall Mall; the apothecaries who provided him with his unending supplies of medicines; the clothiers and shoe-makers whose goods figure so largely in the *Diary;* and the shops of mathematical practitioners, instrument-makers, and apothecaries, where he often spent long hours in collaborative work. As surveyor, he worked in various parts of the city, taking "views" and supervising building jobs, frequently undertaking domestic architectural projects on a private basis. On days when a Cutlerian or Gresham geometry lecture was required, these would be prepared in the mornings, and given, sometimes almost to the bare walls, at two or three in the afternoon. Later in the afternoon, on Wednesdays or Thursdays, when the Royal Society met, Hooke was on call to exhibit the experiments or read the experimental discourses he had prepared at home. Afterward, he and some of the Fellows would repair once more to a city coffeehouse, where they had further experimental and mechanical discourse. From 1674, many late nights were given over to astronomical observations in a "turret" constructed over his Gresham rooms. He rarely went to bed before two or three o'clock in the morning. Sundays followed only a slightly different pattern for Hooke: he generally stayed at home until quite late, and it was a day frequently used to write up his diary or to put his papers and notes in order. And, in the event, the coffeehouses were open even on Sunday evenings. Holidays seem to have affected Hooke's quotidian pattern minimally, though New Year's Day was sometimes used to draw up his financial accounts and his birthday was generally an occasion for reflection and taking maudlin stock of his miserable life thus far. Hooke does not appear to have been familiar with the category "vacation."[2]

This is, of course, only the most sketchy picture of how Hooke spent his days. It does not portray the full range of his activities nor does it take adequate account of significant variation at any given time in his life or of changes over the years.[3] Nevertheless, even this sketch allows one to pick out three related features of that pattern that bear upon the question of Hooke's contemporary identity. First, there is the extreme heterogeneity of his daily activities. Hooke not only engaged in a very wide variety of work activities, he also moved through highly disparate social worlds in the course of doing so. Second, there is the relationship of dependence that informed much, though not all, of Hooke's work. A great deal of what Hooke did during a day's work was done at the behest of others, in accordance with their general or specific directions. That relationship of dependence was usually signaled by the exchange of money for services, as with his work as curator of the Royal Society. By contrast, it would appear that the area of work in which Hooke had the most independent interest and autonomy was that involving the invention of mechanical and optical devices: lenses, lens-grinding machines, telescope sights, clocks and watches, and, not least, his "thirty several" contrivances for flying that preoccupied him throughout his life and whose secrets he took to his grave.[4] Third, there is the social and cultural significance of Hooke's quotidian physical movements through London. Those movements amount to an active circulation between the private and the public and back again, instanced by the difference between the place where he lived and worked, on the one hand, and the places where he discussed and discoursed, on the other.

One conclusion based on this evidence stands out. Hooke was recognized as a person dependent upon others, a person of at best compromised freedom of action, of ambiguous autonomy, and of doubtful integrity. That is to say, his contemporaries might not generally recognize Hooke as a gentleman. At most, his entitlement to the status and attributes of a gentleman was considered as problematic. This is not, of course, a conclusion that will come as a revelation to anyone at all acquainted with the details of Hooke's life. Nevertheless, I want to take the trouble to establish the point, to show what it was about Hooke that made his standing problematic. What connections were there between the gentleman and the experimental philosopher? How did the attribution of gentlemanly standing and conduct figure in the moral economy of the English experimental community during Hooke's life? In the course of discussing these matters I liberally help myself to certain comparisons between the pattern of Hooke's life and that of a major claimant to the title of experimental philosopher in mid- to late seventeenth-century England, his colleague and patron Robert Boyle.

The Private World of Robert Hooke

From 1664 until his death in 1703, Hooke played his life out in and around the suite of rooms he occupied as professor of geometry at Gresham College in Bishopsgate Street.[5] Hooke never traveled abroad; in fact, he left London and the home counties only a few times after he took up permanent residence there. Unlike Boyle, Hooke was not sent on the Grand Tour of the Continent. His direct experience of the physical world was limited compared to that of many of his philosophical colleagues. Indeed, there were few of Hooke's philosophical colleagues who traveled less than he did—Isaac Newton being his main rival in this respect. Similarly, his social world—the range of his acquaintance in the Republic of Letters—was narrow when compared, for example, to Boyle's or to Henry Oldenburg's (whom he succeeded as Royal Society secretary in 1677). Hooke's was not only a London life, it was a life overwhelmingly centered on Bishopsgate Street and its immediate environs. When the Royal Society met at Gresham, all Hooke had to do was to take the experiments he had prepared in his rooms to adjoining or nearby public rooms in order to show them to the Fellows.[6] He read his geometry and Cutlerian lectures in the reading-hall of the college, just behind his quarters. He tended the society's apparatus and the objects in its repository located in the West Gallery. From about 1666 he seems to have had his own "operatory" in his rooms, and, from 1674, he had his turret.[7] When, for seven years, the society met at Arundel House in the Strand, Hooke was put to considerable inconvenience through having to haul sometimes large and awkward experimental devices, like the air-pump, a mile and a half through the streets of London or possibly taking them to the Thames for water-transport to the quay at Arundel House. It was an imposition Hooke resented, and, when Gresham College became available to the society again, one of the considerations that moved them to return was "the conveniency of making their experiments in the place where Mr. Hooke, their curator dwells, and . . . the apparatus is at hand."[8]

Hooke worked where he lived. In the seventeenth century distinctions between places of habitation and places of intellectual labor were not standard. Neither professors nor private gentlemen typically were obliged to leave places of residence in order to produce philosophical knowledge.[9] Oxford, Cambridge, and Gresham professors thought, discoursed, wrote, and occasionally experimented in their lodgings or in attached spaces, and virtuosi like Samuel Hartlib, Boyle, Richard Towneley, Henry Power, and many others maintained laboratories or observatories in or near their houses. In the mid- to late seventeenth century, Gresham College had degenerated from a place given over primarily to public instruction to one in which a variety of private persons lodged, some with no connections to educational purposes, some of dubious character.[10] Nevertheless, by the time Hooke had become established as the Royal Society's curator of experiments, his rooms at Gresham had developed into what was arguably the most important site in England for the performance of experiments. This, together with Boyle's laboratory in Pall Mall, was where experimental *work* was overwhelmingly done, not in the public rooms of the Royal Society where that work was displayed, discussed, and discoursed of.

The rhetoric associated with the new experimental program stressed the public character of proper scientific activity. Nevertheless, Hooke's rooms and workshops constituted, in practice if not in principle, a relatively private place in the economy of seventeenth-century English science. Contained within the Hooke household at Gresham and living with him, there were his various housekeepers, domestic servants, and, from about 1672, his niece Grace Hooke. Technicians, such as Henry Hunt, Thomas Crawley, Denis Papin, and others, also lived with Hooke during their periods as his paid assistants.[11] It was not, therefore, by any means a solitary life. It was, however, a life relatively isolated and insulated from the public life of his philosophical colleagues and associates in the Royal Society. There were few philosophical friends who were frequent visitors to Hooke's rooms, and fewer still who dined with him in his rooms. Hooke's closest friend in the Royal Society fellowship, and the man he entertained most frequently in his lodgings, was the Hartlibian émigré Theodore Haak, who was thirty years older than him. He played chess with Haak, dined with him, and, perhaps uniquely, never fell out with him or, at least, never recorded that he did so in his *Diary*. Christopher Wren, possibly a distant relation, was also often in Hooke's rooms, though how many of those visits exclusively concerned mutual architectural business is unclear. John Aubrey was on good enough terms to use Hooke's rooms as his postal address when he was in London, and spoke of his friend in glowing, though improbable, terms ("a person of great suavity and goodnesse").[12] Other philosophers who sought and gained routine access to Hooke's quarters included his Gresham colleague and personal physician Jonathan Goddard, Abraham Hill, John Hoskins, Walter Pope, Daniel Colwall, Jonas Moore, and Nehemiah Grew. From time to time, Hooke recorded that the Royal Society's Council or a group of key members met, and even dined, "here."[13] During the period in the mid-1670s when acute concern developed over the society's experimental lassitude, informal clubs of the most serious and active Fellows were accustomed to meet initially in Hooke's rooms and later in various coffeehouses. These included William Brouncker, William Croone, Haak, Thomas Henshaw, Abraham Hill, William Holder, Hoskins, Francis Lodwick, Jonas Moore, and Wren. There is a single reference to Newton visiting Hooke's rooms.[14] If, however, attention is shifted from philosophers to instrument-makers, mathematical practitioners, builders, and the like, a different picture emerges. Hooke spent an enormous amount of time with them, often in his rooms working on mechanical and optical projects, often dining with them.

His early eighteenth-century biographer Richard Waller said that Hooke was accustomed to a "rather Monastick Life," that he lived "like an Hermit or Cynick."[15] Waller presented Hooke as someone who cared little for the conventions, customs, and corporeal rewards of the world. He said, and the *Diary* tends to bear him out, that Hooke slept little and erratically, that he worked hard (often "continuing his Studies all Night, and taking a short Nap in the Day"), and that his temperament ("Melancholy, Mistrustful and Jealous") was not one that suited him to a life of conventional sociability.[16] This was not an uncommon presentation for early modern intellectuals, both sacred and secular. Isaac Newton presented himself similarly, and was understood to stand outside the normal ambit of society's conventions.[17] The presentation of the philosopher's persona as hermit was a way of understanding not only who the philosopher was and what might be expected of him, but also a way of warranting his claims to knowledge. A man so abstracted from the world was a man free of the hold of its idols and in immediate contact with reality, divine or mundane.[18] However, there seem to be problems with characterizing Hooke as hermit. The pattern of his life seems, at a glance, to be neither obviously monastic nor private.

Yet there is a significant contrast discernible between the quotidian pattern of Hooke's life and that of some other notable experimental philosophers in London. For example, while Boyle's London laboratory was a place of pilgrimage, for both English and foreign philosophers, Hooke's was visited scarcely at all by philosophical travelers. Boyle was celebrated, both during his lifetime and upon his death, for his openness of access to the "curious of all nations."[19] Such accessibility was identified, by Boyle himself and by others, as a defining characteristic of the new experimental philosopher. His laboratory, unlike those of the alchemists, was to be a place of public resort and collective witnessing. Easiness of access was also, in the seventeenth century, a defining characteristic of a gentleman. Boyle, like Hooke, worked where he lived, and the obligation to hospitality was one that he acknowledged even though it lay heavily upon him.[20] By contrast, so far from being sought out by visiting philosophers and the "curious of all nations," Hooke was rarely even mentioned by those visitors who thronged to Boyle's company. The personal relations that subsisted between the two reflect their relative public standing. Hooke was a constant guest at Boyle's table, or, rather, at the table of Lady Ranelagh, Boyle's sister (with whom Boyle lodged). There were long periods during which Hooke recorded dining at Boyle's house at least once a week. By contrast, there is no convincing evidence that Boyle ever dined at Hooke's table, nor that he visited Hooke's rooms more than once or twice during the period covered by the *Diary.*[21] Hooke and the philosophical world came to Boyle; Boyle and his philosophical world did not come to Hooke. This pattern of movement was understood in the seventeenth century to be a visible sign of the relative standings of the persons involved. In the most influential seventeenth-century English guide to the code of the gentleman, Henry Peacham said of an individual who was our social superior that "We must attend him and come to his house and not he to ours."[22] Indeed, if we wish to be precise about seventeenth-century gentlemanly usage, it might be better to say that Hooke did not *have* a "home." His lodgings were a fit place to work and, on some occasions, to talk work; they were not a place fit to receive and to entertain gentlemen.

The relative privacy of Hooke's place of habitation and work is underlined by the typical pattern of his daily movements through the streets of London. Hooke's experimental work was, as I have already noted, conducted overwhelmingly in his lodgings. This was the place where, in contemporary parlance, experimental "trials" were performed, and this was not a place much frequented by Hooke's philosophical colleagues. However, on leaving his Gresham rooms, Hooke entered a highly public domain. The coffeehouse was a major (arguably the major) site at which the outcomes of experimental trials were made known, their significance assessed, relevant information, books, and materials exchanged. Indeed, the coffeehouse was occasionally even a place where experimental trials were conducted.[23] The active core of Royal Society Fellows often resorted to the Crowne Taverne in Threadneedle Street, around the corner from Gresham College, while Hooke's "clubb" migrated between Joe's, Garaway's, and Child's coffeehouses, latterly meeting at Wren's house. The Restoration London coffeehouse was a highly democratic institution.[24] It was a place of open entry, largely stripped of the patterns of deference and the segregation of social worlds that obtained outside its doors. While the coffeehouse welcomed all comers (except

women, of course) and mixed them together promiscuously, the great courtiers, the high aristocracy, and the morally squeamish tended to shun them.[25]
Hooke loved coffeehouses, even if he was unsure of the safety and value of
either coffee or chocolate. By contrast, Boyle avoided them. Hooke, who
reliably recorded his meetings with Boyle and the company he met at coffeehouses, gives us no certain evidence that Boyle ever visited a London
coffeehouse.[26]

When Hooke moved from the coffeehouse to the meeting rooms of the
Royal Society, he entered upon another sort of public stage. Here Hooke
met with the gentlemen and philosophical colleagues who paid his salary
and directed his experimental efforts. As curator, he performed for them the
discursive and manipulative tasks he had been engaged to do. Despite much
rhetoric associated with it, the society was not a place of promiscuous public access. Nevertheless, this audience constituted the relevant public for the
experimental trials that Hooke performed at home. This is where Hooke
"shewed" the experiments, that is, displayed them as reliable producers of
matters of fact; where he read "discourses" narrating experimental trials
performed at home; and where those "shews" and "discourses" were considered and assessed by the Fellows. This perpetual circulation between
Hooke's rooms and the meeting place of the Royal Society, between the
relatively private and the relatively public, was a necessary process in the
making of experimental knowledge. It was a circulation insisted upon by
those who engaged Hooke's services.[27]

Hooke as Philosophical Servant

Hooke's *Diary* provides abundant evidence of his acute sensitivity to social
rank and to the patterns of deference that expressed and maintained social
hierarchies. Even in this private document, Hooke took pains to refer to
his friends and acquaintances by their proper designations. The society's
president, William Brouncker, was almost invariably referred to as "Lord
Brouncker"; John Wilkins as "Lord Chester"; Seth Ward as "Lord Sarum";
William Petty as "Sir W. Petty"; Jonathan Goddard as "Dr. Goddard";
George Ent as "Sir G. Ent"; and so on. Even as close a friend as Theodore
Haak was generally "Mr. Haak," while as vexing a patron as John Cutler was
"Sir J. Cutler." Boyle was, of course, most commonly designated "Mr. Boyle,"
though the honorific was exceptionally dropped when Hooke was angry
with him or for reasons not evident in the *Diary*.[28] By contrast, his rival

Henry Oldenburg was almost always just "Oldenburg" (when he wasn't "lying dogg," "villain," or "huff" Oldenburg), and Hooke's various technicians and craftsman-associates tended to be designated informally: "Tom," "Tom Hewk," "Harry" (tending toward "Mr. Hunt" when he became economically independent), (Richard) "Shortgrave" (occasionally "Mr."), (Thomas) "Tompion," (Thomas) "Crawley," (Denis) "Papin" (or "Young Pappin").[29] Until 14 November 1673, when Hooke noted that his friend Christopher Wren had been knighted, the *Diary* invariably designated "Dr. Wren." On his next appearance in the *Diary* on 16 December 1673 he became "Sir Ch. Wren," and so he almost invariably remained in Hooke's usage. Similarly, Hooke carefully noted and observed the translation of "Mr. J. Hoskins" into "Sir J," in 1676 and that of "Mr. J. Moore" into "Sir Jonas" in 1680.[30]

Hooke knew and cared where his friends and acquaintances were located on the social map. He also showed signs that he cared deeply about his own place on that map. On the one hand, he displayed standard patterns of deference to those who were his undoubted social superiors, but, on the other hand, he became agitated when he felt that he was not being treated in a manner appropriate to his real standing or worth. In Hooke's case, eternal vigilance seems to have been the price of maintaining his integrity. What was rightfully his had ceaselessly to be made publicly evident, insisted upon, fought for. Fairness could not be taken for granted; anyone might at any time turn into a cheat, a spy, or a traitor; conspiracies might be hatched against his interests; snubs and incivilities lurked around every corner. Lady Ranelagh, for example, employed Hooke over the years to renovate her houses in Pall Mall and Chelsea. In Hooke's view she dealt with him as a mere tradesman, and he periodically bridled at such treatment: "Dind at Lady Ranalaughs," Hooke recorded in 1674: "Never more." When Hooke had been working on Lady Ranelagh's Pall Mall house for some time, he finally erupted: "At Lady Ranalaughs, she scolded &c. I will never goe neer her againe nor Boyle." Within a week he was back at Boyle's and on speaking terms with both him and his sister.[31]

As curator of the society's experiments from 1664 to 1677, Hooke was employed to do the Fellows' bidding. Geoffrey Keynes only marginally overstated the case when he described Hooke as "the Society's dog's-body."[32] When, however, he succeeded Oldenburg as co-secretary (with Nehemiah Grew), Hooke dearly expected better treatment and more autonomy. He was vigilant that he be dealt with appropriately. At the meeting of 13 December 1677, he recorded that he "Read notes Distinctly. Grew placed at table to take Notes. It seemed as if they would have me still curator, Grew Secretary." The next month Hooke was outraged by the fun-loving new president, Sir Joseph Williamson, who suggested, "Ironically," that the hunchbacked secretary wanted a higher chair.[33]

His precise role and function within the Royal Society and the philosophical community generally remained a source of uncertainty and trouble to Hooke through much of his life. Initially, he publicly accepted the identity of philosophers' assistant. Hooke had been accustomed to a deferential relationship with gentleman-philosophers since his student days at Oxford, where he was an impecunious chorister and "servitor to a Mr. Goodman" at Christ Church. Probably while still a student, Hooke entered into remunerated assistantships, first with Thomas Willis, then with Boyle, with whom he lived and worked at Deep Hall from about 1657.[34] In Hooke's first publication of 1661 the dedication to Boyle was unrestrained even by the hyperbolic standards of the genre: he feared that the "Minuteness" of his text would make it "a Present very unfit for so great a Personage" as his master.[35] Boyle was the Sun, the source of light in Hooke's life. But if Hooke accounted himself "minute" with respect to his gentleman-benefactors, he reckoned himself considerable with respect to those he viewed as his inferiors. In *Micrographia* Hooke tellingly placed himself in a condition riven with social tension and ambiguity. He identified himself as a master of technicians and a technician of masters: "all my ambition is, that I may serve to the great Philosophers of this Age, as the makers and grinders of my Glasses did to me; that I may prepare and furnish them with some *Materials,* which they may afterwards *order* and *manage* with better skill, and to far greater advantage."[36]

Hooke continued in Boyle's employment at least until 1662, when Boyle recommended him to the Royal Society as their curator, and there are some reasons to believe that Hooke was paid by Boyle until 1664, when he began to acquire alternative sources of income.[37] Yet the deferential relationship with Boyle continued intact after that time. Neither in *Micrographia* nor in subsequent publications did Hooke ever claim authorship of devices and findings to which many historians think he had a "right": "Boyle's airpump," "Boyle's Law," "Boyle's theory of colors."[38] There is only fragmentary evidence that Hooke's relations with Boyle had a remunerative basis after he began his work for the Royal Society and Gresham College: Boyle made a personal contribution to Hooke's turret in 1674 and Hooke designed and constructed a new laboratory for Boyle in 1676–1677. As late as

1678, he recorded that he was coming to Boyle's presence on command.[39] And it is evident that Hooke continued to perform major services for Boyle in obtaining and delivering instruments, books, and medicines, and in acting (together with Oldenburg) as intermediary between Boyle and a host of printers, engravers, builders, and other craftsmen. For all that, the New Year's Eve 1676 summing up of monies owed by and to him shows nothing relating to Boyle.[40]

The terms of Hooke's engagement with the Royal Society made his dependent position clear. He was strictly charged to supply each meeting "with three or four considerable Experiments" and also to perform whatever other experiments were suggested by Fellows. He was, therefore, as J. A. Bennett has demonstrated, unmistakably an employee at the outset of his career, and, even though his circumstances altered somewhat over time, he could never assume that he would be uniformly treated as a colleague.[41] When Hooke was admitted to the Fellowship ("to come and sit amongst them"), it was on different terms than those that then applied to the rest of the Fellows. His membership fees were waived, and he was paid to lodge at Gresham for the purpose of looking after the society's growing repository and to carry out the society's work. Hooke was, therefore, both in a collegial and in a dependency relationship with the other Fellows. The sanctioned mode of dealing with a colleague stood in contrast to that of dealing with a servant. Hooke's position was, therefore, deeply ambiguous. Which mode of conduct would he be confronted with in any given circumstance? Was he wholly a colleague or wholly a servant? Was he a free agent in experimental matters, or was he the directed instrument of others' free action? How did he see himself, and how did he present himself to others?

For more than fifteen years, certainly the most experimentally active years of Hooke's life, his daily work was largely subject to the will of others. These others manifested no doubt about their entitlement to set the terms of Hooke's scientific work and to chastise him when he failed to give satisfaction. In the mid-1660s, for example, Sir Robert Moray chided Hooke (through Oldenburg) for "his slackness" and complained about the time Hooke allegedly frittered away in his mechanic activities: "I easily beleeve Hook was not Idle, but I could wish hee had finisht the taskes lyet upon him, rather then to learn a dozen trades."[42] The language used by his Royal Society colleagues and masters to direct Hooke's work has been widely noted by other historians. After an initial period lasting no more than a few months, when the *Journal* records that Hooke was treated like other Fellows and "desired" to perform his experiments, by early in 1663 he was being increasingly "directed" and "ordered" to do so.[43] These usages persisted through the 1660s, during which period there are no more than a handful of references to any other Fellow—other than Oldenburg, also an employee—being "ordered" to do anything. By contrast, usage with reference to Boyle scarcely varied: requests for the experiments or discourses to him uniformly took the form of "desires." And, of course, when Oldenburg died in 1677 and Hooke succeeded him as secretary, the form shifted over entirely to "desires," together with the *Journal* portraying a much more prominent and aggressive role for Hooke's experimental and discursive activities. Doubtless, these usages, and their change over time, are partly a function of who was taking the minutes. But it is evident that they also reflect Hooke's standing vis-à-vis the other Fellows, and their perception of his dependence. As late as December 1675, Hooke was content to record in his *Diary* that Brouncker "ordered" him to do experimental work.[44] And one need only thumb through the *Journal* to see what a matter of routine it was for Fellows of the society to cause Hooke to prosecute concerns other than those he was most interested in.[45]

This is not to say that Hooke accepted his dependent status without reservation—at any stage of his career. For example, the boundary between the identity he acknowledged and that of a mere mechanic was one that Hooke carefully policed, especially after his succession to the secretaryship of the Royal Society. Evidently feeling himself badgered by excessive demands for experimental entertainments, Hooke explained why he had not, and would not, "trouble the Society at their meetings with a confused enumeration of experiments" of any given type. His job, Hooke said, was to innovate and to illustrate those innovations; the repeated display of a well-working experimental apparatus was "only the work of a labourer or operator to perform, when once the instruments were contrived, and the method chalked out."[46] While he acknowledged without serious question the right of his legitimate superiors to direct his labors, he bridled at unseemly treatment from those he either did not know or whom he considered to be mere equals or inferiors, particularly when those perceived slurs were committed in the gaze of Hooke's masters. The case between Oldenburg and Hooke is the most spectacular instance of this, though Hooke's dealings with foreign practitioners like Johannes Hevelius are also instructive. Oldenburg was not only the society's other major retainer, he was also the other important, and equally long-serving, recipient of Boyle's patronage. Hooke could not abide him, and used no discretion in dealing with him in public or recording his private sentiments about him. Oldenburg was, in Hooke's view, a common thief and traitor. He had betrayed Hooke's rights in his inventions to foreigners. More important, he constituted the major threat to Hooke's standing among the philosophical colleagues in the Royal Society. He refused, in Hooke's opinion, accurately or fully to record Hooke's experiments and inventions; he fomented trouble between Hooke and colleagues with whom Hooke had no quarrel.[47] By 1676, Hooke had become so vexed at the "Grubendolian Caball" in the Royal Society that he "Resolved to Leave Royal Society," typically abandoning his resolution as soon as it was made.[48] Hooke's transactions with Newton are also indicative of the way he assessed his relative standing in the philosophical world. He appears, like others in the Royal Society, to have known little about who Newton was on their first engagement. As keen as Hooke was initially to insist on his priority and even superiority to Newton, he was equally content to withdraw to a more defensible position as soon as Newton insisted on it and leading Fellows of the Royal Society took Newton's side. Making peaceable noises, Hooke explained to Newton that it was Oldenburg who had been wholly responsible for the troubles between them.[49]

How did Hooke deal with his own servants? How did those dealings compare with the way in which other experimental philosophers dealt with their servants, and even with Hooke himself? And what do those dealings tell us about Hooke's identity? The contrast in these respects between Hooke and Boyle could not be more extreme. In the whole of Boyle's published work and letters I can find no more than six occasions when his employed technicians were mentioned by name. These technicians include Hooke himself, Denis Papin, and John Mayow (once only, and then by initials).[50] This is not because Boyle did not employ technicians; indeed, he employed a great number, some of them resided with him, and their role in his experimental work was of enormous importance. It appears, rather, that Boyle's technicians and various assisting presences were largely invisible to him. Naming them would be one way of acknowledging their agency in the work over which Boyle presided. Boyle evidently saw no reason to acknowledge their presence and role in that work.[51]

Hooke dealt with his technicians according to an entirely different pattern, a pattern that tells us as much about who Hooke was as it does about who they were. Unlike Boyle, Hooke treated his technicians on the model of craft apprenticeship. Boyle was in no way concerned to train his technicians to do what he did; in a sense they could not do what he did because they could not be who he was.[52] Hooke *was* vitally concerned with such training. In 1675, he told Aubrey, who was recommending a young man for employment with Hooke, what his terms were. He wanted a full commitment on the young man's part to live with Hooke for seven years. He was to have adequate lodgings and a plain but reasonable diet. In return, Hooke would teach him how to do what Hooke did: "to fit him for the doing my business." Hooke reminded Aubrey what a success he had made of Henry Hunt, who was considering taking a position at £150–200 a year. The training Hooke offered was both informal (watch-and-do) and formal. There were occasions in the late 1670s when most of his audience for his lectures at Gresham was made up of his current and past technicians and other independent mechanics.[53] Indeed, Henry Hunt succeeded Hooke as the Royal Society's curator, Papin also became curator after serving Hooke, and several others became independent craftsmen. Hooke was therefore in fact exactly what he portrayed himself to be in *Micrographia*—a master of technicians. They were not his equals, but they were youths who might eventually become, ideally should become, his equals.

In Hooke's *Diary,* his various resident and nonresident technicians are major presences. Far from being invisible, as in Boyle's narratives, Hooke's technicians are seen to figure hugely as named presences in the structure of his working day. They were, it is true, rarely alluded to in Hooke's published works. Here the pervasive use of the first-person singular accurately reflects, as it does not in Boyle's practice, the extent to which Hooke acted as his own technician, while Hooke, unlike Boyle, was concerned in a material way with the establishment of his innovative priority. But his awareness of technicians' work and identity is well displayed in the *Journal* of the Royal Society, where his accession to the secretaryship is immediately marked by changes in the *Journal*'s conventions: now there are repeated named allusions to the society's operator (Henry Hunt) and chief clerk (Michael Wicks).[54] Those he engaged to live and work with him were, as Margaret 'Espinasse pointed out, treated as members of the family. He dined with them, rowed with them, and made up with them, exactly as he did with the women with whom he had sexual relations, though what one ought to make of the reference to an occasion when a technician, Thomas Crawley, "Slept by" Hooke is unclear.[55]

Robert Boyle and the Christian Virtuoso

From early in his philosophical career Robert Boyle labored to establish the identity of the new experimental philosopher. What sort of person was this? How did he go about producing knowledge that was true, potent, and safe? How could his identity be made publicly visible as a surety for the knowledge he produced? In Boyle's vision, the new knowledge was to be made by a new sort of practitioner, working in new sorts of social spaces. This practitioner had, as an urgent practical business, to be characterized and modeled. The new experimental natural philosopher required a template, and from this template copies could be multiplied. Throughout his career, Boyle offered two sorts of pattern upon which proselyte experimentalists might model themselves and their practice. One was textual. In *The Christian Virtuoso* and related tracts composed from the 1650s, Boyle delineated the identity of the new philosopher and located his practice in existing and in as-yet-uncharted cultural terrain. The other template was corporeal. Boyle constructed his own life as a visible exemplar of Christian virtuosity. The authority of Boyle's textual depictions was understood to reside in the real moral character of the author.[56]

Boyle's portrayal of the experimental philosopher was substantially novel. The pattern he traced publicly contrasted the new role with a number of existing roles, for example, that of the combative professor, the secretive and selfish "chymist," the overconfident mathematician, the facile and speculative "wit," and the tawdry mechanical "wonder-monger" or "juggler." On the other hand, in constructing the experimental character Boyle practiced moral *bricolage*, pointing to and recombining the moral characteristics of roles that were very widely understood in seventeenth-century English society. Put simply, Boyle modeled the experimental philosopher on the recognized patterns of the devout Christian and the English gentleman.[57]

First, such a man was said to be personally uninterested in the material rewards that might flow from genuine natural philosophy. Although proper science would undoubtedly yield useful outcomes, the Christian virtuoso set himself against Mammon; his concern was solely with the truths whose evidences God left in the natural world; making that truth manifest was his ambition. Boyle said that the "genius and course of studies" of "an experimentarian philosopher . . . accustoms him to value and delight in abstracted truths . . . such truths as do not at all, or do but very little, gratify mens ambition, sensuality, or other inferior passions and appetites." Indeed, experimental study was an effective antidote to sensuality: the only personal goal of the Christian virtuoso was to "entertain his understanding with that manly and spiritual satisfaction, that is naturally afforded it by the attainment of clear and noble truths." The Christian virtuoso was a moral hero. His work, Boyle said, satisfied him "of the vanity of the world, and the transitoriness of external, and especially sinful engagements." The Christian virtuoso set himself against the search for lucre and against unwonted secrecy; he was open and generous with his findings and inventions.[58]

Second, the Christian virtuoso was humble. Modesty and the rejection of presumption were both the ideal "temper of mind" for the practice of experimental philosophy and the natural outcome of a proper engagement with God's creation. Reading God's Book of Nature engendered in the Christian virtuoso "a great and ingenuous modesty of mind." It was an activity designed to give its practitioners a "well grounded . . . docility" serviceable to religion.[59] Third, the Christian virtuoso was a man of honor and he dealt honorably with his philosophical colleagues. Honor was an integral part of experimental social relations. It was unavoidable that the practitioner accept on trust both the "historical experience" represented by the testimony of past philosophers and the vicarious experience represented by the testimony of present-day philosophers whose experiments could not, in principle or in practice, be physically replicated.[60] This meant that the practice of experimental philosophy and the solidarity of the experimental community were founded upon trust. The Christian virtuoso was obliged to deal with other authentic philosophers as honorable men, and he must give other philosophers the visible signs that they could and should treat him as an honorable man.[61]

Fourth, and fundamentally, the authentic experimental philosopher was a devout Christian; he displayed himself as such, and he identified his work as a form of religious practice. Boyle described the experimental philosopher as a "priest of nature" and compared his laboratory to a place of divine worship.[62] The Christian virtuoso devoted equal study to the Book of Scripture and to God's Book of Nature. Boyle said that experimental philosophers ought to be "assiduous studiers of the Scriptures," and, of course, that experimental study afforded "divers motives to piety, and incentives to devotion."[63] Boyle offered himself as the pattern of a Christian virtuoso. His was not a Christianity of mere belief; it was one of active practice, as his texts, his pattern of worship, his engagement with evangelical work, and finally his will made clear. It was said of him that he never mentioned the name of God, which he did with great frequency, without an audible pause in his discourse.[64]

Finally, the experimental philosopher was *independent*. He relied upon the authority of nature, not upon the authority of other men. He displayed no deference to reputation or standing, going on "the visible testimony of nature." The experimental philosopher's freedom of action, the freedom to say what he witnessed and believed to be the case, was a precondition for the production of reliable knowledge. Boyle largely assumed the condition in which the Christian virtuoso was able to act independently of the opinions, reputations, and power of other men. But apologists for the new scientific practices explicitly discussed the necessity of independent free action. Thomas Sprat, for example, while advertising the social heterogeneity of the Royal Society's membership, deemed it essential that "the farr greater Number are Gentlemen, free, and unconfin'd." Neither the model of master and servant, nor that of master and pupil, was appropriate for the experimental community. How, Sprat asked, could the philosopher come to his own conclusions and give his own witness "in the presence of one, whom he *fears* and *reverences*?"[65] Boyle's identification of the experimental natural philosopher can be economically summed up. The authentic experimental philosopher was a Christian gentleman. Gentility in conduct and piety in belief were the proper postures in which to undertake experimental study, just as the experimental study of nature reinforced the attributes of a gentleman and a Christian.

Boyle's portrayal was locally potent. Neither his associates during his lifetime nor his eulogists after his death missed the point: Boyle himself was the Christian virtuoso. He was reckoned to be the very paragon of a Christian gentleman, who brought his piety and gentility to the altar of nature and who extracted from the study of nature further inducements to right religion and genuine morality.[66] If Boyle was the Christian virtuoso, who, then, was Robert Hooke? How did the pattern Boyle constructed bear upon Hooke's identity? How did it affect Hooke's practice as an experimental philosopher and the career of knowledge Hooke did so much to make?

"The Greatest Mechanick This Day in the World"

Hooke did not present himself to his contemporaries as unconcerned with lucre and the material rewards that might flow from a life in experimental natural philosophy. Indeed, within a year of making the acquaintance of the

Honourable Robert Boyle and his friends at Oxford, Hooke was concealing alleged mechanical secrets from them and negotiating patent rights by which he might make "a considerable advantage."[67] It was a practice he persisted with. Throughout his career, Hooke kept technical secrets from his colleagues, and made sure that they knew he was doing so. Hooke's *Diary* massively testifies to the extent of his association with Mammon. He was vigilant, indeed he was at times genuinely obliged by others' turpitude to be vigilant, in ensuring that he obtained what was owing to him. Hunter has meticulously documented Hooke's problems with Sir John Cutler, involving hundreds of pounds.[68] But Hooke was also anxious about much smaller sums. He kept careful records of even relatively petty amounts of cash lent to intimate friends: "Lent Mr. Aubery 5sh., which with the former made 40sh."; and three months later, "Lent Mr. Aubery 3sh. which maketh 43."[69] Hooke's work in providing Boyle with books and scientific instruments was also a continuing source of financial anxiety. In 1673, he sent Henry Hunt to convey a new microscope to Boyle. Boyle gave Hunt 5 shillings: "twas worth 20sh," Hooke noted.[70] Five years later, Hooke recorded the problems he had getting Boyle to pay up for books supplied: "Got from Boyle, Lana booke, also the 6sh. and 6 pence not without much asking for."[71] Similarly, Hooke scrupulously noted acts of petty generosity when rightly extended to himself or wrongly to others. In 1674, he recorded that he had dined at Wren's: "He would not let me pay."[72] In 1676, he was evidently overwhelmed when Haak treated him to chocolate worth £1 5s. 6d., but was annoyed when Oldenburg "was excused from paying" his share of the dinner bill.[73] When Hooke entertained friends in his rooms he observed the consumption of claret with the carefree generosity of an Edinburgh accountant.[74]

Yet we now know that Hooke was not a poor man, and we are reasonably justified in assuming that his associates also knew that. Through his surveying and architectural work in the rebuilding of London, Hooke had, as Aubrey said, "gott a great estate." By the 1670s, Hooke had stuffed a chest in his Gresham rooms that by his death contained "In ready money," "old money," "gould and silver" over £8,000, that is, enough to keep his household going at its normal levels of consumption for most of another lifetime. Nor, having amassed a sizeable fortune, could Hooke bring himself to act upon intermittently expressed intentions to endow a lectureship and laboratory for the Royal Society.[75] There is nothing inherently deplorable in such behavior, nor, of course, is it of any interest in the present context to pass moral judgments on a historical actor. No doubt, Hooke came to be what he was through the interaction of an innate temperament and a unique set of environmental circumstances. Nevertheless, there were structural patterns in his culture that were available to him as ways of modeling and justifying his developing conduct, and that were available to others as ways of locating and understanding such behavior.

Aubrey described his friend as "the greatest mechanick this day in world."[76] The quotidian pattern of Hooke's life definitively reveals how important the identity of mechanic was to him. It is clear that the greatest proportion of Hooke's working day was devoted to mechanical and architectural activity. Bennett has rightly criticized 'Espinasse for claiming that Hooke's work on scientific instruments "must be regarded as by-products of a constant preoccupation with the basic general problems of science." Instead, Bennett has stressed the importance of Hooke's identity as a mechanic and the conceptual significance in Hooke's philosophical work of that identity.[77] In fact, for Hooke the real and potential economic significance of mechanical innovation and architectural work was far greater than that associated with Gresham and the Cutlerian lectures, Royal Society curatorial work, or the authorship of predominantly philosophical texts like *Micrographia*. We know how lucrative Hooke's architectural work was, and we also have solid evidence of his expectations from mechanical invention. The contract drawn up in 1657 looked for thousands of pounds from the chronometer designed to solve the longitude problem.[78] Toward the end of his life, Hooke was actively engaged in a project for a joint-stock company involved in glass-making.[79] And there are traces in the *Diary* of intermittent negotiations with politicians like Sir Joseph Williamson over patents on his inventions.[80]

Hooke's mechanic work was therefore central to his perceived social identity and to his economic position. It was one thing for a son of the Earl of Cork to portray himself as unconcerned with personal reputation, priority, or the ownership of philosophical goods, and it was quite another thing for Hooke to do so. There is no sign that Hooke labored under a code of conduct that obliged him to display openness and humility. He did not so much "violate" one code as operate normally under another. If his priority in inventing time-keeping devices was challenged by Christiaan Huygens, or if the necessity of telescopic sights was denied by Hevelius, Hooke's uniform response was vigorously to defend his interests, insisting that his originality and proprietorship be publicly acknowledged. Equity required it. His sensibilities in such matters were informed not by the patterns of the Christian virtuoso but by those of the trades. In his Cutlerian lectures Hooke referred continually to the proprietary problems faced by mechanical inventors. Spies and traitors were lurking everywhere in Hooke's world, especially in the Gresham College lecture room. Oldenburg was, of course, thought of as a professional spy, but Hooke's watchfulness extended to members of his own household. At one time he even noted "Grace a spy."[81] Secrecy was not a regrettable and intermittent retreat from the free communication that characterized the ideal community of experimental philosophers; it was an absolute necessity in order to secure Hooke in his authentic rights. It seems that at least some of the Royal Society's discussions of the advisability of secret and closed meetings were instigated by Hooke. He clearly relished the secrecy of his "New Philosophicall Clubb" of 1675–1676 whose members "resolvd upon Ingaging ourselves not to speak of any thing that was then revealed *sub sigillo* to any one nor to declare that we had such a meeting at all."[82]

When Hooke apparently felt that his interests were materially engaged in experimental or mechanic dispute, his dealings with real or imagined adversaries were unrestrained. Frequently, if not invariably, Hooke refused to deal with his antagonists as men of honor, men whose words might be relied upon. In the most public possible way, Hooke declined at various times to accept the testimony of Oldenburg, Hevelius, Huygens, and Newton about, so to speak, what they knew and when they knew it. In the 1620s, Henry Peacham spelled out the code governing the word of a gentleman: "We ought to give credit to a noble or gentleman before any of the inferior sort."[83] Indeed, by refusing publicly to credit a person's testimony, one was understood to be contesting his entitlement to the standing of gentleman. By contrast, both Boyle and the Royal Society's other major servant Henry Oldenburg well understood the necessity of refraining from any public suspicions about the factual status of experimental testimony originating from gentleman-philosophers. Any public withholding of trust in these matters would, Oldenburg affirmed, "certainly prove very destructive to all philosophical commerce."[84] Conversely, anyone who refused publicly to accept the word of a gentleman advertised the dubiousness of his own credentials. It does not appear that Hooke behaved as he did because he was unfamiliar with the code. Among the books that Hooke recorded he owned and lent was one called *The Rules of Civility*.[85]

If the pattern of Hooke's behavior was deplorable in a Christian gentleman, it was widely considered to be nothing exceptional among tradesmen. Contemporary social guides to the code of English gentility stressed the contrast between the openness, the generosity, and the reliable truth-telling of the gentleman and the secretiveness, the "sordid interests," and the duplicity of the tradesman and the merchant. Some commentators condemned what they saw as increasing associations between the English gentry and trade. Tradesmen were widely said to be "a baser sort of people"; the practice of trade was incompatible with the honor of a gentleman.[86] It is interesting in this connection that we have so little evidence that Hooke's conduct was condemned by his associates. Sir Godfrey Copley said that Hooke was a miser and that "he hath starved one old woman [house-keeper] already"; Thomas Molyneux called him "the most ill-natured, self-conceited man in the world . . . pretending to have had all other inventions when once discovered by their authors to the world"; Gottfried Leibniz accounted Hooke's illegitimate claims to priority "unworthy of his own estimate of himself, unworthy of his nation, and unworthy of the Royal Society"; Robert Moray complained of the "folly" of Hooke's secretiveness about his mechanical inventions and the "inconvenience" he thereby caused to others; and Oldenburg (albeit without mentioning Hooke by name) told Boyle that "Some body of ye [Royal Society] . . . hath too slender thoughts of all what comes from abroad of a philosophicall nature, or is done by strangers."[87] However, in the main, Hooke's gentleman-associates seem not to have found the overall pattern of his behavior worthy of significant remark. Gentlemen did not behave like that, but tradesmen, merchants, and mechanics notoriously did.

When Boyle described the ideal experimental philosopher as a Christian virtuoso, he did not conceive of Christian obligation in a merely conventional sense. The Christian virtuoso was publicly to display his piety—in his discourse, in actions designed to spread and support right religion, in the social forms of observance, and in his publicly visible moral deportment. The Christian virtuoso was to be a moral paragon. His character vouched for the authenticity of his knowledge. Whether Hooke was perceived by his contemporaries as a moral paragon is doubtful. This is not the place to deal with Hooke's erratic, if not erotic, sexual life. In the event, we have the confident assurance of Lawrence Stone that Hooke's "sexual drive was far below that of the average Western man today. The central interest of his life lay not in women—not even in Nell or Grace—but in his scientific, technological and architectural pursuits."[88] In fact, it is extremely difficult to establish what about Hooke's private life was known to others, and it is only the public perception of his private life that matters in this connection.

On the one hand, I have found no significant or detailed allusion to Hooke's sexual morality by any of his contemporaries. (It would appear that he shared a minor interest in pornography with the Duke of Montagu, since it was he who first showed Hooke the "Naked [woman picture]," and a week later apparently lent it to Hooke for home consumption. This picture was still in Hooke's cellar when he died.)[89] On the other hand, it is hard to believe that the nature of Hooke's relations with his various housekeepers and his young niece was unknown to those like Boyle who kept such an anxious eye on standards of public morality.

The matter of Hooke's Christian observance and its public display is more straightforward, even though it too is hindered by difficult problems of evidence and interpretation. While God is a pervasive presence in the scientific texts of Robert Boyle, He is elusive in the published works of Robert Hooke. The invocation of the Deity is most notable in Hooke's early *Micrographia,* and it may be relevant that this was a work submitted by the new curator for the approval of his corporate masters. A significant number of *Micrographia*'s invocations argue the special case for microscopic (and telescopic) skill in the culture of natural theology. Hooke claimed, for instance, that "the Wisdom and Providence of the All-wise Creator" are as evident in the minute parts of the fly and the moth, "which we have branded with a name of ignominy, calling them Vermine," as they are in larger animate bodies visible to the naked eye. Those possessed of mechanical skill in devising instruments that extended the empire of sense were, Hooke said, very valuable to the advertised goals of the philosophical enterprise, including its theological goals.[90] In later texts, particularly those deriving from the Cutlerian lectures, there is far less invocation of the Deity and little evidence of serious commitment to natural theology, though it could plausibly be argued that this might be expected from the nature of these works. Some writers have attempted to make Hooke out as a secularist-before-his-time, particularly with reference to his geological work. This would be overstating the case. His *Lectures and Discourses of Earthquakes* made repeated reference to the Noachian deluge, and there is no expression of disbelief in the reality of the events described in Genesis. Nevertheless, it is noteworthy that many of the allusions to the Flood in this work occurred in the context of arguments *against* its adequacy as an explanation of the present distribution of fossils. The Flood, in Hooke's view, was too brief an event to account for the finding of fossils on mountaintops; naturalistic processes like earthquakes, which raised the ocean floor into mountains, were more plausible explanations.

Given seventeenth-century discriminations between the natural and the supernatural, Hooke clearly preferred to invoke the former.[91]

Waller strove manfully to present Hooke as a pious Christian, though the effort of doing so seems to show. "He always," Waller said, "exprest a great Veneration for the eternal and immense Cause of all beings, as may be seen in very many Passages in his Writings." According to Waller, Hooke never made any inventions or solved any philosophical problems without "setting down his Acknowledgment to the Omnipotent Providence," though whether the pervasive divine expletives in Hooke's *Diary* indicate Christian piety, as Waller claimed, is doubtful.[92] On suffering through a performance of *The Virtuoso: "Vindica me Deus"*; when his left nostril "looked black," or when port made him sick: *"Mirerere mei Deus"*; when he became exasperated with city bureaucracy: *"Libera me Domine."*[93] Nor is it certain, despite Waller's testimony that Hooke "was a frequent studier of the Holy Scriptures in the Originals," that Hooke was, indeed, a regular reader of the Bible. Compared to the work of Robert Boyle, there are few scriptural allusions in his writings, and, according to Feisenberger, while there were several Bibles in Hooke's personal library, there was "comparatively little theology" represented, "usually the largest section in a seventeenth-century library." Unlike Newton's or Boyle's libraries, Hooke's contained a number of profane works: French plays and works with louche titles like *The Practical Part of Love* and *Merry Drollery or Jovial Poems.*[94]

Hooke's *Diary* gives no evidence that he was either a notable church-goer or Sabbatarian. His Sunday routine was, to be sure, somewhat different from that of weekdays, though these changes might have been substantially dictated by structural patterns in Hooke's society. Hooke typically spent most of the Sabbath at home, doing his *Diary,* writing notes from the Royal Society's meetings, "rectifying" proofs, and working in his operatory. It tended to be a day he received visits from his mechanic friends, rather than paying visits to his philosophical colleagues, who were, certainly in Boyle's case, more observant than Hooke. Nor was Sunday a day to miss the coffeehouse, where Hooke would go in the late afternoon or evening. There are only a handful of notes in Hooke's early *Diary* that unambiguously establish his presence at places of Christian worship. In 1677, Boyle apparently dragged him to chapel when Hooke called on a Friday; in 1680, Hooke was visiting Lord Conway in Oxford and heard a sermon from a parson whom, he decided, was "a Sycophant or worse"; in April 1678, he recorded that he "heard Dr. [Gilbert] Burnet about providence," in December, on "spirits and against the Pope," and in February 1680, on "peace." In 1676, he noted that the "St. Helens parson rayld against Philosophers &c." So far as the conventional social forms of Christian worship are concerned, that is about all the evidence we have that certainly establishes Hooke's attendance through the 1670s.[95] There are periodical allusions to Hooke's participation in theological discourse, but little to establish what his views were. He condemned *The Virtuoso* as an "Atheistical" play as well as a "wicked" one; and he deprecated a Royal Society colleague as an "enthusiastick quaker," while there are occasional inconclusive hints that Hooke may have had heretical inclinations. There is apparent interest in Kabbalah and Mosaic philosophy; there is a reference to Hooke's private philosophical group as a "Rosicrucian" society (though the allusion is most probably to alchemy). His encomium to John Wilkins in *Micrographia* suggests an attachment to a theory of pure and primitive Christianity; and even Waller was unwilling to hold up his orthodoxy to close scrutiny: "If he was particular in some Matters, let us leave him to the searcher of Hearts."[96] If Hooke's virtuosity was Christian in inspiration, it was a very private Christianity. There is little work deriving from the core of the Royal Society whose public presentation was as disengaged from theological aims as Hooke's.

Integrity, Independence, and Experimental Testimony

There were no qualities more important for the Christian virtuoso to possess than integrity and independence. He had to have the integrity to wish to be a truth-teller, and he had to have the independence reliably to tell the truth. The experimental philosopher told the truth because there were no forces acting upon him that might make him need to tell an untruth and because there were no considerations that could compromise or damage him if he told the truth. It was said that this integrity and independence distinguished the new experimental philosopher from existing practitioners. The peripatetics were slavish followers of the word of the ancients; those bred up in Schools and accepting the moral economy of Schools followed their masters' authority and lived in fear of them; and the alchemists kept secret whatever legitimate knowledge they possessed because they wished to gain advantage from it.

The qualities of integrity and independence were shared between the ideal experimental philosopher, the ideal English Christian, and the ideal English gentleman. The pious Protestant went on the authority of no other man, no priest, and no pope. He inspected the evidence of God's Books for himself, and he inspected his conscience. Scripture enjoined him to tell the truth, and Protestantism encouraged him to give active witness to what he conceived to be the truth. The English gentleman was also characterized by his independence and integrity. The capacity for free action was, indeed, a defining feature of an English gentleman in the sixteenth and seventeenth centuries: he was a man so bred and so positioned in the economic and social orders that he was free to do and to speak as he wished, subject to civil law and the law of God. By contrast, the merchant and the tradesman might neither be free to tell the truth nor might they desire to do so. Insofar as anyone in a position of economic and social dependence was subject to the will of his master, he was not able to give his own free opinion. The way in which dependence compromised free action was importantly discussed in the debates over the franchise during the civil wars and Interregnum. Both sides to the Putney Debates, for example, assumed that those who sold their labor to another had so compromised their integrity and independence that they could not legitimately participate in voting. Whatever was said by those who sold their labor could not reliably be ascribed to them. They might speak as they did because it was their master's will.[97] In addition, the merchant and the tradesman might not tell the truth because they might not wish to do so in certain circumstances. Considerations of gain and advantage could decide the tradesman knowingly to say what was not true or to keep silent about what was true. Edward Chamberlayne (a Fellow of the Royal Society) identified the reason that tradesmen "in all Ages and Nations have been reputed ignoble"; it was "the doubleness of their Tongue, without which they hardly grow rich." And even Daniel Defoe wrote at length on the prevalence and necessity of the "trading lie" among those engaged in selling goods and services.[98]

For the gentleman, however, truth-telling was not only the result of his capacity for free action; it was an obligation, freely assumed, that was acknowledged to lie heavily upon him. A popular seventeenth-century guide to gentility specified that nothing could "disparage or lay a deeper aspersion upon the face of *Gentrie*, than to be taxed for fabulous relations."[99] Francis Bacon acknowledged the differences among men in their tendencies to tell truth. Poets lied for "pleasure" and merchants lied for "advantage." The gentleman was a truth-teller because he was bound by a code of honor that enjoined him not to lie to other gentlemen. The violation of this code was a source of "shame": "There is no vice that doth so cover a man with shame as to be found false and perfidious." In the end, for a gentleman to tell an untruth was a sign of irreligion and a blot on his honor: "Montaigne saith prettily, when he inquired the reason why the word of the lie should be such a disgrace, and such an odious charge, saith he, 'If it be well weighed, to say that a man lieth, is as much as to say that he is brave towards God and a coward towards men. For a lie faces God, and shrinks from man.'" A gentleman, by contrast, was brave toward men and humble toward God.[100]

The importance in science of a code that allows practitioners to discern who is and who might not be a reliable truth-teller is rarely appreciated. This is partly an inheritance of seventeenth-century empiricist rhetoric that stressed direct engagement between an individual and natural reality. If direct experience is the paradigm of knowledge-making, the role of testimony and trust would seem to be negligible. Yet the experimental program of the seventeenth century, like empiricist practice generally, was inescapably founded upon the social relations that constituted trust. While the ultimate justification for a claim to empirical knowledge was said to be an act of direct witnessing, it was widely understood in the seventeenth century that one could not, as a practical matter, insist upon direct experience in order to constitute one's factual knowledge. Testimony was necessary, and its quality had to be assessed. The testimony of credible witnesses was to be preferred to that of less credible witnesses. The maxim seems banal, but it was a potent resource. In general, everyone in a local society understood who was creditworthy and who might not be. The imputation was structural; it did not depend upon detailed knowledge of individuals' characteristics. Hooke himself tried intermittently to codify the rules for assessing testimony, but in doing so he achieved little more than a transliteration of the informal code.[101]

The Incredible Robert Hooke

I conclude by displaying some philosophical consequences of attention to the social identity of practitioners. Why certain knowledge-claims and testimony are credited is partly a function of who makes the claims and who gives the testimony. I show some relations between Hooke's perceived position on the social map and problems he encountered in making scientific knowledge. I argue that these problems grew out of ambiguity in Hooke's identity. That ambiguity can be conceived as the gap between who Hooke was understood to be and the identity of a Christian gentleman. Insofar as

Hooke was seen as a mechanic, as a dependent instrument of others, and as a seller of services and goods, certain characteristics were attributed to him that constituted troubles for his role in the community of experimental philosophers. The most significant of those troubles were encountered in the reception of his scientific testimony.[102]

Throughout his career as curator of experiments and even after he stopped serving in that formal capacity, Hooke was the Royal Society's major experimental performer. He was the person, far more than Boyle, who actually possessed manipulative skill. Without him (and such other skilled personnel as Richard Shortgrave and Henry Hunt), the experimental work of the Royal Society would have collapsed. He knew how to build the machines and how to make them work. In this respect, no Fellow was his equal. Despite that, Hooke's masters and colleagues in the Royal Society reserved the right publicly to withhold trust from his experimental testimony. Occasionally, they did actually decline in public to credit it; more commonly, they laid conditions upon the acceptance of Hooke's testimony not imposed on Fellows generally.

In the early 1660s, Christiaan Huygens claimed to have observed the so-called anomalous suspension of water, the failure of a column of water, when purged of air, to descend in the Torricellian apparatus when moved into the receiver of an air-pump. This was a finding that, if genuine, appeared to threaten the conceptual basis of Boyle's pneumatics. Boyle reckoned that the water should descend; if it did not, this was probably because the receiver of the air-pump in which the tube had been placed was leaking. Late in 1662 and early in 1663, the newly appointed curator was ordered to replicate the experiments described by Huygens. Through 1663, Hooke appears constantly to have disappointed and irritated his masters by producing what they took to be experimental "failures."[103] The judgment of whether or not anomalous suspension existed as an authentic matter of fact was informed by judgments of Hooke's skill in constructing and operating the pump, in particular his skill in making the pump tight.[104] Those who rendered judgments of Hooke's skill generally lacked the relevant skill themselves. They proceeded on the basis of their *knowledge* of what phenomena a well-working pump ought to produce, and they asserted the right of knowledgeable agents to define the meaning of skilled agents' work.

It was not uncommon for Hooke's testimony about the outcomes of experimental trials performed in his own operatory to be contested when this conflicted with the expectation of knowledgeable colleagues. Early in 1663, for instance, the *Journal-Book* records that "Mr. Hooke made the experiment of condensing air by the pressure of water; but the trial not agreeing with the hypothesis, it was ordered to be repeated at the next meeting."[105] In 1672, the Royal Society was considering the question whether air was generated or consumed by burning. Success or failure in these experiments had to be defined in relation to some theory or expectation about the resulting measurement. Hooke's colleagues reserved the right to define whether or not his experimental work had succeeded. Indeed, when he eventually reported "success," the *Journal-Book* referred cautiously to the experiment "he said, he had made," and members of the society were delegated to act as witnesses.[106] Repeatedly, Hooke's masters and philosophical colleagues simply assumed the right to identify when Hooke had or had not performed a competent experiment. On failure, Hooke was instructed to take the experiment away until it worked properly, and only then to show it in public. Moreover, Hooke was frequently obliged to make good his testimony about experimental trials by displaying the operations in public. In the 1680s, as Stephen Pumfrey has shown, Hooke was being taken to task "for not performing his experiments publicly." Martin Lister referred to a set of magnetic experiments "which I recommend to farther trial, because Mr. Hooke owned he could not make them succeed in private trial, accusing the too soft temper of the drill; and therefore he is desired to order better (if it can be) to be made that we may not break off in uncertainties, but have the experiments tried before us."[107]

As with all directions he received from those he recognized as his superiors, Hooke generally tended to accept his orders without significant demurral. There are, for all that, occasional indications that he resented the liberty with which his testimony was doubted. In 1667, Hooke was one of the major experimenters in the society's vivisectional work on respiration. He was clearly irritated that his report of experimental success had not been credited by his philosophical colleagues:

> I did heretofore give this *Illustrious Society* an account of an Experiment I
> formerly, tryed of keeping a Dog alive after his *Thorax* was all display'd by
> the cutting away of the *Ribs* and *Diaphragme*; and after the *Pericardium*
> of the Heart also was taken off. But divers persons seeming to doubt of
> the certainty of the Experiment (by reason that some Tryals of this matter,
> made by some other hands, failed of success) I caus'd at the last Meeting
> the same Experiment to be shewn in the presence of this *Noble Company*,

and that with the same success, as it had been made by me at first . . . This I say, having been done . . . the Judicious Spectators [were] fully satisfied of the reality of the former Experiment.[108]

It would be incorrect to claim that refusal to credit Hooke's experimental testimony, or even its public qualification, was a routine occurrence. It was not, and one can hardly imagine how the Royal Society could have arranged its affairs if Hooke's testimony had not been generally accepted. The point is that the withholding of trust was acknowledged to be a very serious act. This trust was withheld only exceptionally. Typically this occurred when the testifying individual was not known to the Fellowship or when he was known, but known to have suspect credentials. A relevant contrast is with the fate of Boyle's experimental testimony. So far as I can discover, there is only ambiguous evidence that an English Fellow of the Royal Society ever withheld trust from Boyle's experimental testimony concerning matters of fact, or even required public replication for his factual narrations to be credited.[109]

Who, then, was Robert Hooke? At the end of the exercise there is still no satisfying simple answer to the question of Hooke's identity. The easy answer—that he was a scientist—becomes even more implausible and historically insupportable. His identity was complex and ambiguous. Some of his associates, some of the time, evidently thought of, and dealt with, Hooke as a mechanic, as a tradesman, as a servant. Insofar as they did so, Hooke's contemporary entitlement to the role and attributes of the experimental philosopher was problematic. Hooke was probably not considered to have the attributes proper to the pattern of Christian virtuosity that was being created and exemplified by his patron Robert Boyle and endorsed by leading figures of the Royal Society. Hooke's experience therefore helps the historian understand some of the seventeenth-century connections between the emerging role of the experimental philosopher and the existing codes of English gentility and Christian morality. There are massive problems of trust and authority that lie largely unacknowledged at the core of empirical science. They are unacknowledged because these problems were practically, not philosophically, solved in the seventeenth century. The word of the Christian gentleman was part of that practical solution.

CHAPTER 10

Who Is the Industrial Scientist?

*Commentary from Academic Sociology
and from the Shop Floor in the United States,
ca. 1900–ca. 1970*

Here are two stories about the identity, condition, and state of mind of the industrial scientist that circulated in America from the early twentieth century to the 1960s or thereabouts. The first story had it that he—almost always "he" during this period—was unhappy, anxious, and maladjusted to his lot. Powerfully socialized as a young man into a unique set of scientific values associated with the university, he took industrial employment because suitable academic research careers were in short supply or because they were just too low-paid to keep body and soul together for those not keen on an ascetic way of life. He found adaptation to incompatible industrial moral economies difficult, sometimes smoldering with resentment throughout his career. As the twig was bent so grew the tree: socialization into such values was strong and consequential. The industrial scientist deeply disliked, if he did not actively rebel against, the violation of scientific values he found in industry: secrecy, regimentation, hierarchy, constraint, and short-termism. The money—for the money was on the whole good—never made up for it. The scientist was being forced into a gray flannel lab coat, and it just didn't fit. That was his central problem and it was a problem that industrial research managers would have to deal with as best they could. This is a story about "conflict of interest," though here the texture and vector of conflict are rather different from what they are understood to be in the present-day American research university. In this story, the emotional pull of the unique scientific ethos is so strong that there are grounds for worry that industry, or indeed governmental laboratories that share some of industry's characteristics, can obtain a sufficient supply of such men, or, when they are recruited, that they can be kept happy and productive. Of course, in the current state of affairs, concerns about "conflict of interest" usually take other forms. Now, university administrators and ethics committees understand that the pull of commercial lucre is so strong that standards of proper academic behavior are seriously liable to corruption.

The historical sources of story number 1 are familiar to sociologists and historians of science. While such sentiments were fairly widely distributed in American culture in the early to mid-twentieth century,[1] the most influential systematic version of this story was elaborated by the late Robert K. Merton in essays of the early 1940s—modified and developed in the 1950s and 1960s by such students and colleagues as Bernard Barber, Norman Storer, Walter Hirsch, and Warren Hagstrom—and institutionalized in the canon of American sociology of science. The "norms of science" into which the scientist was socialized became, as Merton said, "internalized," where they formed the scientist's "conscience" or "super-ego." Even though Merton was then understandably preoccupied with such threats to scientific integrity as those posed by the Nazi idea of "Jewish physics" or the Soviet concept of "bourgeois genetics," his initial description of the "institutional ethos" of science also remarked upon the tensions between scientific and commercial values. In science, Merton pointed out, *Die Gedanken sind frei*, and the "rationale of the scientific ethic" whittles down "property rights in science" to "the bare minimum" needed to secure recognition and esteem to the originator of a scientific idea. Science is public knowledge or it is not science at all, and the very idea of secrecy offends those who have internalized its values: "The communism of the scientific ethos is incompatible with the definition of technology as 'private property' in a capitalistic society . . . Patents proclaim exclusive rights of use, and often, nonuse. The suppression of invention denies the rationale of scientific production and diffusion."

And Merton here specifically alluded to late nineteenth-century litigation between the federal government and Bell Telephone Company that established the inventor's "absolute property" in his invention and his right to withhold crucial knowledge of it from the public. For mid-twentieth-century scientists, Merton wrote, this was a "conflict-situation," and, while different scientists were more or less uneasily responding to it in different ways (by taking out patents, by becoming entrepreneurs, or, indeed, by advocating socialism), nevertheless the fundamental friction arose from a conflict of values between pure science and commerce. Conflict was always likely to appear whenever science came into contact with institutions whose values differed from those needed for the pursuit of certified objective knowledge and that attempted to enforce "the centralization of institutional control."[2]

From these early statements of the norms of science, there emerged what seems to be a *prediction* about what empirical research would eventually show, if, indeed, systematic empirical research was deemed necessary to confirming such a matter-of-course state of affairs: scientists socialized into this value system would suffer the "pain of psychological conflict" when presented with situations that required or encouraged them to behave in ways that violated the norms they had acquired. To avoid or free themselves of this "pain," it was "to their interest" to conform to the ethos in which they had been socialized. Should the internalized "pure science sentiment" be put under pressure by "other institutional agencies" committed to the application of knowledge and concomitant organizational control, the result would be "the persistent repudiation by scientists of the application of utilitarian norms to their work . . . The exaltation of pure science is thus seen to advance and threaten the stability and continuance of scientific research as a valued social activity."[3] So it might be deduced that, for both psychological and social-functional reasons, scientists would vigorously resist assimilation into the value-system of commerce—such was the strength of the psychological grip of scientific values and such was the functional dependence of science on the embrace of these values. The scientist in industry, or in the military, would be an unhappy, awkward, and possibly even a disloyal figure, in constant conflict with commercial values and organizational structures. As a result of scientists' unique pattern of socialization, their personalities were such as would not tolerate organizational constraints: scientists were too fiercely independent and mindful of their individual integrity, too skeptical, too hostile to authority structures, too loyal to science, and too disloyal to local organizational values. Such persons would pose a major problem for the smooth running of commercial organizations.[4]

The work of Merton, Barber, Storer, and, to a lesser extent, of Hagstrom was largely programmatic, but by the 1960s academic sociologists—of whom the most prominent were William Kornhauser at Berkeley and Simon Marcson at Rutgers—were producing extended quasi-empirical studies of the "scientist in industry," centering on the problems of normative "conflict" and assessing the extent to which both industry and scientists themselves had been obliged to come to some sort of in-practice "accommodation" between value systems that were in such fundamental in-principle conflict.[5] In 1956, the founding of the academic journal *Administrative Science Quarterly* (*ASQ*) proceeded from a desire for a "general theory of administration,"[6] while a number of contributions to early issues—notably the work of Merton's student Alvin Gouldner on "locals and cosmopolitans" in organizational life—

fit within the general framework that diagnosed conflicting loyalties between the scientist and other employees of commercial organizations.[7]

Certainly by the 1950s, the conflicted and unsatisfactory position of the scientist in American industry and in governmental research facilities appeared to commentators as an enormous practical problem in the context of the Cold War: how to recruit such people in numbers sufficient to respond to the massive Soviet threat? The low birth rates of the Depression era combined with the burgeoning demand for scientists and engineers to drive the Cold War arms race, with the result that the "science-gap" and the "engineer-gap" preceded the "missile-gap" in American anxieties during the 1950s and 1960s.[8] Few publications from this period dealing with recruitment and with allied problems of retention, creativity, morale, and the efficient use of research workers omitted to justify the salience of such problems to the Soviet threat, and federal government agencies, led by the Office of Naval Research (ONR), were prime sources of funding for social science work in this area. Thomas Kuhn's paper on "The Essential Tension," which was the precursor to key ideas in his seminal 1962 *The Structure of Scientific Revolutions,* was given in 1959 to a conference on scientific creativity that included contributions from officials in the Pentagon's Advanced Research Projects Agency, the Air Force Personnel and Training Research Center, as well as the Dow Chemical Company.[9] Already by the late 1940s, the Steelman Report to the President's Scientific Research Board devoted much attention to problems of recruitment, creativity, and morale, and worried *inter alia* that scientists socialized into a preference for the freedom, autonomy, openness, and collegiality of academic settings might constitute a substantial obstacle to scientific and technological mobilization against the Soviet threat.[10]

In this Cold War context, while academic social scientists elaborated a picture of the industrial scientist made-unhappy-through-early-socialization, they had allies in university departments of industrial relations and in business schools, as well as in strands of nonacademic cultural commentary. So, for example, the Harvard Business School professor Charles Orth wrote in 1959 that corporate managers completely fail to understand "the kind of people who work in the world of science and the influence on their values of the special training which made them scientists." Socialized in academic environments, scientists all want to do basic research: they are accustomed to, and value, "an independence of thought" and a "permissive, low-pressure atmosphere": they care deeply about the good opinion of their disciplinary colleagues and less about the approval of their corporate superiors. That's just what scientists are like, whether the cause is academic socialization or the selective recruitment into a scientific career of special personality types (Orth didn't especially care): "the resultant personality matrix of the scientist is familiar to anyone who has associated at all closely with these men." Put these sorts of people in industry, and "the fact that they think and behave differently makes life difficult for practically everyone in an industrial organization who must deal with them. Businessmen find it hard to understand almost everything such men say and do." Indeed, scientists don't enter industry "because they like the idea of working for an industrial organization"; they are drawn to it against their own inclinations, because they need the facilities, resources, and "somewhat larger salaries" that industry can offer and that academia cannot. The scientist "typically does not really want to work for an organization."[11]

Here Orth acknowledged indebtedness to the most widely distributed 1950s picture of the disastrous, but necessary, engagement between scientists and industry, William H. Whyte's *The Organization Man.* The centerpiece of Whyte's book was a set of three chapters on "The Organization Scientist," which identified the pressures brought on scientists in American industry to conform to industrial values, work conditions, and structures of authority—pressures that were well on their way to eroding the nation's capacity for technological and commercial innovation at just the historical conjuncture when those capacities were most needed. For Whyte, as for Merton, the crux of the matter was a conflict of values between science and those institutions called upon to support and enlist science, and in particular a failure on the part of sustaining institutions to comprehend the unique values that alone would encourage scientific geese to lay their utilitarian golden eggs. In the relevant sections of *The Organization Man,* the target was an obtuse, and ultimately self-destructive, unwillingness on the part of industrial managers to recognize authentic scientific values and to accommodate those values in the organizational life of the industrial research laboratory. The fundamental argument, Whyte said, was that "between the managerial outlook and the scientific there is a basic conflict in goals." The current "orgy of self-congratulation over American technical progress" attributed it overwhelmingly to "the increasing collectivization of research." But this was a mistake that was storing up enormous trouble for the national future. If America effectively organized the individual researcher out of existence, it would eventually put an end to creativity, and, by extension, to America's ability to compete, commercially and militarily.[12]

We are, Whyte wrote, currently lectured by bureaucrats and industrial managers on "how the atom bomb was brought into being by the teamwork of huge corporations of scientists and technicians." Only occasionally does someone have the good sense "to mention in passing" that the creative impulse for the atomic bomb was an individual, unorganized, and spontaneous act of genius—indeed, something that "an eccentric old man with a head of white hair did back in his study forty years ago." American industrial management was working remorselessly to "mold the scientist to its own image; indeed, it saw the accomplishment of this metamorphosis as the main task in the management of research." If it succeeded, Whyte judged, it would be committing suicide, for "every study" has demonstrated that the "dominant characteristic of the outstanding scientist" is "a fierce independence" that will not tolerate control, interference, or collectivization: "In the outstanding scientist . . . we have almost the direct antithesis of the company-oriented man." American industry's aversion to the necessary eccentricity of genius was, for Whyte, signaled by a Monsanto recruiting film in which the voice-over announced "No geniuses here; just a bunch of average Americans working together."[13]

So that is story number 1, or at least a plausible enough précis of it for present purposes, since the story is still so familiar. There is, however, quite a different story—more precisely, a set of stories—about the identity, condition, and state of mind of the American scientist-in-industry during the twentieth century, and these stories are not at all well known among academic social scientists and historians. The literature from which these stories emerge is seldom cited, and, from what I can tell, scarcely read, by relevant sociologists and historians, with the result that the sensibilities represented there are rarely, if ever, confronted by academic writers. And probably the major reason why this literature is little known, and why, therefore, its lessons are rarely if ever engaged with, is that this commentary emerges not from the academy but from industry itself, or, more specifically, from those research managers and administrators whose job it was, from early in the century through the 1960s, to recruit, manage, coordinate, and motivate the work of scientists in industry.

Sources for story 2 can be found in a wide variety of places. From the early 1950s, industrial consortia and managers established journals of their own in which to trade experiences about problems encountered in the new, and fast-changing, world of commercial research: *Industrial Laboratories* in 1949, then *Research Management* in 1957. The *I[nstitute of] R[adio] E[ngineers] Transactions on Engineering Management* intermittently published practical commentary on problems of research management in issues from 1955 to the early 1960s, as did *Chemical and Engineering News, The Journal of Industrial and Engineering Chemistry, Mechanical Engineering, The Technology Review, Personnel* (the journal of the American Management Association), and similar periodicals. Conferences and publications on research management convened and sponsored by such Big Business firms as Standard Oil of New Jersey, Standard Oil of California, and the business consortium, the Industrial Research Institute, are further sources for such stories; and, from the 1920s through the 1960s, a small number of the more reflective research executives published books on the subject of organizing industrial research facilities and administering scientists. I call these sources "shop-floor" commentary, even if there must be some pertinent differences in experience and perspective between research managers and the research workers they managed.

In marked, if unsurprising, contrast to the academic commentary on research management, shop-floor writing displayed no interest whatever in making points of general sociological interest, in using passages of research management as "case studies" for any other purpose than coming to some more-or-less robust findings about recurrent problems in and about the industrial laboratory and in proffering some more-or-less plausible practical solutions to those problems. So, for example, a research director at RCA in the early 1950s rejected the pertinence of general theories of administration in terms typical of practical administrators, insisting that the requirements and problems of research management "will vary among different units of industry" and will even depend upon the "differing outlooks towards research" taken by "individual research administrators."[14] As an agricultural research administrator succinctly put it in 1926, "It is a condition and not a theory that confronts us."[15] This material is not academic in tone or appearance: there are rarely, if ever, any footnotes or literature references—for writing in this genre that is contemporary with, or subsequent to, the work of Merton and his followers, it is as if such work did not exist—and the evident purpose is not apologetic, defensive, or even celebratory, but, rather, in the spirit of trading "war stories" among congenial colleagues. The question here is always practical management, not sociological generalization: how can one make industrial scientists more productive, more creative, happier, more likely to stay with the firm? what forms of organization work best for the industrial research laboratory, or for particular types of laboratories and in different industries and in firms of certain sizes? how does one attract the most able research workers and how can one recognize the signs of ability in potential recruits? That is to say, while the shop-floor literature cannot possibly be confused with academic social science, it is focused on the same social arrangements and relations as the academic genre. It is warrantably *about* the same social world, and, therefore, one would expect that problems identified as central in the academic tradition ought to preoccupy writers of the practical literature as well. Unhappy, value-conflicted industrial scientists were, in part, what academic theory predicted, but the same unhappy scientists, and the same grounds for their unhappiness should be enormous practical problems for research managers and their corporate associates.

However, with vanishingly few exceptions—exceptions that may dissolve on further investigation—unhappy industrial scientists, made unhappy by the strength of their socialization into unique academic values, *just do not exist* in the commentary of research managers and allied executives.[16] And some of the rare reflections on the transition-process produced by practicing research workers themselves draw attention to disorientation in leaving the *more* "regimented" world of thesis-research, in which you had just one supervisor to satisfy, for an industrial laboratory in which the lines of control were not nearly so clear. True, orientation to commercial outcomes was not something that the newly minted Ph.D. was accustomed to, but those who chose industry might either embrace or accept those goals, and "in general," as one newly recruited electrical engineer said, "industrial research is well-respected in the academic world."[17] It is not that the industrial research laboratory is seen as problem-free: far from it; story 2 is about nothing *but* problems and their possible solutions—problems of recruiting, remunerating, retaining, motivating, organizing, and directing the labors of research workers. It was very widely recognized that research workers moving from university to industry may go through processes of adaptation, but such adaptation was usually seen in terms of getting people and their families settled in (schools for the kids, canasta-partners for the wives, etc.); getting research workers familiarized with organizational culture, routines, and expectations; and sometimes, indeed, getting them accustomed to a more team-orientated style of work than was typical in academia.[18] Nor is it that members of industrial research facilities were necessarily seen as "one big happy family"—though such attitudes were sometimes expressed—or that serious tensions were not recognized to exist between companies' research functions and such of their other arms as accounting and production.[19] In shop-floor commentary, the industrial research laboratory was full of tensions, just as its place in overall corporate culture continued to be problematic throughout much of the twentieth century. But to recognize such tensions and conflicts was much the same sort of thing as it was to recognize endemic tensions and conflicts between, say, firms' production and marketing divisions or, on the production floor, between supervisors and skilled workers.[20] Just as these sorts of tensions and conflicts were acknowledged and dwelt upon in internal business commentary, so the research managers whose writings make up story 2 were obsessed by the organizational problems of the industrial research laboratory. It is just that the persistent and consequential problem of socialization so precisely and persistently identified in the academic literature did not exist in shop-floor commentary.

Indeed, there are important and pervasive strands of such commentary that portrayed the quotidian realities of industrial research in ways that make academically predicated role-conflict problematic. Doing major violence to the heterogeneous nature of this literature, I draw attention here to just a few of these realities and associated values. First, consider the related questions of autonomy and planning. Autonomy of research work is always a relative matter. If the university was an institution marked by the value it placed on intellectual autonomy—and, indeed, a number of research managers vigorously insisted that it was, and that this marked a fundamental difference between academia and industry[21]—nevertheless the substantial reality was that some scientists might choose industry because they would in fact be freer to do the work they wanted. In 1905, Willis Whitney recruited William Coolidge from MIT, and, more famously, in 1909, Irving Langmuir from the Stevens Institute of Technology, to conduct fundamental research at General Electric—if not precisely free as birds, then certainly freed from heavy teaching loads and academic colleagues' lack of interest in research, with the resultant commercial bonanza from improvements in the tungsten-filament incandescent light-bulb and the incidental benefit of the 1932 Nobel Prize in chemistry for Langmuir.[22] The chemist Wallace Carothers, later the discoverer of nylon, reckoned Harvard to be the "academic paradise" for teaching, but Du Pont was able to woo him away in 1928 because, as David A. Hounshell and John Kenly Smith say, "he really did not like to do it. He preferred research." Carothers cared about the higher salary that industry offered, but worried about the potential loss of "the real freedom and independence and stability of a university position." However, weeks after his move to Du Pont, he had no regrets, writing to a friend: "Already I am so accustomed to the shackles that I scarcely notice them . . . Regarding funds, the sky is the limit . . . Even though it was somewhat of a wrench to leave Harvard . . . the new job looks just as good from this side as it did from the other."[23] Reflecting on university conditions for research in 1916, Charles Steinmetz, chief consulting engineer for GE, wrote about the "false commercialism," which figured professors' "output" as their pedagogical production rate, and which "wasted the universities' best assets, its professors": "Thus we find in our colleges men who had shown themselves capable as investigators to do scientific research work of the highest order, overloaded with educational or administrative routine, and deprived of the time for research work. Private industries rarely commit such crimes of wasting men on work inferior to that which they can do: industrial efficiency forbids it."[24]

Many commentators, in both academia and industry early in the century, reckoned that research, even of the fundamental sort, might not have a future in the American university because of its primary commitment to teaching and because of its poverty of resources to support research. Autonomy doesn't mean much—then or now—if you can't get the time or funds to do the research you want to do. Of course, industrial research managers made no bones about the fact that the scientists in their employ were free only within limits, and that those limits were ultimately defined by their company's commercial goals. Recognition of that constraint was consistent from the origins of industrial scientific research. In 1919, the physicist Frank Jewett of science-friendly AT&T remarked that "The performance of industrial laboratories must be money-making . . . For this reason they cannot assemble a staff of investigators to each of whom is given a perfectly free hand."[25] And a few years later, his colleague John J. Carty insisted that "Unless the work promises practical results it cannot and should not be continued." Corporate scientists must ask themselves, and be required by their superiors to answer, the fundamental question "Does this kind of scientific research pay?"[26] And in the 1950s, the director of research at Minneapolis-Honeywell was one of many in his position underlining the requirement that "*the results of research must pay.*"[27] Research administrators at Owens-Illinois Glass Company noted in the 1960s that there was no point in directing fundamental research if commercial benefit was not reasonably to be expected of it. Research executives were firm in their position: there was no reason for a company to support fundamental research, so to speak, for its own sake, nor for industrial research workers' enjoying the freedom to pick any kind of problem they pleased: glass companies could not be expected to support employees' fundamental research in oceanography.[28] Creative people were, of course, always likely to be tempted into "fascinating but irrelevant side alleys," and then it was the supervisor's task to get them back on organizational track, but not before checking that the byways really were devoid of commercial potential.[29] At the same time, there was no reason for research workers to imagine a conflict of agendas where none necessarily existed: as James Fisk of Bell Labs put it, "Because a man considers the needs of the organization that employs him, there is no reason to think that this makes for any lesser contribution in the scientific sense."[30] "Our fundamental belief," Fisk wrote, "is that there is no difference between good science and good science relevant to our business," and by the late 1950s Bell Labs had the Nobel Prizes to support their claim.[31]

Managers commonly acknowledged it to be in the nature of genuine research that the unexpected sometimes turned up, and that these unexpected outcomes might be commercially consequential. Research worthy of the name was always to a degree unpredictable, and expressions of that sort of sentiment were absolutely standard among industrial research managers. In 1950, Kenneth Mees and John Leermakers of the Eastman Kodak Research Laboratory insisted that "It is really not possible to foresee the results of true research work," and in 1958 a GE administrator writing about "Free Inquiry in Industrial Research" defined research as "systematic inquiry into the unknown," drawing from that bland definition the conclusion that the "detailed course of a scientific inquiry" is always subject to unpredictable vicissitudes, and, consequently, that "a certain amount of freedom on the part of the investigator" is implied by the very idea of research.[32] Even if such a thing were possible, it was very widely understood that the firm would get few benefits from research whose outcome *was* wholly predictable. Moreover, it was appreciated that some (not all) research workers were the sorts of people from whom one could not get the best without finding ways to give them their heads. In the ongoing, and highly politicized, debates over the extent to which industrial research, or research of any kind, could and should be *planned,* some research managers made a distinction between the possibility and even necessity of planning what they called the "function" of research over a number of years—that is, organizing commitments to it and its place in corporate activities—and planning the "act" or "conduct" of research, in which considerable freedom of action was deemed simply necessary.[33] David Noble's survey of early American industrial research correctly cites occasional corporate rhetoric pointing to the desirability of tightly managing the creative process, while eliding any distinction between this rhetoric and shop-floor realities and ignoring a large body of managerial rhetoric that frankly acknowledged *limits* to any such control.[34]

Research managers defended their own autonomy by persuading their bosses—to the extent they could—that the research function could *not* be held to the same system of cost-benefit accountability as other corporate activities. On the one hand, they wanted, and sometimes obtained, financial and temporal flexibility. Kenneth Mees at Eastman Kodak secured a commitment of ten years' worth of research funding from George Eastman in 1912; a consortium of oil company executives counseled against expecting returns from industrial research in less than five to seven years; and one survey of industrial research managers in the late 1960s found expected average "payback" times from R&D of about four years—with big firms having a more generous time-horizon than smaller ones—though another survey by the management consultancy firm Booz, Allen & Hamilton noted with alarm that less than a quarter of American large companies had *any* formal method for evaluating research, still less calculating a payback time.[35] In 1916, Mees wrote that "Those with the most experience of research work are all agreed that it is almost impossible to say whether a given investigation will prove remunerative or not."[36] And Charles "Boss" Kettering, when he was recruited from National Cash Register by Alfred Sloan to be director of research at General Motors, made sure that Sloan understood "You can't keep books on research, because you don't know when you are going to get anything out of it or what it is going to be worth when you get it."[37] RAND Corporation economists in the late 1950s drew importantly upon such sentiments in arguing vigorously *against* imposing rigorous cost-benefit regimes on research and development.[38]

On the other hand, allowing research workers a significant amount of company time to do their own research is definitely not the late twentieth-century invention of the Silicon Valley high-tech and biotech "knowledge economy." Mees encouraged such autonomous research in the 1910s, and a survey of industrial practices in 1950 found that an allowance of 10 to 20 percent—that is, a half-day to a day a week, with company resources to match—was then common, even in American "smoke-stack" industry.[39] The president of Dow Chemical said that he had "learned that if a research laboratory is to produce results, the men must be allowed the freedom to be a bit crazy," and the American Cyanamid research director quoting this remark approved up to 20 percent paid company time for research personnel "to work out their own ideas."[40] For very highly prized scientific employees, that free time could amount to considerably more: the terms that won William Coolidge for GE in 1905 included one-third—in other versions one-half—of his time to continue an existing personal research project.[41] When Du Pont tried to recruit the organic chemist Louis Fieser from Bryn Mawr in 1927, Fieser was struck by the research freedom on offer: "I never expected to go into industrial work but the thing which makes a decision so difficult in this case is that I don't have to sell my soul at all; they even said I could bring my quinones along and continue my present work."[42] And a 1952 survey reported that, for some individuals, in some industrial organizations, "no plans are made as to how they shall spend their time."[43] While David Hounshell is surely right in identifying this sort of research latitude as a recruiting and retention tactic for some "academic elitists" who saw industrial research "as a poorer career option than that offered by a university or a private basic research institute," he does not claim that all research workers felt that way, and there is abundant evidence of *commercial* justifications for considerable degrees of industrial research freedom.[44]

Eastman Kodak's Kenneth Mees became famous in management circles for his celebration of research *disorganization*. The epigraph to his influential 1920 book on *The Organization of Industrial Scientific Research* boldly stated that "There is danger in an organization chart—danger that it be mistaken for an organization," the source of which was not a scientist fed up with corporate bureaucracy but one of the founders of American technical management consultancy, Arthur D. Little.[45] Disorganization, and the research autonomy consequent on recognition that planning in these areas was naturally constrained, were justified as conditions of commercial functionality. Mees was quite hard-headed enough to insist that "the primary business of an industrial research laboratory is to aid the other departments of the industry," and that its central responsibility was to contribute to the corporate bottom-line,[46] but he rejected the distinction—central to the writings of the British anti-planning Society for Freedom in Science—that it was only "pure" science that was incompatible with planning. Commenting on early writings by Michael Polanyi, Mees said that "I take issue with [the] description of applied science as a field in which freedom of science might conceivably be undesirable. I have been engaged in applied science for forty years, and in that period I have come very definitely to the conclusion that the prosecution of applied science in its most efficient form is identical with that of pure science. I don't think for a moment that it is desirable that applied science should be directed *except in times of emergency.*"[47] Mees's own much-quoted aphorisms include: "When I am asked how to plan, my answer is 'Don't,'" and "No director who is any good ever really directs any research. What he does is to protect the research men from the people who want to direct them and who don't know anything about it."[48] And the only remark by Mees that finds its way into the reference books is his most eloquent condemnation of research control: "The best person to decide what research work shall be done is the man who is doing the research. The next best is the head of the department. After that you leave the field of best persons and meet increasingly worse groups. The first of these is the research director, who is probably wrong more than half the time. Then comes a committee, which is wrong most of the time. Finally, there is the committee of company vice presidents, which is wrong all the time."[49]

This sort of sensitivity to the pathological consequences of attempts to control research was, according to the president of Bell Labs in the 1940s and 1950s, something that "All successful industrial research directors know . . . and have learnt by experience": the "one thing a director of research must never do is to direct research, nor can he permit direction of research by an supervising board."[50] The same Minneapolis-Honeywell research director who insisted on the bottom-line criterion for judging research results concluded his piece in *Industrial Laboratories* by conceding that, while "the amount of freedom or control of research projects is probably the most difficult question in its administration . . . I tend toward the principle expressed by Thomas Jefferson for government: 'The least government is the best government.'"[51]

In practical terms, such sensibilities translated into some tolerance for the spontaneous and unpredictable emergence of novel research agendas among industrial research workers who enjoyed significant freedom to formulate their courses of work. In his own laboratory, Mees had to warn a scientist who developed an interest in high-vacuum pumps and gauges that this was work in no way compatible with the concerns of a photographic company, and one lesson drawn from this story was that scientists in Eastman Kodak's research laboratory could *not* do just anything they wanted. But Mees then repeatedly went on to relate that when he saw what a splendid technology was being developed, he secured the resources to spin off a distinct commercial laboratory, one that ultimately became a highly profitable vitamin-producing joint venture with General Mills. Here, as elsewhere, the lesson that Mees wanted research managers to learn was that autonomy was after all good business. There was no alternative to a high degree of autonomy, and attempts to be unremittingly hard-headed were ultimately self-defeating.[52] Intriguingly, in the very same year that Whyte's *Organization Man* appeared, with its eloquent condemnation of industry's commitment to crushing scientific genius under the dead weight of teamwork and management control, *Fortune* magazine—for which Whyte once wrote—ran a piece celebrating changed industrial sensibilities toward basic research and its demands for individualism, under a title—"geniuses now welcome"—that gave the lie-oblique to Whyte's story.[53]

A second basis for industrial scientists' supposed discontent, and consequent role-conflict, was the secrecy that was a necessary feature of commercial research. And, indeed, no research manager commenting on what is now called intellectual property thought that scientists could possibly be allowed freely to publish valuable trade secrets. Their work was understood to belong to the company, and it was for the company to decide what could or could not be published in the open scientific literature.[54] Scientists joining Du Pont, for example, were obliged to sign non-negotiable agreements that all "inventions, improvements, or useful processes" made while in the company's employ were the "sole and exclusive property" of Du Pont, and they agreed not to "disclose or divulge confidential information or trade secrets."[55] However, in practice many research managers vigorously endorsed the commercial prudence of a quite free publishing policy and argued for the barest minimum of secrecy. The aphorism "When you lock the laboratory door, you lock out more than you lock in" comes not from Robert Merton or Michael Polanyi but—in the same temporal and cultural context—from "Boss" Kettering at General Motors.[56] The free flow of technical information, or, at least, the freest flow compatible with broad corporate interests, was widely, if not universally, acknowledged in these circles as a net benefit to all parties.[57]

Research managers had to advertise their laboratories as workplaces attractive to the most talented research workers, many of whom were found in academia, and there was no better way to do this than to encourage their scientists to participate in professional society meetings and to publish in the same journals as their academic disciplinary colleagues. So in 1948 the director of research at Sylvania wrote that "The reputation of the organization whose research men do publish their findings will be enhanced according to the caliber of the work done"; an oil company research executive agreed that encouraging research workers to publish their results—subject, of course, to "adequate patent protection"—brings "substantial indirect value to a company, which gets a reputation for being willing to have its men present papers and for being progressive. It is not just a matter of humoring the man, but can usually be justified from the company viewpoint"; a vice president of Union Carbide noted that "Forward-looking companies have come to realize that they can attract better men if they are willing to permit publication at the earliest moment compatible with a proper regard for economic advantage"; and a research manager at the General Electric Research Laboratory pointed out (without qualification) the likelihood and advantages of reciprocity: "Any laboratory that must do good basic scientific research must also encourage open publication of research results because this policy insures access to the work of other laboratories." For these, and many other, reasons, a chemical company research director insisted that "a liberal company policy on publication" was simply a "must": "Only a blind or short-sighted management curbs or prohibits publication."[58]

Even at Du Pont—a company that some academic scientists reckoned to be unduly secretive—official policy was massaged by research directors who considered that a liberal publication policy was just necessary for attracting first-rate chemists and maintaining their morale. So in the late 1920s two leading Du Pont research managers recruited scientists by assuring them that "the work of these [fundamental researchers] shall be published almost without restriction," and, as Hounshell and Smith document, that open policy with respect to Du Pont's fundamental research was substantially realized.[59] In addition, publication was understood as a vehicle for interesting disciplinary communities to take up your problems, whether in their relatively pure or relatively applied forms. The more people working away in your area, the better for you. (And that is one, sometimes unappreciated, reason why Edward Teller at Livermore in the 1950s urged the freest possible dissemination of information concerning thermonuclear weapons.) Patentable findings, of course, had to be legally protected, and, while patents had the desired effect of securing property rights, they were understood also as a way of ensuring that the findings were in the public domain, thus satisfying, through a different route, those scientists who craved publication and consequent recognition from their peers. An RCA research director writing about recruitment problems noted that publication *was* important to young scientists, but so were patent and incentive awards, as each of these was a sign that the company recognized individual contributions.[60] Moreover, in fast-moving fields, the trick was not to secure advantage by locking up all possible intellectual property but by keeping one step ahead of the competition. Secrecy in such fields was of little concrete value. And a Sylvania research manager noted that whatever necessity there might be, so to speak, to embargo publication was usually only "temporary": commercial commitment to publication could and should remain as a principle and as a substantial reality.[61] Such a widely accepted principle coexisted with the practical possibility that publication might, at any moment, be prohibited or delayed for commercial reasons—and industrial research workers understood this very well. Some potential recruits to industry bridled at this state of affairs, but many others did not: Leonard Reich writes that even with significant "restrictions on communication and publication, Whitney had little trouble finding researchers willing to work at GE," with such a prestigious and well-equipped institution as MIT being a rich hunting-ground.[62]

Finally, and most interestingly from the point of view of academic theories of socialization, a number of industrial research managers were worried that their newly recruited academic scientists became *too quickly and too totally accepting of the values and research agendas of what they took to be corporate, as opposed to academic, culture.* At Eastman Kodak, Mees judged it very important that personal credit for research be given to individuals and publication be under their names. "The publication of the scientific results obtained in a research laboratory is quite essential in order to maintain the interest of the laboratory staff in the progress of pure science . . . When the men come to a laboratory from the university they are generally very interested in the progress of pure science," Mees wrote, "but they rapidly become absorbed in the special problems presented to them, and without definite direction on the part of those responsible for the direction of the laboratory there is great danger that they will not keep in touch with the work that is being done in their own and allied fields. Their interest can be stimulated by journal meetings and scientific conferences, but the greatest stimulation is afforded by the publication in the usual scientific journals of the scientific results which they themselves obtain."[63] The practical problem pointed to here was not the strong and persistent socialization into academic values predicated by sociologists but its opposite—the matter-of-fact willingness of research workers trained in universities to *abandon* such putatively distinct values. And it was that spontaneous abandonment that concerned research managers like Mees—not for moral or ideological, but for wholly practical, reasons.[64]

One could go on painting this sort of picture of American industrial research from the beginning of the twentieth century to the 1960s or so. It would be a picture that, minimally, makes the academically predicated role-conflict problematic and, if it were taken to an extreme, would argue the total illegitimacy of traditionally established contrasts between institutional values, just as it would open up the possibility of a more "evolutionary" than "revolutionary" account of widely imitated practices in contemporary American knowledge-intensive industry. But even at this superficial level of detail the evidence presented here licenses some brief speculations about the relations between strands of rhetoric and institutional realities in this area.

First, despite this evidence, I cannot see any reason to dispute *all* versions of academic theories taking as their task the identification of distinctive values attached to universities and to commercial undertakings. If the question is one that asks what values *distinguish* academia from industry, then I can see no better response than that which points to academic values clustering around disinterestedness, autonomy, spontaneity, and openness, versus commercial values centering on concrete economic outcomes, organization, planning, and the control of intellectual property. In academic institutions, it might plausibly be said, the "Mertonian" values can be publicly celebrated as institutional essence, while in industrial research they are more often asserted tactically as reminders to the uninformed that research is, to a great extent, an uncertain business, not to be subjected to the accountability regimes of other corporate activities. Yet, a theory of ideal-typical *differences* between institutional cultures is one thing and a description of quotidian realities in complex institutional environments is quite another. Those in the practical business of managing research enterprises have tended to acknowledge the intractable problems of distinguishing between these institutional environments, for theorizing essential differences has been of little concern to them. A paper company research director, given the task of speaking about such things in the mid-1960s, threw up his hands in despair: "Neither industrial nor academic research is, as it were, monochromatic; by whatever criteria we choose for judgement, both yield very broad spectra." He found intra-group differences at least as great as inter-group differences, and thought it "quite possible that there are no such unique species as industrial *vis-à-vis* academic research."[65] Just as nominalism or particularism in these matters seems the natural attitude of the practical actor, so theorizing essential differences has an affinity with the ideological work of distributing value across the institutional landscape. So among academic commentators, most especially in the social sciences, there has been a persistent tendency to elide a distinction between ideal-typifications and accounts of institutional realities. The task of celebrating and criticizing institutional kinds is thereby made so much easier.

So far as possible, apples should always be compared with apples and not with oranges. It may well be that the best response to the questions "What makes university science different from industry?" and "What are scientists like?" is that offered in story 1. For all that, certain obstinate facts remain: (1) by the middle of the twentieth century the majority of academically trained American scientists did not work in universities, and neither industrial nor government laboratories had any problems recruiting as many as they wanted, even if there was always a struggle for the best and brightest;[66] (2) American universities, certainly at the beginning of the century, were not globally regarded as natural homes for research: most were under-resourced; most had a primary commitment to teaching;[67] many experienced cultural, political, and religious pressures that seriously compromised any notion that universities, as such, were communities of free, open, and suitably re-sourced inquirers.[68] With due respect, the University of Southern Mississippi is not MIT, and, as we have seen, General Electric was well able to attract distinguished researchers away from even MIT because there was more effective freedom of action at the company.[69] Free action in research is a matter of material resources and time as well as of rhetoric and ideals. A comparison between conditions at Harvard and at the scientific laboratory of a small chemical company might sustain story 1, but many highly perti-nent comparisons between particular institutions of higher education and particular industrial research facilities definitely do not. And it is good to remember the restrictions on free research problem-choice that were, and remain, endemic in even the best universities. As a one-time member of a sociology department, for example, I was free in principle to conduct re-search in any area I liked, but I could not expect, and I did not receive, credit from my colleagues for work they could not recognize as sociology, nor was I institutionally well placed to obtain support for any research I might pro-pose to do in psychology or economics or particle physics. Moreover, as a research director for the Carrier Corporation noted in the early 1960s, "I know of some cases where competent Ph.D. candidates left the universities of their choice [for industry] because they either were not allowed to work on a topic picked by them, or they could not get as advisor the man they wanted in their own field."[70] These, and other, autonomy-restricting states of affairs are not at all uncommon in academia, nor are they inconsistent with saying that universities have an in-principle association with the ideal of research freedom.[71]

Second, some academic sociologists, and most explicitly those involved in the early *ASQ,* were in the business of elaborating what they called "a general theory of administration," and in that cause both ideal types and the most robust possible generalizations of organizational similarities and differences were central. It is in such a theoretical context that questions like "What makes the academy different from industry?" take on special salience. The shop-floor commentary, by contrast, was not about offering an opposing theory of administration, nor, even, of asking "What makes the academy similar to industry?" Such commentary took problems as they presented themselves to managers concerned to cope with their own quotidian circumstances and to learn from others' experience—if and when such experience seemed warrantably pertinent. The research managers, so to speak, played Simplicio to Galileo's Salviati: they wanted to know how to deal with *this* research laboratory; they were not much interested in specifying "the nature of science," "the nature of the university," or "the nature of industry." There is, of course, no escaping the science studies principle that "it's rhetoric all the way down"—how else would you have access to past organizational realities except through the heterogeneous commentaries of various participants and observers? But one reason why I feel somewhat more confident in the empirical grip of the shop-floor version is that there is no evident commitment to theory-building and no evident apologetic and justificatory flavor to such commentary. If the scientists were being awkward for the reasons predicated by the academic sociologists, I feel reasonably confident that the research managers would have said so.

So there are, indeed, problems with the empirical adequacy of the academic story, problems similar to those pointed out in the late 1960s and early 1970s by such British sociologists of science as Barry Barnes, Steven Box, and Stephen Cotgrove, and even by such renegade Merton students as Norman Kaplan, whose work in this area now appears outstandingly sensitive and prescient.[72] In this connection, criticism of academic theorizing culpably disengaged from empirical realities can be joined to a historical appreciation of the circumstances in which such a story emerged and secured some local credibility. On the one hand, an academic social science strongly committed to establishing its scientific *bona fides* was actively seeking theories of great predictive power, just as it was looking for typologies that allowed practitioners to sort institutions into their species and genuses. To that extent, I consider that much academic social science that dealt with "unhappy industrial scientists" was in large part (not wholly) *deducing its objects from theory.* At this distance, one has to take on trust the interview-responses recorded by academic sociologists documenting substantial "role-conflict" among industrial scientists in the 1960s, but it is not impossible to give alternative plausible interpretations to some of this evidence, and one wonders why a post–World War II survey of scientists' attitudes saw no problems of bias in posing such questions as this: *"Aside from money considerations,* where do you think a person can get most satisfaction from a career in science—in the Federal Government, in an industrial laboratory, in a university, or somewhere else?"[73] Nor is it any part of my argument that all, or even a certain high percentage, of industrial scientists were happy as clams at high tide—this would be a remarkable state of affairs for members of any organization, much less organizations with such a high degree of organizational uncertainty as industrial research laboratories. Nor is there any reason to ignore occasional—though scarce—expressions of unhappiness among industrial scientists drawing upon repertoires that circulated in the cultural neighborhood and that might lend some credibility to story 1. However, what is undeniable is the remarkable gap between the world as seen by academic sociologists in mid-century and the same bit of the world as seen by research managers.

Moreover, the cultural and political conjunctures from which this academic commentary emerged had characteristics that made the story about role-conflict especially appealing to American social scientists and humanists. If natural scientists and engineers in the Cold War period had institutional options—many could jump from the university to industry or government if they wished—the great majority of social scientists and humanists did not.[74] The university was their natural home in a much stronger sense than it was the natural scientists' home. And, as we well know, during the Cold War that home, and the values that were most cherished by the humanists, were under serious attack. Many of the social scientists and their allies did indeed defend themselves against McCarthyism and the militarization and commercialization of the academy, but, more to the point, some emerged as the most articulate defenders of *science,* whether or not they identified what *they* did *as* science. So, for example, in 1958, Merton's colleague Paul Lazarsfeld wrote that college professors as a body tend to be naturally selected from a "permissive" and "liberal" environment: "Once he is on campus, economic and social circumstances also militate against a conservative affirmation by the academic man." But for that minority who came into the academy from a different background, there might be problems adjusting: "For a young person coming from a business background, a teaching career involves a break in tradition. The business community has an understandable affinity with the conservative credo, with its belief in the value of tradition and authority, its corresponding distrust of people who critically scrutinize institutions like religion and the family, and its beliefs in the social advantages of private property and the disadvantages of state interference in economic affairs."[75]

Role-conflict was thus naturalized by some sociologists as a feature of the academic identity, whether humanist, social scientist, or natural scientist. The message was in effect "We are all in this together," and Oppenheimer's struggle in the 1954 security hearings was assimilated to their own. If you wanted to fight external control, and, of course, many humanists and social scientists during the Cold War did *not*, then damaged scientific allies were nonetheless important allies. The sociologists developed a picture of the nature of the scientific community that invited scientists to consider themselves allies of their nonscientific colleagues. That picture was *inter alia* a resource for healing emerging fissures in elite American universities. It was a cause as noble as it eventually proved to be futile.[76]

PART IV

The Body of Knowledge and the Knowledge of Body

In 1826 the French gourmand Brillat-Savarin wrote "Tell me what you eat and I will tell you what you are," and in 1863 the German materialist Ludwig Feuerbach announced that "You are what you eat." The contemporary familiarity of the saying is something of an obstacle to appreciating the historical meanings attached to that sort of sentiment. In the early twenty-first century, we understand that health and disease importantly flow from patterns of eating and drinking—through the transformation of one set of chemicals into another—but in the traditional dietetic culture that remained strong and consequential from Antiquity through at least the eighteenth century alimentary practices were an important component of a *self-making system*. Dietetics was as much moral as it was medical; what was taken to be *good for you* occupied the same cultural terrain as what was considered *good*. Take, for example, the prudential counsel of *moderation,* and consider how this prudence made *asceticism* available for the bodily performance of other-worldliness, including intellectual other-worldliness. For these reasons alone, dietetics is a compelling cultural object for historians interested both in the making of the modern self and in the interplay between bodies of academic expertise and lay knowledge.

CHAPTER 11

The Philosopher and the Chicken

On the Dietetics of Disembodied Knowledge

This work, though it deals only with eating and drinking, which are
regarded in the eyes of our supernaturalistic mock-culture as the lowest
acts, is of the greatest philosophic significance and importance . . . How
former philosophers have broken their heads over the question of the
bond between body and soul! Now we know, on scientific grounds,
what the masses know from long experience, that eating and drinking
hold together body and soul, that the searched-for bond is nutrition.

LUDWIG FEUERBACH, review of Jacob Moleschott's
Theory of Nutrition (1850)

A story is told—and much repeated—about Sir Isaac Newton when he was
living in London toward the end of his life:

> His intimate friend Dr. [William] Stukel[e]y, who had been deputy to
> Dr. [Edmond] Halley as secretary to the Royal Society, was one day shown
> into Sir Isaac's dining-room, where his dinner had been for some time served
> up. Dr. Stukel[e]y waited for a considerable time, and getting impatient, he
> removed the cover from a chicken, which he ate, replacing the bones under
> the cover. In a short time Sir Isaac entered the room, and after the usual
> compliments sat down to his dinner, but on taking off the cover, and seeing
> nothing but bones, he remarked, "How absent we philosophers are. I really
> thought that I had not dined." [1]

Here is another story, circulating among modern academic philosophers,
and it is about another, and much later, Cambridge philosopher. In 1934
Ludwig Wittgenstein came to stay with his friend Maurice Drury at a cot-
tage in rural Ireland, and, as Drury relates,

> Thinking my guests would be hungry after their long journey and night
> crossing, I had prepared a rather elaborate meal: roast chicken followed by

suet pudding and treacle. Wittgenstein rather silent during the meal. When we had finished [Wittgenstein said], "Now let it be quite clear that while we are here we are not going to live in this style. We will have a plate of porridge for breakfast, vegetables from the garden for lunch, and a boiled egg in the evening." This was then our routine for the rest of his visit.[2]

In 1945 his American former student Norman Malcolm visited Wittgenstein in his Whewell Court rooms at Cambridge. Malcolm relates that Wittgenstein prepared supper:

The *pièce de résistance* was powdered eggs. Wittgenstein asked whether I cared for them, and knowing how he valued sincerity, I told him that in truth they were dreadful. He did not like this reply. He muttered something to the effect that if they were good enough for him they were good enough for me. Later, he related this incident to [Yorick] Smythies, and (according to Smythies) Wittgenstein took my distaste for powdered eggs as a sign that I had become a snob.[3]

That was wartime. Afterward, when Wittgenstein lived in Dublin, "he would go to Bewley's Café, in Grafton Street, for his midday meal—always the same: an omelette and a cup of coffee." What especially pleased Wittgenstein was that he became so well known at the café that he did not have to utter a word to order his food: it just came: "An excellent shop: there must be very good management behind this organization."[4]

My concern here is not to do with a late-Wittgensteinian solution to the problem of chicken-egg priority. Nor is it the moral of these—and a series of strikingly similar—stories that those who love wisdom do not love chicken: there is no reason to suppose that there is some special philosophic foulness that attaches to chicken. Rather, the point made by those telling these stories is publicly to say something of consequence about the special constitution of individuals who give themselves up wholly to the pursuit of truth. These are *stipulations* about the bodies of truth-seekers. The chicken is both real and figurative—made into symbolic capital for the quality of knowledge. It is, so to speak, epistemological chicken. And what these stories stipulate is that the truth-seeker is someone who attains truth by denying the demands of the stomach and, more generally, of the body. That is one way in which it is said that the individuals in question *are* truth-lovers—that is, *philosophers*—and one way available to philosophers to be recognized as such. And, if (as is likely) there is now a distinct sense of the bizarre in discussing truth in relation to the stomach, it is that very oddness-of-association that is my topic of inquiry. Why is it that the belly is conceived to stand at the opposite pole to truth?[5]

These stories—and many like them—are unusually widely distributed and persistent over a broad sweep of Western culture. I find them fascinating and important, and I want to tell a few more of them as I go on. My fascination with these stories proceeds partly from a puzzle I sometimes encounter in conversation with academic colleagues in philosophy and in the history of ideas. They occasionally say that, in contrast with some social historians and sociologists of knowledge, their concern is with "disembodied knowledge," with knowledge *itself*, rather than with its embodied production, maintenance, and reproduction.[6] Such locutions are standard, well institutionalized in a range of academic practices, and rarely contested. Yet, to tell the truth, I have never seen a "disembodied idea," nor, I suspect, have those who say they study such things. What I and they have seen is embodied people *portraying* their disembodiment and that of the knowledge they produce or the documentary records of such portrayals. These portrayals are the topic in which I am interested here. How are they done? With what cultural materials are they accomplished? To what ends? I start with a prejudice: it is that the portrayal of our culture's most highly esteemed knowers and forms of knowledge as disembodied has been one of the major resources we have had for displaying the truth, objectivity, and potency of knowledge.[7] These stories, and the performances they describe, constitute that portrayal. They are stories about the meager and the physiologically disciplined bodies of truth-lovers.

My particular interest has been with early modern natural philosophers and the stories attached to their bodies. And I will briefly rehash some familiar stories attaching to Robert Boyle, Henry More, Isaac Newton (again), and Henry Cavendish. But the stories are, indeed, *attached*, since they were associated with other truth-lovers in other, and much earlier, settings. So I want to get to Newton et al. by way of settings from which emerge our earliest knowledge of such stories.[8] And finally I want to suggest that these stories no longer attach to present-day truth-seekers in quite the same way. The career of such stories, I speculate, tracks the development of modern conceptions of knowledge and the knower in a perspicuous way.

The Ascetic Ideal and Its Classical Tropes

We are dealing here with a *trope,* one of very great antiquity and pervasiveness, a trope that has been consequentially attached in a range of settings to those who are said to be authentic lovers of truth.[9] Possibly, the original of the trope is found in Plato. In the *Phaedrus,* Socrates tells a charming story about the origins of the race of cicadas. Once upon a time, before the Muses were called into being, cicadas were human beings. And when the Muses were created, "some of the people of those days were so thrilled with pleasure that they went on singing, and quite forgot to eat and drink until they actually died without noticing it. From them in due course sprang the race of cicadas, to which the Muses have granted the boon of needing no sustenance right from their birth, but of singing from the very first, without food or drink, until the day of their death." And when the cicadas die they report to the Muses "how they severally are paid honor among mankind, and by whom." These people—dancers, singers, historians, and the like—are the blessed of the Muses. But of these some are specially blessed: "To the eldest, Calliope [Muse of epic poetry], and to her next sister, Urania [Muse of astronomy], they tell of those who live a life of *philosophy* and so do honor to the music of those twain whose theme is the heavens and all the story of gods and men, and whose song is the noblest of them all."[10]

At the very end of his life, Socrates made clear the special affinity between the cicada's way of life and that of the philosopher. Sentenced to death, Socrates argued against those of his friends who would have him flee Athens and avoid the hemlock. In the *Phaedo,* Socrates brings Simmias round to the view that of all men the philosopher is one who, rather than fearing death, should embrace it. The argument proceeds by way of the role of the body, its desires and requirements, in the philosopher's search for truth. Socrates: "Do we believe that there is such a thing as death?" Simmias: "Most certainly." Socrates: "Is it simply the release of the soul from the body? Is death nothing more or less than this, the separate condition of the body by itself when it is released from the soul, and the separate condition by itself of the soul when released from the body? Is death anything else than this?" Simmias: "No, just that." Socrates: "Well then, my boy, see whether you agree with me . . . Do you think that it is right for a philosopher to concern himself with the so-called pleasures connected with food and drink?" Simmias: "Certainly not, Socrates." Socrates went on to establish that the philosopher is a different sort of person from the ordinary run of humanity: he "frees his soul from association with the body, so far as is possible, to a greater extent than other men." And if the philosopher's disembodiment is the condition for his hope to attain truth during mortal life, so death, which is the final freeing of the soul from the constraints of the body, is not to be shunned but welcomed: "Surely the soul can best reflect when it is free of all distractions such as hearing or sight or pain or pleasure of any kind—that is, when it ignores the body and becomes as far as possible independent, avoiding all physical contacts and associations as much as it can, in its search for reality." In "despising the body and avoiding it, and endeavoring to become independent—the philosopher's soul is ahead of all the rest . . . If we are ever to have pure knowledge of anything, we must get rid of the body and contemplate things by themselves with the soul by itself." In this way, the practice of philosophy during life was the imitation of death, since both philosophy and death act to free the soul from its bodily prison: "True philosophers make dying their profession."[11]

The ancient Greek association between the truth-lover's way of life and the denial of the body was widespread and *mutatis mutandis* persistent. Diogenes the Cynic was advertised as a philosopher who cared so little for fleshly and material rewards that, when asked by the great Alexander what thing he might desire of him, he requested only that Alexander should "stand out of my light."[12] Stoic philosophers, content with water and plain bread, able to miss their dinner without complaint or even without noticing, were celebrated for the simplicity of their diet. Epicurus, whose identification of pleasure as the goal of life was much misunderstood, was "thrilled with pleasure in the body, when I live on bread and water," and commanded a friend to "[send] me some preserved cheese, that when I like I may have a feast."[13] The Greek seeker after truth was recurrently said to eat only enough to keep life going. To eat more than a bare minimum, or to yearn after delicacies, was to compromise the philosopher's ideal self-sufficiency. The condition for truth was an austere dietetics.

Pythagoras and his followers were famous for their abstemiousness. Legend had it that they routinely performed "an exercise of temperance": "There being prepared and set before them all sorts of delicate food, they looked upon it a good while, and after that their appetites were fully provoked by the sight thereof, they commanded it to be taken off, and given to the servants." Later commentators made much of Pythagorean vegetarianism and the prohibition against eating beans. Both animal flesh and beans produced noxious effluvia that corrupted the body and rendered it impure and unfit for intellectual activity.[14] Accordingly, a frugal diet was not only a display of dedication to knowledge and an emblem of a person who cared little for its pleasures and needs, it might also be understood as a physiological condition for putting the body in a fit posture for the intellectual and spiritual quest. (As Ludwig Feuerbach much later punned, "Der Mensch ist, was er iszt.")[15] Broadly Pythagorean sentiments persisted into the later Roman Empire, Plotinus and his pupil Porphyry arguing strenuously for abstemiousness and vegetarianism for all, but especially for those intending to live a philosophical life: "Abstinence from animal food . . . is not simply recommended to all men, but to philosophers." Porphyry's tract commending vegetarianism was written against a philosophical friend who took up flesh-eating on his conversion to Christianity. You are what you eat, and those who consumed flesh fed their animal natures while they poisoned their souls.[16]

The Ascetic Ideal and Its Christian Tropes

After Jesus wandered in the desert for forty days and nights, "he hungered" (Matthew 4:1–2; Luke 4:1–2). Satan's first temptation was not power but food: "If thou art the Son of God, command that these stones become bread. But he answered and said, It is written [quoting Deuteronomy 8:3] Man shall not live by bread alone" (Matthew 4:3–4; Luke 4:3–4); "Is not the life more than the food and the body more than the raiment?" (Matthew 6:25). When the disciples wondered that the Rabbi did not eat, "he said unto them, I have meat to eat that ye know not" (John 4:30–32). For the faithful, Jesus himself was "the bread of life: he that cometh to me shall not hunger, and he that believeth on me shall never thirst" (John 6:35). Paul lectured the Corinthians: "Meats for the belly, and the belly for meats; but God shall bring to nought both it and them" (1 Corinthians 6:13).[17]

The early Christian idiom for expressing the relationship between the denial of bodily wants and the attainment of spiritual knowledge is probably more familiar than that of Greek Antiquity, and fine recent historical work has yielded new understandings of the ascetic culture produced by the Egyptian monks of the third and fourth centuries. Peter Brown, for example, has corrected dominant modern assumptions about the temptations Christian hermits and anchorites took themselves to the desert to confront and surmount. For St. Anthony, the desert was "a zone of the non-human," and, for this reason, Brown writes, "the most bitter struggle of the desert ascetic was presented not so much as a struggle with his sexuality as with his belly. It was his triumph in the struggle with hunger that released, in the popular imagination, the most majestic and the most haunting images of a new humanity . . . The titillating whispers of the 'demon of fornication,' much though they appear to fascinate modern readers, seemed trivial compared with [the obsession with food]."[18] (In the Middle Ages, the skin disease erysipelas was known as St. Anthony's blush, because, as one legend has it, the anchorite saint blushed every time he was obliged to eat.) The Desert Fathers regarded eating as a matter of both shame and spiritual danger: "The body prospers in the measure in which the soul is weakened and the soul prospers in the measure in which the body is weakened."[19] Another legend tells of a friend, concerned for the health of the hermit Abba Macarius, bringing him a bunch of grapes. Macarius was unwilling to indulge himself and sent them to another hermit, who then passed them on to still another, until at last they came back to Macarius, uneaten.[20] Here the religious life of the mind appears not just disembodied but specifically disemboweled.

The ascetics of Late Antiquity tended to conceive of the human body as an "autarkic" system. In ideal conditions, and, tellingly, before Adam's original sin—it was food, after all, that brought him down—the body was thought capable of running "on its own heat." It needed just enough food to maintain that heat. It was only "the twisted will of fallen men" that gorged the body with surplus food, and it was this dietary surfeit that produced the excess energy manifested in "physical appetite, in anger, and in the sexual urge." The passions, including that of sexuality, were in part epiphenomena of dietetics: food before sex. Brown writes that "in reducing the intake to which he had become accustomed, the ascetic slowly remade his body. He turned it into an exactly calibrated instrument. Its drastic physical changes, after years of ascetic discipline, registered with satisfying precision the essential, preliminary stages of the long return of the human person, body and soul together, to an original, natural and uncorrupted state." In Genesis (1:29) the Lord said that "I have given you every herb . . . and every tree . . . and to you it shall be for meat." From the early Christian era well into the eighteenth century and beyond it was debated whether Adam and Eve were vegetarians and whether they ate only raw foods; whether this was the natural diet of prelapsarian humans; whether the Fall from Grace altered the human constitution so that we now required flesh and cooked foods; and, importantly, whether fallen humans might restore their pure state, and their pristine and powerful intellectual capacities, by a pure and primitive diet.[21]

In the early Christian era, St. Augustine was perhaps the most influential voice advertising the disciplined body as the condition for spirituality. The Jews feared certain foods, while to the Christian all foods were equally clean or unclean: "It is not the impurity of food I fear but that of uncontrolled desire." God taught Augustine "to take food in the way I take medicines. But while I pass from the discomfort of need to the tranquillity of satisfaction, the very transition contains for me an insidious trap of uncontrolled desire." Moreover, the variety of fleshly pleasures offered by the variety of foods was a snare. Routine consumption of *the same* foods—for Augustine as for Wittgenstein—was a way of ensuring against the "tumult of the flesh" and "bringing the body into captivity."[22] That was the human condition: to be human was not only to err but to eat, and, in eating, people inevitably fed those animal wants that had the potential to corrupt the soul.[23] In this way, the Eucharist Host and Communion wine expressed not only particularly Christian worship but also the general human predicament, until such time as bread was replaced by the Bread of Heaven. After the Resurrection, there would be no need to eat in order to prevent decay.[24]

By contrast, Jewish traditions of asceticism, and ascetic warrants for knowledge, were relatively poorly developed. Immediately after the Old Testament's most eloquent commendation of decorum—"To everything there is a season"—the aged Solomon wrote that there is nothing better for men "than to rejoice, and to do good so long as they live. And also that every man should eat and drink, and enjoy good in all his labour, is the gift of God . . . Go thy way, eat thy bread with joy, and drink thy wine with a merry heart; for God hath already accepted thy works" (Ecclesiastes 3:1, 12–13; 9:7–8).[25] In the twelfth century, the Spanish-Jewish physician Moses Maimonides worried about the effects, both on pious Gentiles and on his own coreligionists, of the heroic asceticism of Christian "saintly ones." In fact, Maimonides said, such abstinence was best understood as a periodic means of "restoring the health of their souls" and as a contingent reaction against "the immorality of the towns-people." The mistake of the ignorant was to think that extreme abstinence was virtuous in itself, "that by this means man would approach nearer to God, as if He hated the human body, and desired its destruction. It never dawned on them, however, that these actions were bad and resulted in moral imperfection of the soul." Aristotelian moderation was identified as the dietetics of both spiritual and civic well-being.[26]

Nor is it the case that pagan and early Christian ethical and medical authorities issued blanket recommendations of severe abstemiousness. From the pre-Socratics through the Hippocratic and Galenic corpus, and the writings of such Stoic philosophers as Epictetus and Seneca, health was seen to flow from observing *moderation*—in exercise, in study, and in diet. Both gluttony and excessive fasting were explicitly identified as recipes for moral and physiological disaster. Let the body serve the rational mind, not the mind the body; in eating, let your aim be to "quench the desires of Nature, not to fill your belly"; "allow thy belly what thou shouldst, not what thou mayest"; "eat to live, not live to eat."[27] Such advice, as well as the physiological schema that justified it, proved remarkably stable over a great span of European history. Tweaked, tuned, and idiosyncratically interpreted by individual writers, balance, stability, and moderation remained the dominant dietetic counsel from Antiquity to the modern period.[28] The lay wisdom of an early modern proverb had it that "he that is ashamed to eat is ashamed to live." Yet the prudent "middle way" to which free civic actors were enjoined created at the same time a way of understanding, and celebrating, the special dietetic self-denial of truth-seekers.[29] The philosopher *was not as other men:* his discipline of the belly was recognized in the culture both as a condition of spirituality and as a badge by which authentic truth-lovers might be identified. A "lean and hungry look," like a specially ascetic way of life, might visibly mark not only the politically risky person— "He thinks too much: such men are dangerous"—but also the exceptionally virtuous and wise man.[30]

Heroic abstinence constituted a potential problem as well as a resource for the developing institutions of Christianity. By Late Antiquity a Church that had assumed substantial responsibilities of civic management was in a different position with respect to gestures of otherworldly disengagement from the one that had once stood on the political periphery. While the solitary ascetic continued symbolically to represent piety in its purest and highest form, such examples could not be effectively offered as a pattern for the ordinary conduct of the whole body of the faithful. The clerical hierarchy increasingly worried about the uncontrollability of individual gestures of heroic asceticism and about the potentially subversive alternative claims to religious authority that such gestures might represent. Orthodoxy was now in a position where its canons formally celebrated heroic asceticism while its institutions reserved the right to counsel a temperate course and to monitor the authenticity and interpretation of individual ascetic gestures. When the bishop lived in a mansion and kept a sumptuous table, personal acts of heroic asceticism might plausibly be treated as subversive criticism. The temperate and highly ordered dietetics of monasticism was one way of managing the problem: the sixth-century monastic Rule of St. Benedict, for example, provided for victuals (excluding "the flesh of quadrupeds" but including a ration of wine) whose nature and quantity were prudently adapted to the local climate as well as to individual brothers' work routines, constitutions, and momentary states of health.[31] Another was the careful surveillance of the heroically abstinent: was this fasting figure genuine or a fraud? was it quite clear that the faster was not motivated by pride? that such abstinence did not testify to an unreasonable and unwholesome *attention* to the demands of the body? that he or she was not diabolically rather than divinely inspired?[32]

When St. Francis of Assisi was ill with a fever, his friends urged him to take a little solid nourishment, only to have his eventual backsliding made into a further spectacular public display of self-abasement. Stripping himself naked, and putting a cord round his neck, he commanded a colleague to lead him into the piazza, where he addressed the people: "You believe me to be a holy man, and so do others who, on my example, leave the world and join the Order and way of life of the brothers. But I confess to God and to you that in this sickness of mine I ate meat and broth cooked with meat . . . Here is the glutton who has grown fat on the meat of chickens."[33] It was just this kind of gesture that might be interpreted as proceeding more from pride than piety. In the fourteenth century St. Catherine of Siena progressed from a diet of bread, water, and raw vegetables (occasionally supplemented by pus from the suppurating ulcers of a cancer victim's breast) to an announcement that she took nourishment only from the Host. Her friends reminded her that Jesus told his disciples to "eat such things as are set before you" (Luke 10:8), and skeptics suspected that she was in fact sustained by Satan. Carefully watched, Catherine nevertheless satisfied her monitors that she could retain no food in her stomach and that "her body heat consumed no energy."[34] In the seventeenth century those set to watch over St. Veronica's fasting observed her periodically to gorge, but this was explained as the work of the devil. Pressure was successfully brought to bear to get her to submit to the regular dietetics of her order, of which she ultimately became abbess.[35] So the dietetic moderation to which the civic actor was pervasively enjoined was also, albeit typically on a more ascetic scale, counseled by the Church to its clerics and to the community of believers. The cultures of both civic and sacred institutions possessed ways of understanding, sometimes approving, and sometimes worrying about, the special moral state and the special epistemic claims of the heroically abstinent.

As the examples of Catherine, Veronica, and many other female saints make plain, the gesture of heroic abstinence was at least as available for holy women as it was for holy men. Caroline Bynum has beautifully described differences (as well as similarities) in medieval male and female gestures of holiness, noting the special significance for women of food and its renunciation. Food was pervasively "a powerful symbol" and was therefore central to interpreting the human condition and its eschatological future, especially in endemic conditions of scarcity: "But food was not merely a powerful symbol. It was a particularly obvious and accessible symbol to women, who were more intimately involved than men in the preparation and distribution of food."[36] Women's bodies were, indeed, the source of life and of food, and their acts of giving birth and nursing could be recruited as powerful analogies of Christ's body. Yet male medical writing from Antiquity through the Middle Ages (and beyond) tended to conceive of the female body as colder and wetter than the male body, more liable to corruption, more *organic*. "Although all body," Bynum says, "was feared as teeming, labile, and friable, female body was especially so." Yet, she notes, "women could triumph over organic process." This meant that dominant understandings of women's bodies could count as an obstacle to female gestures of spirituality (and female entitlements to spiritual knowledge), while at the same time they gave grounds for regarding the gestures of the heroically abstinent woman as specially powerful.[37]

Temperance and Its Early Modern Meanings

From Antiquity to the early modern period, medical texts consistently counseled the prudent person to adopt a dietetics of moderation. Yet strands of sixteenth- and seventeenth-century culture contested the meaning of the temperate life, debating how it was that ancient philosophers had lived and how the modern wise person ought to live. The dominant notes in texts written for a genteel readership remained the prudential commendation of a temperate and moderate course of life and an associated condemnation of fashionable excess. The English humanist Sir Thomas Elyot closely followed

Galen in listing the qualities of different foods and their effects on persons of varying temperaments. Gross meat made gross bodily juices, and, while the roast beef of Olde Englande offered suitable victuals for laborers and for others of coarse constitution, "it maketh grosse bloude, and ingendereth melancoly." (Hare too was proverbially said to be "melancholy meat" but capon was recommended for those whose complexion was that way inclined.) Simply prepared things were best; the simultaneous consumption of a variety of meats was to be avoided; gluttony and drunkenness were worst of all. Abstinence, however, might itself be dangerous, and its practice too must be observed in moderation. After all, both Plato and Galen (St. Paul was not mentioned here) recommended using "a little wine for thy stomach's sake." Excess in abstinence might be conducive to melancholy.[38]

The mid-sixteenth-century homespun advice manual *La Vita Sobria* by the centenarian Venetian gentleman Luigi Cornaro became the most widely circulated early modern tract celebrating dietary temperance. Like Elyot, Cornaro denounced the routine gluttony of modern patrician society. He ate in quantity "only what is enough to sustain my life": bread, broth (with perhaps an egg), no fruit, all sorts of fowl, veal (but no beef), some fish—all flesh being taken in moderation. While the temperate life was "pleasing to God," its justification here took a largely secular form: this is the way one ought to live if one desired health and a robust old age. The lives of ancient philosophers—Plato, Isocrates, Cicero, and Galen—were recruited as patterns of dietary restraint, but nothing about this version of temperance made it unfit for those "in service of the State" or for the ordinary civic actor: "I am nothing but a man and not a saint."[39]

Montaigne's late sixteenth-century skepticism was targeted at dietary as well as at philosophical *systems:* "My way of life is the same in sickness as in health; the same bed, the same hours, the same food serve me, and the same drink. I make no adjustments at all, save for moderating the amount according to my strength and appetite. Health for me is maintaining my accustomed state without disturbance. It is for habit to give form to our life, just as it pleases."[40] "There is no way of life so stupid and feeble as that which is conducted by rules and discipline," and one who attempted to eat and drink by the book was no less liable to go wrong than one who sought to regulate belief and action by the book. The "most unsuitable quality for a gentleman" is "bondage" to system. One should not decline to follow local dietary custom because it conflicted with systemic medical principles:

"Let such men stick to their kitchens." In dietary matters, one should conform to rules tested by experience but not be "enslaved" by them. There was indeed a vice of "daintiness," of taking "particular care in what you eat and drink," but that vice might be equally manifest in vigilant temperance or in gormandizing fastidiousness. Over a lifetime, one's sense of pleasure adapted one's stomach to its usual fare, and radical change was always likely to do more harm than good. And even if long life was promised to those who would radically amend their dietary habits, "Is it so great a thing to be alive?"[41]

Francis Bacon's posthumously published *History of Life and Death* (1636) worked subtle but consequential changes both on the dietetic culture handed down from Antiquity and on Cornaro's program of systematic temperance. He agreed that the Pythagoreans and the Church Fathers were unusually long-lived and that their abstemious dietetics was substantially responsible for that longevity. And he claimed that "light contemplations" had similarly beneficial effects in prolonging life: "For they detain the spirits on pleasing subjects, and do not permit them to become tumultuous, unquiet, and morose. And hence all contemplators of nature, who had so many and such great wonders to admire, as Democritus, Plato, Parmenides, and Apollonius, were long-lived." By contrast, what he called "subtle, acute, and eager inquisition shortens life; for it fatigues and preys upon the spirits."[42] Yet Bacon dissented from the ascetic tradition that causally associated dietary abstemiousness with intellectual good: "It is certain also that the brain is as it were under the protection of the stomach, and therefore the things which comfort and fortify the stomach by consent assist the brain, and may be transferred to this place."[43] Most important, Bacon adapted traditional injunctions toward dietary moderation, generally preserving the *form* commending the dietary Golden Mean while altering its content and prescriptive meaning. "Frequent fasting," he announced, "was bad for longevity"; and experience showed that great gluttons "are often found the most long-lived": "Where extremes are prejudicial, the mean is the best; but where extremes are beneficial, the mean is mostly worthless." Diets that were too spare were to count as extreme, with all the effects on body and mind that flowed from excess. The gentlemanly actor in society was placed in a position where occasional surfeit was a routine and civically prescriptive fact of life, and accommodating oneself to such circumstances was both politically and dietetically prudent: "With regard to the quantity of meat and drink, it occurs to me that a little excess is sometimes good for the irrigation of the body; whence immoderate feasting and deep potations are not to be entirely forbidden."[44]

So early modern culture worked with, and ingeniously reworked, dietetic traditions ultimately inherited from pagan and early Christian cultures. Sixteenth- and seventeenth-century advice was firmly linked to its ancient sources by the recommendation of prudent moderation and temperance for those wishing to live a healthy, happy, and productive life in society, even as the meaning of what it was to observe a temperate dietetics was modified according to differing conceptions of *how* and *where* the good life was to be lived and according to differing conceptions of *who* the philosopher and the prudent person were. When humanist writers urged a relocation of the ideal life of the mind from cloistered to civic settings, dietary advice was part of that attempted cultural transformation. If study and philosophizing were to be legitimate activities within a civic setting, contributing to civic concerns, then the dietetics of the legitimately learned should be substantially similar to that of the prudent civic actor. At the same time, this attempted respecification continued to offer ways of understanding, and even appreciating, the austere dietetics of the otherworldly thinker.

By the Renaissance and early modern period, Greek and Roman theories of the humors, temperaments, and complexions had been developed into important reflective understandings of what scholars and philosophers "naturally" were like and, in turn, what effects the life of truth-seeking wrought upon their bodies. In the late fifteenth century, the neo-Platonist Marsilio Ficino wrote influentially about the melancholy to which learned people were especially prone, by virtue of their natural constitutions (disposing them toward the philosophical life) and by virtue of the effect their habits had upon humoral balance.[45] In the early seventeenth century, *The Anatomy of Melancholy* (1628) by the Jacobean scholar Robert Burton codified and distributed a picture of scholarly melancholy: you could identify those who unremittingly pursued truth by their bodily "temper," their countenance, their situation and way of life. The philosophical body was different from the civic citizen's body. Dedication to truth was physically inscribed upon it. Bodily form and mode of life were visible as ways of recognizing a philosopher, and these were also ways by which those meaning to present themselves as philosophers might effectively do so. These, then, are the cultural traditions against which stories about early modern philosophers should be understood. The stories that attached to seventeenth- and eighteenth-century natural philosophers emerged from traditions attaching them to spiritual intellectuals, with much the same meaning for portraying the status and value of knowledge.

The Dietetics of Early Modern Philosophy

Aristotle wondered why men of genius tended toward melancholy, and Seneca asked why God afflicted the wisest men with ill-health.[46] These questions continued to circulate in the seventeenth century and beyond. The natural philosopher Walter Charleton announced that the "finest wits" are rarely committed to "the custody of gross and robust bodies; but for the most part [are lodged] in delicate and tender Constitutions."[47] Dead White Males, that is, were generally Sick White Males. And in seventeenth-century English natural philosophy Robert Boyle was widely recognized as such a one. Poised between the role of the gentleman and that of the Christian scholar, Boyle (and his friends) reflected upon the state and meaning of his special bodily constitution and way of life. Few contemporary commentators on Boyle omitted to mention, and to draw out the cultural significance of, Boyle's disengagement, abstemiousness, and physical delicacy. For some it represented melancholy, the badge of "a hard student," while others contested his identity as a melancholic on the grounds of its incompatibility with gentlemanly civic obligations. John Evelyn saw Boyle's fragility as a form of refinement and even as a kind of strength: his body was "so delicate that I have frequently compared him to a chrystal, or Venice glass; which, though wrought never so thin and fine, being carefully set up, would outlast the hardier metals of daily use."[48] The funeral sermon preached over Boyle's corpse by his friend Gilbert Burnet carefully rejected the charge of scholarly melancholy. Boyle was too much the gentleman for that: "To a depth of knowledge, which often makes men morose . . . Boyle added the softness of humanity and an obliging civility." At the same time, Boyle exercised rigorous stoic bodily control, neglecting all display of "pomp in clothes, lodging and equipage." And, tellingly, over a course of more than thirty years "he neither ate nor drank, to gratify the varieties of appetite, but merely to support nature."[49] Arguably, everyone listening to that sermon in St. Martin-in-the-Fields recognized the trope. The meaning of this stipulation proceeded from its resonance with a culturally pervasive sensibility causally associating types of bodies and qualities of minds.

Boyle's contemporary, the Cambridge philosopher Henry More, devoted much attention to the care of his own body and to the dietetic regulation of other philosophers' bodies. More's Platonism here expressed itself in the view that all individuals have within themselves a "Divine Body, or Celestial Matter," the state of which depended upon the management of dietetics and passions.[50] The care of the philosopher's special mind involved the special care of his special body: "There is a sanctity even of Body and Complexion. which the sensually-minded do not so much as dream of."[51] More's early eighteenth-century biographer announced that

> he was of a singular Constitution both for Soul and Body: His very Temperature was such as fitted him for the greatest Apprehensions and Performances; especially when by his Temperance, and most earnest Devotion he had refin'd and purified it. A rich "Æthereal sort of body for what was inward" (to use here his own Pythagorick phrase) he had even in this Life; that is to say, a mighty Purity and Plenty of the Animal Spirits, which he still kept up lucid and defaecate by that Conduct and Piety with which he govern'd himself.[52]

Constitutionally endowed with this purity of animal spirits and a warm complexion, More's way of life distilled his natural inclination to the search for truth. He "had always a great care to preserve his Body as a well-strung instrument to his Soul." He said that his body "seem'd built for a Hundred Years, if he did not over-debilitate it with his Studies." A disciplined body was made to serve the philosophical will, for philosophizing was heroic work. By his abstemiousness he had "reduc'd himself . . . to almost Skin and Bones; and was to the last but of a thin and spare Constitution." At the end of his life, More said that there were two things he repented: the first was that, although he had the means to afford it, "he had not lived [at Cambridge] as a Fellow-Commoner," and the second was that he had "drunk Wine."[53]

More's remarkable correspondence with Lady Anne Conway is well known to historians as a rich source for English Cartesianism and theology, but it is also almost uniquely informative about the dietetics of early modern philosophical bodies. Incessantly, More and Conway (as well as her husband and her brother John Finch) exchanged recipes for the diet that would best adjust and manage the bodily heat necessary to high philosophical inquiries, while preventing that heat from flaming over into pathological enthusiasm and "phrensy." This was a task requiring the most painstaking management of the quantity and quality of food and drink, as well as attention to the fine adjustment of consumption in relation to momentary bodily state and the precise nature of intellectual labor. Early modern culture understood thought, emotion, and diet as elements of a reciprocally interacting casual system: just as diet could influence mood and cognition, so the forms and content of intellectual activity could affect humoral balance and dietetic requirement.[54] "Too much small beere and fruit" damped the body's heat; wine and roasted meats stoked its fire. To know Anne Conway was to know that her complexion was warm and to know the risks and capacities that attached to such a complexion. More counseled Conway "to eat such kinde of meat as begets the finest and coolest blood, and to abstain from all gross food, which many times is the most savoury, but breeds melancholy blood," while her brother warned her against *overdoing* a cooling diet: "Take heed of overcooling your selfe for your temper being naturally hott to take perpetuall cool thinges is to cure not your disease but to disturb your temper."[55]

And here again considerations of gender have epistemic pertinence, as pervasive understandings of the female complexion (colder than the male) could provide a general basis for explaining women's absence in philosophic enterprises while the same humoral scheme allowed a heightened appreciation of Anne Conway's special individual constitution.[56] In warmth of complexion and its bearing on the capacity for and nature of philosophical speculation, More and Conway recognized each other (despite male/female difference) as similarly endowed, facing similar predicaments. The advice More gave to his warm woman friend was advice he took for himself. The dietetic counsel conveyed in their letters and (presumably) in their face-to-face conversations *tuned* each other's philosophic thermostats, while Conway worried that her humor might "prove infectious."[57]

Probably the richest seventeenth- and eighteenth-century sources for portrayals of the disembodied philosopher attach to the person of Isaac Newton, and stories of his disengagement and otherworldliness echo into the twentieth century, painting some of our culture's most vivid pictures of the special body whose mind is wholly given over to truth. The legendary and the portable status of these stories is an index of their topicality. George Cheyne's *Natural Method of Cureing Diseases* (1742) noted that, in order to "quicken his faculties and fix his attention," Newton "confined himself to a small quantity of bread."[58] Other contemporaries observed both Newton's abstemiousness and forgetfulness of food—as in the chicken story retold at the outset. He "gave up tobacco" because "he would not be dominated by habits." In London, his niece remarked that Newton "would let his dinners stand two hours": "his gruel, or milk and eggs, that was carried to him warm for supper, he would often eat cold for breakfast."[59] In Cambridge, an amanuensis related that he often went into Newton's rooms and found his food untouched: "Of which, when I have reminded him, he would reply, 'Have I?'" "His cat grew very fat on the food he left standing on his tray." Still another (contested) report testified that Newton, like the Pythagoreans, abstained from meat.[60]

In the 1930s L. T. More collected these and other stories—"which the world has so often heard"—and gave them new life. More wrote about "Sir Isaac's forgetfulness of his food when intent upon his studies"; "He took no exercise, indulged in no amusements, kept no regular hours and was indifferent to his food." And More drew out the significance of Newton's abstemiousness, abstractedness, and solitude for his identity and that of the knowledge he produced: "It is little wonder that his contemporaries have passed on to us the impression that he was not a mortal man, but rather an embodiment of thought, unhampered by human frailties, unmoved by human ambition . . . Passion had been omitted from his nature."[61] More recently, Richard S. Westfall's depiction of "a solitary scholar" identified Newton's disembodiment as that "of a man possessed" by love of truth, wholly other, not responsive to his body's needs, not *there* in his own body.[62]

Later in the eighteenth century, Henry Cavendish became a popular attachment for similar stories, and these stories too were prominently told and retold by his biographers, to similar ends. They were struck by the frugality and disengagement of Cavendish's way of life, and this despite his enormous wealth: "A Fellow of the Royal Society reports, 'that if any one dined with Cavendish he invariably gave them a leg of mutton, and nothing else.'" He was so shy of human contact that in his own house he ordered his spare meals by leaving a note for the housekeeper upon a table. George Wilson's mid-nineteenth-century portrayal of Cavendish's body was as telling for the bodily features omitted as for those to which it drew attention: "he did not love; he did not hate; he did not hope; he did not fear . . . [A]n intellectual head thinking, a pair of wonderfully acute eyes observing, and a pair of very skilful hands experimenting or recording."[63]

The Dietetics of Modern Philosophy

I started by juxtaposing Newton and Wittgenstein, so suggesting—intentionally—that the trope portraying truth-lovers' ascetic bodies persisted into late modernity. And so it did. In 1925 the fictional medical scientist Max Gottlieb (maximum God-love) explained to Martin Arrowsmith (in Sinclair Lewis's novel) just why the scientist was not as other men:

> To be a scientist [says Dr. Gottlieb]—it is not just a different job, so that a man should choose between being a scientist and being a . . . bond-salesman . . . It makes its victim all different from the good normal man. The normal man, he does not care much what he does except that he should eat and sleep and make love. But the scientist is intensely religious—he is so religious that he will not accept quarter-truths, because they are an insult to his faith.[64]

In 1949 the iconic scientific intellectual of the twentieth century specified the *constitutional* difference between those who lived for truth and those who lived for the belly:

> When I was a fairly precocious young man [Albert Einstein wrote] I became thoroughly impressed with the futility of the hopes and strivings that chase most men restlessly through life. Moreover, I soon discovered the cruelty of that chase, which in those years was much more carefully covered up by hypocrisy and glittering words than is the case today. By the mere existence of his stomach everyone was condemned to participate in that chase. The stomach might well be satisfied by such participation, but not man insofar as he is a thinking and feeling being.[65]

And the stories of bodily abstraction attached to Wittgenstein as one of this century's most celebrated philosophers focus importantly upon his "otherness" in terms instantly recognizable from the classical and early Christian traditions. So some twentieth-century thinkers—like the Desert Fathers and seventeenth-century natural philosophers—*could be* depicted (to use Franz Kafka's striking phrase) as "hunger artists."[66] Some, but not all, or, I think, even very many.

Some caveats against misunderstanding my interpretation of these stories about disembodied truth-lovers. First, it is obviously *not* the case that these depictions attach uniquely to scientists and philosophers. Insofar as the tropes specify the constitution of truth-lovers, they attach to those who secure whatever body of knowledge is represented in the relevant local culture as a repository of truth and value, whether it be religious, scientific, philosophical, or artistic. It ought therefore to be understood that by focusing upon the bodies of early modern and modern scientific truth-lovers I mean to draw attention to the ways in which pervasive tropes locally attached to specific, highly valued forms of culture.

Second, it is also evidently not the case that the trope of disembodiment is without what might be called a "countertrope." Whenever and wherever the trope of disembodiment works to specify proper knowledge an opportunity is created for its purposeful denial or modification. So in Antiquity some sects of philosophers (for example, the Cynics) played with a carnal presentation, and did so as a way of marking out their philosophical practice from that of the dominant tribes of philosophers.[67] And, as I will shortly note, late modern philosophical voices importantly analyze, interpret, and reject the very idea of disembodiment as the condition for making and recognizing truth. Here I want to say that the presentation of disembodiment just has the character of a cultural *institution,* against which critical voices stake their position, not that disembodiment is the only way of presenting and warranting truth. The sociable, merry, and moderately gormandizing philosopher of the eighteenth century—a perspicuous instance of which is "le bon David" Hume—makes a statement about the nature and placement of philosophic knowledge whose meaning is understood against the background of a dominant ascetic ideal.[68]

Third, I want to acknowledge both the possibility and, within limits, the legitimacy of a "realist" psychological and sociological way of talking about the disengagement and ascetic discipline of intellectuals. It might, for example, be plausibly said that abstraction, solitude, and self-denial simply *are the conditions* for innovation or for producing knowledge of a certain character. Truth-lovers are "just like that"—by temperament, or are made so by their way of life. And in this connection I am well aware of recent psychological and psychiatric causal inquiry into creativity, innovation, genius, and mental health. Such realist claims may be legitimate within their own causal idiom, though their legitimacy within that idiom cannot count as a sufficient explanation of why these stories circulate and persist, and one has to be careful not to take at face value the historical anecdotes that provide some of their evidence. For example, manuscript evidence indicates that, at the very same time stories about Newton's abstemiousness circulated so widely, deliveries to his London household for a single week showed "one goose, two turkeys, two rabbits, and one chicken." A contemporary observed that Newton had grown so fat in later life that "when he road in his coach, one arm would be out of the coach on one side and the other on the other."[69]

And the officially ascetic monks of the abbey of St. Riquier in the twelfth century are known to have received yearly from their tenants seventy-five thousand eggs, ten thousand capons, and ten thousand chickens.[70]

However the case may turn out about "real" philosophical bodies and their "real" dietetics, historical engagement with the stories that speak about them, about their meanings and uses, and about the conditions of their circulation, has its own legitimacy and interest. Such stories are culturally significant public presentations and stipulations. They testify at once to the constitution of knowledgeable bodies and to the status of bodies of knowledge; they represent norms for philosophical knowledge and the philosophical knower. And stories about the normative way of life for the truth-lover could, and did, stably coexist with massive evidence that the ideal might not (always or usually) be realized. That is just the nature of norms in relation to the behavior they both describe and prescribe.

Finally, I want speculatively to explore the possibility that, despite gestures at Einsteinian and Wittgensteinian portrayals, the topic of disembodiment has rapidly been losing its sense and force in late modern culture. While the trope of the absent-minded professor continues with some currency, the very idea that the truth-lover is "not as other people" and, particularly, that he (and now, importantly, she) secures knowledge through denying bodily and material wants seems to many naive or quaintly outmoded. On the one hand, much modern sociology of science was founded on the claim that no special temperament or motives distinguished the scientist from the ordinary run of humanity, while, on the other hand, some of the most popular "realistic" portrayals of the modern scientist (for example, James Watson's *The Double Helix*) secure their public credibility as realistic through free confession of scientists' concern for fame, power, money, and sex. The very idea that Dr. Grant Swinger, Professor Morris Zapp, or, indeed, the author of this essay would ever pass up a pot of money, or a nice chicken, in the quest for truth is currently risible.[71]

Increasingly, I suggest, heroically self-denying bodies and specially virtuous persons are being replaced as guarantees of truth in our culture, and in their stead we now have notions of "expertise" and of the "rigorous policing" exerted on members by the institutions in which expertise lives. Expertise and vigilance, and the warrants for truth these offer, are, of course, no new things in late modernity: the ancients too had the ability to recognize expertise. But they—and, I think, intellectuals through the nineteenth century—had other conceptions of knowledge apart from expertise: conceptions of virtuous and sacred knowledge attached to special persons inhabiting special bodies.[72] So, in an eggshell, the suggestion is that the career of the ascetic ideal in knowledge follows the same career as the notion of sacred knowledge and its warrants. Late modern culture appears to be conducting a great experiment to see whether we can order our affairs without a sacred conception of knowledge, and, thus, without a notion that those who produce and maintain truth are any differently constituted, or live any differently, than anyone else. That is the sense in which it might be thought that all knowledge has the character of expertise: experts don't know *differently;* they just know *more.* W. B. Yeats said that "the passions, when we know that they cannot find fulfilment, become vision."[73] Expertise is not vision.

By the 1880s, strands in philosophy itself took a decisive turn against the ascetic ideal, notably in the work of Friedrich Nietzsche and his followers. In *The Gay Science* Nietzsche meant to acquire "a subtler eye for all philosophizing to date." The philosophical tradition against which he revolted inscribed disembodiment at its pathological core:

> Every ethic with a negative definition of happiness, every metaphysics and physics that knows some finale, some final state of some sort, every predominantly aesthetic or religious craving for some Apart, Beyond, Outside, Above, permits the question whether it was not sickness that inspired the philosopher . . . Whether, taking a large view, philosophy has not been an interpretation of the body, and a *misunderstanding of the body.*

The thrust of Nietzsche's criticism of the ascetic ideal was that the pathologies of philosophy "may always be considered first of all as the symptoms of certain bodies."[74] Nietzsche's tactics were well judged: *if* one means to subvert existing conceptions of transcendental philosophical knowledge one *should* proceed by way of an attack on the ascetic ideal. What Nietzsche could not know, and what his intellectual heirs still cannot clearly visualize, is the shape of a society that has dispensed with those conceptions of knowledge and the knower that lie at the heart of the ascetic ideal.

CHAPTER 12

How to Eat Like a Gentleman

Dietetics and Ethics in Early Modern England

A well-behaved stomach is a great part of liberty.

MONTAIGNE (quoting Seneca)

Two Sorts of Books

Consider two genres of books that were common in early modern England. One genre was the popular medical text. This sort of book was produced by medical practitioners and, quite often, by non-medical men who for various reasons reckoned they had something worth saying on the subject. The general purpose of such works was to extend medical knowledge and to recommend courses of action to preserve health, cure disease, or prolong life. You could act on many, if not all, occasions as your own physician, and this genre told you how to do it, or at least reminded you of the value of what you might be presumed already to know. The stress in these books was not so much on diagnostics and therapeutics as it was on dietetics—regimen and hygiene in their broadest aspects, and not just on what to eat and drink. This reflected the contemporary center of gravity in medical culture, and it also picked out a domain of action in which the maintenance of health was very much in your own hands, importantly taking for granted readers' economic ability to exercise choice about their diet. There were lots of these kinds of books about, and even though we have little reliable knowledge of their circulation, ownership, and uses, we can be fairly sure that the average educated person was familiar with some of them.[1]

Another genre of popular books was comprised of practical ethical tracts, including so-called courtesy books.[2] These were written by gentlemen great and small (or by "gentlemen's gentlemen"—tutors, governors, and companions) for other gentlemen who might appreciate a reminder of what the social game was all about, for those who wanted to be recognized as gentlemen, for those who desired that standing for their children and wished to breed them accordingly, for those who had cultural goods to sell to gentlemanly society, or for those who, for a variety of reasons, wished to know how gentlemen *did* behave or how it was thought they ideally *should* behave. English courtesy books instructed readers about the authentic bases of gentility, and although they generally acknowledged the fact that birth and wealth counted for much, they overwhelmingly stressed (alone or in various combinations) the role of virtue, education, piety, and easy good manners as the proper entitlements to gentlemanly status. Humanism and Puritanism each had their proprietary views of what the English gentleman ought to be, and what was wrong with what he then was. These books told you how to live a virtuous life; how to behave in a polite, prudent, and civil manner; how properly to raise your sons; how to pass muster and sometimes, if necessary in a period of mask and mobility, just how to pass. There were lots of these books around too, and they were often inventoried in the emerging gentleman's "library" of the seventeenth-century English country- or townhouse.[3] Samuel Pepys—a tailor's son, but a well-connected one, and very much on the rise, always curious about how people behaved in circles above his—was an avid consumer of such books.[4] And John Aubrey's practical thoughts on the education of gentlemen's sons recommended the reading of the better courtesy books for instilling in the young what he called "mundane prudence."[5]

Manners and medicine do not seem, on the surface of things, to have much to do with each other. Books telling you how to behave like a gentleman might be presumed very different sorts of things than books telling you how to preserve health and live long. And, indeed, from all sorts of pertinent points of view, the two genres *are* distinct: in frankly medical texts you are not going to find rules for when to "take the wall" and when to bare your head, and a courtesy book or essay in practical deportment is unlikely to contain instructions about whether boiled or roasted meats are most suitable for your atrabilious temperament. Yet there is an overlap in substance between the two genres, and it is a telling one: that which was considered dietetically *good for you* was also accounted *morally good*. The relationship between the medical and the moral was not merely metaphorical; it was constitutive. In doing what was good for you, you were doing what was good: materially constituting yourself as a virtuous and prudent person, giving symbolic public displays of how virtuous and prudent persons behaved, encouraging such behavior in others, fulfilling the noblest aspect of your nature as a human being. The medical and the moral occupied the same terrain, figuratively in the case of cultural modes, literally in the quotidian management of the body and its transactions with the world.

This cohabitation, and the consequent substantive overlap between medical culture and moral commentary, has been noted many times before, both for early modern England and more generally for premodern medicine. In the 1930s, Ludwig Edelstein summarized the fundamental dictum of ancient dietetics: "He who would stay healthy must . . . know how to live rightly."[6] Owsei Temkin described the assumption of Galenic medicine that "A healthy life is a moral obligation . . . Health . . . becomes a responsibility and disease a matter for possible moral reflection." He beautifully depicted what it meant in Antiquity to say that the philosophic life of virtue depended upon the medical care of the self, and he recounted Galen's view that "The disposition of the soul is corrupted by unwholesome habits in food and drink, and in exercise, in what we see and hear, and in all the arts."[7] Temkin observed: "The coupling of right diet and virtue was an essential part of Galenic philosophy. Proper regimen balanced the temperament of the body and its parts, and with them the psychic functions. Correct and incorrect diet could determine health and disease, and because it was under human control, the choice of diet gave a moral dimension to health and sickness."[8] And on this subject Michel Foucault's account of the moral nature of ancient dietetics is little more than an expansion of insights secured by such scholars as Edelstein, Temkin, Sigerist, and a pioneering generation of students of ancient medicine.[9] More recently, Keith Thomas has briefly but perceptively written about the constitutive relationship between medicine and morality in early modern English popular medical texts, noting that "their advice coincided closely with the conventional morality of the day. Indeed, the precepts they offered were as much ethical as medical."[10] That link has also been thoroughly treated by historians dealing with the popular medical literature of later periods and in non-English settings.[11]

My aims here are modest. I want to add more depth and detail to our current understandings of the early modern English connections between dietetics and morality, but, more important, I want to do this by approaching that relationship from a different direction than has been customary. Supposing that you were a moral philosopher, or a historian of practical ethics, what picture of medicine would you get if you looked at historical sources central to your discipline? What does dietetic advice look like when one encounters it not in explicitly medical tracts but in the literature of practical ethics written by and for gentlemen, instructing them how to live a virtuous, prudent, and effective life? What are the social-historical circumstances that mold the dietetic counsel one finds there? What broad agreements and what contests were there about the shape and content of this advice during the early modern period? What does the resulting picture show about the relationship between the layperson and the professional, between common sense and expertise? And, finally, what can this different angle of engagement suggest about changing relations between health and virtue in the early modern and in the late modern constitutions?

Nothing to Excess: Dietary Moderation and Gentlemanly Health

No English practical ethical text of which I am aware omitted passages of medical advice, and dietetics made up by far the biggest portion of that medical advice. The dietetic counsel one finds in this literature is remarkably stable over time and setting; it reeks of prudence and robust common sense; it is skeptical of extremes, innovations, one-size-fits-all courses of action, and claims to external special expertise; but its very banality and cultural robustness is what makes it so deeply interesting.

In 1531, Thomas Elyot's *The Governor* commended temperance in all things and sobriety in diet: surfeit was bad for you, engendering "painful diseases and sicknesses."[12] Thomas Gainsford's *Rich Cabinet* passed on the proverbial form: "Temperance in diet and exercise, will make a man say; a figge, for *Gallen & Paracelsus.*" That is to say, a temperate diet—like the apple—keeps the doctor away.[13] When King James VI of Scotland (later James I of England) wrote to instruct his infant son and heir Henry how to live like a prince, he too warned against "using excesse of meate and drinke; and chiefelie [to] beware of drunkennesse."[14] James Cleland's *The Instruction of a Young Noble-man* (1612) followed the king's counsel closely: "it is the preservation of health not to be filled with meate; & when a man eateth more meat then his stomacke is able to digest he becommeth sicke."[15] In the 1630s, Henry Peacham's *Complete Gentleman*—also much influenced by royal views—said the same: the gentleman was to "be moderate" in regard of his health, "which is impaired by nothing more than excess in eating and drinking (let me also add tobacco-taking). Many dishes breed many diseases, dulleth the mind and understanding, and not only shorten but take away life."[16] Gilbert Burnet's *Thoughts on Education* even pointed to proto-eugenic reasons for shunning "all wasting intemperance, and excesse": "Since the minds of children are molded into the temper of that case and body wherein they are thrust, and the healthfulness and strength of their bodies is suitable to the source and fountain from whence they spring, it clearly appears that persons wasted by drunkenness or venery must procreate unhealthful, crazy, and often mean-spirited children."[17]

Even strongly "Puritan" courtesy texts criticized contemporary dietary excess for practical medical reasons. Restoration gallants made debauchery their profession, but they would surely pay for their pleasure: "The *Table* is the *Altar* where they *sacrifice* their *Healths* to their *Appetittes;* and *Temperance* to *Luxury.*" Gentlemanly infatuation with exotic and expensive foods was proverbial: "*What's farre fetch't and deare bought is meat for Gentlemen.*"[18] Jean Gailhard's *Compleat Gentleman* (1678) cautioned parents to accustom their children early to "sobriety and temperance in their diet." They should be bred to approach the table "not so much to please their palate, as to nourish their body . . . for exuberancy of food causes surfeits, which do endanger their life . . . Plain food is more nourishing, and less hurtful, than that which is accounted more exquisite; because the palate is pleased with it, though it be otherwise with the stomach."[19] And when John Locke, writing more as a household governor than as a philosopher, or even as an Oxford physician, composed his 1693 tract on the education of gentlemen's sons, he too advocated a "plain and simple diet." Little meat, much bread, few spices, small beer only: drunkenness, gluttony, and gormandizing to be avoided at all costs. The English ate far too much meat, and to this intemperate habit Locke imputed "a great Part of our Diseases in *England.*"[20] The Third Earl of Shaftesbury, whose early education was entrusted to Locke, later commended temperance and a moderate diet in his *Inquiry Concerning Virtue.*[21] And so said virtually all the English courtesy and practical ethical writers from the mid-sixteenth to the early eighteenth centuries.

English moralists' commendation of dietary moderation had a national bite to it. Continental as well as some local voices of temperance reckoned that the English were tucking into far too much beef and swilling down far too much ale, and that this was bad for their health. English writers of practical ethical texts, one might think, had a vested interest in criticizing contemporary dietary excess, and in judging that things were now much worse than they had been in the English past. Yet these dietary jeremiads were often plausibly specific in their condemnations of *new* habits of gentlemanly intemperance, and present-day historians broadly agree with them about the facts of the matter. Anna Bryson, for example, describes the English court dinner "as a competitive exercise in conspicuous consumption"; Lawrence Stone reckons that the massive consumption of flesh—reaching mountainous proportions at the court of James I—was a possible cause of a gentlemanly plague of bladder and kidney stone; and Roy Porter and George Rousseau document eighteenth-century appreciations of a causal link between an epidemic of fashionable gout and increased English intake of meat and strong drink.[22] It was common—in England but also on the Continent—to blame new fashions in excess on the influence of foreigners. Peacham said that the English once had a national reputation for sobriety, but that the imitation of Dutch and German drinking-habits had ruined all that. Englishmen were now, Peacham judged, unhealthier, weaker, and even shorter than the victors of Agincourt, and this was due to overeating and overdrinking.[23] Seventeenth-century ethical writers identified this as an age of "luxury": not just dietary excess, but "delicacy," exoticism, variety, and complexity were often assigned to the influence of effete, debauched, and Papist France and Italy.[24] And when these authors pointed to the early Stuart and Restoration court as the center of such unwholesome practices, so the criticism of dietary sophistication and excess became an element in one of the major *political* conflicts of the seventeenth century: Court versus Country ideologies.

The English defenders of Good Olde Roast Beef, and lots of it, were not, of course, bereft of a response, insisting that such fare was physiologically appropriate to their damp and chilly climate; that it bred stout, hot-blooded heroes; that such straightforward and lavishly portioned victuals were suited to honest English natures; and, at a level less often surfacing in print, simply approved dietary abundance as a mark of gentlemanly hospitality, generosity, and gusto.[25] "Gluttony was honourable," Roy Porter writes of eighteenth-century English gentlemanly society: "handsome eating was a token of success, and hospitality admired. Englishmen tucked in and took pride in their boards and bellies."[26] That great eighteenth-century gourmand, Samuel Johnson, showed just what the dietary moralists were up against when he famously announced that "He who does not mind his belly will hardly mind any thing else." Dr. Johnson at table was not a pretty sight: "while in the act of eating, the veins of his forehead swelled, and generally a strong perspiration was visible. To those whose sensations were delicate, this could not but be disgusting."[27]

Preaching and practice in ethical matters commonly diverge. So the first Stuart king—whose court was in fact renowned for its culinary extravagance—formally commended a plain and simple diet. Nor was there anything very novel about the terms in which the English ethical writers criticized intemperance: dietary surfeit, as well as excessive variety and elaboration, had been widely identified as unhealthy and unwholesome since Antiquity, and ancient medics and moralists pervasively accounted *theirs* an age of unhealthy excess.[28] As a general rule, the Golden Age of moderation tends always to lie in the past. Moreover, while English condemnations of dietary excess were shaped by local conditions, texts much read by the English that were written by foreigners and that primarily addressed foreign settings added their support. Almost no book-writer in any early modern cultural context said that frequent gorging or boozing was good for your body. Erasmus's Christian Prince was to flee from "excessive drinking and eating."[29] Castiglione blandly said that it was "well known" that the ideal "Courtier ought not to profess to be a great eater or drinker."[30] And the conferees of Stefano Guazzo's *Civil Conversation* "all agree[d] in blaming and condempning of them, who never cease to fill their bellies up to the throate, and whose love and lyfe consisteth in spending their time in eating and drinkeing, and in riotous and excessive gluttonie."[31] Even the cynical duc de La Rochefoucauld pointed to the prudence of dietary moderation: "We would like to eat more but fear we shall be sick."[32]

When educated Englishmen read the ancients—whether in Latin, Greek, or English translation—they saw the continuity of the counsel of moderation. If they conceded ancient authority—and, despite the so-called moderns of the Scientific Revolution, they almost all did—then they saw in that great continuity further warrants for dietetic wisdom. Educated Englishmen could, and did, get their counsel of dietary moderation from ancient ethical and political tracts as well as from the medical writings of Hippocrates, Galen, Celsus, Pliny, and Orabasius. They could find temperance recommended in the moral writings of Aristotle, Plato, Cicero, Plutarch, and Seneca, and, indeed, many early modern ethical tracts were little more than palimpsests of such ancient sources.[33] Absolutely everywhere that bookish counsel was offered to early modern gentlemen, the Road to Wellville was signed by the Golden Mean.

The Dietetics of Virtue: Moderation and Mastery

Three sorts of Good Things happened if you observed the dietary Golden Mean. First, you preserved your health and obtained all the desiderata that depended upon health; second, as you followed the path of moderation, so you displayed your wisdom and virtue and acquired a valuable public reputation; third, you created the material conditions in your own body for enhanced virtue and wisdom. The first Good Thing is possibly still familiar to early twenty-first century readers even if its grandmotherly common sense sets it against the particularity and physiological detail of current voices of medical expertise and the rampant food-faddism of the culture that often pretends to owe its authority to medical expertise.

But the other two Good Things are rather less familiar to late moderns and need explication. Dietary moderation was a display of wisdom and prudence for several reasons. If you observed moderation, you also *showed* that you cared for your health. For a private person this was personal prudence—even when glossed by a religious idiom that made the human body God's temple—but for a public person, other matters were at stake.[34] Gentlemanly, and especially courtly, eating and drinking were overwhelmingly *public* acts, and they were public acts saturated with meaning. You tended to be *observed* as you ate and drank—at court, in the household, or in public eating and drinking places; communal eating and drinking constituted social order, displayed social order, and sent finely tuned social messages back and forth among the diners and drinkers. The "pledging of healths" followed strict rules of precedence and might carry messages of desired changes in precedence. The offering of choice hunks of meat, the manner and order in which these were offered, and the conditions under which one was obliged to accept, or allowed to decline, offered morsels, were acts rich in hierarchical significance.[35] And, as we now understand through the work of Norbert Elias and his followers, the "civilizing process" of bodily control that is supposed to have done so much to configure the modern social agent was particularly visible on those occasions when gentlemen and aristocrats met to eat and drink together.[36] James I underlined for his son the political importance of resisting any temptation to eat privately: "Therefore, as Kinges use oft to eate publicklie, it is meet and honorable that ye also do so, as wel to eschew the opinion that yee love not to haunt companie, which is one of the markes of a Tyrant, as likewaies, that your delighte to eate privatlye, be not thought to be for private satisfying of your gluttonie, which ye would be ashamed should be publicklie seene."[37] A good behaviour at Table," Gailhard wrote, "is a strong proof of a good Education." If you want to be treated like a gentleman, don't eat like a pig.[38]

King James's *Basilicon Doron* showed acute sensitivity to the obligation toward temperance that bore specially upon a prince. As people observed the king at table, so they inferred his true nature from his visible conduct. Accordingly, James advised his son to let his table-behavior make a powerful display of self-control: "One of the publikest indifferent actiones of a King, & that manyest (especiallie strangers) wil narrowly take heed to, is, his manner of refection at his Table and his behaviour thereat." Keep your dishes simple; eat of them with restraint; and never allow yourself to succumb to gluttony or drunkenness, "whiche is a beastlie vice, namelie in a King."[39] (And if King James himself had not gone to The Globe to see Shakespeare's *Henry IV,* then many of those who read *Basilicon Doron* had vividly in mind Prince Hal's youthful revels with the gluttonous John Falstaff, and his cool rejection of his fat friend on assuming the throne: "Make more thy grace and less thy body hence.") These were all practical matters: on the one hand, a prince (or, indeed, any other public person) who showed that he cared little for his health also showed that he cared little for those people and public enterprises that depended upon him—his capacity for reliable, rational, and effective action. That is why the French physician Laurent Joubert said that princes had a special obligation to their health and to the dietary moderation that would secure health. First, the prince "must serve as a true model for his subjects and deputies and must be of a great perfection, more divine than human in sobriety [and] countenance." Whatever the prince does, his subjects will emulate. Second, the prince has much to do and he may be called to decisive action at any moment. For practical reasons he cannot allow himself to be ill or incapacitated by surfeit of food or drink. Third, the prince must execute policy over extended periods of time. A long and healthy life is the material condition for effective policy, for securing succession, and for ensuring the safety of the state.[40]

Such injunctions toward dietary moderation could be, and often were, conveyed in a secular medical idiom: eat moderately and you will live long and healthily. Or they might be cast in ripely rhetorical economic terms: Lord Burghley counseled his son toward a "plentifull hospitality, but one kept well within the measure of thine owne estate . . . for I never heard nor yet knew, any man growne poore by keeping an orderly Table," and Josiah Dare condemned those "Epicures and Belly Gods [who] gulch down their Estates by gulps, till in the end they come to be glad of a dry Crust . . . The Purses of such Prodigals may be said to be poor by their great goings on, while their Bellies may be said to be rich by their great comings in."[41] But the practical ethical literature more often spoke of temperance in frankly moral and religious language. And here it was said that the overwhelming fault of dietary excess was that it gave proof that the appetitive and bestial had gained sway over the rational, spiritual, and, therefore, uniquely human part of your nature. That sensibility was utterly stable from the ancient Greek moralists to the seventeenth-century English ethical writers. Pagan or Christian, High or Low Church—it made little difference. Human beings were Great Amphibians—hybrid creations, partly animal and partly divine— and it was understood across a broad sweep of European culture that the self was a field of contest between rational will and appetitive desire. Accordingly, he who succumbed to excess—in food, drink, venery, or emotion—displayed a failure of rational control. The glutton or drunkard was more beast than human being. Indeed, he was *worse* than the beasts, "for they doe never exceed the measure prescribed by nature, but man will not be measured by the rule of his owne reason."[42] And the same hierarchy of control also followed the contours of social rank and order: the gentleman showed his entitlements through control of the appetites, while the absence of authentic gentility was displayed by the absence of restraint.[43] Dietary excess, a French ethical writer noted, "is the vice of brutish men," and Lord Burghley said that he had "never heard any commendations ascribed to a drunkard more then the well bearing of his drinke, which is a commendation fitter for a brewer's horse or a drayman, then for either a Gentleman or Servingman."[44] Authentic "generosity"—that is, in early modern usage, the virtuous essence of gentility—"teacheth men to be temperate in feeding, sober in drinking."[45]

Almost all practical ethical texts said the same sort of thing. *The Courtier* made the secular observation that temperance "brings under the sway of reason that which is perverse in our passions."[46] Elyot followed Plotinus's commendation of temperance as that which "keep[s] desire under the yoke of reason" and which permits us "to covet nothing which may be repented."[47] The Puritanical *Gentile Sinner* gave the same advice in a religious idiom: "The *Gentleman* is too much a *man* to be *without* all passion, but he is not so much a *beast* as to be governed by it." Temperance gives him "*Empire* over *himselfe,* where he gives *Law* to his *Affections,* and *limits* the extravagances of *Appetite,* and the insatiable *cravings* of *sensuality.*"[48] Sir Walter Ralegh's advice to his son quantified the measure of drink along a scale leading from well-being and virtue to disease and vice: "the first draught serveth for health, the second for pleasure, the third for shame, the fourth for madness."[49] Richard Lingard's avuncular counsel to a new graduate commended dietary moderation because it "discover[s] you to be your *own Master;* for he is a miserable Slave that is under the Tyranny of his Passions: and that Fountain teeming pair, *Lust* and *Rage* must especially be subdued."[50] William de Britaine's prudential guide to how to get on in Restoration society traced the causal link between political control and the display of self-control: "He who commands himself, commands the World too; and the more Authority you have over others, the more command you must have over your self."[51] And Locke's practical educational tract said "that the Principle of all Virtue and Excellency lies in a Power of denying our selves the Satisfaction of our own Desires, where Reason does not authorize them." Virtue could be acquired by practice, so dietary temperance should be practiced early.[52]

The practical ethical literature was therefore almost unanimous in its commendation of dietary moderation as a *mark* of virtue. More fundamentally, however, temperance, or deliberate moderation, was considered a virtue in itself: so said the ancients, and so said early modern ethical writers. Castiglione observed that "many other virtues are born of temperance, for when a mind is attuned to this harmony, then through the reason it easily receives true fortitude, that makes it intrepid and safe from every danger, and almost puts it above human passions."[53] Elyot wrote that the other virtues followed temperance, "as a sad and discreet matron and reverent governess," preventing excess in all other ways of being.[54] The royal herald Lodowick Bryskett noted that "Temperance [is] the rule and measure of Vertue, upon which dependeth mans felicitie," and cited Platonic authority for the view that temperance was "the guardian or safe keeper of all human vertues."[55] Thomas Gainsford wrote that "Temperance is the protectrix of all other vertues," and Richard Brathwait agreed that "no vertue can subsist without Moderation," the foundation and root of all other virtues.[56] King James gave royal warrant to the ancient hierarchy that made temperance the "Queene of all the reste" of the virtues: without self-command one could not realize any virtuous end. If virtue consisted in the Golden Mean, then temperance was literally the master-virtue.[57]

The six Galenic "non-naturals" were those behaviors presumed to be under volitional control whose rational management constituted the practice of traditional medical dietetics. The usual list of non-naturals current in the early modern period included one's exposure to ambient air (the sort of place you decided to put your dwelling or spend your time), diet (in the strict sense of meat and drink), sleeping and waking, exercise and rest, retentions and evacuations (including sexual release), and the passions of the mind.[58] There is no more concrete sign of the common terrain occupied by practical ethics and practical dietetics in early modern England than the fact that several ethical texts explicitly structured their counsel through a list of the non-naturals, while others did so implicitly or diffusely. Rationally managed moderation of the non-naturals just was virtuous and prudent, and no special indication was, or needed to be, given that consideration had here moved from moral onto medical terrain. That is because no such cultural shift had in fact occurred.

So, for example, Peacham's practical observations on contemporary English mores listed those things upon which both our health and our ability to do civic good principally depended, "which are air, eating, drinking, sleep and waking, moving and exercise, and passions of the mind: that we may live to serve God, to do our king and country service, to be a comfort to our friends and helpful to our children and others that depend on us, let us follow sobriety and temperance, and have, as Tully saith, a diligent care of our health, which we shall be sure to do if we will observe and keep that one short, but true, rule of Hippocrates: 'All things moderately and in measure.'"[59] And Locke's influential educational text started out with a series of counsels that rigorously followed the traditional list of non-naturals.[60] Note especially that the control of the passions—or, as we would say, of the emotions—counts clearly as a key item in ethical discourse. Moralists have always counseled the control of anger and of avarice, and so they did in early modern England. Yet the place of the passions as an item in the list of Galenic non-naturals also establishes their place at the very center of medical dietetics, regimen, and hygiene.[61] Looked at from the point of view of explicitly ethical writers, temperance in the non-naturals seemed the soundest moral advice, while explicitly medical writers were similarly struck by the coincidence between what was good for you and what was good. In 1724, George Cheyne's *Essay of Health and Long Life* announced that "The infinitely wise *Author* of *Nature* has so contrived *Things*, that the most remarkable *Rules* of preserving *Life* and *Health* are *moral Duties* commanded us, so true it is, that *Godliness has the Promises of this Life, as well as that to come.*"[62]

Temperance is a virtue; following the dictates of temperance leads you to the Golden Mean in the observation of all other virtues; and, finally, dietary temperance constitutes the moral-physical conditions for virtuous thinking and acting. Virtue is literally circular. The circle is closed by widely shared notions—again, continuous with Antiquity—about how diet causally influenced the operation of the mind. Francis Bacon felt that "It is certain . . . that the brain is as it were under the protection of the stomach, and therefore the things which comfort and fortify the stomach by consent assist the brain, and may be transferred to this place."[63] Other writers were more concerned to stress how dietary excess had the potential to corrupt the mind. *Sine Cerere et Baccho friget Venus,* was the old adage, common in both medical and civic circles.[64] Excess in food and drink, especially of gross food and strong drink, fed both sexual desire and anger. That is why practical ethical writers could say that "Gluttonie and Drunkennes [are] the mother of al vices." The passions could not be effectively controlled, nor could the mind reason clearly, when the fires of desire and rage were stoked by dietary excess. "What operation can a minde make," Cleland asked, "when it is darkened with the thicke vapours of the braine? Who can thinke that a faire Lute filled ful with earth is able to make a sweet Harmonie? . . . No more is the minde able to exercise anie good function, when the stomacke is stuffed with victuals. How ought Noble men then, whose mindes are ordained to shine before others in al vertuous and laudable actions, stop the abuse of abhominable *Epicurisme?*"[65] Dietary excess was bad for your body, but the gross blood and vapors bred by excess also "dulleth the mind and understanding."[66] By contrast, Peter Charron said of temperance: "Neither is it serviceable to the bodie onely, but to the minde too, which thereby is kept pure, capable of wisdome and good counsell."[67] But when your mind was clouded by dietary excess, then your rational ability to control excess was compromised. The virtuous mind/body circle induced by temperance then became truly vicious. By the 1730s, the fashionable physician George Cheyne, building on Hippocratic and Galenic dietetic ideas, had developed an elaborate and systematic theory of the dietary causation of melancholy—"the English malady"—a major cause of which was excess and to which a sovereign remedy was a severe and formulaic "lowering diet."[68]

What the Mean Meant: Specifying Dietary Moderation

Like all the Aristotelian virtues, dietary moderation was poised between two vicious extremes. The practical ethical literature, however, overwhelmingly concentrated on the vices of excess, commending temperance in opposition to gluttony, drunkenness, delicacy, and overelaboration. The authors of such tracts were appearing in the person of the moralist, and the audiences they had in view were those gentlemanly and aristocratic strata who had the resources to indulge themselves and who, in moralists' opinion, were in fact now indulging themselves on a spectacular scale. Nevertheless, there was also a minor theme in early modern ethical writing that picked out the vices attending the ascetic extreme.

Dietary asceticism was well known in early modern society; those who endorsed and embraced it spoke from some of that society's most authoritative platforms; and its cultural significance was widely understood within gentlemanly circles. Even so, dietary asceticism was very rarely advocated by ethical writers addressing themselves to civic actors, and, indeed, the practical dangers and social inconveniences of asceticism were sometimes spelled out. Asceticism was seen as strongly linked to the character of spiritual intellectuals and to patterns of disengagement and private contemplation that shaped their lives. Such asceticism and disengagement might be accorded high cultural value in civic society, but moralists generally warned gentlemen to avoid these practices. They were just not suited to the lives led by civic actors. They offended against gentlemanly obligations to generosity; they counted as a disagreeable display of self-indulgence and prissiness; they blocked the quotidian rhythms of gentlemanly social interaction; and, they might constitute both practical and social risks to the adaptability and mobility central to gentlemanly life.

If practical ethical writers would not have their readers be gluttons and drunkards, neither did they approve asceticism.[69] Robert Burton's *Anatomy of Melancholy* excoriated excess—ancient and modern—at length, but more briefly noted the mischief wrought on their bodies and minds by those going to the other extreme: "too ceremonious and strict diet, being over precise, Cockney-like, curious in their observation of meats . . . just so many ounces at dinner . . . a diet-drink in the morning, cock-broth . . . To sounder bodies this is too nice and most absurd." This was uncivil and unsound, but there were other dangers: monks and anchorites were well known to have driven themselves mad "through immoderate fasting."[70] Henry Peacham was one of several practical ethical writers who warned against going too far in the *avoidance* of dietary excess: "Neither desire I you should be so abstemious as not to remember a friend with a hearty draught, since wine was created to make the heart merry, for 'what is the life of man if it want wine?' Moderately taken, it preserveth health, comforteth and dispereth the natural heat over all the whole body, allays choleric humors, expelling the same with the sweat, etc., tempereth melancholy, and, as one saith, hath in itself a drawing virtue to procure friendship."[71]

To eat and to drink like a gentleman was, then, to eat and drink both temperately and reasonably. So said virtually all the practical ethical writers of the period. The commendation of routine intemperance by any author pretending to prescribe a virtuous and prudent life is almost inconceivable. The Golden Mean was so thoroughly institutionalized in both ethical and medical canons that its denial would count as a violation of good sense and decency. The sensibility of the carnivalesque did indeed reject temperance, as did exercises of cultural subversion or inversion, but these rejections serve to underline just how central temperance was to early modern ethical and medical traditions, to orthodoxy, and to common sense.[72] Moderation was therefore a great cultural prize, and because it was such a prize, there were contests for giving its counsel specific content and for defining what moderation meant.

That content and meaning could and did vary in early modern England, yet it is noteworthy how variation, and even conflict, occurred while holding stable much or even all of the prescriptive form that counseled dietary moderation and that identified moderation as both morally good and medically good for you. The views of Francis Bacon are particularly pertinent in this connection. Bacon wrote a lot about medicine. Like several other "modern" natural philosophers of the Scientific Revolution, he considered that the medical profession was in a sorry state, and that physicians' relative inability to prevent disease, to cure disease, and to extend human life was largely owing to deficiencies in physiological knowledge. "Medicine," he judged, "is a science which hath been . . . more professed than laboured, and yet more laboured than advanced."[73]

Bacon was specially unimpressed with the state of medical dietetics. Traditional advocacy of the dietary Golden Mean had become, to a degree, trite and unreflective, and it had never been informed by an adequate stock of valid empirical knowledge. Adherence to the Mean was still to count as prudence, but one must properly understand where the Mean was located and what were the nature and consequences of extremes. So Bacon—both in his essay "Of Regiment of Health" and in his much longer tract on "Life and Death"—appropriated Celsus as authority for a respecification of dietary moderation. In the essay, Bacon wrote that "*Celsus* could never have spoken it as a *Physician,* had he not been a Wise Man withall; when he giveth it, for one of the great precepts of Health and Lasting; That a Man doe vary, and enterchange Contraries; But with an Inclination to the more benigne Extreme: Use Fasting and full Eating, but rather full Eating; Watching and Sleep, but rather Sleep, Sitting, and Exercise, but rather Exercise; and the like."[74] And in his philosophical work on longevity and health, Bacon similarly said that "Where extremes are prejudicial, the mean is the best; but where extremes are beneficial, the mean is mostly worthless . . . We should not neglect the advice of Celsus, a wise as well as a learned physician, who advises variety and change of diet, but with an inclination to the liberal side; namely that a man should at one time accustom himself to watching, at another to sleep, but oftener to sleep; sometimes fast and sometimes feast, but oftener feast; sometimes strenuously exert, sometimes relax the faculties of his mind, but oftener the latter."[75] Bacon was here disputing that dietary extremes—just like that—*were* vicious and damaging. Often, if not always, moving toward one extreme might be in itself more beneficial than moving toward the other—eating a lot was better than fasting, sleeping a lot was better than staying awake—and, when this was manifestly the case, then the point of prudence was shifted toward—though not reaching—the beneficial extreme. Bacon practiced what he preached: his seventeenth-century biographer noted that Bacon followed "rather a plentiful and liberal diet, as his stomach would bear it, than a restrained."[76] And his nineteenth-century editors observed that "He could make nothing of a great dinner. He said, 'if he were to sup for a wager he would dine with a Lord Mayor.'"[77] Bacon not only wrote but ate like a Lord Chancellor.

It was, Bacon recognized, fashionable to associate dietary abstinence with longevity, and perhaps there was indeed a causal relationship of this sort: "It seems to be approved by experience that a spare and almost Pythagorean diet, such as is prescribed by the stricter order of monastic life, or the institutions of hermits . . . produces longevity." But there was no absolute certainty in the matter—later in the same tract Bacon wrote that "Frequent fasting is bad for longevity"—nor was it evident that longevity was the only relevant consideration. Bacon cited evidence that, among those "as live freely and in the common way"—that is to say, civic actors—"the greatest gluttons, and those most devoted to good living, are often found the most long-lived." Nor was he aware of any solid evidence that rigorous observance of the so-called "middle diet" contributed to longevity, even though it *might* be a prudent way to health. If you really want to live according to the exact rule of moderation, then you are going to have to do it with very great care, more care than the public actor may wish to, or may be able to, devote to such things. More care, indeed, than it might be worth.[78]

Celsus did indeed point out the medical benefits of a varied diet and way of life, but Bacon went his ancient authority one better. The Mean might be defined in terms of its momentary location between dietary extremes. In this case, the counsel of moderation would be: drink two glasses of wine a day; never drink ten a day; and there's no good in having any wine-free days. Alternatively, the Mean might be redefined by what we would now call a statistical distribution over time of daily behaviors. And in this case, the voice of temperance would say: you may drink ten glasses of wine in a day, but don't make a habit of it. Bacon, it appears, meant to respecify dietary moderation along the latter lines. He concluded his discussion on how aliment operated on the body by offering advice at odds with dominant medical and moral counsel: "With regard to the quantity of meat and drink, it occurs to me that a little excess is sometimes good for the irrigation of the body; whence immoderate feasting and deep potations are not to be entirely forbidden."[79]

Here Bacon was articulating, and giving philosophical and gentlemanly cachet to, a dietary sensibility that evidently ran deep in lay culture, however much it was disapproved by physicians, priests, and most moralizing authors. The vomiting and purging induced by dietary excess was considered to cleanse the system, to get rid of accumulated crud and noxious substances, and to give the body a healthy catharsis. As was common in early modern gentlemanly society, Bacon himself often "took physick" for these purposes, but apparently only his personal recipe for a maceration of rhubarb in a little white wine and beer to "carry away the grosser humours of the body." Describing himself as one "that have been ever puddering in physic all my life," Bacon acted largely as his own physician and he dosed himself in moderation, but with very great attention to detail.[80] Montaigne, whose essays Bacon much admired, similarly wove together medicine and manners in commending the occasional binge: "He will even plunge often into excess, if he will take my advice; otherwise the slightest dissipation will ruin him, and he will become awkward and disagreeable company."[81] Sir Thomas Browne's later compilation of commonly received errors recorded the view "That 'tis good to be drunk once a moneth, is a common flattery of sensuality, supporting it self upon physick, and the healthfull effects of inebriation."[82] Laurent Joubert deplored the popular saying that "there are more old drunkards than there are old physicians," while also identifying ancient authority—Celsus again—for the advice that you should sometimes eat to surfeit.[83] And John Aubrey's life of Thomas Hobbes recorded the philosopher as saying "that he did beleeve he had been in excess in his life, a hundred times; which, considering his great age, did not amount to above once a yeare. When he did drinke, he would drinke to excess to have the benefitt of Vomiting, which he did easily . . . but he never was, nor could not endure to be, habitually a good fellow, i.e. to drinke every day wine with company, which, though not to drunkennesse, spoiles the Braine."[84] In the eighteenth century, such lay approval of occasional excess still troubled physicians, who did not appreciate Bacon giving it further credibility. James Mackenzie's *History of Health* (1760) acknowledged that it was popularly, but falsely, attributed to Hippocrates that "getting drunk once or twice every month [w]as conducive to health."[85] In any case, such counsel was pervasive. It represents some of the sentiments that physicians were up against when they commended a rigorous observance of dietary moderation.

"To Live Medically Is to Live Miserably": The Dietary Vicissitudes of the Active Life

Physicians were piqued by the respecification of moderation as occasional excess, but gentlemanly society nevertheless had excellent reasons for choosing to ignore the physicians. Bacon's essay on regimen aphoristically summed up the relevant consideration: "In *Sicknesse*, respect *Health* principally; And in *Health, Action*."[86] The early modern public actor—the gentleman, the courtier, the politician, the diplomat, the merchant, the soldier—came down firmly on the action side on the ancient debate between the *vita contemplativa* and the *vita activa*. In natural philosophy, Bacon's modernizing reforms were meant to reshape intellectual inquiry to fit the exigencies of political and economic action.[87] So, in medicine, Bacon reckoned that the legitimate test of medical practice was its ability to enhance the capacity for action. And in no case should medical counsel withdraw otherwise healthy men from the active sphere. If occasional feasting and boozing were central to the public life—and in early modern England they spectacularly were—then it could not possibly be a point of prudence or of morality to embrace medical counsel that removed public actors from those scenes in which public action occurred and in which social solidarity was made and subverted. The sort of self-indulgent discipline that was acceptable for sequestered scholars, monks, or retired gentlemen was not proper or permissible for the civic actor. La Rochefoucauld declared that "To keep well by too strict a regimen is a tedious disease in itself."[88] And when the proverbial voice similarly said that "To live physically"—that is, according to the commands of physick— "was to live miserably," two things were meant: first, that it was not pleasant for your body; second, that it was a socially unacceptable way of living.

The rigorous dietetics of moderation had to be tempered by other important ethical concerns, and, indeed, the meaning of moderation might even be re-specified so as to align the notion of temperance with these other concerns. That is why Bacon's early biographer, giving the details of the Lord Chancellor's program of self-medication, felt obliged to insist that "he did indeed live physically, but not miserably."[89]

If you were actually ill, then, of course, you should summon your physician and accept his best advice. It was important to your part in the active life that you got better just as quickly as you could. But if you were not actually ill, there was no reason to submit yourself to the severely ordered regimens prescribed by medical counsels of moderation. This principle was widely known among early modern medical and moral writers as the Rule of Celsus. In the original, the Rule went like this: "A man in health, who is both vigorous and his own master, should be under no obligatory rules, and have no need, either for a medical attendant, or for a rubber and anointer. His kind of life should afford him variety."[90] The Rule of Celsus—the rule of no rule—had all the appeal of common sense; its dictates appeared both to accommodate prudential considerations and to fit with much of what counted as reliable physiological knowledge. Over a period of time—both lay and medical voices said—your body got accustomed to your usual diet and to the usual rhythms of your life. You submitted yourself at your peril to abrupt dietary change, or to the rigorous rule of dietary system. That is the sense of Seneca's maxim that "A well-behaved stomach is a great part of liberty": if you have a healthy stomach, then you are not legitimately subject to any rules but those of your own normal patterns of life. Integrity is a circumstance of good digestion. Nor was it just lay public actors who fell in with the Rule of Celsus. It was such a prize that it was cited approvingly by a number of early modern physicians who elsewhere displayed their predilection for the sovereignty of expert dietary system. In these connections, physicians gave themselves considerable flexibility in specifying when you were in fact healthy or, appearances tending to deceive, actually ill or at imminent risk of becoming so.[91]

More to the point, ethical writers dwelt extensively on just why it was that "living physically" was neither proper nor practical for the man who meant to act effectively on a public stage. King James wanted his son and heir to appreciate the link between dietary adaptability and effective rule. The consideration here was wholly political. In a face-to-face society, the showing of legitimate condescension—noblesse oblige—commonly took place at table, and this is where table manners might merge with matters of state. For physiological as well as prudential reasons you had to get used to what was on offer wherever you had to be. That is why James admonished his son to "eate in a manly, round, and honest fashion," and to get accustomed "to eate of reasonable rude and common-meates, aswel for making your body strong and durable for travel, as that ye may be the hardliner received by your meane subiects in their houses, when their cheere may suffice you." Dietary flexibility was good for you and it was good politics: "your dyet [should] bee accommodatte to your affaires, & not your affaires to your diet." When in Rome, eat as the Romans.[92] An effective prince could not allow dietary squeamishness, or his physicians' orders, to keep him away from where the action was, nor could he afford to give offense to potential allies and valuable followers by declining to eat, and visibly to relish, what was offered. Then as now, declining a proffered dish or drink might be taken as an act of social disengagement.[93]

Less exalted moralists fell in with the king's counsel, often citing ancient warrant for dietary flexibility. Peacham celebrated the example of the Emperor Augustus, who "was never curious in his diet, but content with ordinary and common viands. And Cato the Censor, sailing into Spain, drank of no other drink than the rowers or slaves of his own galley."[94] Locke advised that one not accustom children to regular mealtimes. A body grown used to such strict order would give trouble when public business necessitated the disruption of routine. And in no sphere of active life was such adaptability as important as in military occupations.[95] Of all gentlemanly roles, that of the soldier required a "body used to hardship," accustomed to whatever "accidents may arrive." The soldier's diet was generally unpredictable and often rough: the stomach on which, Napoleon said, an army marches had therefore to be a robust and compliant organ, tempered by the vicissitudes.[96]

So there were many reasons—the typical early modern mélange of the medical, the moral, and the prudential—why the publicly acting gentleman should not live according to expert, externally imposed, dietary rule. To be a slave to system was not civil; it was not prudent; and it was possibly even unnecessary to legitimate interests in preserving the health required to act effectively in society, and to do so until a ripe old age. The doctors' concerns were not necessarily your concerns, nor should they be. Despite their pretensions, they didn't know it all, and they might not even know what was really pertinent to your own health and your own freely chosen way of living. This was very much Montaigne's view, enormously influential in shaping Bacon's opinions on these matters, and more generally, both in French and through John Florio's translation, that of late sixteenth- and seventeenth-century English gentlemanly society.

Like many other late Renaissance and early modern gentlemen, Montaigne preferred his own experience of his body, and its responses to regimen, to the expert advice and the artificial systems of the physicians. The intellectual basis of being able to act as "your own physician" was adequate self-knowledge. No one could know one's body and its responses to food, drink, and patterns of living as well as oneself, but the price of such knowledge was painstaking attention. A prudent man should acquire and value such dietetic self-knowledge, and Montaigne accounted himself such a man: "I study myself more than any other subject. That is my metaphysics, that is my physics."[97] The essay "Of Experience" was composed when Montaigne was fifty-six years old—well past the age at which (as the old saying had it) a man should no longer need a physician.[98] Montaigne here deliberated upon what he had learned, how it bore upon his own proper regimen, and how he then stood in relation to external medical expertise.

"Of Experience" was an eloquent expression of Montaigne's well-known skepticism about external expertise and a vigorous defense of the moral and practical integrity that such skepticism assisted. In form, it is an essay about the general superiority of prudence to systemic pretension; in substantial content, it is in fact a dietary tract about the management of the Galenic non-naturals. If you were a prudent person, then over the years your dietary routine had been informed by the patterns of your body's responses to food and drink, and, in turn, your body had grown accustomed to that dietary routine: "I believe nothing with more certainty than this: that I cannot be hurt by the use of things that I have been long accustomed to." Your appetite was a pretty reliable guide to what was good for you. If you liked it, it probably liked you. "I have never," Montaigne said, "received harm from any action that was really pleasant to me . . . Both in health and in sickness I have readily let myself follow my urgent appetites. I give great authority to my desires and inclinations. I do not like to cure trouble by trouble . . . My appetite in many things has of its own accord suited and adapted itself rather happily to the health of my stomach . . . It is for habit to give form to our life, just as it pleases; it is all-powerful in that; it is Circe's drink, which varies our nature as it sees fit." Even if your physicians urgently advised it, radical alterations in long-established customs could be bad for you: "Change of any sort is disturbing and hurtful." Montaigne was no fool: when his body began speaking to him in unaccustomed ways, he listened. When sharp sauces started to disagree with him, he went off them and his taste followed suit; when ill, wine lost its savor and he gave it up. But in ordinary circumstances, habit, having become a second nature, was to be respected, and Montaigne—falling in with Galenic tradition—thought that you should be careful making abrupt changes in dietary routine unless they were absolutely necessary. Still, soldier and public actor that he once had been, Montaigne did not portray himself as a slave to habit: "I have no habit that has not varied according to circumstances"; "The best of my bodily qualities is that I am flexible"; "I was trained for freedom and adaptability." Vicissitudes of life were one thing; the sudden changes that flowed from adopting doctors' expert systems were quite another.[99]

Montaigne saw no reason not to consult with physicians when he was genuinely ill, even if he suspected that, in general, they knew little and could do less: "The arts that promise to keep our body and our soul in health promise us much; but at the same time there are none that keep their promise less."[100] Expert physicians disagreed among themselves, and, so, if you didn't like the dietary advice that one offered, you could always pitch one doctor's favored rules against another's: "If your doctor does not think it good for you to sleep, to drink wine, or to eat such-and-such a food, don't worry: I'll find you another who will not agree with him." For that reason alone, you might as well do what you thought best, or nothing at all. The curative power of nature was, in any case, probably more effective than the art of any doctor: "We should give free passage to diseases; and I find that they do not stay so long with me, who let them go ahead; and some of those that are considered most stubborn and tenacious, I have shaken off by their own decadence, without help and without art, and against the rules of medicine. Let us give Nature a chance; she knows her business better than we do."[101]

If you put the conduct of your life under the care of physicians, Montaigne too thought they would make you miserable. Forbidding this and forbidding that, the doctors un-man you and, ultimately, they un-do you: "If they do no other good, they do at least this, that they prepare their patients early for death, undermining little by little and cutting off their enjoyment of life."[102] It was a widely shared general ethical principle that a man who was a slave to system was less than a man. He who was ruled by others, or by a book of rules, was no free actor; he lacked the integrity central to gentlemanly identity.[103] Montaigne's essay took that general moral case and made it specific to dietetics. Change was physiologically and morally good— better and more possible for youth than age: "A young man should violate his own rules to arouse his vigor and keep it from growing moldy and lax. [There] is no way of life so stupid and feeble as that which is conducted by rules and discipline." "The most unsuitable quality for a gentleman," Montaigne declared, "is over-fastidiousness and bondage to certain particular ways." Not to do, not to eat, or not to drink what was going wherever you were was "shameful": "Let such men stick to their kitchens. In anyone else it is unbecoming, but in a military man it is bad and intolerable; he . . . should get used to every change and vicissitude of life." By all means, listen to those who may have authentic medical expertise, but do not give up your freedom of action in so doing. Montaigne said that he knew of, and pitied, "several gentlemen who, by the stupidity of their doctors, have made prisoners of themselves, though still young and sound in health . . . We should conform to the best rules, but not enslave ourselves to them."[104]

One established role for philosophical knowledge in relation to medicine was its ability to guide the physician in preventing and curing disease. Skeptical of the ability of current knowledge to achieve these ends, Montaigne embraced another long-established role for philosophy: it could and should reconcile us to the inevitable circumstances of our mortal condition. To live like a man, you must learn to suffer like a man, and, finally, to die like a man: "We must meekly suffer the laws of our condition. We are born to grow old, to grow weak, to be sick, in spite of all medicine . . . We must learn to endure what we cannot avoid." And when it comes time to die, then one should rather die philosophically and well than to live on stupidly and miserably: "Is it so great a thing to be alive?"[105]

The physicians didn't like such talk; they rarely do. Traditional medical dietetics did indeed stress the importance of self-knowledge and the dietetic practice that depended upon this self-knowledge. But physicians also insisted upon the necessity of being consulted in disease, upon the acknowledgment of their expertise in pronouncing whether you really were well or ill, and upon their role in supervising and guiding your self-knowledge. So, for example, Montaigne's contemporary Laurent Joubert countered the apparently fashionable idea that there were other legitimate gentlemanly values than looking after one's health, or that submission to expert medical system was anything but right and wise. It was, the physician wrote, very prudent to live young as if you were old, for then you really would attain vigorous old age. Or, as the proverb had it, "young old, old young." And if princes and public actors objected that they were busy and that it was better for them "to be loose and not observe any rules, schedules, or system," then Joubert settled for the best that he could get, and more than they were usually accustomed to giving. Such people should nevertheless observe "the strictest codes and rules [they] can possibly manage to apply, as much as . . . circumstances will allow."[106] The Jesuit Leonard Lessius also knew his target when he wrote of the reluctance of civic actors to take the best medical advice: "The Wills and Humours of Men (we know) are stubborn and uncontroulable, and their Appetites too ungovernable to admit of any violent Restraints. Men (we see) will, at least the generality of them, eat and drink, and live according to the ordinary Course of the World, and indulge their sensual Appetites in everything to the full. Thus comes it to pass, that all their other Care and Diligence concerning these Physical maxims, or Prescripts, in the End produce little or no benefit at all." They are fools to live like this and they show their foolishness in blaming physicians for what they have brought upon themselves, "leav[ing] all entirely to Nature, and Event. To live physically they hold (according to the old Proverb) is to live miserably; and they look upon it as a very great Unhappiness for a Man to be dieted, to be denied the free Use (perhaps) of an insatiable Appetite, or Desire."[107]

In the eighteenth century, even the great dietary doctor George Cheyne recognized the strength of gentlemanly moral resistance to medical rule: "The Reflection is not more common than just, That he who lives *physically* must live miserably. The Truth is, too great Nicety and Exactness about every minute Circumstance that may impair our Health, is such a Yoke and Slavery, as no Man of a generous free Spirit would submit to. 'Tis, as a *Poet* expresses it, *to die for fear of Dying.*" Yet Cheyne made his considerable living by peddling dietary advice to those well-heeled gentlefolk he judged were, if not actually ill, then in imminent danger of becoming so. And therefore he tempered his appreciation of polite moral resistance by appealing to another set of ethical sensibilities also acknowledged in gentlemanly society: "But then, on the other Hand, to cut off our Days by *Intemperance, Indiscretion,* and guilty *Passions,* to live miserably for the sake of gratifying a *sweet Tooth,* or a brutal *Itch;* to die *Martyrs* to our *Luxury* and *Wantonness,* is equally beneath the Dignity of *human Nature,* and contrary to the *Homage* we owe to the *Author* of our Being. Without some Degree of *Health,* we can neither be agreeable to *ourselves,* nor useful to our *Friends.*"[108] Intemperance was at once religiously sinful, socially uncivil, and personally imprudent.

Expertise, Integrity, and Common Sense: Early and Late Modern Dietetics

If you were a poor person in the early modern period, you would act as your own physician largely because you could not afford the services of a medical expert. And if you were radically inclined, or if your material interests were at stake, you might assert the adequacy or superiority of self-treatment as a way of breaking up the medical profession's corporate power and political privileges. But if you were a free-acting gentleman in this period, neither of these considerations typically had much to do with pervasive assertions of medical self-knowledge or with skepticism toward medical expertise. Rather, you acted very substantially as your own physician because you wanted to, especially in circumstances where self-knowledge was central to medical assessments and where medical advice bore upon the fabric of everyday life and upon the identity and integrity of the self. Dietetic self-knowledge and self-care were marks of mundane prudence and moral integrity.

The management of the Galenic non-naturals constituted the prescriptive part of early modern medical dietetics, but it also made up a substantial part of the early modern cultural practices that established personal identity and social worth. This is just a way of rephrasing the observation that medical dietetics inhabited the same cultural terrain as practical morality. And yet that cohabitation could and did give rise to conflict as well as to consensus— conflict between medical expertise and gentlemanly common sense and conflict between the goals of physicians and those of public actors. Insofar as gentlemen were not professionally qualified, they were, like the "vulgar," laypeople with respect to physicians. But their knowledge was not so easy to condemn as that of the vulgar, nor was it so easy medically to objectify them and their "conditions," nor, again, to dispute their definitions of the situations in which medical counsel might or might not have pertinence or potency.[109] Early modern gentlemen, that is to say, were laypeople of a very special sort. They had a voice, arguably more audible in literate culture than that of the physicians whom they occasionally employed. Gentlemanly prudence could not be dismissed as *mere* common sense or as meretricious "low" knowledge; gentlemanly self-knowledge was hard to gainsay; and gentlemanly goals formed a framework for evaluating physicians' advice whose legitimacy could be challenged only with great difficulty. When early modern gentlemen concurred with what physicians counseled, their assent was consequential, and when they did not, their dissent caused potentially serious problems for the credibility and the social grip of medical expertise.

What has become of that early modern dietetic common domain? One could argue that it dissolved long ago. To a considerable extent, the cultures of moral discourse and of medical expertise have gone their separate ways, though it would be very wrong to describe the divorce as absolute.[110] The adage "You are what you eat" survives as a vestige of a largely lost dietetic culture, while the modern biochemistry of food metabolism gives the formula a renewed charge of credibility at the cost of a fundamental change in meaning. So far as the medical profession is concerned, dietetics no longer exists as a discrete subject in the curriculum. The "dietician" is widely understood to be someone arranging institutional meals for maximum nutritional content at minimum cost; the "nutrition scientist" studies metabolic pathways typically at many removes from the offer of practical counsel; and the heaps of dietary advice that polysaturate the common culture tend overwhelmingly to pick out the virtues or vices of specific food items in relation to specific conditions or diseases. None of these bears anything but a lexical relationship to the early modern culture of dietetics. The counsel of dietary moderation may be hard to discern in the contemporary culture of medical expertise and in what the laity seems to expect of that expertise. Quackery—defined as identifying simple explanations and remedies for complex conditions—is in the ascendant, and one could plausibly describe present-day lay attitudes to diet, disease, and health as an incoherent assemblage of discrete quackeries. The medical profession has almost wholly given up the role of counseling individuals on their way of life, save with respect to disease-specific conjunctures (for example, exercise and a low-fat diet in relation to coronary artery disease; stress-relief in connection with hypertension). Conversely, patients can rarely effectively insist that the advice of medical experts should be weighed against the range of their own life-goals. The dietetic voice of moderation, insofar as it is audible at all in late modern culture, tends to come from other sources than medical experts.[111]

Early modern gentlemanly acknowledgment that the dietetic counsel of moderation might itself have to be qualified by the demands of civility has similarly lost much of its force. If the character of the gentleman no longer exists, nevertheless the scenes of public life and social interaction that gave rise to Montaigne's and Bacon's commendation of dietary decorum still do. Yet some years ago I gave a dinner-party for eight that required—on medical and on ideological grounds—the preparation of four different menus. As to drink, two people would take no red wine; two others would take no wine at all; and one would not drink German wine. Both health and ideology effectively trumped civility.

Scaled up, stripped down, and generalized, these observations about the career of dietetics are just familiar truisms about contemporary culture. They bear a family resemblance to academic cultural-theoretical truisms about modernity that plausibly talk about cultural specialization and differentiation, about the dominance of expertise, about secularization, about the hollowing out and debasement of a once common culture, and about the decline of civility. But, like many truisms about the nature of the modern condition, they tend to mistake the part for the whole, an aspect of how our society is changing for an adequate description of what it is. Indeed, dietetics— interestingly poised at the intersection of self and non-self, of the scientifically descriptive and the morally prescriptive—offers a perspicuous site for reconsidering how one might go about describing how we live now.

It is worth just pointing out the conditions in which, and the extent to which, early modern dietetic culture remains intact in the early twenty-first century. Much of our personal identity and social worth is still asserted, established, and contested through the personal management of the Galenic non-naturals and through others' observations of how we manage them. To that extent, it might be said that medicine and morality continue their cohabitation. Gluttony, drunkenness, sloth, promiscuity, and repeated outbursts of road-rage are still likely to count against assessments of individual character, and, although the counsel severely lacks trendiness, adherence to the Golden Mean probably remains a prudent course of action, even now, for anyone wishing to win friends, to influence people, and, for that matter, to decide what and how to eat and drink. Even after hard reflection, and with due deference to the physicians, the psychotherapists, the social workers, and the personal trainers, it is difficult to know what better advice one could possibly give. Of course, we no longer *know* that these practices have a coherent medical identity, and therefore, for us, they do *not* have a medical identity. Those of us who have passed through physiology courses in school or university are unlikely to credit the humoral framework that linked the emotions to diet and that offered a conceptual framework explaining *why* moderation was good for you.

So dietetic culture, one might say, does survive in late modernity, but as a dispersed set of fragments, ripped free from its original moorings in a medical understanding of the self and without any substantial anchorage in contemporary medical expertise. Like many other allegedly "premodern" modes, the dietetic culture that once bound medicine and morality is possibly better described as submerged in the layered streams of late modern life rather than as dissolved in a unitary, and universally credible, medical expertise. We late moderns resemble Montaigne in many things, not least in our ability to pick and choose which, if any, among the many dissenting voices of medical expertise we will listen to and act upon. Expertise without credibility is nothing at all. And, while the counsel of dietary moderation is indeed hard to hear in modern culture, it is far from impossible to hear it, and, on hearing it, to be reminded of its sheer common sense and unassailable authority. As I finished this essay, the *New Yorker* brought me Julia Child's irresistible, and only slightly modernized, expression of the ancient Rule of Celsus: "A low-fat diet . . . What does that mean? I think it's unhealthy to eliminate things from your diet. Who knows what they have in them that you might need? Of course, I'm addressing myself to normal, healthy people . . . What I'm trying to do is encourage people to embrace moderation—small helpings, no seconds, no snacking, a little bit of everything, and above all, have a good time."[112]

PART V

The World of Science and the World of Common Sense

A distinguished scientist has recently asserted, with serene confidence, that the very idea of science is opposed to that of common sense: if something fits with common sense, he says, then it just isn't science, and that's one way you can tell what is science and what is not. This sort of claim has a long history: one of the characteristic gestures of the Scientific Revolution was the assertion that the ultimate physical and causal structure of the world was inaccessible to common experience and common modes of accounting, while emergent notions of a special and powerful Scientific Method juxtaposed that method to the faulty reasoning of the common people. Science was one thing; the Errors of the People were quite another. That sentiment is pervasive, and it is, therefore, an object worthy of serious historical attention, even if few scholars see the need to spell out what is to be taken as common sense and what as science. How did it happen that scientific knowledge was set in opposition to lay knowledge? With what consequences for conceptions of reliable knowledge and the roles of those possessing it? Who could have that reliable knowledge and how could one recognize its authentic possession? Could, indeed, the distinction between science and common sense be rejected? A historical study of proverbs—short linguistic genres circulating among the common people—is one way of thinking seriously about the science/common sense distinction. What are proverbs, and how do their generalizations and applications compare to those of expert science? Another way of putting some new historical life in the science/common sense distinction is to take a close look at practices in which scientific expertise—abstract and fundamental knowledge of nature's processes—is brought to bear upon *specifics,* in the example of medicine, particular sick and suffering bodies. Is it, as Galileo's Simplicio critically suggested, that the abstract lives in the ideal world and the concrete in the real world? If so, medicine—in its aspects as science and art—is a perspicuous site in which we might concretely address the relations between scientific expertise and whatever counts as common sense, once again transforming resource into historical topic.

CHAPTER 13

Trusting George Cheyne

Scientific Expertise, Common Sense,
and Moral Authority in Early Eighteenth-century
Dietetic Medicine

Historically, the physician's role has been precariously poised between expertise and common sense. Insofar as the physician presented himself as an expert, and asked to be accepted as such by those who engaged his services, what he had to say about the causes, course, and possible cure of your condition, as well as the maintenance of health and the attainment of long life, had to be different from mere common sense. You might be skeptical of the good he actually did you (many were not), but insofar as you recognized that the physician was an expert, you ascribed to him special and superior knowledge about how human bodies worked in health and disease. The physician's expertise might be understood to flow from his vast stock of experience with sick and well people—knowledge of the same sort you had, but in larger store—or that expertise could be constituted by his possession of knowledge not available to you as a layperson, such as knowledge of the hidden structures and processes of your body, its aliment, and the external environment that impinged on your body.[1]

Yet medical expertise was tempered by the knowledge of the people who paid for those expert services. At any time from Antiquity through at least the early twentieth century, much of the basic conceptual vocabulary of medical science and art was held in common by medical experts and laity, and it is not easy in many cases to say to whom such vocabulary authentically "belonged." In early modern England, for instance, essentially everybody who interacted with physicians (and many of those who did not) understood what was meant by having a bilious disposition or a scorbutic humor, by being vaporish, by suffering from a tertiary or quaternary fever. They knew what sorts of things clysters and electuaries were; they knew what rhubarb and the Jesuit bark were and what they were good for; they knew why they were bled, what results were supposed to be produced by bleeding, and why they might be directed to be bled around the equinoxes and solstices. Such a common culture coordinated interaction between physicians and patients; it made such interaction meaningful; and it was used by the laity to understand what was happening to them and why. Physicians used it to make their expertise manifest to laypeople, and, if you wanted to contest the efficacy of medical directions and interventions, the vocabulary of the common culture provided the resources to do that too. Physicians and patients both accepted that laypersons knew a lot about their bodies and the conditions of their health and disease. Even physicians acknowledged that the laity were, to a very large extent, experts on themselves as medical objects.[2]

This essay is meant as a contribution to understanding some pervasive features of engagements between common sense and expertise. First, I briefly note some historically general circumstances attending dietetic medicine as a culture in which medical expertise and morally textured prudence occupied common terrain. Second, I treat a particular, historically specific site of engagement between expertise and prudence: the case of George Cheyne (1671–1743), a fashionable dietetic doctor who, as a prolific author, also made repeated claims to *dernier cri* scientific expertise. What did that expertise look like both in his published work and in his quotidian medical practice? How did these claims to scientific expertise figure in securing the doctor's authority and patients' trust in his counsel?

"Every Man His Own Physician": Owning Expertise

It was commonly said that after thirty years of age—in other versions after forty or fifty—every man should be his own physician. He then knew enough about his body and its vicissitudes to treat himself, to know which foods and activities agreed with him and which did not, to guess plausibly enough about the course of those common illness that afflicted him.[3] Montaigne's skepticism about medical expertise was matched by his confidence about dietetic and therapeutic self-knowledge: habit and constitution molded themselves to each other and everyone who was not a fool came eventually to know best where his own shoe pinched.[4] Physicians themselves quoted this common wisdom, so acknowledging that there was no *necessary* threat to professional expertise contained in the notion that patients knew quite a lot about themselves as medical objects. Indeed, George Cheyne cited this wisdom on the opening page of his first systematic dietetic text.[5]

Moreover, early modern writers recurrently cited the ancient Rule of Celsus, which held that people who were in ordinary good health should have no need for a physician or put themselves under the constraint of medical rules.[6] The proverb "He who lives physically lives miserably" identified the moral and practical dangers of subjecting yourself unnecessarily to the discipline of medical expertise, and lay commentators referred to the common-sensical authority of the Rule of Celsus to argue against the tyranny of those physicians who constantly asserted their authority to order you to live in ways that were unpleasant, inconvenient, and (in skeptics' view) unwarranted by any substantial risks to well-being and longevity.[7] Physicians themselves sometimes invoked the Rule to show that they acknowledged the moral and pragmatic limits of aggressive professional expertise: asserting your expertise didn't have to mean that you had taken leave of your common sense.[8]

So far as prognosis and diagnosis were concerned, laypeople might reasonably come to know enough about common illnesses, the early signs of their appearances, and the course they tended to take in their bodies, to reckon themselves possessed of relevant expertise. And even in therapeutics laypersons might sometimes legitimately juxtapose their expertise to that of physicians, as they might come to acquire sufficient familiarity with common drugs and procedures to know which worked on them and which did not.[9] Apollo's oracle said "Know thyself," and many early modern laypeople thought they did know themselves at least as well as, if not better than, any expert physician could.

If this was the case in diagnosis, prognosis, and therapeutics, it was an even stronger sentiment in that important part of medicine called dietetics (regimen or hygiene). That is because management of the non-naturals constituted such a significant portion of quotidian life. How you arranged eating and drinking, evacuations, sleeping and waking, exposure to airs and other environmental features, exercise, and how you managed your emotions constituted a big part of who you *were* and of your recognized social standing. In dietetics, that is, medicine pitched its tent on ground already densely occupied by moral common sense.[10] Medical counsel toward temperance— "nothing too much" was the other dictum carved on Apollo's temple at Delphi—made such common sense because it was at the same time a cherished article of moral prudence. Gluttony was, for example, bad for you, but it was also just *bad;* temperance, after all, was one of the classical virtues. So when dietetic physicians said that you ought to observe the Golden Mean, they spoke with the joint authority of medical expertise, of common sense, and, sometimes, of Divine Law. Cheyne wrote that "The infinitely wise *Author* of *Nature* has so contrived *Things,* that the most remarkable *Rules* of preserving *Life* and *Health* are *moral Duties* commanded us, so true it is, that *Godliness has the Promises of this Life, as well as that to come.*"[11] This was a powerful combination, and its advice was difficult to deny.[12]

From Antiquity through the early modern period, dietetic expertise counseled moderation with very great cultural stability and uniformity. Perhaps that is why dietetics has attracted so little historical attention: its advices seem banal and it is not a culture that changes very much over a great sweep of history. There appear to be no real *ideas* at play, certainly nothing as headily intellectual as the changes in medical theorizing ushered in with the Scientific Revolution. Of course, the ancient natural philosophy of the elements and the doctrine of natural place underpinned the counsel of moderation, and the same ideas shaped the vocabulary of humors, complexions, and temperaments that allowed physicians to understand the complicated relations between individuals, aliments, environment, and medically directed measures. But both the stability of those ideas and their joint-ownership by experts and laypeople have seemed unattractive to scholars who conceive of the history of medicine as the history of its novel ideas.[13]

This joint-ownership of dietetic culture could give physicians great authority, just so long as what they advised counted as common sense. However, the same cultural sharing also presented them with problems in asserting their expert authority, just because their counsel might appear as little *else* than common sense, or even, where it departed from temperate prudence, as *less* than common sense. From the physician's point of view—though not, of course, from the patient's—dietetics held out limited possibilities for cultural and social distinction. As dietetics was such an important part of the physician's role, this was the predicament wrestled with in much of the late Renaissance and early modern literature on popular "medical errors." Proverbial common sense advertised itself as containing all you really needed to know of physic: "Piss clear and make a fig at physicians"; "He who pisses, sleeps, and wags well has no need of Doctor Bell"; "Kitchen physic is the best physic"; "A good cook is half a physician"; "Use three Physicians still; first Doctor *Quiet,* Next Doctor *Merry-man,* and Doctor *Dyet.*"[14] This was the sort of stuff opposed by Laurent Joubert and other medical critics of "popular errors": the "ungrateful common people" are always likely to give themselves and their commonsense beliefs and practices the credit that rightfully belongs to learned expertise, just as they "forget rather easily the benefits they receive [from physicians' care] and retain in memory the most insignificant mistakes." They attribute to "all nature's doing" or to "good soups" what was really done by physicians with their rational expertise, their uniquely effective dietetic advice, and their well-judged therapeutic interventions.[15]

Medical Expertise and Micromechanism

The late seventeenth and early eighteenth centuries witnessed radical changes in the cultural conditions of medical expertise and its relations with common sense. The micromechanism of René Descartes, Pierre Gassendi, Robert Boyle, Giovanni Alfonso Borelli, and, above all, Isaac Newton testified to an invisible realm radically different from that posited by the Aristotelians or by commonsense actors. When the corpuscles of micromechanism took the place of the four elements, there were new implications and opportunities for physiological and medical expertise.[16] Iatromechanism and iatromathematics were platforms from which the advanced physician could speak for an invisible realm publicized as more securely founded than that which it was bidding to supplant, and, moreover, whose natural philosophical champions were rapidly turning into powerful cultural allies. Francis Bacon had complained that there would be no progress in medical practice until medicine acquired reformed natural-philosophical foundations.[17] With Newton's work, such foundations were finally considered to be available, ready to be exploited by physicians mathematically and philosophically able to do so. Now that such physicians definitively knew the microstructures and micromechanisms of aliment and the body they could intervene and advise with radically improved effectiveness. Micromechanism ambitiously promised the maintenance of health, the cure of disease, and the prolongation of human life.[18]

The "rational physician" traditionally distinguished himself from the vulgar "empirick" because he alone systematically grasped the fundamental underlying causes of health and disease. Among learned and fashionable physicians in early to mid-eighteenth-century Britain, the display of iatromechanical expertise was a powerful vehicle for cultural product-differentiation. So the Scottish iatromathematician Archibald Pitcairne advertised the professional advantages that would flow from adopting Newtonian principles: the "infamous Mark of *Uncertainty*" would be erased from medicine, and "the Honour of our Profession" would no longer be at "the Mercy of the Vulgar."[19]

And Richard Mead announced that soon "*Mathematical Learning* will be the Distinguishing Mark of a Physician from a Quack."[20]

That is how George Cheyne started his medical and literary career, well narrated in Anita Guerrini's now-indispensable recent biography. In the 1720s, John Woodward satirically counseled professionally ambitious doctors newly arrived in London "to make all the Noise and Bustle you can, to make the whole Town ring of you if possible: So that every one in it may know, that there is in Being, and here in Town too, such a Physician."[21] When Cheyne removed from Edinburgh to London in 1702, the instrument of his "Noise and Bustle" was *A New Theory of Continual Fevers,* an aggressively iatromathematical tract inspired by his mentor Pitcairne.[22] The display of Newtonian expertise was one way to make yourself known; association with London's literary and philosophical fast crowd was another. Within months of his arrival in London, Cheyne secured election to the fastest philosophical club of all, the Royal Society. A great talker, and, at the time, a very great trencherman and drinker, Cheyne cruised the coffeehouses and the taverns, where he "found the *Bottle-Companions,* the *younger Gentry,* and *Free-Livers,* to be the most easy of *Access,* and most quickly susceptible of *Friendship* and *Acquaintance.*"[23]

After a rocky start, Cheyne eventually established himself not only as a fashionable physician—shuttling between London and his main base in Bath—but also as one of England's most influential medical authors. His clientele included Alexander Pope, John Gay, Beau Nash, Samuel Richardson, the Methodist Countess of Huntingdon, Robert Walpole's adolescent daughter, Catherine (who died under Cheyne's care of something resembling anorexia nervosa), and the Earl of Chesterfield (who passed Cheyne's advice ineffectively on to his "dear boy"). Cheyne's published medical advice was favorably quoted in *Tom Jones;* Samuel Johnson commended his books; and John Wesley's *Primitive Physick* copied out whole sections of Cheyne's work.

Historians of medicine sometimes say that around 1720 Cheyne "repudiated his youthful mathematical brashness and excessive Newtonian enthusiasm."[24] In this "second phase" of his career, he refashioned himself into a dietetic doctor, centering his attention on chronic conditions and prudently counseling "moderation in diet and drink."[25] Certainly, dietetics did become the focus of Cheyne's publishing career, and this did represent a considerable change from earlier writings that elaborated micromechanical accounts of the human body relatively disengaged from streams of practical hygienic counsel.[26] General dietetic advice was the meat of such works as his *An*

Essay of Health and Long Life (1724), *An Essay on Regimen* (1740), and *Natural Method of Cureing the Diseases of the Body* (1742), while his popular *Essay on the Gout* (1720) and the celebrated *The English Malady* (1733) commended a largely dietary regime in dealing with both conditions.

By the 1730s and 1740s, the whole polite British world was talking about Cheyne and his diet. And while Cheyne did indeed counsel moderation for people in normal good health—even articulating a version of the Rule of Celsus[27]—the diet for which he was famous was a severe "lowering" regime, suited to the valetudinary, the sedentary, the studious, and the otherwise fine-nerved, who, he warned, were risking their lives by persisting with a normal course of food and drink: "I advise . . . all Gentlemen of a *sedentary Life,* and of *learned* Professions, to use as much *Abstinence* as possibly they can," *in extremis* descending to an exclusive regime of asses' milk and seeds.[28] Many people swore by Cheyne's lowering diet, announcing that it had saved their lives; others swore *at* it, considering it bizarre, unbalanced, rigidly doctrinaire, unnecessary, impossible to maintain, and probably totally ineffective.[29] (By 1740, Cheyne was sufficiently aware of his identification with the weird milk-and-seed diet that he took pains publicly to deny that he recommended it to any but those in desperate need of its cooling effects.)[30] In this primarily dietetic work, Cheyne may well have moved away from his earlier iatromechanical and iatromathematical "enthusiasms," but he did *not* "repudiate" his claims to micromechanical expertise.

So Cheyne appeared in the person of an expert author, subject to all of the interests, irritations, conventions, and constraints of the authorial life in early eighteenth-century Britain. At the same time, he appeared in the person of a fashionable physician, purveying face-to-face and epistolary diagnoses, prognoses, dietetic counsel, and therapeutic prescriptions to patients with whom he was often intimately familiar. A large number of letters survive from Cheyne to two of his patients in the 1730s and 1740s: the printer and novelist Samuel Richardson (1689–1761) and Selina, Countess of Huntingdon (1707–1791).[31] It is this sort of evidence, considered side by side with public displays of scientific expertise, that allows one to reconstruct some of the complex and often edgy interrelations between medical theory and practice, expertise and prudential common sense, public statements and private counsels, the general and abstract and the particular and concrete, in early to mid-eighteenth-century Britain. What did iatromechanical expertise look like when confronted with an individual sick patient? How did such expertise figure in quotidian medical practice? And how was this expertise implicated, together with other features of Cheyne's knowledge, life, and character, in securing the credibility of his claims and the authority of his practical advice?

Medical Expertise and the Invisible World

The overall framework of Cheyne's iatromechanism is not radically different from that of other early eighteenth-century British Newtonian physicians. The human body, he wrote, is "nothing but a *Compages* or Contexture of Pipes, an *hydraulic Machin.*" [32] The "elasticity" of the body's solids was a pretty durable paternal inheritance—not easy (though not impossible) to alter over time. But the juices came from your mother, and these were readily modifiable by your way of life, and, especially, by your food and drink. That is the general philosophical explanation of why the primary task of the dietetic physician was to tend the juices: if these could be mended by expert dietetic advice, "*they* will in time . . . rectify and confirm the *Solids* into their proper Situation and *Tone.*"[33] In health, the fluids enjoyed free passage through these "Pipes"; diseases resulted *inter alia* from obstruction to that flow. The expert rational physician could mathematically model blood flow and could demonstrate the precise causal links between the configurations of alimentary particles and their effects on fluid flow. That expertise was what allowed him to give good advice: "*Art* can do nothing but remove Impediments, resolve Obstructions, cut off and tear away *Excrescences* and Superfluities, and reduce Nature to its primitive Order; and this only can be done by a proper and specific *Regimen* in Quantity and Quality."[34]

The "*Grand Secret*" of health and long life was, in principle, quite simple: it was "to keep the Blood and Juices in a due State of Thinness and *Fluidity,* whereby they may be able to make those Rounds and Circulations through the animal Fibres, wherein Life and Health consist, with the fewest Rubs and least Resistance that may be."[35] But if, through age and improper diet, the fluids become "*viscid, thick* and *glewy,*" the circulation slows and ultimately stops, producing first disease, then death.[36] *The English Malady* commenced with a causal explanation of "Chronical Distempers" in general, the paramount cause being a "*Glewiness, Sizyness, Viscidity,* or *Grossness* in the Fluids."[37] The "*best* Blood" was the "thinnest and most fluid Blood," as it "most easily circulates thro' the *capillary* Vessels, which is the most solid Foundation of good Health and Long Life."[38]

One was to understand, therefore, that Cheyne's advice to eat this and not to eat that, to take the drugs and embrace the other therapeutic measures he prescribed, and to adopt specific regimens of exercise proceeded from his deep and systematic knowledge of the invisible world, and that the quality of this expertise was in large part vouched for by its derivation as a deduction from Newtonian natural philosophy and mathematics. Cheyne's special contribution was to put flesh on this deductive skeleton, to identify the micromechanical structures of particular aliments and body parts. His books also assured readers that he could bring this fleshed-out framework to bear on the management of particular sick bodies. Indeed, he wrote that he had repeatedly and successfully done so, most spectacularly in the extraordinary thirty-page "The Case of the Author" appended to *The English Malady.* In curing himself by largely dietetic means, Cheyne announced that he had cured by far his hardest case.[39]

Cheyne's books identified a number of ways to keep the blood and other bodily juices thin, sweet, and flowing. Exercise was important; so were air, blood-letting, and the judicious use of drugs, but "It is *Diet* alone, proper and specific *Diet,* in *Quantity, Quality* and *Order,* that continued in till the Juices are sufficiently thinn'd, to make the Functions regular and easy, which is the sole *universal Remedy.*"[40] "A *thin, fluid,* spare and lean Diet" made for thin and free-flowing juices.[41] Here Cheyne clearly had the force of analogy on his side: the ultimate particles of thin and fluid aliments were themselves thin and fluid, while gross and sharply flavored foods and drinks were made up of large and angular particles, likely to scrape the vessels and to deposit an obstructing crust on them. The authority of micromechanical expertise thus helped itself to the very analogy between visible and invisible that was formally denied by the distinction between primary and secondary qualities.[42] As a dietetic physician, Cheyne acknowledged the obligation to identify the vices and virtues of specific aliments, transforming micromechanical expertise from abstract pronouncements into particular counsels.

Consider Cheyne on water-drinking. "*Pure Water,*" he wrote, is "the only *Beverage* designed and fitted by Nature for long *Life, Health* and *Serenity.*"[43] To drink water as your sole dietary liquid "is the only Preservative, I am certain, known or knowable to Art."[44] If you started with a course of exclusive water-drinking when young, and persisted with it, you "would live probably till towards an *hundred* Years of Age."[45] Water is the "true and universal *Panacea,* and the *Philosopher's Stone.*"[46] Despite the confident promise of longevity, this was radical advice, seriously unpopular in free-toping Georgian England. Cheyne appreciated that it could be justified only on the strongest philosophical grounds, and these micromechanical grounds were duly supplied.[47] The ultimate particles of water were so fine and smooth that this was the "*sole* Fluid that will pass through the smallest animal *Tubes* without Resistance."[48] The microstructure of water was what made it such a good solvent of vascular obstructions and such an effective vehicle for keeping the juices flowing. Cheyne understood that water in its naturally occurring states also contained dissolved in it "a little fine vegetable *Earth, Salt* and *Sulphur*"—the "smallest and finest" of such particles—and, accordingly, possessed powerful nutritive as well as therapeutic properties.[49]

All aliment contained in various combinations the principles of sulphur ("from whence Spirit and Activity"), salt ("hard angular" and highly attractive particles), water ("from whence alone Fluidity"), and earth ("the base and *Substratum* of these others").[50] Cheyne averred that "it is past all Doubt in Philosophy, and in philosophical *Chem[istr]y,*" that animal foods were richer in the first two principles while vegetable foods were richer in air, water, and earth.[51] From "undeniable Experiments," it was philosophically known that the first two principles are "the most active, *energic,* and *deleterious,* and tend more, by their Activity, to the Division, Dissolution and Destruction of the Subject, than those others when they enter in any great Proportion."[52] Of these invisible states of affairs, "there are so many and convincing Demonstrations, that none can have any Doubt of it, that has the least Acquaintance with natural *Philosophy.*"[53] Expertise was *that* sure of the matter: "Infinite *Experiment,* and the best *natural Philosophy,* confirm to a Demonstration, that those Substances, which have least of *Salt* and *Sulphur,* of *Spirit, Oil,* and hard pungent Particles, and most of soft Earth, Water and Air, are the fittest to circulate, and be secreted through animal Tubes, create least Resistance to the motive Powers, tear, rend, and wear out the *Tubes* themselves least, and form less obstinate and powerful Obstructions, in the smaller Vessels."[54] Water was not only good in itself, but it might, over time, wash away the incrustations caused by the micromechanical structure of animal foods and strong liquors, restoring the vessels to their naturally elastic and healthy tone.[55]

Expertise in advising regimens of food and drink was therefore said to proceed from privileged knowledge of the micromechanical realm, and such philosophically informed regimens were of central importance to Cheyne's professional practice. Yet Cheyne never confined himself to dietetic advice, and the same expertise in testifying about the realm of the invisibly small was also brought to bear on his bleeding and drugging practices. So consider

Cheyne's views—to be sure, not unique to him—on the properties and medical virtues of mercury. The therapeutic virtues of mercury also proceeded from its micromechanical structure. The particles of mercury are the smallest of any known fluid; they are the most perfectly spherical, the heaviest, and the most extremely attractive and repulsive. It follows that mercury is, of all substances, the most easily raised by heat; it possesses the greatest momentum; and it is the most able "to pass through all *animal* Substances, which are lax and porous."[56] Physiologically, this meant that mercury was uniquely suited to break up viscid and gluey accretions bunging up the vessels. And from these physiological capacities the specific medical uses of mercury deductively followed: mercury was the most powerful medicine against scurvy, palsies, gout, and, indeed, all the "*chronical* Distempers caus'd by Excesses."[57] Even so, knowing all this about mercury was not sufficient to its safe and effective curative use: mercurial medicines had to be prepared with exquisite care. Its degree of fineness was crucial to mercury's action either as a powerful antidote or as a dangerous poison, so the physician had to be responsible for selecting, supervising, and validating the work of the preparing apothecary or chemist.[58]

While the authorial display of micromechanical expertise identified Cheyne as a rational physician, the inclusion of case histories (notably in *The English Malady*) and discussions of the causes and cures of common chronic illnesses advertised the concrete pertinence of his knowledge and his effectiveness in matters of pressing concern to existing and prospective patients. *Natural Method of Cureing Diseases,* for example, showed Cheyne's philosophical expertise and proprietary regimen powerfully at work in a long list of chronic distempers, including rheumatism, dropsy, gout, colic, sciatica, the stone, and menstrual complaints. In the case of scurvy, Cheyne told readers what the disease was in micromechanical terms and how his method worked to cure or alleviate it. The manifestations of scurvy were various and diffuse, including "a habitual white, or *foul crusted* Tongue"; a reddish sediment in the urine expressed during the night; and a long series of "*Hysteric* and *nervous Symptoms,*" such as alternating chills and burning sensations in the extremities, dermatological eruptions, vomiting, interrupted sleep, thirst on waking, depression, convulsions, and flatulence.[59] The micromechanical cause of all this was, again, "viscid Juices," the saturation of the blood and other body fluids with "*saline, sulphurous,* or *firy* Particles,*" obstructing the circulation, the perspiration, and, ultimately, blocking the viscera and the nervous system.[60] Accordingly, the appropriate method for dealing with scurvy, or a tendency toward it (a "scorbutic habit" or "humour"), centered on the dietary thinning of the fluids, for example, by drinking "sweet Cow-whey" and a "light white-meat *trimming* Diet," with little or no fermented liquors: "Living on Milk and Vegetables . . . will keep this Distemper long under."[61] Mercury and phlebotomy could be judiciously used to aid the thinning process, while rhubarb, aloes, or other emetics might be employed to free up the bowels. By this method, Cheyne wrote, he had rarely failed "of a perfect Cure, or a notable Relief," except in cases very far advanced.[62] You had to consult Cheyne early enough, or even his formidable expertise might not save you.

Intimate Relations: Cheyne and His Patients

So far this story has remained in the public realm, describing the philosophical and practical expertise that Cheyne published in the books written from the 1720s until close to his death in 1743. What configurations of expertise, prudence, and authority are evident when one shifts attention to the relatively private sphere of Cheyne's medical practice? Here the letters to Samuel Richardson and the Countess of Huntingdon are major sources of evidence.[63] Both patients were well known to Cheyne: he treated various members of the extended aristocratic family over many years, and he had extended business relations with Richardson as printer of several of his medical books. The social pleasantries exchanged in the letters establish intimacies that possibly passed the norm in early eighteenth-century doctor-patient relationships, while the correspondents each got the mode of civility appropriate to their social standing—fawning deference to the Countess and a kind of affectionate condescension to the tradesman-printer, whom Cheyne praised for an integrity rare in people of his sort and who benefited from the good doctor's prescriptions for effective novel-writing.[64]

Both patients also stood in an intimate relationship with Cheyne because both were made to understand that their conditions were similar to those suffered by the doctor himself. Symptomology was evidently agreed between physician and patients: with few exceptions, Cheyne accepted not only their reports of signs and feelings, but the vocabulary they used to report those signs and feelings.[65] At the start of the correspondence, Richardson suffered from the classic complex of signs that Cheyne designated the "English Malady" (hypochrondriasis, "the Hyp," or "the nervous Hyp"): that combination of vertigo, paroxysms, "Giddiness and Lowness," and "terror and confusion" that Cheyne's *English Malady* so vividly described in his own case.[66] Paramount in Lady Huntingdon's sufferings was severe and persistent constipation, but she also reported extreme itchiness (possibly hemorrhoids, as the correspondence gets coy on the matter), an erysipelous skin condition, flatulence, gripes, colic, hot flashes, occasional unspecified complaints about her eyes, menstrual irregularities (she was continually pregnant during the period of her correspondence with Cheyne), and a series of "distracting, sinking nervous complaints."[67] And, while some of these complaints were attributed to Lady Huntingdon being constantly "a-breeding"—one of the few conditions from which Cheyne did not claim his own "crazy Carcase" suffered—he traced the ultimate roots of many others to an underlying disposition that was said to be remarkably similar to his own.[68]

Cheyne's medical counsel to both Lady Huntingdon and Richardson was very various. It definitely included aggressive drugging, bleeding, and the active management of all the non-naturals. So, over the course of ten years, a small selection of drugs Cheyne regularly directed Lady Huntingdon to take included Anderson's (or the Scotch) pill (a mild aperient), Jesuit's bark (quinine, a febrifuge but apparently widely taken for a diffuse range of complaints), rhubarb (a common purgative), senna (a cathartic and emetic), ipecac (emetic, diaphoretic, and purgative), spirit of lavender, laudanum and other opiates (for pain and sleeping problems), Glauber's salt (sodium sulphate, a purgative and laxative), cream of tartar (potassium bitartrate, an emetic), "cinnabar of antimony" (cinnabar usually designated a mercuric sulphide, but Cheyne warned of the dangers of "active mercurial medicines" in her case)—often elaborately compounded to the doctor's precise instructions—and washed down with lashings of mineral water, Bristol or Pyrmont by preference.[69] Richardson was dosed with enormous quantities of mercury in various forms (as purgative and cathartic), the Jesuit bark, squill (a botanic diuretic), ipecac, hiera picra, aloes (botanic purgatives), ethiops mineral (probably a sulphide of mercury), asafoetida (an antispasmodic), spirit of niter (for the relief of flatulence), "tincture of soot" and powdered steel (purposes unknown), various extracts of spruce and fir, often compounded together, and, as with Lady Huntingdon, taken with Spa or Pyrmont water. (Cheyne disapproved Richardson's preference for Tunbridge.)[70]

Both patients were directed frequently to be bled, for diagnostic as well as therapeutic purposes. And both were prescribed the precise forms of light exercise: Lady Huntingdon to ride abroad (when her pregnancies permitted), while Richardson (who loathed any form of exercise) was evangelized on the virtues of the then-fashionable hobby- (or chamber-) horse: "It is certainly admirable and has all the good and beneficial Effects of a hard Trotting Horse except the fresh Air."[71] Cheyne pursued a vigorous regime of vomits, induced both chemically and manually, but, perhaps surprisingly, these were commended not so much for lightening the load on the stomach as for the *exercise* they afforded the body. His proprietary method of "Thumb Vomits" was minutely detailed to Richardson (the Countess was spared this method): "the Virtue lies in the Exercise, the Throws and Pumpings of the Cavities; 40 or 50 Kecks is more Exercise of the whole Body than half a Dozen Miles Coaching."[72] Daily thumb vomits "work the whole Man and shake every Fibre and Gland which cannot be otherwise reached."[73] Lady Huntingdon was diffusely counseled to maintain a cheerful frame of mind and an optimistic view of her case, while Richardson was offered explicit expert advice on how to manage the emotional terrors of the authorial life (writer's block and bad reviews): "Now as to yourself I never wrote a Book in my Life but I had a Fit of Illness after."[74]

Although Cheyne evidently struggled with Richardson to get him to take the recommended exercise, in the main the letters offer no evidence that the prescribed bleedings and polypharmacy met with much resistance from the patients. For all the elaborateness of Cheyne's prescriptions, and for all the obvious inconvenience and unpleasantness of having to spend so much of your life on the water-closet, his practice here seemed well within local norms.[75] The matter was quite otherwise with his program for food and drink. Here Cheyne was obliged to confront a serious conflict between his dietary counsel and the demands of common sense, civility, and appetite. And here the patients did indeed resist ("I find . . . you go on timorously, grudgingly, and repiningly," he told Richardson),[76] and Cheyne had to bring to bear all the resources at his command to secure his authority and their assent to his advice.

In both of the cases at hand, Cheyne wound up ordering his patients to adopt a radically lowering milk-and-seed diet. At the start of their correspondence, Cheyne was not seriously concerned about Richardson's condition: "All your Complaints are vapourish and nervous, of no Manner of Danger."[77] Accordingly, Cheyne saw no reason to urge on Richardson anything more than "that general Temperance I have so often recommended to you and which I know you pursue."[78] Abstinence from wine and fermented liquors was good, but moderation in all food and drink would do nicely:

"One Dish of plain Fish or Flesh at Dinner, at Supper a Toast with another Half Pint of Wine and Spaw Water with a Bit of Cheese, and the ordinary Breakfast";[79] there might even be "Times and Seasons when a little Indulgence in Chicken . . . may not only be convenient but necessary."[80] Why not "Half a Chicken in Quantity of any fresh tender Meat (any Thing else to fill Chinks you please)"?[81] But already by the summer of 1741 the doctor was beginning to be seriously concerned that Richardson had taken a "Plunge" and was now in immediate danger of a fatal apoplexy. Cheyne would not answer for his patient's life if a radically restricted regimen was not followed precisely: "You had as good shoot yourself as alter your Diet."[82] The doctor's advice became more extreme, and he ordered Richardson onto a totally vegetarian and wine-less diet. By the spring of the next year, Richardson was encouraged to embrace the pure milk-and-seed regimen, asses' milk for preference: "at Dinner Rice Pudding, and at Night Watergruel or Milk Porridge."[83] Hunger was the best medicine, and Richardson had to learn to live with gnawing hunger and a miserable diet or, Cheyne assured him, he would not live at all.

Lady Huntingdon was also started out on a "cool and tender regimen," but nothing too severe: lashings of sweet cow's whey and water-drinking provided the base; ripe fruit was all right ("as much as you please") and salads too, if they agreed; at dinner, she was permitted to rise as high as "Chicken, partridge, a little white fish, lobster, cray fish, lambstons [lamb's testicles?], veal feet, [and] jellies."[84] Pregnancy further warranted an easy regime: Lady Huntingdon was reminded that she had two to feed and was told, accordingly, to "Let your appetite be your rule."[85] But after a year and a half of this treatment (including active drugging) her condition did not respond, and Cheyne first urged her to carry on with the temperate diet and then, with a show of sympathetic reluctance, to lower it further—alternating a light main meal of white meats with one of milk-and-seeds: "Milk is the only certain and infallible remedy" for her condition;[86] "Live as much as you can on milk and milk meats";[87] no fermented liquors; "a bit of chicken now and then for a relishing of spirits."[88] The total milk-and-seed diet might yet be avoided if only she would follow Cheyne's directions religiously. A year later, her complaints persisting, Cheyne ordered her to be lowered still further: "I think milk and rice, sago, barley, or bread the best emulsion and diet for you."[89] Asses' milk would, of course, be best, but the Countess evidently stuck at that.[90] From then on, Cheyne's letters fine-tuned Lady Huntingdon's diet, constantly urging her to keep on course and promising an ultimate return to ordinary victuals when the low regime had at last worked its healing effects. Normal life could then resume.[91]

Trusting George Cheyne

Cheyne had to work hard at getting his patients to adopt his diet and to stick to it. Both his side of the exchanges and contemporary satire indicate the extent of lay resistance: as influential as he was, many eighteenth-century English readers found Cheyne's dietary prescriptions ludicrous, impossible, unlikely to do anyone any good—while much of his medical correspondence with these patients is constituted by a continual battle for practical and moral authority. What resources were available to him to secure this authority and patients' assent? Why would you agree to *do* that to yourself when so much about Cheyne's regime violated dietetic tradition, contemporary common sense, and moral prudence?

The first thing that might be done by a physician counseling such extreme measures would be to represent this kind of diet as *itself* a form of temperate common sense, to respecify an apparent extreme as a prudentially sanctioned Golden Mean. So when Cheyne was advising Richardson only to reduce his consumption of meat and wine, he said, "You may keep the golden Mediocrity by thus trimming";[92] when he urged him on to "A Fleshless and Wineless Diet," he commended it as "a just Medium between a Common Animal and Wine Diet, and a Milk Diet";[93] and when, finally, he directed Richardson to adopt the asses' milk regime, Cheyne shifted attention to the total *amount* of aliment to be consumed, describing that as "a just Mediocrity."[94] The prudential Golden Mean could also be reckoned through a social calculus. Even with the lowering measures pressed on him by Cheyne, Richardson was assured that his diet "will still be fuller and higher, more nourishing and salutary than Nine Parts of Ten in England can have";[95] "All below Farmers scarce taste Animal Food Six Times a Year . . . and yet one Tenant is generally supposed to out-live Three or Four Landlords at an Average. These have few or no natural Distempers except epidemical Ones."[96] Throughout the early modern period, the prudential Golden Mean remained such a powerful cultural resource that there were always contests for the rights to its blessings.[97]

Extreme dietetic advice might also be smoothed in its course by embedding itself in the rhetorical forms of common sense. Counsel that everybody recognized as prudential wisdom was hard to gainsay. The Hippocratic aphorism stipulating that "desperate diseases require desperate remedies" had long passed into common usage, applied in a wide range of nonmedical as well as medical contexts. And when Cheyne was confronted by patients' reluctance to follow what even he acknowledged to be extreme measures, he asserted his expertise by telling them that their condition had become so serious that mere temperance was no longer enough. "Extreme cases must have extreme cures," he wrote to Lady Huntingdon,[98] and Richardson similarly was motivated to compliance by being told that "a desperate Disease must only have a desperate Remedy."[99] Proverbial common sense was pervasively used as an emollient wrapping for bizarre and unpleasant expert counsel: "you must take Care of the Brute else he will be at last too hard for the Man";[100] "The Disease must in a Manner be starved";[101] "He who is in the Fire should get out as fast as he can";[102] "Custom is no Reason."[103]

Other means for asserting his dietary authority responded to more skeptical sentiments among his patients. When Cheyne evidently felt that Lady Huntingdon and Richardson were wandering off-message, he pointed to his track-record in many previous cases, to healthy and happy patients who were vibrantly living testimony to the efficacy of the dietary method, and, even, to a primitive form of statistical evidence. So he reminded Lady Huntingdon of the pertinent case of the cleric who first inspired him to his present dietetic method. This man had totally cured himself of epilepsy by living for twenty-two years "on a total rigid milk diet"—not even any bread, fruit, or vegetables; then, when he was persuaded by his family and friends to resume a "higher, tho even, temperate, diet," his former condition reasserted itself, and he "perished miserably under it."[104] On a more positive note, the Countess was put in mind of two patients suffering from the same scorbutical humor that afflicted her, who, keeping to Cheyne's diet, were now "big . . . healthy and gay." He enclosed in his letter to Lady Huntingdon a copy of a letter he had received from "a vegetable [eater] of 72 years"—"a considerable person in the House of Commons"—who had thus recovered from "mortal agonies."[105] Family members also treated by Cheyne were now looking "fresh, clear, and plump" by virtue of their milk and vegetable diet,[106] and there were several Cheyne-advised low-living neighbors, now "gay as a bird," whom Lady Huntingdon could visit, and whose good health gave witness to the power of the method.[107]

To Richardson, Cheyne enclosed a personal letter from a satisfied patient, and also a statistical summary of his vast experience in such matters: "In at least 30 Years Practice" in cases such as Richardson's,

> and all the other incurable Cases by Drugs or Doctors in which I have treated some Hundreds and of which many are yet alive over all the Dominions of His Majesty and some Abroad I do not remember to have lost above Three, and they were too far gone and died in the Beginning of their Course. One or two more I have lost by being over persuaded by eminent Physicians to alter their Method, but never one that I had the most remote Reason to ascribe their Failure to their Diet, nor do I think the Thing possible. Some indeed do not recover to that high, athletic Health some strong young Beef-eaters enjoy but that is because they have had originally broken, tender, debilitated Constitutions from their Fathers or began [the method] too late in Life . . . but never one who continued in it 2 or 3 Years that did not live out the natural Duration of their Lives and went on infinitely easier than they did before they entered upon it. I have Letters every Post from some one or other such from all Parts of the Kingdom.[108]

However, in these intimate epistolary settings, the invocations of statistics appear as a less-central feature of credibility management than the assertion of a moral compulsion. Cheyne's exchanges with his patients took place on a moral and emotional field: following the doctor's expert but difficult advice became a mark of personal virtue. Again and again, his patients were urged to be *courageous*. When they showed signs of reluctance to go on with his radically abstemious dietary regime, they were told to summon up the courage to persevere, until they were "brought to perfect health and gayety at last."[109] To Richardson: "Courage! you will come to laugh at your own Fears";[110] "take Courage";[111] you must have the courage to stand against both your animal appetites and the ridicule of your friends.[112] Patients needed patience: the cure takes time, and courage is the virtue required to let the medicine work its inevitable curative effects. As Cheyne rather unfortunately told Lady Huntingdon, "the *lightest* and the *least,* tho slow, is as certain as death."[113] This virtue was not gendered: Lady Huntingdon was applauded in just the same way as Richardson for the "courage to go on steadily";[114] for "her resolution and courage to enter upon such a course of self denial";[115] for her stout refusal "to be sneered, ridiculed, or frightened" out of her regime.[116]

The doctor encouraged his patients, and, in turn, they were applauded for their displays of dietary courage: the relationship was a mutual tuning of the virtues. They owed dietary persistence to themselves (of course), but also to their friends and family. Richardson was promised material benefits as the undoubted result of his compliance: he would get to "a moderate, Active, gay Temper and Habit and write Books without End, as I have done, and grow rich as a Jew and settle all your Family to your Heart's Content."[117] Both Richardson and Lady Huntingdon were repeatedly told that their restoration to health, and therefore their persistence in Cheyne's regime, were moral obligations they owed to those who loved and depended upon them. Here the virtue of courage was joined to that of fidelity.

His patients' return to health was also enmeshed in a network of reciprocal moral obligations with their physician. Cheyne worked hard—possibly as hard as any British physician of his time—to produce and sustain such obligations. His patients were repeatedly given to understand that he had suffered what they were suffering. Practically speaking, this meant two things: that his experience was not merely theoretical, and that the efficacy of his method could be vouched for by his own now-healthy, but once seriously deranged, body. Time and again, Cheyne reassured Lady Huntingdon and Richardson that his condition was the same as, or similar to, their own.[118] If Richardson complained of a pain in his ears, Cheyne "had it often" and knew what cured it;[119] when the printer was bothered by "Startings, Twitching, and Cramps" making him apprehensive, Cheyne sent him a complicated prescription, and assured him that "I took it myself some Months" with good effects;[120] when Richardson had a worrying "Plunge" early on in the radically lowering diet, the doctor consoled him with the information that he himself had experienced such a relapse, even a year-and-a-half or two years into the regime: but that was "16 or 17 Years ago, and now at 70 you know I am tolerably well."[121] To the lay injunction "Physician, cure thyself," Cheyne had a robust response: he had done so.

Both Richardson and Cheyne were fat—at about 450 pounds Cheyne was spectacularly so ("overgrown beyond any one I believe in Europe")[122]— and, while obesity was not then in itself a medical complaint, it brought about physical inconveniences that might become medically consequential. Physician-patient intimacy reached its height when Cheyne, encouraging Richardson to keep on his low diet, shared with the printer a disgusting obesity-induced disorder that "is a Secret to all the World except to my own Family": "my Guts fell out through the Cawl where the Spermatic Vessels perforate it made a Kind of Wind Rupture"; diet and thumb vomits cured that too.[123] Again, specific hernias and child-bearing apart, Cheyne's ability to feel your pain was not notably gendered. The English Malady afflicted men and women alike, and Cheyne both understood and sympathized with Lady Huntingdon's underlying scorbutic humor and "erisipelatous" outbreaks: "I have had more of that distemper than anyone I ever heard, I believe above 40 times, and it was the principal reason why I entered upon a milk and vegetable diet . . . and just now, by indulging too freely in high, rich vegetables, growing too fat, using too little exercise . . . I have suffered the most universal erisipelas ever was known."[124] Cheyne bound himself emotionally to his patients because he had suffered their pain, but also because he himself had at times fallen away from "the dietetical Ghospel" just as they were in imminent danger of doing.[125] Only a sinner knew the true value of salvation.[126]

Cheyne also gave his patients to understand that he cared for them very much. His relationship with them incorporated expertise, but it was never merely instrumental. The letters between doctor and his patients exchanged civilities, inquiries after the health and doings of friends and family members, notifications of gifts about to be sent and of gifts gratefully received. From Lady Huntingdon's family there came occasional presents of venison (only partly foul);[127] from Richardson repeated gifts of oysters (only sometimes spoiled).[128] Cheyne recommended servants to Lady Huntingdon and sought personal favors from her for his brother.[129] Repeatedly, he assured both patients of his deep and abiding concern for their well-being. He cared for them, and loved them, as he did himself, and he would not commend to them any measures but those most certain of success. On this "I will venture my life and reputation," he told Lady Huntingdon;[130] "As to your life, I could venture mine a thousand times for the security of it."[131] Urging Richardson to adopt his dietetic regime, Cheyne would "venture my Life" on a successful outcome,[132] and, later said that he would "go to Death for it."[133] Cheyne told Richardson that he loved him as he would a family member, and he would no more give bad advice to Richardson than he would to a brother, son, or father.[134] In such a relationship, the only interest Cheyne could conceivably have was in his friend's well-being. And so he told Richardson "If I have either Honour, Honesty, Friendship, or Virtue I would not suffer a Man who trusts me, is my Friend on whom I have no Views, to run Risques."[135] To Lady Huntingdon he wrote that if she failed to improve on his regime "I am a cheat and a deceiver"; if she had "any trust to give me," would she really suggest that someone who loved her would tell her a damaging lie?[136]

Cheyne asked, even demanded, to be trusted. He wanted, of course, to be trusted as an expert, but, more than that, he wanted to be trusted as someone who would bring his expertise to bear on his patients' case—conscientiously, prudently, heroically (if necessary), and with as much skill and art as he would use in his own case. It is one thing to acknowledge expertise; it is another to accept that expertise will be diligently brought to bear on your behalf. Cheyne asked to be trusted even in the extremities of his dietetic method, and the reciprocal ties of moral obligation gave adequate grounds for doing so. It was not easy to withhold such trust from a friend, and especially to do so in intimate relations.

Where is natural philosophical expertise in all this? In fact, displays of micromechanical knowledge figure *not at all* in these exchanges. Cheyne *never* instructed Lady Huntingdon or Richardson about the ultimate particles of their juices or aliment or about the corpuscular causes of their conditions. Occasionally, the doctor reasserted his superior knowledge of the underlying causes of their various symptoms. In Lady Huntingdon's case she was several times reminded that all her complaints proceeded from "that sharp scorbutic humour you brought into the world with you," and that made her ailments so difficult to alleviate quickly using dietetic methods.[137] Similarly, he told Richardson that his problems resided in the solids and not in the juices: "I take it your Solids are loose, flabby, and soft though fresh and sound like untwisted Silk Threads"; through poor diet and habits early in his life, the "original lax Membranes and Vessels" were filled too full, "and they being somewhat broken are not sufficiently strong and elastic to force out the perspireable Wind and Steams which being retained perpetuate on the Membranes."[138] In only one instance in these exchanges did Cheyne even gesture at knowledge at a more fundamental ontological level. Sometime in 1742, Richardson sought the opinion of another physician.[139] This physician recommended a course of chemical medicines that Cheyne had always opposed. Accordingly, Cheyne reminded Richardson of his superior natural philosophical knowledge: "I have studied Chemistry and read most of all the Rational and Philosophical Chemists, but never could make any Thing of them that I could rely on, and even despise Boerha[a]ve for his wild Brags of some of his chemical Medicines which I have ever found false on frequent Trial . . . I never saw a chemical Medicine of any Kind that I could not *over-match* with a natural and simple one."[140]

On the other hand, when Cheyne was confronted with outbreaks of special obstinacy, he was willing to remind both patients that he himself was the author of *books* in which the natural philosophical foundations of his regimen were systematically set out. The Countess was intermittently informed of Cheyne's books in which such cases as hers were systematically addressed. In 1737, he told her that he was about to complete another book, then to be titled "the Universal Remedy," which Cheyne believed, "will hit my lord and your ladyship's taste."[141] He requested permission to dedicate the book—eventually appearing in 1740 as the *Essay on Regimen*—to Lord Huntingdon, which he did do, though at the time the two had evidently never met. This book (printed by Richardson) set out "the true principles and theory of medicine, natural and moral philosophy." At the same time, Cheyne was aware that his noble patrons might not be especially interested in such things: the book was "chiefly designed for learned and philosophical men, [so] I fear there might be some impropriety" in begging leave for this dedication.[142]

Richardson too was occasionally referred to the systematic principles set out in Cheyne's published work, but possibly with a somewhat greater expectation that these books might actually be consulted.[143] By 1742, Cheyne was expressing a degree of special irritation that Richardson showed little familiarity with the doctor's treatises, since it was Richardson, after all, who saw them through the press: "It is a surprize to me that you, [who] have printed 3 or 4 of my Books wherein all the Turns, Symptoms, Nature and Cause of nervous Disorders are narrated and accounted for, should seem to know as little of the Affair as if you had never seen them."[144] Even the very literary man who had printed Cheyne's expert works appeared to have little need of their formal and systematic expert counsel.

Expertise in Action

What sort of thing is this expertise that laypeople might recognize, desire, and, possibly, evaluate? One kind of expertise can be called *prudential*.[145] Such expertise is possessed, for example, by a thoracic surgeon who has done very many coronary bypass operations; it belongs to a restaurateur who has opened, and closed, lots of restaurants; to a car mechanic who has seen hundreds of faulty Ford transmissions; to a marriage counselor who has accumulated experience of the tensions in a May–December relationship; and, of course, by your grannie in regard to all sorts of things that

depend on a large experience of life and its vicissitudes. So prudential expertise must not be thought of as just the property of commonsense actors: very
highly trained professionals may possess such expertise and it may, indeed,
be definitive of their role and authority. However, the nature of prudential
expertise is that it *need not* pretend to flow from knowledge of underlying
processes reckoned qualitatively different from, or superior in kind to, lay
knowledge. You could be such an expert if you'd "been around the houses"
as much as the acknowledged prudential expert. It's accumulated experience, and the judgment informed by that experience, that matters here. A
great surgeon might be a poor physiologist, but that need not matter.

There is another kind of expertise that might be called *ontological*. This
expertise bases its claims to authority on the possession of special knowledge about the underlying or hidden structures of the world or of the domain in question. This knowledge is argued to be different in kind from that
held by lay actors or by prudential experts in the same domain, or, it might
be said, that ontological experts alone penetrate behind appearances to hidden realities. As is the case with all such formal distinctions, there is no
reason to think that the types of expertise map very neatly onto actual social
roles: the prudential expert may happen to know something about the hidden world, and the ontological expert may well possess considerable prudential knowledge. It's a rare theoretician who is totally devoid of practical
common sense. Nonetheless, the modes of expertise are analytically distinct.
It is not impossible to imagine the one without—or almost without—the
other: the marriage counselor who has never been married and who takes
her theories off the shelf; the business executive who makes no claims to
knowledge in the areas of rational decision theory or the academic sociology
of organizations.

Such a distinction between types of expertise seems to make some obvious sense in the history of medicine. Indeed, it maps easily onto empirically
consequential battles over social roles and their attendant values. If in the
eighteenth century you pretended to be a "rational" physician, your identity
and worth flowed from some version of ontological expertise whose power
was contrasted with the "empiric's" inadequate, superficial, and unreliable
merely prudential knowledge.[146] And, although the "rational" physicians
tended to write most of the books, the empirics were not without a come-
back: prudential knowledge was soundly based on things that mattered—
particular sick bodies; claims to ontological expertise were unreliable and
unverifiable. Ontological medical expertise was no new thing in the late

seventeenth and early eighteenth centuries, but the corpuscular and mechanical philosophies of the Scientific Revolution provided it with new forms and bases for cultural authority. Cartesianism and Newtonianism gave ontological expertise a novel language to assert itself, a language that identified both its own foundational status and its divorce from the categories and vocabularies of everyday experience. That's what the Galilean distinction between primary and secondary qualities was about: where the ultimate structures of the physical world were concerned, appearances were deceiving. You could not *see* the ultimate particles of roast beef or of blood vessels, for example; you had to take their characteristics on trust from an expert who had securely deduced their existence.

We now have a pretty good understanding of how such displays of ontological expertise figured in the professional contests of English Restoration and Augustan medicine. Iatromechanism and iatromathematics were used, *inter alia,* as professional *displays:* when you wrote a book about the micromechanical structure of the human body and its aliments, you could establish your authentic standing as a rational physician and even your superior rationality vis-à-vis other physicians. That was good for you in the professional community, and there was a reasonable expectation that standing in the professional community might translate into a lucrative practice. Fashionable patients might be presumed, after all, to want intellectually, as well as socially, fashionable doctors.[147]

Just as the micromechanism of the natural philosophers was identified as a radical break with Scholasticism, so iatromechanism could provide an intellectual license for radically new medical *practices.* George Cheyne, buttressed by the cultural authority of the new ontological expertise, took on a drastic reconfiguration of dietetics, perhaps the most stable and traditionally entrenched of all domains of medical practice. It was in large measure because Cheyne advertised his new ontological expertise that he took the risk publicly to defy dietetic tradition and common sense. The counsel of temperance might give way to an extreme milk-and-seed diet because, in some professional circles, ontological expertise was accounted such a powerful cultural commodity.

So far as the laity are concerned, it's probably banal to say that they want their physicians to be experts. But how medical expertise is parsed, between its prudential and ontological forms, and how it is given content, are historically and sociologically contingent matters. There is little evidence that Cheyne's patients cared much, if at all, about his public displays of ontological expertise. Such displays were largely for the benefit of other medical men and philosophers, and if his patients cared about such things at all, it was indirectly—through hearing from friends and family that Cheyne's ontological views were highly esteemed by pertinent physicians and philosophers.[148] There is more reason to think that some *modern* middle-class patients do expect their physicians to possess such cutting-edge expertise, or even that some patients reckon they can and should assess such ontological expertise, as occurs, for example, in sophisticated patient support groups. Nevertheless, even educated patients concerned that their physicians should possess ontological expertise may continue to feel a degree of discomfort with doctors just out of medical school where such expertise has its natural home and is most concentrated. More to the point, all patients care about medical expertise—of whatever sort—not as a theoretical matter, but in deeply personal terms. They want their doctors to possess and deliver *relevant* expertise. That's to say, they want medical expertise to be brought to bear *on their cases*—humanely, conscientiously, and effectively.

The patient's practical task is to assess the credibility of relevant expertise. In principle, that's a very hard thing to do, since, by definition, the laity themselves possess neither form of expertise. (If they did, they would be experts themselves.) This means that laypeople have to look for the visible marks of expertise. If it's prudential expertise, this large pertinent experience has got to be vouched for by some visible or audible warrants (for example, those of age, manner, and commonly expressed opinion). If ontological expertise happens to be an issue, then some other visible or audible warrants might be looked for (perhaps the visible marks of learning, the occupation of a role or an institutional habitation widely known for such expertise, and, again, general opinion). So in this connection another distinction is indicated—this time between *formal* and *informal* channels for knowing about expertise. In these materials, the formal channel is represented by Cheyne's books, and the informal channel by intimate relations, both epistolary and face-to-face, with particular patients.

If we accept that patients are interested in *relevant* expertise, what are the capacities of the different channels for informing themselves on such matters? The formal channel cannot do that very well, just because the book is not written *for* you and it is not exactly *about* you. It might be written for people of your sort in general, or, as Cheyne told Lady Huntingdon, it might be written for another type of audience altogether, a readership of "learned and philosophical men." You really want to know two things together:

whether this doctor is an expert (in any sense) and whether his or her expertise will be conscientiously and sensitively brought to bear on you—that is, whether he or she is a relevant expert. This is where the informal channel is so powerful. To use Internet terms, the informal channel has a very high "bandwidth," the more so in face-to-face interaction (including "bedside manner"), and to a lesser, but still significant, extent in epistolary exchanges between already familiar persons. Lots of information can be conveyed through the informal channel—for example, via inflections of mood and tone, by communications geared specifically to you, and by finely tuned reactions to responses received. And by definition, the informal channel is relevant because you are one of the parties to exchanges in it.[149]

Cheyne was a virtuoso in using the informal channel. His letters did not have to be used as tokens of his relevant expertise, because in large part they were the thing itself. The letters not only *said* that he cared about his individual patients; they were a major way in which such caring was instantiated. As a way of establishing his authority, Cheyne's masterful use of the informal channel worked pretty well. Far more than the public literary display of ontological expertise, this was what enabled Cheyne to get some of his patients, some of the time, to follow a dietetic regime of asses' milk and seeds, and so to fly in the face of tradition, appetite, and common sense.

CHAPTER 14

Proverbial Economies

*How an Understanding of Some
Linguistic and Social Features
of Common Sense Can Throw Light
on More Prestigious Bodies
of Knowledge, Science for Example*

Learned expertise describes and commends itself as it describes and con-demns vulgar knowledge.[1] This state of affairs is pervasive at the present time and it belongs to a long historical tradition. Scarcely any canonical text of the Scientific Revolution, for example, failed to applaud proper concepts and methods by way of a flattering contrast with the uninstructed ways of the common people. The failings of vulgar knowledge were legion, but two defects were considered paramount among them: its tendency to remain trapped in the world of misleading superficial appearances and its unreflec-tive tolerance of logical untidiness, of incoherence, or even of contradiction. Superficiality, and the unreflectiveness that generated it, were just what the great philosophical modernizers had to overcome. False belief was a popu-lar illness in pressing need of learned therapy.[2]

Sometimes the learned pointed their fingers at common linguistic forms in which vulgar knowledge was cast and that revealed its superficiality and incoherence in a particularly clear way. Proverbs, and similar folkish expres-sions, often served the turn. These were brief descriptive and prescriptive generalizations that the common people were known to value and routinely employ. They evidently spoke about how things were in nature ("Great oaks from little acorns grow") or in human affairs ("New brooms sweep clean"), or they explicitly prescribed how prudent people ought to behave ("Look before you leap"). Quite commonly, proverbs contained metaphors that folded prescriptive elements about human action into apparently descriptive generalizations about nature ("The early bird gets the worm"), or they talked about human nature by drawing on metaphors from human-animal interactions ("Lie down with dogs; rise up with fleas").[3] But to the learned eye a simple inspection of such sayings revealed their inferiority to the propositions and prescriptions of learned expertise. So the seventeenth-century physician and moralist Thomas Browne contrasted expert reason with vulgar irrationality. The people, he said, were "unable to wield the intellectual arms of reason," so they tended

to betake themselves unto wasters and the blunter weapons of truth; affecting the grosse and sensible waies of doctrine, and such as will not consist with strict and subtile reason. [So] unto them a piece of Rhetorick is a sufficient argument of Logick, an Apologue of Æsope, beyond a Syllogisme in Barbara; parables then propositions, and proverbs more powerfull then demonstrations. And therefore they are led rather by example, then precept; receiving perswasions from visible inducements, before intellectual instructions.[4]

With some notable academic exceptions, this broad learned characterization of proverbial common sense continues in currency. In the late nineteenth century, the logician Alfred Sidgwick announced that "Proverbs . . . are frequently employed in arguing by indistinct resemblance. It is the slackness with which any 'striking' analogy will commonly pass muster that leads at all times to the use so freely made of proverbs. To assume that some case comes under some well-known proverb, without a shadow of evidence to show that it does so beyond what may be gathered from the crudest superficial inspection, is still in many quarters a favourite practice."[5]

Philosophers tend to dislike proverbs for the same reason they tend to dislike metaphorical reasoning (and other forms of indexical expressions): both are undisciplined and both are supposed to embody imprecise and superficial modes of inference, leading to inexactitude and error.[6] In social science, too, proverbs are occasionally used as a foil to expert knowledge: in modern textbooks folk generalizations about how people tend to behave are shown to be both shallow and incoherent, needing repair by learned expertise, and the inadequacies of proverbial common sense are offered to students as major inducements to take academic social science seriously.[7] The learned recurrently talk about proverbs as they address themselves to common sense and its standing vis-à-vis formally instructed expertise, notably including philosophy and the sciences, both natural and social. For this reason alone, proverbs are a pertinent site for interpreting the pervasive contrast between expertise and common sense, and for suggesting some new ways of thinking about that contrast.

A few caveats, qualifications, and explanations should be made at the outset:

1. All intellectual traditions generate their subversive elements, and there are well-known counter-instances to generalizations about learned contempt for ordinary reasoning and associated proverbial forms. In philosophy, Montaigne, David Hume, William James, John Dewey, Ludwig Wittgenstein, Richard Rorty, and some practitioners of "ordinary language philosophy" have criticized their own discipline for defining its role as the repair of ordinary cognition and language-use; in social science, the phenomenologists, the symbolic interactionists, and the ethnomethodologists have performed much the same function with respect to "objectivist" sociology; and in the natural sciences there is a strand of thought that rejects the prevalent contrast between common sense and the supposedly special methods of scientific expertise. For all that, there are scarcely any better sources than these internal criticisms for documenting dominant learned tendencies to condemn common sense and to offer expert repair of common modes of reasoning and judging: that dominance is conceded and described even as it is criticized.

2. For a host of reasons, it will not do simply to equate proverbs with common sense. For one thing, proverbs are linguistic items, often propositional, and not all everyday knowledge is linguistic or, insofar as it is linguistic, propositional. Proverbs can, however, be usefully treated as markers of common sense, not least because the learned themselves have traditionally used them for that purpose, and their usage is something one wants to inspect and interpret. Using proverbs as a counterpoint to elements of learned expertise has the evident advantage of equitable comparison: insofar as the learned have treated their knowledge as propositional, it is apposite to offer an interpretation of the propositional aspect of proverbs.

3. When scholars first began collecting, printing, and commenting on proverbs—in the late Renaissance and early modern period—there was a great debate about whether such things were authentically folkish or whether they were of ancient learned origin, achieving wider distribution as they descended the social scale. In the sixteenth and seventeenth centuries a general learned approval of proverbs was associated with a view that their genealogy did indeed trace back to learned authorship. The eighteenth century saw a polite and learned backlash against proverb-use associated with a

growing tendency to treat such expressions as folkish through and through. For some time, the practical consensus among relevant modern linguists and folklorists has ascribed proverbs overwhelmingly to the common people. Even where ancient learned usage can be established, many scholars are now reluctant to take that as conclusive evidence of learned authorship: Socrates, Cicero, and Juvenal might well have been expressing learned sentiments in then-current folkish linguistic forms.[8]

4. So far as common sense itself is concerned, it is useful to retain the phrase as an importantly institutionalized marker—sorting, bounding, and evaluating what are taken to be different sorts of knowledge, cognition, people, and practices—while retaining the most open mind possible about common sense's coherence, identity, and standing with respect to supposedly different learned forms of cognition and practice. I follow learned characterizations of common sense, and of proverbs as one of its elements, in order finally to express skepticism about the sorting, bounding, and evaluating traditionally marked by those same characterizations.[9]

I argue that many learned condemnations of proverbs are not merely wrong but interestingly misdirected and misconceived. They tell us little about what proverbs are or about how proverbs work in naturally occurring settings, and they set up a contrast with learned knowledge that makes it hard to understand what *that* knowledge is in *its* naturally occurring settings. If we want to get learned knowledge right, that is, we can make a contribution by trying to get proverbial common sense right. Our culture has historically tended to seize-up with anxiety when asked to give an account of the cognitive and linguistic processes of highly valued science, mathematics, philosophy, and associated learned practices. When we are not using it merely as a foil to learned expertise, it is easier to engage with common sense in a relaxed and naturalistic frame of mind. I am primarily interested in giving an account of scientific knowledge and related practices. But if we take a detour by way of common sense and some of its linguistic forms, when we meet up with science again we see it from an unaccustomed angle. I want here just to show some paths that might be taken to achieve this changed angle of vision; I paint the resulting revised picture of scientific expertise only with the broadest of brushes. Nevertheless, I have reason to think that taking this detour might be useful and interesting to students of science and other expert practices.

The first substantive section briefly characterizes proverbs while pointing out problems associated with any attempt to give them an exact and coherent definition. This section goes on to consider some structural features of proverbs that help us to appreciate their grip on the mind, their ability to circulate undeformed, and the real epistemic and moral value some people have seen in them. I note how a proper appreciation of proverbs' often metaphorical character allows one to understand their semantic and referential scope, while showing that *translation* from proverbs' metaphorical base to situations-at-hand is unnecessary. The second section argues the importance of considering proverbs as features in naturally occurring scenes of action, spoken by certain kinds of people, with a view to judging and acting properly in those specific scenes, and against the backdrop of all the knowledge participants bring to those scenes. I call such scenes *proverbial economies,* and I show the insufficiency of treating proverbs solely in their propositional aspect. A proverbial economy, in my usage, is a network of speech, judgment, and action in which proverbial utterances are considered legitimate and valuable, in which judgment is shaped and action prompted, by proverbs competently uttered in pertinent ways and settings: that is to say, a cultural system in which proverbial speech has the capacity of making a difference to judgment and action. In the third section, I comment on the unfoundedness of viewing proverbial common sense as unreflective: if you wanted to treat proverbs solely in their guise as propositions-about-the-world, you could retrieve a host of such propositions that make serious trouble for learned criticisms of common sense as trapped in the superficial world of mere appearance. The fourth section mobilizes evidence that much proverbial common sense is not only reflective but even "fashionably" relativistic about knowledge-claims, social conventions, and cultural authority. In the fifth section, I dispute a traditional learned condemnation of proverbs as self-contradictory and, for that reason, worthless. Again, I argue that this charge enjoys local plausibility only by virtue of misconceiving the object of attack—as a body of proverbial propositions rather than a scene of speech and action. Finally, I revisit the contrast between learned knowledge and proverbial common sense. In very general terms, what does this contrast look like at the end of the exercise? What possibilities for understanding science and other formal bodies of learned knowledge are opened up or assisted once we have taken a sideways look at some linguistic and social features of common-sense-in-action? I draw attention here to the *heuristics* of modern expert practices, often embodied as maxims and "technical proverbs," that are significantly involved in the making, transmission, and justification of expert bodies of knowledge.

What Are Proverbs and How Do They Work?

From the earliest learned engagement with proverbs to the inquiries of present-day academic folklorists, sociolinguists, and anthropologists, there has never been notable agreement about how to define proverbs and how to distinguish them from other, formally related, short linguistic genres. Some individual scholars seemed (and seem) confident in their ability to define and distinguish a range of such items—adages, aphorisms, apophthegms, clichés, commonplaces, dicta, epigrams, exempla, gnomes, maxims, precepts, saws, sayings, sententiae, and tags—but no definitional scheme seems ever to have escaped learned criticism.[10] Archer Taylor, the premier twentieth-century scholar in the area (in terms of art, a parœmiologist), despaired of any structural definition and fell back on competent tacit knowledge: "The definition of a proverb is too difficult to repay the undertaking; and should we fortunately combine in a single definition all the essential elements and give each the proper emphasis, we should not even then have a touchstone. An incommunicable quality tells us this sentence is proverbial and that one is not. Hence no definition will enable us to identify positively a sentence as proverbial. Those who do not speak a language can never recognize all its proverbs."[11] Moreover, definitions that seek to pin down proverbs by their structural linguistic characteristics give the game away, achieving clarity at the price of pertinence.

Proverbs are not fixed natural kinds: different sorts of people define them differently, depending on their purposes and points of view. One might as well say that proverbs are what these different people have said they are and, for the modern learned, that proverbs are what you find in proverb dictionaries. There are, however, some widely quoted definitions which, in their family resemblances, capture much of what is relevant in present connections. It is relatively uncontroversial to say that proverbs are short sentences, and their brevity is one traditional way folklorists and linguists have of distinguishing them from such more expansive "short genres" as the aphorism and the apophthegm. But how short is short? Probably "Short enough to remember and for a lot of people to use in a linguistically stable form." As it happens, few proverbs found in standard compilations are longer than ten or twelve words. But brevity is not considered enough to make a proverb, and many commentators conjoin brevity and some notion of pithiness—pointing to the ability of a genuine proverb to distill experience, to say something worthwhile and important in an unusually economical way and, moreover, in a manner that marks it off from the flow of ordinary speech. So the Restoration collector Thomas Fuller famously said that proverbs were "much matter decocted into few words."[12]

Other criteria specified that proverbs, properly so called, have a homespun, metaphorical character—indeed, Aristotle defined proverbs as "metaphors from one species to another."[13] They are supposed to be a kind of referential poetics, often explicitly compared in that way to the precise and literal propositions of learned speech and writing. Proverbs are said to draw upon familiar, everyday experience—for example, about how birds and dogs behave—but they make their meaning through metaphorical extension to human situations and to other natural situations of interest to human beings. And, while many locutions found in proverb compilations are indeed figurative in this way, others are not, speaking about what's in the nature of priests and cooks, women and men, the young and the old, just as they are.[14] Some attempts to characterize proverbs, and to mark them off from aphorisms, insisted upon their antiquity: the origins of proverbs were either lost in the mists of time or they descended from respected ancient authors, but, in any case, they were not supposed to be the kind of thing that you could now just make up on the spot, claim authorship of, and put into general circulation. There is nothing new in the notion that, as it were, "the age of proverb-making is past."[15]

In an overwhelmingly oral culture—such as that of sixteenth-century England, where scholarly proverb-collecting became an important activity—proverb-like sayings were widely used, by both the learned and the common people. Their form and pithiness gave them great mnemonic and rhetorical force and, when properly used, they secured easy recognition and, often, assent. Their value flowed partly from the primacy of orality and the way that proverb-citing could give the written text some of the authority that then powerfully resided in the oral and the face-to-face. Scholars' and gentlemen's commonplace books were chock-full of them, sometimes arranged under appropriate "heads" or topics, testifying to the value placed upon them and ensuring their easy retrieval for occasions of argument, pleading, instructing, or entertaining. Some modern scholars trace changing learned evaluations of proverbs to the increasing dominance of literate over oral modes of communication, as well as to the declining plausibility of assigning proverbs to specific ancient learned authorship.[16]

Suppose one accepts that proverbs do belong to "the people" and that a search for their authorship—learned or otherwise—is generally bound to fail. What would this mean for proverbs' identity and authority? In this view—which is the modern learned consensus, and which may well have been dominant among the common people themselves—proverbs express the condensed experience of nameless hosts of knowing ancestors. Barbara Herrnstein Smith nicely characterizes the proverb as a "saying" rather than a "said" or a "says": it is "speech without a speaker, a self-sufficient verbal object rather than a verbal act, an utterance that asserts itself independently of any utterer—continuously, as it were, or indeed eternally."[17] The "they" in the "they say" commonly prefacing proverb-utterances is generic ancestral wisdom, not a set of nameable authors. That is one reason why a proverbial economy should be pertinent material for anyone taking a sociological, or indeed a historical, view of how knowledge acquires authority. So one commentator perceptively speculates that the denigration of proverbs is testimony to the epistemological individualism of the modern learned classes: "Perhaps there is now something unacceptable in the very notion of collective wisdom: more to the modern individualist taste is Wilde's quip that 'a truth ceases to be true when more than one person believes it.'"[18] Whatever is thought of the virtues and vices of vulgar knowledge has tended to be thought of proverbs as well. If you don't think much of prudence, practical reasoning, rules of thumb, tradition, and situated knowledge, you probably won't think much of proverbs either.[19]

From the Restoration through the nineteenth century a list of "six things required to a proverb" was pervasively cited: the first five items on the list are unexceptional—a proverb should be short, plain, common, figurative, and ancient—but the sixth comes as something of a shock against the general run of learned opinion: the proverb is said to be "true."[20] What evidence could there be to support such an apparently perverse claim? And, in general, how might one go about appreciating the considerable authority and grip of proverbial expressions? First, consider the means by which proverbs arrest attention and grip the mind. A range of linguistic characteristics has the effect of setting the proverb off from the normal run of speech or even of writing. Proverbial expressions, as Erving Goffman might have said, "break frame."[21] Metaphor is one way in which proverbial out-of-the-ordinariness is secured, a striking figure allowing the proverb to break free of the supposed literalness of ordinary discourse. In this sense, metaphorical proverbs mark what is being said as special in the same way that poetry does: the juxtaposition of the homespun and familiar with novel situations, the special extension of meaning between the one and the other, and the sense that language has "gone on holiday" invite special notice.

Still other structural linguistic characteristics work to secure for the proverb what I call *mnemonic robustness*—the capacity of the proverb to seize the mind, to be easily remembered and retrieved, and to resist deformation as it circulates in the culture and over time. Aids to this mnemonic robustness include rhyme ("A stitch in time saves nine"), alliteration ("Many men, many minds"), semantic symmetry, parallelism, or inversion ("Better a lean peace than a fat war"), surprising contrasts ("A shlimazl falls on his back and hurts his nose"), and so on. Indeed, robustness is perhaps too weak a term for talking about how proverbs resist deformation. It is very important to say a proverb just right. If you say "You can conduct a donkey to the pasture, but you cannot make him consume grass," or even "A new broom cleans efficiently," you will probably be corrected by those who know the proper form, and the persuasive or communicative effect sought for will be lost, even though the message conveyed by your mistaken form is, from a certain point of view, "the same." While there are well established variant forms of particular proverbs—some people say "Stolen fruit is sweet" while others say "Stolen apples are sweetest"—you are supposed to use an established form just as it is—no paraphrase will do—and in this respect proverbs are a form of ritual utterance. The linguist Thomas Sebeok, writing of the "charm" (or magical incantation) of a Uralic traditional culture, notes that its effectiveness "depends on its literally exact citation, and, conversely . . . any departure from its precisely set mechanism may render the magic wholly ineffective."[22] And, of course, the same may be said of religious professions, blessings, and the formulaic incantations of childhood cultures studied by the Opies.[23] Although proverbs are uttered in, and take their sense from, specific occasions, linguistically they stand apart from, and above, the specificities of those occasions. Their stability, the ethnomethodologist Harvey Sacks noted, "can be something independent from any occasion of use." Modifying proverbs at will, like summarizing or paraphrasing their message, would result in something that at once lacked the authority and the "frame-breaking" character of the proper form: it's not done.[24] Those who accept Bruno Latour's association of power with that which is immutable and mobile should be interested in how proverbs resist deformation while traveling.[25]

Consider also the *reference* of proverbs and what it is that they counsel with respect to their objects of reference. Proverbs are orientated toward experience. They report on accumulated experience, human and natural; they make those reports efficiently available to people who mean to act in the world; they recommend courses of action in light of experience; and therefore—jarring as it may seem to say so—proverbs represent a widely distributed form of *expertise*. The "expert" is, after all, someone who has relevant experience, and expertise is that embodied experience. That is to say, proverbs have both a representational and a pragmatic component. They are about the world, but not about it mainly as an object of contemplation.[26] Sometimes they comment upon action taken, drawing or inviting conclusions so that future actions should be better informed, as if to say "Well, what can you expect?" or "That's what those sort of people will do." "That's the way these things turn out": "You play with matches, you get burned"; "The squeaky wheel gets the grease"; "You can't take trouts with dry breeches."[27]

Kenneth Burke thought that if you correctly understood how proverbs work you could arrive at a better appreciation of literature in general. Proverbs were a kind of "medicine" or therapy: it's just a lot easier to say that kind of thing about proverbs than about *King Lear*. Why not, Burke asked, "extend the analysis of proverbs to encompass the whole field of literature? Could the most complex and sophisticated works of art be considered somewhat as 'proverbs writ large'?" Proverbs, Burke recognized, are indeed about experience (in both natural and human domains), but what they do is to name experiential "'type' situations," and often to counsel how one is to act in these type situations. One orientates to activity as one sees what kind of situation one is in. So proverbs "are *strategies* for dealing with *situations*. In so far as situations are typical and recurrent in a given social structure, people develop names for them and strategies for handling them."[28] Or, as Harvey Sacks put it, proverbs are used "to make events noticeable, perhaps to make their ordered character noticeable."[29]

The matter can be put more strongly than that. Since "the same" situation is potentially construable (noticeable) in different ways, proverbs are resources for *creating* scenes of observation and action, for making situations recognizable as situations of a certain kind. Confronted, for example, by discussions about military tactics, is the gist of the situation summed up, and its proper purpose identified, by a proverbial pronouncement on the risks of ambition and the misplaced search for certainty—"The best-laid schemes of mice and men gang aft agley"—or by one stressing the importance of detail—"Look high and fall low"? Each proverb can count as a pertinent way of identifying "what is going on here," of picking out and directing notice to its salient features. And, as each names and configures the situation differently, so each offers resources for acting in it differently, so *making it* the situation proverbially named. This is how proverbs help to constitute their referential realities. The metaphorical component of many proverbs also provides resources for creating situations through naming them. When we address some passage of human behavior by saying "Birds of a feather flock together," we can make available whatever is known about birds for understanding and orientating to kinds of human beings.[30] But what it is specifically about birds that is relevant to the human case at hand is not exactly defined. In Aristotelian terms, proverbs belong to the process known as deliberation—the taking of decisions about what to do, what may be brought about by our own efforts, in the realm of the more or less and of the contingent—where absolute certainty is neither available nor rationally to be expected. They belong to the complex circumstances of life-as-it-is-lived, not to the idealizations of philosophy or science.[31] Or, in Stephen Toulmin's vocabulary, proverbs are aids to action not in the domain of Reason but of reasonableness.[32]

What is the experience about which proverbs speak? Like scientific theories and laws, proverbs are generalizations.[33] This bears upon—indeed it is another way of pointing out—proverbs' often figurative content. The "cock" in "Every cock crows on his own dunghill" is both a generalized rooster and a generalized human being, while the "every"—as opposed to "some" or "many"—marks the fact that a generalization of wide scope is being offered. The "dirt" referred to when we say "Throw dirt enough and some will stick" similarly can be a fabricated story of sexual misconduct or a concocted accusation of scholarly dishonesty, but it is rarely garden soil, even if the properties of soil may have some relevance to the reference at hand. This is just a way of pointing out proverbs' enormous semantic reach. Proverbs evidently speaking about what's in the nature of birds, brides, and brooms may nevertheless find use in an unpredictably wide range of domains or situations. And as they find those applications, so the proverbs' reference subtly changes. Put another way, philosophical suspicion of proverbs is wholly justified: they are a logical empiricist's nightmare.

Nonetheless, too much should not be made of the move from proverbs' metaphorical base to the particularities of the situations in which they find application. You do not have to know very much, if anything, about the ways of chickens or the adhesive properties of garden soil to use the relevant proverbs properly and pertinently. It was only a few years ago that I understood why one might want to look a horse in the mouth: I'd never done it myself, and where I grew up—not exactly horsey country—no one else did either. However, I insist, for many years before that, I was able to understand, and properly use, the proverb "Don't look a gift horse in the mouth." That is because I was wholly familiar with a large number of occasions of competent usage, and I understood—as part of competently knowing what was happening in these scenes of competent usage—what intention was expressed in the saying. Reiterated usage builds up the reference, and, while translation from the farmyard domain may occur for some users, on some occasions, such translation is not at all necessary.[34] That is presumably why we still properly say—and find sense in—such proverbs as "The exception proves the rule," "You can't make bricks without straw," and "Strike while the iron is hot," even though many of those who competently say and hear such things have no experience with the techniques of brick-making and blacksmithing, and have never known the historical sense of "proving" as "trying" or "testing" that puts the seventeenth-century meaning at 180 degrees from its present-day sense.[35] If enough situational context is available, and enough experience with competent usage is on tap, you don't have to translate from metaphorical proverbs' literal aspect reliably to grasp their meaning here and now.[36] A theory about how proverbs originated, and how they circulated and signified in their original settings, is not necessarily adequate to account for how they circulate and signify in other contexts. Even while resisting deformation in their utterance, proverbs escape any such semantic discipline. And that is a mark of their referential power.

Of course, some currently used metaphorical proverbs may continue to draw upon familiar experience. Most of the educated classes probably still know just enough about lubricating mechanical gadgets to understand literally why "The squeaky wheel gets the grease," and you don't have to be an ornithologist to know that many birds of the same species just do tend to hang out with each other. In such cases, it can make some sense to talk about an *external* mode of generalizing experience: in order to understand metaphorical proverbs we are supposed to move outward from their manifest content (for instance, wheels and birds) to their situationally intended reference (complaining academic colleagues and clannish members of the "-ology" down the corridor). But metaphorical proverbs also importantly display an *internal* generalizing disposition. Take the proverb "A rolling stone gathers no moss," and take for granted that its situational reference is usually competently understood not to be rocks but instances of human behavior.[37] Perhaps knowledge of stones *is* involved in competent proverb-use: for some people, on some occasions, it might well be. But what one has to understand here is not something to do with features of that little-oval-gray-stone-with-pyrite-speckles-that-you-trod-on-in-the-desert-last-week. The stone of proverbial reference is not a historically specific stone, just as its rolling behavior is not a historically specific event. Anyone who presumed otherwise would be judged incompetent, at least pedantic, possibly even mad, and that is why proverbs are sometimes used in clinical tests of mental illness.[38] Rather, insofar as mineral concretions are involved in the reference of this proverb, what's talked about is *stones* and what it is in their nature to do, and about what tends to happen to them as they do what it is in their nature to do.[39]

In Aristotelian terms, what happens to the proverbial stone is that which happens in a certain way "for the most part," and, therefore, proverbial references to stones, or to pertinently similar aspects of human behavior, are to be treated as true "for the most part." There are all sorts of complexities and contingencies involved in real-life situations (natural or human); and it is only in the mostly irrelevant ideal-world that the essential nature of things is invariably and precisely expressed. The competent proverb-user appreciates that there may be stones in the world that roll and gather some moss; stones that do not roll and nevertheless gather little or no moss; and, indeed, people who are constantly on the go and yet who are boringly habit-bound. Instances may be found that contradict the proverb's experiential generalization, without compromising its truth. This is how generalizing proverbial statements of wide validity ("Every cock will crow on his own dunghill") or non-metaphorical proverbs ("Like father like son") can coexist, without contradiction or damage to their validity, with knowledge of specific counter-instances to the generalization. Indeed, the physical fact that it is almost *never* "darkest just before dawn" detracts in no way from the validity of its proverbial application, as the proverb, for example, captures something of the psychological trajectory of human despair. Here is yet another way in which proverbs are epistemically powerful.

Given that the generalizations expressed in proverbs are true "for the most part," or "in the sense in which they are competently intended," it is not pertinent—it is, indeed, a violation of decorum—to treat proverbially expressed generalizations either as invitations to systematic inquiry (as prolegomena to empirical study of metamorphic stones and their environmental accretions) or as vulnerable to empirically observed counter-instances.[40] Harvey Sacks noted academics' general tendency to treat proverbs solely as propositions about their explicit referents, "and to suppose then that it goes without saying that the corpus of proverbs is subjectable to the same kind of treatment as, for example, is scientific knowledge."[41] Bruno Latour specifically contrasts the referential "softness" of proverbs to the "hardness" of scientific propositions. As an example of the distinction, he offers a mother telling her son "An apple a day keeps the doctor away." The awkward son replies by citing scientific studies that contradict the empirical validity of the proverbial generalization. What impresses Latour about the exchange is that the proverb's claims to truth cannot stand the test of the resources the son brings to bear on it: the son's "hard" language-game mobilizes resources that the mother's "soft" one can't cope with. Latour observes that the proverb is not used, as scientific propositions are, as an argument to win a counter-argument and that it does not have the strength to do so. That's right, of course, but the power of the mother's proverbial utterances is linked to the fact that she will, in this case, accept no counter-argument from nutritional research. The son doesn't win because of the "hardness" of his scientific speech; he just winds up looking silly or insolent. The upshot of any such exchange is not the triumph of scientific over proverbial propositions; it is a failure in the son's sense of decorum, as the mother herself would fail were she to interject that proverb into a formal exchange between expert nutritionists.[42]

Sacks observed that "one of the facts about proverbs is that they are 'correct about something,'" and Aristotle rather irritably wrote that "just because they are commonplace, everyone seems to agree with them, and therefore they are taken for truth."[43] That reputation is a real epistemic advantage, and one index of this advantage is the manner in which proverbs are often used. They are recurrently employed as a kind of coda—a way of summing-up or bringing matters to a head or conclusion, often performing a function similar to Dr Johnson's "And that's an end of the matter." *The*

Economist's celebrated house-style continues to use proverbs abundantly to the same effect—summing up a line of description and evaluation and occasionally at the same time deflating the balloons of fancy economic or management theorizing.[44] Unless the setting is the extended proverb-exchange common in traditional societies, proverbs are not invitations to continued conversation on the topic at hand; they are, rather, conversation-stoppers, signaling that perhaps it's time to move on from this particular subject, that this is, for the moment at least, the last word on the matter.

Proverbial Economies

Many proverbial sentences can be thought of as rule-like propositions, used to regulate judgment and counsel action in a range of situations. So it is pertinent to compare them with other rule-like propositions we know about and how these other propositions work. There is a sense in which proverbs are *truer* than those propositional rules that function in economies that do invite inquiry as to their validity. Sacks offered as a comparison the law and its rules: "[There] if you invoke a rule by reference to precedent, the occasion of using it can provide the occasion for reconsidering the rule to see whether, not only in this instance but in general, it ought to obtain for anything. So that a rule introduced to govern a situation in a law case can be changed altogether." And so Sacks saw no reason to resist the apparently odd conclusion that "even a strict precedence system such as [the law] doesn't have objects as powerful and as limitedly attackable as proverbs."[45] Part of the sensible oddness here consists in the observation that a weak Popperian object—a proposition that resists falsification and that therefore is widely supposed to have no proper epistemic virtue—is among the most useful and robust elements of our culture. What is supposedly lost in refutability is gained in adaptability.

Whether metaphorical or not, proverbs are generalizations that nevertheless take their meaning as they are invoked in particular situations of judgment and action. The proverb thus appears in two guises—first, the generalizing proposition or prescription, treated as true, held in common, and circulated as linguistically stable; and, second, its highly variable semantics and references, deriving from the multiplicities of situations to which it is applied and which determine its meanings. Proverbs link the general and the particular. They make *this* instance of judgment or action understandable or legitimate, in light of statements about what it is in the nature of things to do, or of people to behave, or of situations to turn out. And, as they get applied to further particular occasions, so the references of the generalizations are themselves modified.

The proverb "Everyone knows where his shoe pinches" can be pertinently brought to bear on lay or expert medical diagnosis, to methodological debates in Anglo-American sociology, to the marketing and design of personal computers, to the distribution of charitable funds, and to innumerable other situations now existing or to unpredictable situations that will come into existence in the future. Such specific situations, as Kenneth Burke observed, "are all distinct in their particularities; each occurs in a totally different texture of history; yet all are classifiable together under the generalizing head of the same proverb."[46] The purpose of using this proverb is to draw attention to an accepted and widely known general rule that is illustrated by the case at hand, or, conversely, how the case at hand comes under the compass of some accredited and familiar generalization.[47]

These observations suggested to Sacks a way of understanding proverbial economies as "atopical" phenomena. The point of the rolling stone that gathers no moss is atopical in that nothing pertinent to its sense and use is to be illuminated by reference to particular mineralogical or botanical findings. It is not a matter here of concrete versus abstract modes of thinking, as if failure to comprehend the proverb's moral message were a failure of the ability to abstract. The proverb-as-proposition is quite abstract just as it is, since it is not to be understood as concerning the behavior of any particular stone. I have already indicated that in vernacular usage proverbs are not competently subject to empirical inquiry about their validity, nor are they as a body monitored for logical consistency or non-contradiction. On the contrary, while they may be misapplied—and thus lack in force—proverbs are held to *be* true, and it is the work of the auditors in a scene to figure out *how* they are true here and now, how they pertinently address judgment and action in this particular setting. This is a very powerful way of organizing bodies of knowledge and action. "In that way," Sacks noticed, "instead of constantly revising a body of knowledge by reference to the discovery that it's not correct here, now, for this, you maintain a stable body of knowledge and control the domain of its use."[48]

It has, indeed, been a notable feature of learned engagement with proverbs to compile them, to make a list of them, to inspect them as a body for coherence and sense, and I will return to that tendency toward the end of this chapter. But the acts of compilation, arrangement, and inspection transform their objects. Proverbs-compiled-in-a-list are not the same things as the vernacular items purportedly compiled. A proverbial economy does not compile its body of short-generic propositions or prescriptions and inspect them in these ways or make their generalizations occasions for inquiry. Were it to do so, then it would be some other kind of epistemic economy. In a proverbial economy, the pertinent judgment does not concern the truth of the proverb—that is largely taken for granted—but the pertinence and productiveness of inserting this particular proposition into this scene.[49]

Against the general learned tendency to treat proverbs as naked propositions-on-a-printed-page, there are some anthropological and sociolinguistic recommendations to conceive them as speeches in situations of use, that is, to come to grips with functioning proverbial economies.[50] As early as 1926, the anthropologist Raymond Firth insisted that "The meaning of a proverb is made clear only when side by side with the translation is given a full account of the accompanying social situation—the reason for its use, its effect, and its significance in speech. It is by nature not a literary product."[51] Anthropologists recognized that the meanings of proverbs were rarely transparent to naïve or non-native auditors just as they were.[52] They may have to be explicated, decoded, seen to be correctly chosen, spoken, and applied. Who does that? Who *can do* that? Aristotle's *Rhetoric* sounded a significant warning to people about to spout these sorts of sayings: "The use of maxims is appropriate only to elderly men, and in handling subjects in which the speaker is experienced. For a young man to use them is—like telling stories—unbecoming; to use them in handling things in which one has no experience is silly and ill-bred: a fact sufficiently proved by the special fondness of country fellows for coining maxims, and their readiness to air them."[53]

As general counsels about how to judge and what to do, they just don't *work* when uttered by the kinds of people who are competently known not to have broad experience and not to have the authority to pronounce about what should be done. In a society that acknowledges the epistemic and moral virtues of embodied age-and-experience, you don't ask your grandmother about her warrants for attaching a proverbial generalization to a particular event—for the very same reason that you don't teach her to "suck eggs," "get children," "sup sour milk," or "grope her ducks"—just because it is her embodied authority that provides adequate grounds for such attachments. And, again, your grandmother's proverbial assertions are not occasions for inquiry into their evidential warrant, just because it is her embodied authority that tells you what adequate evidence is.

Similarly, an African ethnographer observes how proverbs can work to reassure or to reconcile: "This is also the function of proverbs in modern Occidental culture, or rather of the 'bromides' with which they have been merged . . . [A] proverb is usually quoted to the disturbed individual by a senior, and it comes as the voice of the ancestors, his seniors *par excellence*."[54] Note the pervasive "My son" prefacing the biblical Proverbs—"My son, keep my words"; "My son, forget not my law"—and the presumption that proverbs are both repositories of value, that their wisdom may not be transparent to the young and naïve, and that they may require explication by the old and experienced. The young or naïve person is apt to choose proverbs badly or to misapply them, so rendering them worthless. So said the Apocryphal book Ecclesiasticus: "A proverb will fall flat when uttered by a fool, for he will produce it at the wrong time."[55] This was Sancho Panza's problem: his proverbs came out all in a jumble, without due recognition of their proper occasions of use. "Look you, Sancho," Don Quixote said, "I do not find fault with a proverb aptly introduced, but to load and string on proverbs higgledy-piggledy makes your speech mean and vulgar."[56] While many proverbs are indeed propositions, the knowledge of how to store, select, and apply proverbs—and therefore of how to give them force—is itself not propositional: it is a skill acquired by experience and, when acquired and displayed, a mark of wisdom. There are no propositions that adequately specify the conditions of proper usage. This sense of decorum is substantially given by age-and-experience, and it is made manifest to others—and hence potentially transmissible—by the example of embodied wisdom. Absent such examples, proverbs in themselves can have little authority. For this reason, a proverbial economy cannot be described without figuring in the culture surrounding embodied proverb-speakers as well as the occasions of proverb-use.

A similar consideration applies to the metaphorical aspect of many proverbs. If the pertinence of a metaphor cannot be subject to proof, how is it made locally persuasive? In this connection, too, a particular speaker in a particular setting has got to point out, if necessary iteratively, just what it is about, for example, rolling stones and moss in general that competently applies to *this* aspect of human behavior at hand. This "matching up" or "correlation" can draw on any and all aspects of the present scene, but one very potent scenic element is, once more, the embodied attributes and authority of the one who speaks, who invokes the proverb, and who adds his or her personal and generic authority to those of the nameless ancestors for whom he or she now speaks.[57] One cannot properly talk about how proverbs are true and pertinent without talking about the capacity of certain kinds of people in certain kinds of scenes to identify what is to count as truth and pertinence. Just as Aristotle said that you should not utter proverbs until you reached a certain age, so he recognized the "character" of a speaker as "almost the most effective means of persuasion he possesses."[58]

Proverbs as Reflective Knowledge

From traditional learned points of view, talking about proverbs—items of commonsense knowledge—as ways of improving judgment and action must seem odd. It suggests a kind of reflectiveness usually associated only with the deliberations and pronouncements of the learned. Indeed, this chapter set out by gesturing at a great tradition in which the learned commended themselves and condemned proverbs by noting that proverbial common sense was endemically unreflective. The common people, it is said, take things just as they seem to be, habitually declining to go beyond superficial appearance to the truth or pattern that lies behind. However, as stable as this imputation has been over the centuries, there is a sense in which it is flatly contradicted by the most cursory inspection of any proverb dictionary. Thousands of proverbs, in all cultures, enjoin just the sort of reflectiveness that is supposedly absent from commonsense knowledge. They counsel the inexperienced and the naïve against vulgar errors of inadequately justified judgment or undisciplined inference. (Even the vulgar have their vulgarians.) Actors in proverbial economies have available to them a stream of advice that counsels against taking things just as they seem to be. Experience advises otherwise: "Every light is not the sun"; "Everyone thinks his own fart smells sweet."[59]

Proverbs of this sort can be called *inference instructors,* and they come in several varieties. One type of proverbial instruction cautions against premature or overenthusiastic inference from particular to pattern. It warns those with a restricted stock of experience that it is unwise to infer from one instance, from short-term patterns, or from local manifestations, to the way things will normally pan out, to the course of nature or the nature of people. Such inference-instructing proverbs, when suitably uttered by suitable people, identify the pitfalls to sound judgment that have been noticed by long experience. Indeed, they tell inexperienced people that they *are* inexperienced, and in what ways, that what they might regard as a sufficient stock of experience and basis for inference are no such things. "One swallow does not make a summer," or "Don't count your chickens before they're hatched," is something the experienced chief financial officer might tell a young biotech researcher convinced that new experimental results warrant an immediate Initial Public Offering, or, that a father might tell a son celebrating early sporting success, or (as "One robin doesn't make it spring") that a skeptical California farmer recently told an agricultural scientist conducting field trials of a new variety of celery.[60] And a mother might say to her adolescent daughter moping over a failed first romance that "There are many more fish in the sea." You think, the mother means, that you'll never be in love again, but you will, probably many times over; no, I can't guarantee it, but that's the way these things tend to work out—usually or for the most part. The proverbial voice here notes that the patterning—and the distorting—influence of a single striking instance may be strong, and it advises prudent actors to recognize its effects: "Once bitten, twice shy"; "The burnt child dreads the fire."[61]

In such ways, naïve persons are told both about what the world is like and about sources of knowledge about the world: the what is contained within the proverb; the persons speaking it in a setting help secure its meaning while, at the same time, they constitute themselves as reliable sources of knowledge about the underlying structure or pattern of the world and as receptacles of collective wisdom. Naïve and unreflective persons are told that they *are* naive and unreflective, that their stock of experience is in fact restricted, and that there are human sources of knowledge available who embody vast stores of experience and prudence. Proverbs of this sort, when suitably uttered, act as vehicles for the transmission of accumulated experience from the old to the young, and, more generally, from the experienced to the inexperienced in any endeavor. They uphold the moral order as they testify to the order of nature.[62]

Despite the torrents of learned commentary deploring vulgar perception and judgment for their entrapment in superficial appearance, yet another large body of proverbs warns against mistaking appearance for reality: "All that glitters is not gold"; "All are not friends that speak us fair"; "You can't tell a book by its cover" or "wine by the barrel"; "Just because there's snow on the roof doesn't mean there isn't fire in the oven"; and, more generally and theoretically, "Appearances are deceptive." Don't be taken in by flash superficiality. Things are rarely what their surface appearances suggest: "Truth lies at the bottom of a well" and "The best fish swim near the bottom." What is merely superficial—however fashionably and fiercely valued—is likely at the end of the day to prove empty or meretricious. Go for the solid and enduring stuff; don't follow the confederacy of dunces. Neither truth nor any social or material goods worth having are easy of attainment. Anyone who thinks so is a fool; anyone who tells you so is a fraud. Again, it takes accumulated experience to know this. Listen to the voice of that experience and learn. You will then be warned of life's recurrent pitfalls, and freed from the painful necessity of making your own mistakes: "Experience keeps a dear school, but fools will learn in no other."

There are, however, proverbial aids to right inference from particular to pattern, and accumulated experience is available to identify these aids and to counsel how they should be recognized, applied, and acted upon. Many proverbs direct notice to means by which one can reliably discern real states of affairs from visible signs: "There's no smoke without fire"; "Nearest the heart, nearest the mouth" (a folk version of the "Freudian slip"); and "The eye is the index of the soul." Still other proverbs speak about the sorts of alterations that are not to be expected, given the nature of things, about how natural and human things tend to work out, usually or for the most part. Take, for instance, "The leopard does not change his spots," and the series of proverbial commentaries on heredity and development that includes "What's bred in the bone will not out of the flesh"; "Like father/mother, like son/daughter"; "The apple falls not far from the tree"; "The child is father of the man"; "Such is the tree, such is the fruit"; and "Blood will tell."[63] Here the message is that things in general tend to go on in the future as they have in the past. Given the nature of things, there are limits to the changes of which people and natural processes are capable, and which it is reasonable to expect of them.

"No tree grows to the sky," and "Whatever goes up must come down." Singular violent occurrences and extremes meet natural tendencies in the opposite direction. Things—natural and human—tend to even out over time. One extreme is counter-balanced by another. Whatever "this" is, this too will pass: "All that is sharp is short." Learn to recognize the violent, the singular, and the extreme for what they are. It is not reasonable to expect that such instances, however remarkable and however much they may grip the imagination and the emotions, offer reliable signs of what is normal, and it is not prudent to plan judgment or action that is predicated on their long continuance or even recurrence: "Lightning never strikes the same place twice." In this way, some proverbs teach at once what is singular or extreme, as opposed to what is normal, and, again, they do so on the basis of accumulated experience, condensed in proverbs and brought to bear on a particular scene by those entitled to speak in the name of such experience: "Every dog has his day"; "After a storm comes a calm"; "The tide will fetch away what the ebb brings"; "Pride goes before a fall"; "What goes around comes around."

Just when you think that things are really set fair, they will turn lousy. Or when you think that things will never improve, they will get better: "Every cloud has a silver lining"; "It's an ill wind that blows nobody any good." Or, again, when you think the ways things are will never change, they do: "It's a long lane that has no turning"; "Sometimes all honey and then all turd." It's not reasonable to persuade yourself that good comes without bad, or that either good or bad can long continue. In general, there's little purity in the world; things come all jumbled up: "Every path has a puddle"; "There's no mirth without mourning"; "No pleasure without pain"; "No weal without woe." Murphy's (or Sod's) Law ("If an aircraft part can be installed incorrectly, someone will install it that way")—made up by the Northrop aeronautic engineer Captain Edward A. Murphy in 1949—has age-old proverbial predecessors: "Nothing is certain but the unforeseen"; "The unexpected always happens"; "The bread never falls but on its buttered side."[64]

Proverbial Relativism

Proverbial voices warn that it is imprudent to take at face value claims to universal, timeless, or absolutely certain knowledge. (The vulgar too have their postmodern moments.) Beware of anyone who tells you that there is a global formula for right judgment or a royal road to right action. Watch out for anyone who claims that generalizations about the real world can hold universally and without exception. Life is too complicated, too rich, and too heterogeneous to support any such assertion. Human intelligence can't compass the jumble of creation or the idiosyncrasy of people, and it's a mark of learned fools that they think their wit is up to the task, that a precise "theory of everything" is at hand or just around the corner. For every sucker born every minute there's a simplifying rationalizer or a moralizing snake-oil salesman who's willing to take the sucker's money. In intellectual claims, as in material goods, proverbial voices say "Caveat emptor": "Comparisons are odious"; "Circumstances alter cases"; "Every like is not the same"; and neat abstractions tend not to hold good when one is concerned with the contingencies and complexities of real-world judgment and action. Always best to be cautious in one's judgments, and circumspect in the scope of one's conclusions: "Almost was never hanged."[65]

As in the natural, so in the moral: there is no one right way to judge and to act that holds good in all places, times, and circumstances. Solomon taught that "To every thing there is a season, and a time to every purpose under the heaven."[66] The voice of the people concurs: "Many men, many minds"; "Other times, other manners"; "When in Rome do as the Romans"; "Being on sea, sail; being on land, settle"; "Horses for courses." There's no accounting for, or disputing, tastes: "One man's meat is another man's poison"; "Chacun à son gout"; "Every shoe fits not every foot." "Live and let live" because "It takes all kinds to make a world." People and their predicaments differ, and so their customs, values, and standards differ accordingly. There's no exact science of such things, except in the excremental purity of the uselessly abstract. And there should be nothing either surprising or troubling about injunctions to moral or epistemic decorum: "Measure not another by your own foot," or, as Sly and the Family Stone put it, "Different strokes for different folks."

So relativism has deep roots in common sense. Many proverbs acknowledge interpretative flexibility and express suspicion of claims to semantic fixity or the sufficiency of propositions to firmly fix meaning: "Everything is as it is taken"; "It is not the matter but the mind." At the same time, such voices do not proceed from the flexibility of meaning to the commendation of postmodern playfulness. Judgment *does* vary from situation to situation, but local standards may be, and legitimately are, obligatory. When in Rome you *must* do as the Romans, even though you are well aware that things are legitimately done otherwise in Florence. There is nothing "mere" or "arbitrary" about such proverbs' view of local custom. Just as folk wisdom testifies to both the variability and the force of local obligations, so it voices skepticism about either the availability or the necessity of transcendental justification. Justification both comes to an end, and is *bottomed*, in local obligation. Household gods are all the gods going, and they are quite powerful enough.[67]

There is no reason to push too far the claim that folk wisdom is globally relativistic in any precise academic sense. There are, in fact, other proverbial voices pointing out, or urging, commonality of standards and finding unity in apparent diversity. "What is sauce for the goose is sauce for the gander" can be used to argue that different categories of people ought to be judged *not* by different but by the same standards. That "There is one law for the rich and another for the poor" counts both as a cynical statement about how things are and as an implicit recommendation that they ought to be otherwise. And, in the case of truth and authority, similar sensibilities are voiced by "Gold speaks"; "Money is the best lawyer"; and "If the doctor cures, the sun sees it; but if he kills, the earth hides it." (Proverbial voices have never spoken well of doctors and lawyers, or, for that matter, millers.) There are also proverbs that direct attention to genuine and pertinent sameness against uninstructed or biased tendencies to see differences and particularities: "In every country dogs bite" or "The sun rises in the morning"; "All the world is one country." Even a king "has to put on his trousers one leg at a time," the same king that "even a cat may look at," and robust subjectivity may turn out to offer a more secure basis for consensus than learned pretences to objectivity: "Hearts may agree though heads differ."

The most pervasive sentiment informing such pieces of folk wisdom is not, of course, some formal relativist *position;* it is, rather, an emotionally charged skeptical, even iconoclastic, deflation of intellectual pretension and moral absolutism. The learned person is likely to prove a fool in ordinary life—and it's ordinary life that really does matter. The pious preacher is quite possibly a hypocrite when it comes to everyday moral action. Those who pretend to speak for God, Truth, and Reality usually turn out to be speaking for their own special interests. Grand rational plans, systems, and abstractions rarely hold good when they are actually put to the test in the real world: "It's easier said than done"; "Talking pays no tolls"; "Fine words butter no parsnips." But putting fine words and rational plans to the test of everyday life *is* the appropriate assay: "The proof of the pudding is in the eating." So intellectuals come off no better than lawyers: "A handful of nature is better than an armful of science"; "A mere scholar, a mere ass"; "Much science, much sorrow." Nor do priests and the pious: "The nearer the church, the farther from God"; "All are not saints that go to church." If your concern is with practical judgments in real life, then common sense is a surer guide than book-learning: "Years know more than books."[68] As Clifford Geertz maintained about "commonsense" sentiments in general, "Sobriety, not subtlety, realism, not imagination, are the keys to wisdom."[69]

Are Proverbs Logically Incoherent?

If you treat proverbs as naked propositions—if you do not bring to bear on their interpretation the culture and contingencies of a setting of use—then they suffer from some quite obvious epistemic flaws. The truth of the proverbial proposition "A fish rots from the head" seems doubly vulnerable: it is susceptible to counter-evidence that might be contained either in your refrigerator or in a university faculty whose malaise is not obviously the dean's fault. Moreover, there are evidently quite a lot of proverbs—one thinks of weather proverbs about Bredon Hill putting on his cap, about Groundhog Day, green Christmases, and white Easters—that are trite, that don't stand up to the findings of modern science, or that continue to circulate in settings where they have little or no predictive value: it's just what you say to have something to say about the weather (though that has its own definite kind of "phatic" purpose).[70] For all the wisdom one might want to acknowledge in proverbial economies, and even for all of proverbs' robustness, the body of proverbs is not *philosophy*, or, to be more precise, not philosophy as it is ideally represented. Proverbial propositions need a lot of situational help in order to be pertinent and true. If proverbial propositions are to be accounted true, then a whole raft of qualifications, reservations, and stipulations about context and contingency has to be noted or taken for granted as part of their assessment and application. But once they are granted that help—that is, once they are seen as part of a working proverbial economy—then they are powerful stuff.

Yet scholars have historically been reluctant to concede proverbs that situational help. Scarcely had the dust settled on the earliest learned efforts systematically to collect such folkish sayings when it was recognized, and loudly trumpeted, that valued proverbs could be found that formally contradicted other valued proverbs. In Tudor and Stuart England, humanists liked to play with the genre known as *crossed proverbs*—assembling proverbs that self-evidently clashed with each other—to make sport, and sometimes also to make a point about the inadequacy and the methodical indiscipline of the common people's way of thinking.[71] More recently, some social psychology textbooks demonstrate the inadequacy of lay reasoning by drawing students' attention to the phenomenon of contradictory proverbs: "A standard ploy is the presentation of a set of maxims, proverbs or bits of folk wisdom as 'common-sense theories' of social psychology. Then, when certain pairs of maxims are shown to conflict (for example, 'Birds of a feather flock together' as against 'Opposites attract'), and the utter senselessness of common-sense psychology has thereby been demonstrated, the writer is free to appraise students of the virtues of the scientific approach to these matters . . . Everyman [is given a] perpetual role as a straw man."[72]

For every apparently sage "Look before you leap," there is an equally valued and opposite "He who hesitates is lost"; "Absence makes the heart grow fonder," but "Out of sight, out of mind." How could rational persons find the slightest value in any such generalizing propositions when the body of them was so evidently vulnerable to contradiction? How could vulgar knowledge be anything but worthless when such contradictions were tolerated, amazingly even going unnoticed until the learned collected them, arrayed them in lists, reflected on them, and pointed out their contradictory nature? Elementary logic soundly teaches us that a proposition cannot be true if its opposite is also true: "Socrates is mortal" is true just on the condition that "Socrates is immortal" is untrue. And so, to the extent that proverbs massively contradict each other, the body of proverbs is incoherent and its members individually unreliable.

Here, again, the learned criticism mistakes its object. In his study of the relationship between cultural change and the development of literacy, the anthropologist Jack Goody noted an interesting feature of scholarly engagement with proverbs. The first thing the learned did was to *make lists* of them. By that simple act, Goody observed, the learned not only removed the proverb from the contexts in which it had its traditional being, but also shifted its identity and altered its epistemic value. By taking the proverb out of its oral, situated, and purposive setting, "by listing it along with a lot of other similar pithy sentences, one changes the character of the oral form. For example, it then becomes possible to set one proverb against another in order to see if the meaning of one contradicts the meaning of another; they are now tested for a universal truth value, whereas their applicability had been essentially contextual (though phrased in a universal manner)."[73] The proverb that is thus contradicted by another proverb is a different thing from the proposition or prescription uttered in context of practical use.

Given that proverbs are strategies for dealing with situations, they may properly be said to contradict only when both the particular situations they typify and the particular attitudes they express about that situation are the same. But just as proverbs-in-an-economy name (and help create) recurrent situations and point out pertinent similarities between them, so concrete situations lumped together for one purpose may differ in any number of respects, and the respects in which they differ may be pertinent for other purposes. "Many hands make light work" contradicts "Too many cooks spoil the broth" just on the condition that, say, all kitchen work is the same with respect to the value of your helping me. But why ever should that be? If I am washing the dishes, if I am fed up with the work, and if my kitchen happens to be big enough, then "Many hands make light work" is what I might say to sum up a situation, to link it in your mind with other warrantably similar situations, and to summon your assistance. However, if I am whisking up egg-whites for my famous flourless walnut cake, no matter how hard the work is, and no matter how large the kitchen, I decline any help you might offer by saying "Too many cooks spoil the broth": this is a one-person job, like others to which (you know) that saying has been applied. The contradiction that seems so evident when proverbs are treated as isolated propositions-in-a-list vanishes like smoke when they are interpreted as utterances-in-a-particular-situation.

Proverbial Common Sense and Science Revisited

I began by sketching a pervasive evaluative contrast made by the learned between proverbs (as tokens of vulgar knowledge) and properly expert forms (the propositions of science and philosophy). The vices of proverbial common sense consisted largely in its unreflectiveness, its referential imprecision, and its incoherence; the virtues of learned knowledge included its refusal to remain trapped in the world of appearance, its clarity, and its logical tidiness. Accordingly, I end by reviewing some of the epistemic virtues of proverbial economies, and then by reconfiguring aspects of the traditional contrast between our most highly valued forms of expert knowledge and the world of proverbial common sense.

Supposing one wanted to say that proverbs are epistemically powerful and worthy: what sorts of virtues could one now mobilize in their favor? First, proverbs' linguistic structures flag that something special is being said— for example, summing up a situation or giving an overall judgment about what is to be done—and these same structures facilitate recall and undeformed circulation—what I have referred to as proverbs' mnemonic robustness. Anything that is so readily retrieved, that travels while remaining relatively unchanged in form, and that is put to such a range of practical uses might be thought rather potent. Defining knowledge as that which is archived may point in the direction of one sort of epistemic virtue; defining knowledge as that which can be easily accessed and put to productive use here and now points toward another kind of virtue.

Second, while proverbs cannot avail themselves (as aphorisms and quotations can) of the authority that may be attached to a prestigious individual originator, they are compensated by speaking with the voice of tradition, the ancestors, and anonymous collective wisdom. No individual or sectional interest attaches to their claims; no special circumstances attending their origins limit their applicability. Third, proverbs-in-use can have a self-referential quality. They tell you what sort of situation you are in, so orientating you toward appropriate action and, in so doing, they help to *create* the cultural realities they describe. Fourth, proverbs-in-use are generalizations about experience and action that are semantically protean and highly adaptable to different situations. They can speak about cocky chickens and cheeky children at the same time, and their potential range of reference is not subject to knowable limits. Their form stays stable, and their truth may be conceded, while both their meaning and their reference change. Fifth, as cherished generalizations, proverbs-in-use are highly protected from refutation by way of empirically available counter-instances. Competent members of the culture understand that proverbial generalizing speech about chickens is not to be negated by awkward facts available from the expert knowledge of poultry science, nor are such competently used metaphorical proverbs subjectable to queries about the appropriateness of chicken behavior to the human case at hand. Thomas Kuhn has shown us how deeply entrenched in expert communal life are the paradigms of a scientific practice, but I suggest that it is far easier for members of the appropriate subculture to negate or modify the formal generalizations of either poultry science or particle physics than it is to dispute the truth or pertinence of a competently uttered proverb in its naturally occurring economy. For these and other reasons, it is hard to challenge the epistemic power of proverbs in those natural economies. What you can dispute, and what has indeed been repeatedly disputed, is the epistemic *value* of proverbs compared to the propositions of such learned practices as philosophy, natural science, medicine, and engineering. How might such criticisms be answered?

Part of the answer has already been given: apples should be compared with other apples, oranges with other oranges, but an equitable and informed comparison is rarely on offer when the learned compare their knowledge and practice with that of the common people. Contrast proverbial propositions with the propositions of formal philosophy or natural science and they tend to come out badly. But even here learned generalizations about proverbs are often flawed by selective inattention to the variety of sentiments expressed in proverbial propositions. I have shown that it is no hard work to assemble a mass of proverbs commending a reflective and skeptical attitude to superficial appearances or to the claims of established authority and, hence, that it would be an easy matter to reconstruct the sentiments, norms, and gross methods of science from the propositions of proverbial common sense. Proverbial propositions are very various. If we mean by common sense the sentiments expressed in its proverbial propositions, then *there is no one direction* in which they collectively point; if, however, we mean by common sense the cognitive capacities employed in an array of everyday reasonings, there is no convincing reason yet offered to distinguish these capacities from those employed in a range of learned activities. More important, in learned evaluative contrasts, proverbs are typically denied the help of their natural scenes of use, and the epistemic virtues of a proverbial economy are transformed into the vices of proverbial-propositions-in-a-list.

The unevenness of the comparative playing-field is more evident than that, for a parodic account of proverbs is typically contrasted not to the real worlds of scientific and philosophical practice but to cosmetically worked-up idealizations of science and philosophy. However, if we take a closer look at a range of modern expert practices, we can begin to notice the role of linguistic forms strikingly like those the folklorists have documented in everyday life. Present-day learned practices also have their proverbs and other mnemonically robust short genres; proverbial economies are present there too. For example, canonical scientific laws and meta-principles frequently avail themselves of the mnemonic robustness of the proverbial form: "Ontogeny recapitulates phylogeny"; "For every action there is an equal and opposite reaction"; "Opposites attract"; "All life/cells/organization from pre-existing life/cells/organization"; "Nature abhors a vacuum"; "Nature doesn't make leaps"; "Remove the cause and the effect will cease." More significantly, proverbial (and similar) forms often express modern technical practices' valued rules of thumb. They identify recurrent predicaments, point out pitfalls, and instruct practitioners how to proceed in different types of situation. Such proverbs only occasionally find their way into textbooks and formal presentations of their practices' knowledge-base, but practitioners would arguably be lost without them. They would find it much more difficult to transmit their cultural heritage from one generation to the next, or, should they wish to do so, to make their principles accessible to cultural neighbors. The case for inarticulable tacit knowledge in science, medicine, and engineering is now well established as a matter of principle. But no legitimate appreciation of the tacit dimension in science should dispute the value of mnemonically robust linguistic genres—what I call "technical proverbs"—in transmitting expert lore within relevant cultures or even across some cultural boundaries. Of course, the understanding of such technical proverbs is dependent upon a prior shared culture but, in practice, such a culture-in-common is sometimes available. Whether or not it is available in specific cases is a matter for empirical inquiry, not for methodological fiat.[74]

Some of these technical proverbs are by tradition established or imported from common usage; others are evidently recent special creations. In biochemistry, for example, it is a maxim not to "waste clean thinking on dirty enzymes"; in population biology, and many allied fields, they say "Statistics is a way of making bad data look good"; an immunologist, discussing the merits of rival hypotheses, reminded readers of the legal and forensic maxim that "Absence of evidence is not evidence of absence." Sara Delamont and Paul Atkinson's recent study of doctoral training in science quotes the wisdom a biochemistry supervisor aims to transmit to his students: "Where . . . an experiment is not working my attitude is 'don't flog a dead horse,'" and David Noble documents the "First Law of Machining"—"Don't mess with success." In English law, judges are prudently counseled "When in doubt do nowt." Stock market speculators are warned that "Many a good mine has been spoiled by sinking a shaft," that "No tree grows to the sky," and that they should "Buy on the rumor, sell on the fact." But in a bull market investors are assured that "A rising tide floats all boats." Designers of search engines say "Big guns rarely hit small targets," and start-up companies having a poor sense of their potential market are sometimes reminded by their advisors that "If your only tool is a hammer, everything looks like a nail." Venture capitalists are told that "There's no premium for complexity"; that they should "Bet on the jockey, not on the horse"; and that "It takes money to make money." Optimists among them say "Go big or go home," while post-dotcom-crash realists say "More companies die of indigestion than starvation." Entrepreneurs are warned against the strategy of "selling vitamins [optional goods] rather than aspirins [necessities]." Human resource managers distill their craft's wisdom about remunerating talented technical staff by reminding CEOs that "If you pay peanuts, you get monkeys." A Florida state emergency planner stated as an expert adage in his business that one should "Run from the water, and hide from the wind." An ecologist summing up his position about Colorado River usage judged that "The fish doesn't drink up the pond in which it lives." A baseball pitching coach encapsulated his expert advice by saying "Challenge early, nibble late," and a Royal Navy officer in the Falklands war referred to a maxim quoted by his public school boxing coach: "If they fight, box them; if they box, fight them." In intellectual property law they say that "Tradition is permission." A linguist commenting on the "Ebonics" controversy said that "A language is merely a dialect with an army." And a now-influential dictum in macrosociology has it that "War made the state, and the state made war."[75]

From Antiquity to the present, medicine has relied heavily on maxims and aphorisms that economically and memorably transmit expert knowledge of probable causes, effective therapies, and the nature of life in the medical profession. Some of the Hippocratic aphorisms have proverbial form—"Life is short, art is long"; "Desperate cases need desperate remedies"; "Cures may be effected by opposites." A few modern medical maxims are metrically catchy: in dermatology, there is only some whimsy in instructing novices "If it is dry, make it wet,/If it is wet, make it dry,/If it is red, make it blue,/If it is blue, make it red,/If all this fails, soak it in warm Pablum." Still others have only brevity, and a certain local vividness, to assist their retention in physicians' unusually well-trained organs of memory. The great clinician Sir William Osler offered some aphorisms that have proverbial warrant—"The glutton digs his own grave with his teeth"—and many others that are just brief, and well-turned, enough to strike home: "Feel the pulse with two hands and ten fingers"; "Depend upon palpation, not percussion, for knowledge of the spleen"; "If many drugs are used for a disease, all are insufficient"; "Pneumonia is the captain of the men of death and tuberculosis is the handmaid."[76] Collections of anonymous medical maxims—terse, but rather less catchy—continue to circulate, especially among students always on the lookout for cribs and abridgments to assist the medical memory under strain: "Headache due to hypertension is generally occipital"; "It is uncommon for vascular disease to be limited to one area."[77] Some modern commentary on the degree of certainty legitimately to be expected in medical practice celebrates the probabilistic character, and proper understanding, of medical maxims: "Maxims that begin with probability, rather than with certainty, are more faithful to the wisdom of the experienced clinician."[78]

Computer programming famously contributes to the general culture "Garbage in, garbage out" and "What you see is what you get." And an early guide to writing BASIC, FORTRAN, and COBOL was entitled *Programming Proverbs,* the cover designed to look like an old American almanac. It was organized as a series of twenty-six glossed maxims for programmers, each "proverb" identifying possible pitfalls or suggesting proven ways of working round them, for instance, "Never assume the computer assumes anything." "As with most maxims or proverbs," the author sagely noted, "the rules are not absolute, but neither are they arbitrary. Behind each one lies a generous nip of thought and experience . . . Just take a look at past errors and then reconsider the proverbs. Before going on, a prefatory proverb seems appropriate: 'Do Not Break the Rules before Learning Them' . . . Experience keeps a dear school, but fools will learn in no other."[79]

We are now in the world of heuristics. The term derives from the Greek, designating the art of discovering, and, as it came into modern English usage, it tended to pick out the cognitive processes and linguistic resources used to solve problems and to render judgments when information is incomplete and when the tools of formal logic and probability theory are either inappropriate or practically unavailable. In the mid-1940s, the Hungarian-American mathematician George Polya surveyed the heuristics of mathematical problem-solving in his classic *How to Solve It*. These heuristic principles, Polya wrote, "are general, but, except for their generality, they are natural, simple, obvious, and proceed from plain common sense . . . They state plain common sense in general terms." Commonly, Polya noted, heuristics take proverbial form, and he concluded his book with a section entitled "The Wisdom of Proverbs." Folkish proverbs, he observed, often capture crucial aspects of mathematical problem-solving, identifying recurrent pitfalls and prescribing constructive action. Indeed, they are used in mathematical culture to transmit lore and warn of dangers. Heuristic proverbs are not perfect: "There are many shrewd and some subtle remarks in proverbs but, obviously, there is no scientific system free of inconsistencies and obscurities in them." And so Polya joined the legions of scholars pointing out their contradictory advices: "On the contrary, many a proverb can be matched with another proverb giving exactly opposite advice, and there is a great latitude of interpretation. It would be foolish to regard proverbs as an authoritative source of universally applicable wisdom but it would be a pity to disregard the graphic description of heuristic procedures provided by proverbs." "It could be an interesting task," Polya said, "to collect, and group proverbs about planning, seeking means, and choosing between lines of action, in short, proverbs about solving problems." In fact, the proverbs Polya listed in this connection were all of folkish origins, but, when suitably glossed and brought to bear on their new mathematical scenes of use, their previous employment among the common people made them no less valuable: "Diligence is the mother of all good luck"; "Perseverance kills the game"; "An oak is not felled at one stroke"; "Try all the keys in the bunch"; "Arrows are made of all sorts of wood."[80]

By the early 1970s, the most influential work on heuristics was being done by the cognitive psychologists Amos Tversky and Daniel Kahneman, focusing not on mathematical or scientific problem-solving but on the heuristics of everyday judging and decision-making. While Tversky and Kahneman formally acknowledged that the employment of such common heuristics in everyday life was "highly economical and usually effective," the overall thrust of their research was to show how often and how seriously their use led to "severe and systematic errors." Everyday heuristics, such as those encapsulated in "the Gambler's Fallacy," were identified as sources of "bias." Better outcomes would be secured if further information about the situation was sought, or if the tools of logic and probability theory were systematically employed, as they would be by trained experts.[81] Some readers of this work drew the lesson that it was best not to involve the common people in consequential political and technological decision-making activities, as their cognitive processes systematically led them into error.[82] The heuristics of everyday life were poor cousins to the methods used by modern experts and, while Tversky and Kahneman only sometimes supplied the linguistic forms commending these principles of judgment, proverbial versions for many such heuristics and, indeed, for their logical opposites, are easily located.

It was not until fairly recently that dissenting voices emerged from within cognitive psychology. Gerd Gigerenzer and his colleagues in the Center for Adaptive Behavior and Cognition have criticized the notion that decision-making—lay or learned—ever takes place in scenes where time is very abundant, where total pertinent knowledge is possible, and where computational capacity is unlimited.[83] That is to say, all human judgment that is actually judgment about real-world predicaments is judgment under uncertainty. The learned are in the same boat as the vulgar. As Gigerenzer and his colleagues write, the greatest weakness of the model of unbounded rationality "is that it does not describe the way real people think." Not even how philosophers think: "One philosopher was struggling to decide whether to stay at Columbia University or to accept a job offer from a rival university. The other advised him: 'Just maximize your expected utility—you always write about doing this.' Exasperated, the first philosopher responded: 'Come on, this is serious.'"

Acting on the Gambler's Fallacy—the belief that, for example, the probability of heads on the ninth toss of the coin is greater than 50 percent after eight tails in a row—will, indeed, lose you money, *just on the condition that the game is an ideal, unbiased one,* but in real-life gambling one is often faced with the decision about "what kind of game this is," crooked or straight, and, if crooked, in what way, what to do about it, and what one's opponents will do in light of what one does. Moreover, what Gigerenzer and his colleagues call "fast and frugal" heuristics are not only surprisingly adequate to tasks at hand, they can also be demonstrably superior to problem-solving techniques that attempt to secure further information, to survey a wide range of possible outcomes, and to compute in a more thoroughgoing manner. The best, as the proverb has it, is truly the enemy of the good. Like Tversky and Kahneman, Gigerenzer's group does not seek to identify the linguistic embodiments of such "fast and frugal" heuristics, but, toward the end of their important book, they offer an intriguing comment on how such heuristics may be acquired: "Simple heuristics," they say, "can be learned in a social manner, through imitation, word of mouth, or cultural heritage." And they note that "cultural strictures, historical proverbs, and the like" are effective ways of transmitting such powerful "fast and frugal social reasoning."[84]

Where, then, is a legitimate contrast between, say, science and common sense? And is there a role for such linguistic forms as proverbs (and related short genres) in any such legitimate contrast? Nothing in this study argues that there can be no legitimate contrasts between various modes of cognition and practice, that all, so to speak, is on a level, and for all purposes. Much, however, cautions against facile assumptions about the domains to be contrasted. What counts as common sense is probably pretty diverse in its attitudes and counsels. Geertz argued that "there are really no acknowledged specialists in common sense," correctly gesturing at pervasive lay suspicion of learned expertise and its pretensions.[85] Yet such a claim seems to rest on an excessively homogeneous, and too systematic, view of any such cultural entity as "common sense." There may be no specialized experts in "common sense," but there are acknowledged specialists in fishing, gardening, and market trading. There are also quite common people who are conceded such expertise as there is in finding and holding a spouse, bringing up the kids, and gauging the credibility of different sorts of folk. As individuals and as members of groups, some common people are conceded to be very experienced in such things; others less so. And just as proverbs' counsels, and proverbial scenes, are very various, so there is little reason to presume the distinctive unity of learned practices or of the forms of reasoning employed in the different moments of any one learned practice.

In Galileo's *Dialogue Concerning the Two Chief World Systems*, the Aristotelian Simplicio, objecting to Salviati's mathematical physics, claimed that "these mathematical subtleties do very well in the abstract, but they do not work out when applied to sensible and physical matters."[86] Bringing Simplicio up to date, and giving him his due, one is tempted to say that all learned practices, when they are dealing with real-world contingencies, implicate judgment under uncertainty. So the Aristotelian might concede that proverbial economies may be attached to physics rather than mathematics, or, in a more contemporary sensibility, that such economies are to be found in engineering, medicine, politics, business, and cookery rather than in pure science, mathematics, philosophy, or logic. Conceding engineering (and its cousins) to the world of proverbial economies is indeed an important step, but it is not enough. As we have learned more about the practice of science, so we have learned to appreciate the extent to which it is like how we understand engineering to be, the extent to which the conduct of science is like a craft or an art. Like the proverbs of the common people, the heuristics of science belong to the domains of the more or less, the usually or for the most part, the *ceteris paribus* and the *mutatis mutandis*. Philosophers have rightly warned us not to seek a "logic of discovery," and in the processes of discovery the role of proverbial heuristics should be uncontentious. But the making and the justifying of ideal worlds are themselves a real-world business. Learned practices for judging and for justifying also involve judgment under uncertainty, and there too proverbial economies may be found. Only in the ideal worlds produced by the real worlds of learned practices is there judgment under total certainty, and only there might one expect to dispense with proverbial economies. But it is good to remind ourselves that no human practitioner has ever yet been to such an ideal world to confirm the expectation.[87]

While learned opinion has historically tended to contrast the cognitive processes used by the learned and the vulgar, it is not impossible to find scientists themselves arguing to the contrary. In the 1850s, T. H. Huxley wrote that "Science is, I believe, nothing but trained and organised common sense."[88] "The whole of science," according to Albert Einstein, "is nothing more than a refinement of every day thinking."[89] Max Planck agreed: "Scientific reasoning does not differ from ordinary reasoning in kind, but merely in degree of refinement and accuracy."[90] And so did J. Robert Oppenheimer

("Science is based on common sense; it cannot contradict it"), the chemist James Conant (science is "one extension of common sense"), and the biologist C. H. Waddington ("Science is, after all, largely common sense").[91] If proverbs belong to the worlds of commonsense practice, they and their short-generic cousins belong also to the worlds of science, and what we can understand about proverbial economies should be available as a resource for anyone wanting to understand real-world scientific practice.

So if we seriously want to make some cultural distinctions, this study of proverbial economies suggests three things: first, that we ditch the traditional straight-up contrast between proverbial common sense and the cognitive resources of learned expertise; second, that we look to differences in how, as Huxley suggested, knowledge economies are organized, how their members interact with each other, and how they relate to their cultures' stock of knowledge; third, that the objects of comparison be individuated: "science" versus "common sense" doesn't work, but why shouldn't we be interested in the differences and similarities obtaining among, for example, accountancy and botanical taxonomy, fly-fishing and neurology, cooking and chemistry? "The devil," as the proverb has it, "is in the details." But, as another proverb says, so is God.

CHAPTER 15

Descartes the Doctor

Rationalism and Its Therapies

Philosophy and Its Medical Point

There are four questions that have historically been put to philosophers and intellectuals that, so to speak, call them to task. "If you're so smart, why aren't you good?" That is a Greek question, presuming that the goal of intellectual inquiry is personal virtue, and that the presence or absence of virtue in philosophers is a legitimate test of the knowledge they profess. "If you're so smart, why aren't you happy?" That is a Greek question too, and a related one: it supposes that wisdom should delight, either because it issues in a better manner of living in the world or because its attainment fulfills a basic human longing. Then, there is, "If you're so smart, why aren't you rich?"[1] That is what has been called "the American question," though versions of it can be traced at least as far back as Francis Bacon.[2] Here the presumption is that true knowledge is materially useful knowledge, and that individual or collective wealth is a reliable sign that genuine knowledge is at your disposal.[3]

The fourth—formally, perhaps, a version of the second and third—is nonetheless distinct in its cultural framing and consequences. It is far less familiar than the first three, yet, I want to show, it was quite important in the early modern period as a way of advertising and gauging the quality of philosophical knowledge. This question was, "If you're so smart, how come you're sick?" or—a variant—"how come you died?" or, less ambitiously, "how come you didn't live to 120?"

Life-and-death matters matter. They matter to every-person and they matter to the learned person as every-person, since even Francis Bacon and René Descartes had vulnerable and, as it disappointingly turned out, mortal bodies about which they cared a great deal—viewed at times as instruments to their extraordinary souls, and at other times just as everybody else views their bodies, wishing that they would be less painful, more serviceable, and more durable. That is one—blindingly obvious—reason why the ability to prevent and cure disease, to alleviate suffering, and to extend human life has recurrently been used as a public test of the truth and power of philosophic and scientific systems, and why the learned too might share in that public assessment.[4]

However, there were reasons particular to the learned classes, and more particular to the great philosophical modernizers of the seventeenth century, why medicine mattered, even to those who were not themselves physicians. In general terms, it was bound to matter because of the close but contested relationship that has historically obtained between natural philosophy and the science of medicine and, through the science, the art of medicine. For Aristotelians and Galenists, knowledge of the hidden makeup and workings of the human body was wholly integrated within an overall philosophy of nature. The humors and the temperaments were just a part of the same system of elements and qualities that allowed philosophers to explain what the natural world was made of and why inanimate bodies moved as they did. For Aristotle, Galen, and their followers "medicine was the philosophy of the body," or, put another way, medicine was philosophy in action, put to the test.[5] And so it remained for both Aristotelians and their critics in the early modern period.

Ars longa vita brevis—that is Hippocratic aphorism number one—but it did not matter how long you had to learn the art of medicine if that art was not securely founded on good philosophical principles. Early modern Aristotelians and Galenists, on the whole, considered that medical art *was* so founded, and that, correcting for the complexity of the human-environment transactions that made for individual health and disease, and for the intractability of patients' behavior (even when expertly advised), medical practice did pretty well. And the fact—as they reckoned things—that it did pretty well was powerful testimony to the validity of medicine's philosophical foundations as well as to individual practitioners' skill.[6]

But if you were a seventeenth-century modernizing and mechanizing critic of traditional natural philosophy, you tended to take a very different view of the matter.[7] Aristotelian natural philosophy was no good—unintelligible, absurd, vacuous—and any practice informed by that philosophy was itself likely to be no good. Claims to medical efficacy by Aristotelians and Galenists were deemed largely bogus. Thomas Hobbes was exceptional among the great seventeenth-century philosophical moderns in *not* developing a frontal assault on the state of medical knowledge and practice, though, to be sure, his explicit applause of the current state of medical knowledge was mainly a rhetorical vehicle to appropriate William Harvey for the mechanical cause and to ironically imply that the London physicians were not as Harveian as they ought rationally to be.[8]

In the main, the rest of the great modernizers took a very dim view of contemporary medicine. They said that contemporary medicine was just that kind of practice and belief that was largely false, mainly barren, and, accordingly, in need of systematic reform. Only when medicine was put on solid natural philosophical foundations would it be of genuine use in healing, alleviating suffering, and extending life. The precise nature of their criticisms differed, but their drift was consistent: we the moderns have reformed natural philosophical knowledge, and a reliable sign and legitimate test of our reformed natural philosophy will be not just a better science of medicine but a vastly more powerful system of medical practice. Why not be your own physician when existing medicine was so faulty and when you possessed the philosophical means for its reform?[9] Moreover, and perhaps owing to the intense individualism of modernizing philosophical culture, thinkers' personal states of health and the lengths of their individual lives were sometimes used to gauge the goodness of their philosophical knowledge—both by themselves and by their supporters and critics.

The range of medical promissory notes issued by modernizers is one of the fascinating byways of the Scientific Revolution, though, from presentist points of view, the notes do not make for edifying reading: they look more like testaments to human folly than to the power of reason. Nevertheless, from an impeccably historicist perspective there are inducements to further consider these medical promises, for they offer an intriguing and perspicuous way of reconfronting and reframing some longstanding questions about the relations between theory and practice, and between tradition and change in the history of culture and of technique. What did it look like when modern philosophic reason in its purest and most rigorous forms became its own physician?

Francis Bacon offered wide-ranging criticisms of the state of medical science and art. "Medicine," he wrote, "is a science which hath been . . . more professed than laboured, and yet more laboured than advanced; the labour having been, in my judgment, rather in a circle than in progression." Unless medicine were refounded upon a purified register of natural fact and a proper philosophy of nature, we could never hope for the effective treatment of disease or the prolongation of life: "the science of medicine, if it be destituted and forsaken by natural philosophy, it is not much better than an empirical practice."[10] Medical practice lacked both evidential discipline and philosophical system. But were it to achieve discipline and system, what benefits might humankind expect? The "prolongation of life," Bacon wrote, is the "most noble" of all the parts of medicine, and he wrote extensively about how vast extension of human life might be achieved.[11]

Robert Boyle was deeply concerned with medical therapeutics throughout his life, and his advocacy of pharmaceutical "specifics" was persistently framed and rationalized in terms of his version of corpuscular natural philosophy. "The physician borrows his principles of the naturalist," Boyle wrote, so he should borrow only the best.[12] If you understood the hidden textures of matter, you would rightly grasp how drugs worked and how to select specific drugs that targeted particular illnesses and conditions. He supposed he knew that better than did the physicians, and, accordingly, that he was more effective than they were in treating both his own and others' diseases. For humors, substitute corpuscular textures and mechanisms: "I think the physician . . . is to look on his patient's body, as an engine, that is out of order, but yet is so constituted, that, by his concurrence with the endeavours, or rather tendencies, of the parts of the automaton itself, it may be brought to a better state."[13] It was commonly said that the Draconian laws "were written in blood"; in Boyle's view this would be more accurately said of certain Hippocratic aphorisms, which, slavishly followed by contemporary physicians, cost the lives of countless patients. Systematic reform in natural history and natural philosophy will allow us to do better: "there may be unthought of methods found, whereby, by ways different from those formerly used by physicians, a man may be much assisted in the whole manner of ordering himself, so as to preserve health, and to foresee and prevent the approach of many distempers."[14]

Infallible Demonstration: Descartes's Medical Project

Descartes's vision of the practical medical goods that would certainly be delivered to those possessed of right philosophical reason was more clear, coherent, and ambitious than that of any other seventeenth-century intellectual modernizer. His vision was philosophically systematic and his commitment was lifelong. The *Discourse on Method* of 1637 announced that a renovated philosophy would make us "as it were, the lords and masters of nature." This mastery would, of course, be desirable "for the invention of innumerable devices which would facilitate our enjoyment of the fruits of the earth and all the goods we find there, but also, and most importantly, for the maintenance of health." Medicine "as currently practiced does not contain much of significant use," and most honest medical practitioners would freely acknowledge the disproportion between what we now know and what remains to be known. But, were medicine to be refounded on proper philosophical principles, "we might free ourselves from innumerable diseases, both of the body and the mind, and perhaps even from the infirmity of old age."[15] Descartes publicly committed himself to achieving those ends, and even announced that he had already discovered a method that would "inevitably" lead him to such practical knowledge, unless, he added, without evident irony, "prevented by the brevity of life." The essay concluded with Descartes's resolution "to devote the rest of my life to nothing other than trying to acquire some knowledge of nature from which we may derive rules of medicine which are more reliable than those we have had up till now."[16]

Descartes came from a medical family—his paternal and maternal grandfathers were both physicians—and it is possible that he studied medicine at Poitiers after he had finished at La Flèche.[17] In 1645, he told the Earl of Newcastle that "the preservation of health has *always* been the principal end of my studies, and I do not doubt that it is possible to acquire much information about medicine which has hitherto been unknown."[18] And there is much evidence that Descartes's theoretical and practical medical concerns arose very early in his philosophical career. One of his first biographers traced Descartes's formulation of medical goals back to 1629, when "not being able to forget the End and Scope of his Philosophy, which was only the benefit of Mankind, he seriously undertook the Study of Physick . . . He imagined that nothing was more capable to produce the temporal Felicity of this World . . . [and] he thought it necessary to find out some way or other to secure Humane Body from Evils that might disturb its Health."[19]

By 1630, in any event, Descartes was in earnest pursuit of a practical therapeutic system founded on his rational reform of philosophy. In that year Marin Mersenne told Descartes that he was suffering from the skin disease erysipelas. Descartes sympathized and told Mersenne "to look after" himself, "at least until I know whether it is possible to discover a system of medicine which is founded on infallible demonstrations, which is what I am investigating at present."[20] Three months later Descartes still had not discovered such a system, but announced to Mersenne that he was then study-ing chemistry and anatomy in the hope that these fundamental studies would help "find some cure for your erysipelas."[21] Nine years passed, and still no medical system based on "infallible demonstrations," but Descartes was un-fazed, convinced that he was on the right lines, having supplemented theory with the dissection of many animal bodies: "I have spent much time on dis-section during the last eleven years, and I doubt whether there is any doctor who has made such detailed observations as I. But I have found nothing whose formation seems inexplicable by natural causes. I can explain it all in detail, just as in my *Meteorology* I explained the origin of a grain of salt or a crystal of snow."[22]

The Description of the Human Body (composed in the winter of 1647–1648 and published posthumously) began with the Delphic injunction to "know ourselves." But the benefits accruing from that self-knowledge were here identified as specifically medical: "I believe that we would have been able to find many very reliable rules, both for curing illness and for prevent-ing it, and even for slowing down the ageing process, if only we had spent enough effort on getting to know the nature of our body."[23] Questioned two years before his death about the *Discourse*'s hints concerning the philosophi-cally mediated prolongation of human life, Descartes's confidence remained unshaken: "it should not be doubted that human life could be prolonged, if we knew the appropriate art. For since our knowledge of the appropriate art enables us to increase and prolong the life of plants and such like, why should it not be the same with man?"[24] So there is Descartes's own testi-mony, repeated throughout his life and in a variety of settings, that medical prophylaxis, therapeutics, and the extension of human life were central goals—even the most cherished goals—of his philosophical reform program.

How did Descartes think these goals could be achieved? And what did he see as the relationship between proper philosophy and practical healing? The preface to the French edition of the *Principles of Philosophy* (1647) introduced the celebrated metaphor of philosophy as a tree: "The roots are metaphysics, the trunk is physics, and the branches emerging from the trunk are all the other sciences, which may be reduced to three principal ones, namely medicine, mechanics and morals."[25] This is how things should be, and, following Cartesian reform, how they would be. But given the poor state of contemporary physics, it was deemed fortunate that matters were now otherwise arranged. So in *The Passions of the Soul* Descartes (or, at least, a close associate) wrote that medicine, "as it's practised today by the most learned and prudent in the art," does not depend on physics: "they are content to follow the maxims or rules which long experience has taught, and are not so scornful of human life as to rest their judgments, on which it often depends, on the uncertain reasonings of Scholastic Philosophy."[26]

In this connection Roger French has noted how Descartes was bidding to fill a space in the university curriculum then securely occupied by Aristotle's *Organon* (or body of logical writings). One of the characteristics of this space, French writes, "was that the arts course naturally led on (for those who wished to follow that path) to medicine. Not only did the arts course generally end with one of the more biological of the Aristotelian physical works, but the whole theory of medicine was based on the principles of Aristotle's natural philosophy: 'where the philosopher finishes, there begins the physician' was a common doctor's defense of his own subject." So if Descartes was going to supplant Aristotle in the university curriculum, a Cartesian medicine was required. It was institutionally and culturally natural for Descartes causally to associate the new philosophy with a new and vastly more powerful medicine.[27]

Descartes's mechanistic physiology was formally elaborated in a number of tracts—the *Discourse,* the *Treatise of Man* (composed 1629–1633), the *Description of the Human Body,* and *The Passions of the Soul* (1649). The heart powered the animal machine. Its motion was the source of all other bodily motions, and its natural heat was the natural source of those motions. The heat of the heart rarefied the blood, "and this alone . . . is the cause of the heart's movement."[28] That is why the explanation of the heart's motions was offered as the key exemplar of Descartes's mechanical philosophy in part V of the *Discourse.*[29] "The cardinal tenet," as T. S. Hall puts it, of Cartesian physiology is the assumption that motion is transferred from part to part by contact, and the original source of bodily motion is the heat of the heart.[30] Having sketched his physiological system in the *Description of the Human Body,* Descartes declared that "this will enable us to make better use both of body and of soul and to cure or prevent the maladies of both." And, given the significance of cardiac theory in Descartes's physiology, he insisted that without accurate knowledge of "the true cause of the heart's motion"—which he considered he now had—"it is impossible to know anything which relates to the theory of medicine."[31]

The gap between the theory of medicine and its practical realization in both hygiene and therapeutics was only occasionally bridged in Descartes's writings. That is, Descartes rarely spelled out exactly how it was that certain dietetic practices or therapeutic interventions worked in mechanical and corpuscular terms or how proper philosophical principles led to new and uniquely effective changes in medical art.[32] Despite this largely unbridged gap between theory and practice, there is much evidence that Descartes thought his philosophical achievements did put him in a position to understand and to manage his own body better than the physicians and to offer advice to friends and colleagues that was superior to available professional alternatives. I will first consider how Descartes acted as his own physician and then detail and interpret the range of concrete dietetic and therapeutic advice he gave his associates.

Descartes His Own Physician

At least twice in his life Descartes explicitly endorsed ancient counsel that one be one's own physician. The wise man (the argument went), taking the advice of Apollo's oracle, ought indeed, after a period of time, to come to know himself better than any physician possibly could. This is what he told the Earl of Newcastle in 1645: "I share the opinion of Tiberius, who was inclined to think that everyone over thirty had enough experience of what was harmful or beneficial to be his own doctor. Indeed it seems to me that anybody who has any intelligence, and who is willing to pay a little attention to his health, can better observe what is beneficial to it than the most learned doctors."[33]

Three years later he repeated the same sentiments to Frans Burman when the young student asked him what foods one ought to eat and how one ought to eat them. We become our own experts on such matters: "So, as Tiberius Caesar said (or Cato, I think), no one who has reached the age of thirty should need a doctor, since at that age he is quite able to know himself through experience what is good or bad for him, and so be his own doctor."[34] Here Descartes was not apparently advertising the special virtues of his new philosophical system; he was merely pointing out that if you were a wise and attentive man, then current medical expertise deserved no deference, and noting that he himself was such a man. Descartes's self-monitoring began at an early age. A sickly and delicate child, he was born (so he later told Princess Elizabeth of Bohemia) with "a dry cough and a pale colour" that he believed he inherited from his mother and that remained with him until he was twenty, "so that all the doctors who saw me up to that time gave it as their verdict that I would die young."[35] As a child and young man, he was used to taking special care of himself. At La Flèche he was permitted to lie in bed until quite late in the morning, partly owing to weakness, partly because he recognized the special virtue of post-waking bed-time for deep meditations.[36] Descartes liked spending time in bed: in 1631 he was still getting a solid ten hours of (dreamless) sleep a night, and he recommended his friends to do likewise.[37] Deep contemplations were hard work, and he was careful to ration philosophizing-time to just a few fully rested hours a day.[38]

Descartes's already acute self-monitoring seems to have been sharpened by a midlife crisis, when, after turning forty, he began to notice "the gray hairs that are coming in a rush": he did not want "to study anything any more except the means of postponing" old age.[39] He told Constantijn Huygens that "I have never taken greater care in looking after myself than I am doing at the moment." Yet he was in an optimistic frame of mind: he had once believed that he was built to last only thirty or forty years, but now he saw the prospect of living "a hundred years or more" if only he could produce that infallible system of medicine he had told Mersenne about seven years before:

> I think I see with certainty that if only we guard ourselves against certain errors which we are in the habit of making in the way we live, we shall be able to reach without further inventions a much longer and happier old age than we otherwise would. But since I need more time and more observational data if I am to investigate everything relevant to this topic, I am now working on a compendium of medicine, basing it partly on my reading and partly on my own reasoning. I hope to be able to use this as a provisional means of obtaining from nature a stay of execution, and of being better able from now on to carry out my plan.[40]

Mersenne said that people were worried about Descartes's health, but Descartes told them not to be:

> by the grace of God, during the last thirty years I have not had any illness which could properly be called serious. Over the years I have lost that warmth of the liver which in earlier years attracted me to the army; and I no longer profess to be anything but a coward. Moreover, I have acquired some little knowledge of medicine, and I feel very well, and look after myself with as much care as a rich man with gout. For these reasons, I am inclined to think that I am now further from death than I ever was in my youth.[41]

Two years later, he still saw physical signs that he would live a very long time: he told Huygens that "my teeth are still so firm and strong that I do not think I need fear death for another thirty years, unless it catches me unawares."[42]

These claims were circulated throughout the learned community. Encouraged by the man himself, at least some commentators were prepared to assess Descartes's philosophy by his visible success or failure in acting as his own physician. In 1637, Huygens begged Descartes to write him just a few lines about "the means to live longer than we now do," and later valued Descartes's judgment enough to consult him about a physician for the Stadtholder.[43] And at some unknown date the visiting English philosopher Sir Kenelm Digby told Descartes that life was too short to waste time in "bare speculations": if you really have reliable physiological knowledge, then the thing to do was to find practical "ways and means to prolong" human life. Descartes evidently replied that he already had the matter in hand: while he (modestly) could not promise "to render a man immortal . . . he was quite sure it was possible to lengthen out his life span to equal that of the Patriarchs," that is to say, to almost a thousand years.[44] The belief that Descartes was after the secret of living much longer, or even that he had already cracked that secret, was not uncommon.[45] So it was a great shock to some of his friends when he died in Sweden, aged just fifty-four. Descartes acted as his own physician even on his deathbed. He declined the ministrations of the Swedish court physicians, preferring his own preparation of tobacco-flavored wine to bring up the phlegm. He resisted being bled and only submitted toward the end, "in a great quantity, but to no purpose."[46]

On hearing the news of Descartes's death, his friend the Abbé Claude Picot said that "he would have sworn that it would have been impossible for Descartes to die at the age of 54, as he did; and that, without an external (*éstrangère*) and violent cause as that which deranged his machine (*dérégla sa machine*) in Sweden, he would have lived five hundred years, after having found the art of living several centuries."[47] Stories spread that Descartes had in fact been poisoned, and, if this tale were credited, then the cause would have been external and violent indeed. But in the main it was not believed. Most commentators seem to have been satisfied that he had died of a "*fiebvre continue avec inflammation de poulmon*," or "*une pleuresie*," or, probably, in plain modern terms, pneumonia.[48] His early death was taken, G. A. Lindeboom says, as "a blow at the reliability of his philosophy." When news of his death reached the Low Countries, a Flemish journal reported "that in Sweden a fool had died who had claimed to be able to live as long as he liked."[49] Queen Christina, whose demands for philosophy lessons at five in the morning were the proximate cause of the fatal illness, even sneered, "*Ses oracles l'ont bien trompé.*"[50]

Descartes Prescribes

Throughout his adult life, Descartes offered friends and colleagues a stream of practical medical advice. Some of this advice was explicitly advertised to follow from his new philosophical principles, while other advice not advertised in this way might nevertheless have been so informed. He did not ultimately deliver Mersenne the powerful medical system based on "infallible demonstrations," but many years after that promise Descartes still felt himself in a position to tell Mersenne exactly how a mutual friend, Claude Clerselier, ought to be treated for epileptic fits. First, Descartes cautioned against blindly accepting physicians' recommendations that the patient be bled: "they are great ones for bleeding in Paris, and I am afraid that when they see the benefits of one blood-letting, they will keep on with the treatment, which will greatly weaken the brain without improving his bodily health."[51]

Mersenne told him that Clerselier's condition began with a sort of gout in the toe, and Descartes replied that "if it is still not better, and he continues to have epileptic fits, I think it would be beneficial to make an incision right to the bone in the part of the toe where the trouble began, especially if he is known to have been injured in that area; there may be some infection still present which is the cause of the trouble, and it needs to be driven out before a cure can be effected." As confident as he was of this advice, Descartes was reluctant to have it known that he was appearing in the person of a professional physician, "interfering with medical consultations . . . So if you think it right to pass on my suggestion to one of his doctors, please make sure that he is quite unaware that it comes from me."[52]

In 1647, Descartes paid a call on the sickly young mathematician Blaise Pascal in Paris, Pascal having moved there from Rouen in order to consult physicians. Descartes wanted to speak to Pascal about the latter's vacuum experiments, but, seeing how weak he looked, Descartes volunteered some medical advice: a lot of bed-rest and a lot of soup (*force bouillons*).[53] Princess Elizabeth consulted Descartes about a range of medical conditions from the start of their correspondence in 1643. In December 1646, she wrote to Descartes about a swelling in her hands, for which court physicians had already been consulted. Descartes advised against accepting their remedies—above all phlebotomy—at least until spring, "when the pores are more open and so the cause can be eliminated more readily." And should the condition persist until spring, "it will be easy to drive it away by taking some gentle purgatives or refreshing broths which contain nothing but known kitchen herbs, and by not eating food that is too salty or spicy." Elizabeth was further warned against meddling with non-Galenical "chemical remedies." Since their corpuscular textures were precisely related to their physiological effects, it was vital that such drugs be prepared with much greater care than one could usually expect from chemists and apothecaries, "for if you make the slightest change in preparing them, even when you think you are doing your best, you can wholly change their qualities, and make them into poisons rather than medicines."[54] These drugs should be used "rarely and with great precautions."[55] On the other hand, there was no reason to "despise good remedies just because they are in common use": in its sorry current state medical expertise was likely to be more dangerous than common custom.[56]

He was similarly concerned about the composition of mineral waters and their exact medical effects. The waters of Spa were said to be all right—provided that they were taken at the proper season—but the so-called miraculous spring at Hornhausen was likely to be dangerous because of its possible chemical composition:

> There are many wretched people who broadcast its virtues, and are perhaps hired by those who hope to make a profit from it. For it is certain that there is no cure for all ills; but many people have made use of this spring, and those who have benefited from it speak well of it, while no one mentions the others. However that may be, the purgative quality in one of the springs, and the white colour, softness and refreshing quality of the other, make me think that they pass through deposits of antimony or mercury, which are both bad drugs, especially mercury. That is why I would advise no one to drink from them. The acid and iron in the waters of Spa are much less to be feared; and because they diminish the spleen and chase away melancholy, I value them both.[57]

The Dietetics of Pure Common Sense

The advice Descartes took for himself seems to have been broadly similar to that he offered others. On the whole, Descartes's personal dietetics and therapeutics do not appear either radical or innovative. Descartes's dietetics were temperate in most respects, and he was skeptical about a whole range of heroic medical therapies that purported to do better than the healing powers of nature. Knowing yourself, and acting as your own physician, meant—for Tiberius, Descartes, and very many early moderns—that you learned, over a long period of time, what foods agreed with you and what did not, what order and quantity they should be taken in, what manner of living, forms of exercise, patterns of sleeping, modes of emotional expression and restraint, and types of intellectual stimulation were best for your individual body. If you needed medicine, then you could call on a stock of personal experience, or even intuition, to judge what was required for your specific condition. That is to say, learning to be a prudent and effective gentleman meant that you fashioned yourself physiologically as well as culturally, and that you learned how best to manage the six non-naturals, not as a general and theoretical matter but as a particular, personal, and practical matter.[58]

"Our own experience," Descartes told young Burman, teaches us "whether a food agrees with us or not, and hence we can always learn for the future whether or not we should have the same food again, and whether we should eat it in the same way and in the same order." Even when we are ill, we should let our appetites be our guide. Every wise person knew this through experience, and it was only learned physicians who counseled otherwise: "perhaps if doctors would only allow people the food and drink they frequently desire when they are ill, they would often be restored to health far more satisfactorily than they are by means of all those unpleasant medicines."[59] While Descartes himself was dietetically temperate, he did not make a fetish of his food. He was neither gluttonous nor abstemious. He was restrained in his consumptions, yet he was not one to fail in his obligations as a gentleman. As Baillet wrote, "His course of Diet was always uniform. Sobriety was natural to him. He drunk little Wine, and was sometimes a whole Month together without drinking a drop yet seeming very jocund and pleasant at table, his frugality not burthensome to his Company. He was neither nice nor difficult in the choice of his Victuals, and he has accustomed his Palat to every thing that was not prejudicial to the health of the Body."[60]

Descartes was no ascetic: "He knew Nature must be supply'd'"; he took moderate exercise; he counted bodily health "the greatest Blessing in this life next to Virtue"; and he systematically explored the dependence of the mind upon the body. "His judgment was, that it was good always to keep the Stomach and other *Viscera* a doing, as we do to Horses," and, while he was not a vegetarian, he did seem to commend a high-fiber diet, namely "Roots and Fruits, which he believed more proper to prolong the Life of Man, than the Flesh of Animals."[61] Why should that which was good for the animal machine not be good for the human machine? So he told Burman that "the best way of prolonging life, and the best method of keeping to a healthy diet, is to live and eat like animals, i.e. eat as much as we enjoy and relish, but no more."[62] His "two grand Remedies" were "his spare regular Diet, and moderation in his Exercise."[63]

Looking on the Bright Side of Life: Managing the Passions

Baillet went on to say that the remedy Descartes favored even before moderate diet and exercise was the proper management of the emotions.[64] And, indeed, by far the most pervasive medical advice met with in Descartes's writings, and the most explicitly reflected upon advice, was the injunction to cheer up. Everyone who consulted him got the advice to cheer up; Descartes said he took it himself, to great effect; and he elaborated the physiological reasons for its efficacy in great detail. Cheering up appears as a characteristically Cartesian remedy, and it connects Cartesian philosophical radicalism to medical tradition in a revealing way.

In 1644, Princess Elizabeth consulted Descartes about her upset stomach. Descartes, of course, recommended moderate diet and exercise, but especially drew her attention to the influence of the soul on the body. Believe that you are basically healthy and you will in general be healthy: "I know no thought more proper for preserving health than a strong conviction and firm belief that the architecture of our bodies is so thoroughly sound that when we are well we cannot easily fall ill except through extraordinary excess or infectious air or some other external cause." Contrarily, fear or expectation of illness can be a self-fulfilling prophecy: there are "people who are convinced by an astrologer or doctor that they must die at a certain time, and for this reason alone fall ill, and frequently even die." Descartes said that he had personally seen that happen to several people.[65]

The next year, Descartes was plainly concerned about Elizabeth's continuing health anxieties. He told her what the remedy was; he assured her that he had taken it himself and that it worked; and, as her philosophical master, he told her why it worked in terms particular to his system. The therapy was "so far as possible to distract our imagination and senses from [sources of distress], and when obliged by prudence to consider them to do so with our intellect alone." Descartes himself had "found by experience" that cheering up had "cured an illness almost exactly the same as hers, and perhaps even more dangerous." This was the "dry cough and pale colour" he had inherited from his mother. But early on he recognized and reflected upon his own "inclination to look at things from the most favourable angle and to make my principal happiness depend upon myself alone, and I believe that this inclination caused the indisposition, which was almost part of my nature, gradually to disappear completely."[66] Descartes, acting as his own physician and moral counselor, had cured himself of a condition that was, for all practical purposes, part of his innate constitution. He had changed his nature by repeated rational acts of will.[67]

The way in which you could cure yourself by avoiding sadness lies at the heart of Descartes's physiology and of his medically consequential system for understanding the relations between body and mind. Sadness (and its opposite, joy)—like wonder, love, hatred, and desire—was understood as a passion of the soul. The passions, for Descartes, were those thoughts (emotions, sensations, or perceptions) "which are . . . aroused in the soul by cerebral impressions alone, without the concurrence of its will, and therefore without any action of the soul itself; for whatever is not an action is a passion." Commonly, however, the term was restricted to thoughts caused by some particular agitation of the spirits.[68] The soul could be made sad by external influences. Descartes's account of how this happened was similar to his theory of sensation in general: the scenes and objects that make one sad cause motions on the body's surface—in the case of saddening visual scenes, motions on the eye and in the optic nerve. These stimulate motions of the animal spirits in the nerves, which motions are immediately transferred to the brain, and thence to the cavity in which the pineal gland is delicately suspended. The consequent movement of the gland is the cause of a sensory perception or of the emotion of sadness in the soul. The bodily feeling of sadness is caused by the body→brain→soul traffic flowing in the reverse direction: the soul's sadness is communicated via the animal spirits in the nerves to the heart, and thence to the different organs and muscles involved in the total bodily experience we call "feeling sad."

Externally caused sadness imprints itself on the brain, storing up perceptions and emotions to be retrieved as the soul wills. If the soul decides to summon up and dwell upon sad thoughts, the whole physiological system involved in the original, externally caused experience of sadness will again be put into play, internal causes now acting as effectively as the original external causes. Conversely, if the soul decides not to summon up sad thoughts but rather to retrieve joyful thoughts, the corresponding physiological processes will be suppressed and activated. And, of course, the soul can decide whether or not to put itself in a position where external sources of sadness are likely to occur, say by deciding not to read a sad-making book. The passions were thus open to volitional management: they were not, as in the Stoic vision, to be denied; they were to be disciplined and rightly directed.[69]

Canonical history of philosophy correctly identifies Descartes as the most systematic dualist of the Scientific Revolution: the ontological distinction between soul and body in Cartesian metaphysical writings is, of course, explicit and repeated. Descartes never abandoned that dualism nor did he ever reflectively compromise it. Nevertheless, it would not be a difficult matter to infer from his actual medical practice a quite different unarticulated metaphysics. It is not just that Descartes's medicine was psychosomatic (or somatopsychic): this is an accepted, and a plausible, thing for historians to say about Cartesian thinking on such matters.[70] Its plausibility proceeds from the intensity of the transactions Descartes's medical practice posits between material body and incorporeal soul. In these connections, interaction is so intense and consequential that ontological integrity is blurred. Susan James's recent study of the emotions in early modern philosophy is a brilliant provocation to recognize the plausibility of a case for Descartes as anti-dualist, and a key passage merits extended quotation:

> The thorough-going interconnection between body and soul therefore ensures that both changes in our bodily states and the patterns of our own thoughts give rise to passion. Throughout our lives, we are subject to delicately inflected and ever-changing emotions which direct and redirect our actions and are a central feature of our experience. The capacity of the passions to stream across the line between body and soul is matched by their capacity to cross the boundary around the body. By virtue of our passions, our thoughts are enormously sensitive to the states of our bodies, which have their own economies of normal and abnormal motions. Physical disease is accompanied by emotional disturbance, and a multitude of tran-sitions in the temperature and pace of the blood coming from the various bodily organs

cause a succession of passions, like clouds. Furthermore, we are continually affected by the world around us and respond emotionally to other people, their gestures, conversations, the books we read, the weather, music, the buildings we inhabit, and a thousand things besides. The fact that so many emotions pass through our porous outer skins ensures that our emotional life is constant and various, often more so than we realize. It serves not only to protect us from harm, but also to connect us to the material realm beyond our bodies.

Note how, in suggesting "that *the body thinks,* and that among its thoughts are the passions," James naturally finds herself in the domain of health and disease, their practical understanding and management.[71] The passions are systemic: having a sensation of sadness involves the soul and all the organs of the body that might be pertinent to the experience: the heaviness or heat of the heart, the sensation of sluggishness arising from the composition of the blood, and the condition of the spleen. Because of the systemic interaction between the soul and the body's solids and fluids, you could not only cure yourself by the rational management of thought, you could also cure yourself by the rational management of the other non-naturals. Diet mattered: aliment was the basis for all bodily constituents, including spirits.[72] Airs and waters mattered, for example, as they bore upon the composition and consistency of the blood: when the blood was in a fine and thin condition, joy was a likely consequence, and when it was too thick and sluggish its movement in the heart tended to produce sadness. Taking the Spa waters (and very judicious phlebotomy) might achieve that thinness, while dry air rendered the blood "more subtle."[73] Sleeping and waking mattered: sleep allows the blood to refresh the cerebral substance, making it more responsive to the movement of spirits.[74] And the active control of the passions mattered a great deal. (Recall that Descartes's theory of the passions developed in a specifically therapeutic medical context: Elizabeth complained about her state of health and Descartes offered medical advice.) The passions, after all, counted as one of the traditional non-naturals. The non-naturals were customarily identified as those things bearing upon the condition of our bodies that indeed were under our volitional control. Descartes, who started out by defining the passions as those things that were not volitional actions, ended up treating them in a therapeutic context in just the same way as the Galenists—something about which you could give medical advice, expecting that the prudent person would rationally act upon it.

Custom and Reason in Descartes's Medical Practice

So, if Descartes were your doctor, this is the sort of advice you would get:

- First of all, once you have got a sufficient stock of experience with your own body, reflect upon and trust that experience, and do not be led by medical experts whose knowledge of your body is manifestly inferior to your own.

Within the terms of that general restriction, some robust sorts of advice included:

- Observe dietary moderation: be neither an ascetic nor a glutton. Go for a high-fiber diet: more vegetables than meat; avoid very spicy and salty foods. Do not drink too much. Variety of foods is good. Soup is very good.
- On the whole, let your appetites be your guide. Your body is probably telling you something; listen to it.
- Take exercise (and, by implication, venery) in moderation.
- Get plenty of bed-rest, and do whatever you know works best for you to secure a dreamless sleep. Don't get up from your bed too suddenly, and don't try to think too hard for long periods or at times that might interfere with your sleep.
- Don't let yourself be browbeaten into acceding to physicians' fashion for heroic bloodletting in particular or for any extreme or bizarre treatments in general. If you have to be bled to thin the blood, let it be in moderation and at the right times of year. Take great care with newfangled drugs, especially those whose composition is uncertain or known to be related to things that can harm people.
- And, above all, cheer up: avoid thinking about things that make you distressed; dwell on pleasant objects and memories; look on the bright side of life.

When the rationalist rubber hit the road—as he reckoned it ought to do—this is what Descartes's revolution in philosophy came down to. He said that he would reform philosophy on foundations wholly new, rejecting everything that he had been taught in the schools, and he said that practical effects on human health and longevity were among the principal aims of his philosophical career. Yet the medical advice he offered and took recognizably belonged to longstanding and approved traditions of dietetics. It is quite true that the medical advice I have retrieved here derives overwhelmingly from personal communications rather than from published texts, but this is just the form in which Descartes chose to proffer practical advice. His texts as well as his informal communications generated expectations of radically new and powerful medical practice, even suggesting that he already possessed elements of this practice. So it is pertinent to see what Descartes's practice consisted of—not as it might have been had he lived long enough to perfect his system, but as it actually was. In his own terms, it is a legitimate question to ask. What were the dietetics and the therapies of Reason?

In the main, it was the kind of advice you might give if you were an orthodox early modern Galenic physician. Even the skepticism about medical expertise—which, in Descartes's case, was very probably part of his proprietary challenge to Scholasticism—resonates with the sort of skepticism that was broadly expressed in civic circles, and which, indeed, was elaborated by the great anti-rationalist Montaigne.[75] Much of what Descartes counseled can be found in standard sixteenth- and seventeenth-century advice manuals, and, indeed, in the early twelfth-century Salernitan verses, the *Regimen Sanitatis Salernitanum*, endlessly reprinted, circulated, and translated through the early modern period and beyond: "Use three Physicians still; first Doctor *Quiet*, Next Doctor *Merryman*, and Doctor *Dyet*."[76] The injunctions to "know yourself" and to "avoid excess" were carved on Apollo's temple at Delphi: they are not only central to the art of hygiene; they are its historical origins. Descartes's free and easy attitude toward the appetites in sickness is a version of what was known as the Rule of Celsus: the rule of no rule, the view that a healthy man should not enslave himself to any particular rule or system, that he should do whatever it is customary to do and subject himself to the varieties and vicissitudes of the life that was customary for people of his sort.[77] The foolishness of the view that there was one right global rule for health followed from an appreciation of constitutional individuality that was also standard in traditional medicine: you should be your own physician, and you should find out what worked best for you, because no one else was just like you. Medicine, if it worked at all, worked upon idiosyncratic individual bodies.

The dependence of bodily health upon states of mind was likewise wholly familiar from traditional medical advice. Disease was understood to be the result of disorder and excess of any kind, disrupting the balance of the humors, obstructing their flow, and producing putrefaction and corruption. Physiological conditions could, as a matter of course, have psychological causes. Maintaining a healthy frame of mind was of vital importance to bodily health, and I have noted that the passions counted among the non-naturals of Galenic medicine, their management understood to be as important to bodily well-being as food, drink, and exercise. Indeed, the precise physiological effects of the passions were even identified in some popular medical texts of the sixteenth century: irritability might show up in the composition of the blood.[78] The notion that a healthy body was in causal interaction with a healthy mind was proverbial: you did not have to know Juvenal's *mens sana in corpore sano* to be exposed to a folk version of the same view, while the moral importance of managing the passions was an article of Stoic conceptions of virtue and gentility that were widely influential in early modern civic circles.[79]

Curing or Enduring?

So the practical medical advice that was advertised as issuing from Descartes's radical reform of philosophy was not itself at all radical. Nor was Descartes's confidence unalloyed in the ability of his rational reform to deliver the promised medical goods. Against the general background of boasts, promises, and dreams, there was an undercurrent of doubt, resignation, and reconciliation to inevitable fate. Even in the hubristic *Discourse,* Descartes said that he was reluctant to "commit myself in the eyes of the public by making any promise that I am not sure of fulfilling."[80] In the *Principles,* he expressed confidence that he had already established true philosophical principles, but that "many centuries may pass before all the truths that can be deduced from these principles are actually so deduced."[81] When, in 1639, Descartes told Mersenne that he could already explain everything about the animal frame in detail and by natural causes, he confessed "But for all that I do not yet know enough to be able to heal even a fever. Because I claim to know only animal in general, which is not subject to fevers, and not yet man in particular, who is."[82]

Nor did Descartes neglect formally to give providence its due. Quite possibly, Father Mersenne brought out the pious and the lugubrious side of him, but, in the same letter in which Descartes told him that he was very well and looked forward to a long life, he also acknowledged Divine pleasure and a traditional role of philosophy in counseling the wise man that, as the proverb says, what can't be cured must be endured: "Should God not grant me the knowledge to avoid the discomforts of old age, I hope he will at least grant me a long enough life and the leisure to endure them. Yet all depends upon his providence. One of the main points in my own ethical code is to love life without fearing death."[83]

In 1646, he told his friend Chanut that he had more success using physics to ground ethics than to ground medicine: "Indeed I have found it easier to reach satisfactory conclusions on [moral philosophy] than on many others concerning medicine, on which I have spent much more time. So instead of finding ways to preserve life, I have found another, much easier and surer way, which is not to fear death. But this does not depress me, as it commonly depresses those whose wisdom is drawn entirely from the teaching of others, and rests on foundations which depend only on human prudence and authority."[84] Descartes had his occasional dark night of the soul, but these eruptions of resigned pathos are a counterpoint to the medical optimism that ran through his entire philosophical career.[85]

Descartes understood, or at least intermittently appreciated, the hybrid status of medicine: as a science and as an art. If the object of medical science is body in general, the object of medical art is a particular ailing body. That is just the distinction Descartes recognized when he lamented to Mersenne that he really did not even know how to cure a fever. What worked in and for an imagined body—"a statue, an earthen machine"—might not work in and for *your* body, which was to be understood as the result of your innate constitution plus your total historical transactions with the environment and your history of managing those transactions, suffering from your particular sort of fever on a cold Monday morning in winter, living in a place with gross air.

The art (or practice) of medicine was, therefore, action under uncertainty. Just as you must act in society without having certain knowledge of consequences, so you must eat without knowing whether this particular food is wholesome or poisonous. Knowledge that foods *of this general kind* have been good for you is not certain knowledge that *this particular dish* will not kill you.[86] As in ethics and social conduct, so in the art of medicine: if you must act under uncertainty, then your best guides are, indeed, custom and, one might say, common sense. So in the *Discourse* Descartes frankly acknowledged that skepticism and the demand for certainty were limited as practical matters. Since you had to act even when absolute certainty was not attained or attainable, you had, as a general matter, to provide yourself with some maxims of prudent conduct, the first of which was to follow "the laws and customs of my country."[87] And this is what Descartes's medical practice really amounted to. After the rationalist and revolutionary smoke clears, the hubristic modernist Descartes and the skeptical traditionalist Montaigne turn out to be as one on the issue of medical practice. Custom is a pretty good guide. Know yourself. Avoid extremes.

Modernist philosophical ambitions and rhetorical fanfares did not find it that easy to shift longstanding traditions of medical practice.[88] And the reasons why these traditions were so durable are interestingly related to the grounds on which Descartes's rationally founded medical advice might enjoy whatever authority and credibility it had in his society. Regimen and dietetics—the management of the non-naturals—were central components of the medical art. But, looked at from another point of view, that management amounted to a substantial part of individuals' quotidian life and of the customs held in common by the community or at least Descartes's sector of it. The body of culture that told you where to live, what to eat, when to sleep, and how to control your passions was wholly integrated into the practices of everyday living and into the everyday moral environment that allowed actions to be reliably accounted good or bad. It was not only *good for you* to follow medical advice counseling moderation; it was just *good* to do so. Correspondingly, your health could be read as evidence that you had followed a virtuous course of conduct while your disease might be understood as the upshot of intemperance and viciousness. As Temkin summed up the Galenic hygienic tradition, "A healthy life is a moral obligation . . . Health . . . becomes a responsibility and disease a matter for possible moral reflection."[89] That state of affairs was still firmly in place in the Renaissance and early modern period. Popular medical texts in the sixteenth and seventeenth centuries (and beyond) articulated moral and social norms as they advised how to live a healthy life and how to cure disease.[90] That is why those traditions of medical practice were so resistant to change: revolutionary change in these practices would be just as difficult to achieve as social and cultural revolution.

But Descartes did *not* challenge the practices of traditional medical art in any significant way. Belonging to the domain of judgment under uncertainty, those percepts and maxims were pretty good guides to follow until the promised system of "infallible demonstrations" was secured, or, put another way, until squadrons of pigs flew past his window in tight formation. In so far as Descartes's medical advice enjoyed local authority, that authority must have been substantially borrowed from the same medical tradition that, as a philosopher, he affected to despise. We are now on stage again with Molière's *Le Malade Imaginaire*. Previously you were to control the passion of anger because it unbalanced the humors. Now reference to humors was deemed unintelligible: instead you were to control the passion of anger because it had a certain effect on the animal spirits which, when communicated to the heart, affected the size of its pores, which unbalanced the consistency of the blood.[91] Mechanical philosophy, like its Aristotelian and Galenist predecessors, provided a language for talking about human beings and their vicissitudes in relation to the rest of the universe. There were all kinds of reasons why certain people might want such a language, and why certain people might find a mechanical language preferable to, even more intelligible than, alternatives.[92] But a unique causal relationship to new and effective medical interventions is not one of them.

PART VI

Science and Modernity

In the 1960s, the anthropologist and philosopher Ernest Gellner wrote that "Roughly, science is the mode of cognition of industrial society, and industry is the ecology of science." The rhetorical force of the aphorism enhanced the general appeal of affirming a substantive link between science and the modern economic order—not just science as an instrumental cause of technological development—a problematic claim in its own right—but science as a widely diffused and coherent cognitive disposition that was, as is now said, co-produced with modern industrial (post-industrial?) society. But it was good that Gellner prefixed the claim with "roughly," since we actually do not know whether it is true, or, if true, in what respects. When, for example, it is said that we in the modern order think in scientific terms, we do not securely know who counts as "we," what it might mean to think scientifically, and what consequences our answers might have for the contemporary authority of scientists and scientific institutions. It is time to go back to the traditional Science-Modernity topic and to take a new look at its terms.

CHAPTER 16

Science and the Modern World

Science Made the Modern World and it's science that shapes modern culture. That's a sentiment that gained currency in the latter part of the nineteenth century and the early twentieth century—a sentiment that seemed almost too obvious to articulate then and whose obviousness has, if anything, become even more pronounced over time. Science continues to Make the Modern World. Whatever names we want to give to the leading edges of change—globalization, the networked society, the knowledge economy—it's science that's understood to be their motive force. It's science that drives the economy and, more pervasively, it's science that shapes our culture. We think in scientific terms. To think any other way is to think inadequately, illegitimately, nonsensically. In 1959, C. P. Snow's *The Two Cultures* complained about the low standing of science in official culture, but he was presiding not at a funeral but at a christening. In just that very broad sense, the "science wars" have long been over and science is the undoubted winner.

In the 1870s, Andrew Dickson White, then president of Cornell, wrote about the great warfare between science and what he called "dogmatic theology" that was being inexorably won by science.[1] In 1918, Max Weber announced the "disenchantment of the world," conceding only that "certain big children" still harbored reservations about the triumph of amoral science.[2] Some years earlier, writing from the University of Chicago, Thorstein Veblen described the essential mark of modern civilization as its "matter of fact" character, its "hard headed apprehension of facts." "This characteristic of western civilization comes to a head in modern science," and it's the possession of science that guarantees the triumph of the West over "barbarism." The scientist rules: "On any large question which is to be disposed of for good and all the final appeal is by common consent taken to the scientist. The solution offered by the scientist is decisive," unless it is superseded by new science. "Modern common sense holds that the scientist's answer is the only ultimately true one." It is matter-of-fact science that "gives tone" to modern culture.[3] This is not an injunction about how modern people *ought* to think and speak but Veblen's description of how we *do* think and speak.

In 1925, Alfred North Whitehead's *Science and the Modern World* introduced the historical episode that Made Modernity, and that had not *yet* been baptized as "the Scientific Revolution": it was "the most intimate change in outlook which the human race had yet encountered . . . Since a babe was born in a manger, it may be doubted whether so great a thing has happened with so little stir." What started as the possession of an embattled few had reconstituted our collective view of the world and the way to know it: the "growth of science has practically recoloured our mentality so that modes of thought which in former times were exceptional, are now broadly spread through the educated world." Science "has altered the metaphysical presuppositions and the imaginative contents of our minds." Born in Europe in the sixteenth and seventeenth centuries, its home is now "the whole world." Science, that is to say, travels with unique efficiency: it is "transferable from country to country, and from race to race, wherever there is a rational society."[4]

The founder of the academic discipline called the history of science—Harvard's George Sarton—announced in 1936 that science was humankind's *only* "truly cumulative and progressive" activities, and, so, if you wanted to understand progress toward modernity, the history of science was the only place to look.[5] The great thing about scientific progress was—as was later said, and often repeated—"the average college freshman knows more physics than Galileo knew . . . and more too than Newton."[6] Science, Sarton wrote, "is the most precious patrimony of mankind. It is immortal. It is inalienable."[7] When, toward the middle of the just-past century, the Scientific Revolution was given its proper name, it was, at the same time, pointed to as the moment modernity came to be. Listen to Herbert Butterfield, an English political historian, making his one foray into the history of science in 1949: "[the Science Revolution] outshines everything [in history] since the rise of Christianity and reduces the Renaissance and Reformation to the rank of mere episodes, mere internal displacements, within the system of medieval Christendom. Since it changes the character of men's habitual mental operations even in the conduct of the non-material sciences, while transforming the whole diagram of the physical universe and the very texture of human life itself, it looms . . . large as the real origin of the modern world and of the modern mentality."[8]

Butterfield's formulation was soon echoed and endorsed, as in this example from the Oxford historian of science A. C. Crombie: "The effects of the new science on life and thought have . . . been so great and special that the Scientific Revolution has been compared in the history of civilisation to the rise of ancient Greek philosophy in the 6th and 5th centuries B.C. and to the spread of Christianity throughout the Roman Empire."[9] And by 1960 it had become a commonplace—Princeton historian Charles Gillispie concurring that modern science, originating in the seventeenth century, was "the most . . . influential creation of the western mind."[10] As late as 1986, Richard Westfall—then the dean of America's historians of science—put science right at the heart of the modern order: "For good and for ill, science stands at the center of every dimension of modern life. It has shaped most of the categories in terms of which we think."[11]

Evidence of that contemporary influence and authority is all around us and is undeniable. In the academy, and most especially in the modern research university, it is the natural sciences that have pride of place and the humanities and social sciences that look on in envy and, sometimes, resentment. In academic culture generally, the authority of the natural sciences is made manifest in the long-established desire of many forms of inquiry to take their place among the "sciences": social science, management science, domestic science, nutrition science, sexual science. Just because the designation "science" is such a prize, more practices now represent themselves as scientific than ever before. The homage is paid from the weak to the strong: students in sociology, anthropology, and psychology commonly experience total immersion in "methods" courses, and, while chemists learn how to use mass spectrometers and Bunsen burners, they are almost never exposed to courses in "Scientific Method." The strongest present-day redoubts of belief in the existence, coherence, and power of Scientific Method are found in the departments of human, not of natural, science.

Moreover, though it may be vulgar to mention such things, one index of the authority of science in academic culture is the distribution of cash, a distribution that seems—crudely but effectively—to reflect public sensibilities about which forms of inquiry have real value and which do not. The National Science Foundation and the National Institutes of Health distribute vastly more money to natural scientific research than the National Endowment of the Humanities does to its constituents. Statistics firmly establish pay differentials between academic natural scientists and engineers and their colleagues in the sociology or history departments, and the "summer salary" instituted by the National Science Foundation early in its career was one explicit means of ensuring that result in a Cold War era when the "scarcity" of physicists and chemists, but not of, say, art historians, was a matter of political concern. These days it is more likely the "opportunity cost" argument that justifies this outcome, even if it means that not just scientists and engineers but also academic lawyers, physicians, economists, and business school professors now command higher salaries.[12] Many scientists and engineers are now the apples of their administrators' eyes because their work brings in government and corporate funding, with the attendant overheads on which research universities now rely to pay their bills. Finally, the ability of university administrators to advertise to their political masters how their activities help "grow the local economy," spinning off entrepreneurial companies, transferring technology, and creating high-paid, high-tax jobs all supports the increasing influence of science and engineering in the contemporary research university. In the 1960s, social and cultural theorists—following Jürgen Habermas—began to worry about what they called a "technocracy," in which decisions properly belonging in the public sphere, to be taken by democratically elected and democratically accountable politicians, were co-opted by a cadre of scientific and technical experts—as the saying is, "on top" rather than "on tap." And even though that worry seems to have been allayed by more recent concern with political interference in scientific judgments, a recent *New Yorker* piece complaining about the Bush administration's attack on the autonomy of science blandly asserted the primacy of science as the leading force of modern historical change: "Science largely dictated the political realities of the twentieth century."[13]

Sixty years after Hiroshima, and more than a century after General Electric founded the first U.S. industrial research laboratory, it is almost too obvious to be pointed out that it is the natural sciences that are now so closely integrated into the structures of power and wealth, and not their poorer intellectual cousins. It is science that has the capacity to deliver the goods wanted by the military and industry and not sociology or history, though there are some obvious qualifications that need to be made—not all the natural sciences do this, and there was a period, early in the post–World War II world, when there were visions of how the human sciences might make major contributions to problems of conflict, deviance, strategic war-gaming, the rational conduct of military operations and weapons development, and the global extension of benign American power. Few observers disagree when it is said that science has changed much about the way we live now and are likely to live in the future: how we communicate, how long we are likely to live and how well, whether the crucial global problems we now confront—from global warming to our ability to feed ourselves—are likely to be solved, indeed, what it will mean to be human.

Some time about the middle of the just-past century, sociologists noted an exponential increase in the size of the scientific enterprise. By any measure, almost everything to do with science was burgeoning: in the early 1960s, it was said that 90 percent of all the scientists who had ever lived were then alive and that a similar proportion of all the scientific literature ever published had been published in the past decade. Expenditures on scientific research were going up and up, and, if these trends continued—which in the nature of things they could not—every man, woman, child, and dog in the United States would be a scientist and every dollar of the Gross Domestic Product would be spent on the support of science.[14] By these, and many other, measures, it makes excellent sense to observe that science *is* constitutive of the Modern World. And so it's hard to say that claims that Science either Made the Modern World or that Science is constitutive of Modern Culture are either nonsense or that they need massive qualification. Nevertheless, unless we take a much closer look at such claims, we will almost certainly fail to give any worthwhile account of the Way We Live Now.

Do we live in a scientific world? Assuming that we could agree on what such a statement might mean, there is quite a lot of evidence that we do not now and never have. In 2003, a Harris Poll revealed that 90 percent of American adults believe in God, a belief which, of course, is not now, and never was, in any necessary conflict with whatever might be meant by a scientific mentality. But 82 percent believe in a physical Heaven—a belief that is—perhaps predictably, just because Heaven is so much more pleasant than The Other Place—13 percent more popular than a belief in Hell; 84 percent believe in the survival of an immaterial soul after death; and 51 percent in the reality of ghosts. The triumph of science over religion trumpeted in the late nineteenth century crucially centered on the question of whether or not supernatural spiritual agencies could intervene in the course of nature, that is to say, whether such things as miracles existed. By that criterion, 84 percent of American adults are unmarked by the triumph of science over religion, which supposedly happened over a century ago. These responses are not quite the same thing as the "public ignorance of science" (or "public misunderstanding of science") so frequently bemoaned by leaders of the scientific community. For that, you'll want statistics on public beliefs about things like species change or the Copernican system. Such figures are available: 57 percent of Americans say they believe in psychic phenomena, such as ESP and telepathy, that cannot be explained by "normal means."[15] Americans are often said to be more credulous than Europeans, but comparative statistics point to a more patchy state of affairs. So 40 percent of Americans said astrology is "very" or "sort of" scientific, while 53 percent of Europeans that it was "rather scientific." Americans did somewhat better than Europeans in grasping that the Earth revolves around the Sun and not the other way round: 24 percent of Americans got that wrong compared to 32 percent of Europeans, and only 48 percent of Americans believed that antibiotics killed viruses compared to 59 percent of Europeans. Unsurprisingly, the "Darwin question" was flunked by more Americans than Europeans in 2001: 69 percent of Europeans, but only 52 percent of Americans, agreed that "Human beings developed from earlier species of animals."[16] A still more recent transnational survey published in *Science* shows that, when asked the same question, Americans yielded the second-lowest rate of acceptance (now 40 percent) of all 34 countries polled—above only Turkey.[17] If you believe the Gallup pollsters, then in 2005 the percentage of Americans who agreed with the more specific and loaded statement that "Man has developed over millions of years from less advanced forms of life [and] no God participated in this process" was 12 percent, encouragingly up from 9 percent in 1999.[18]

Whitehead's *Science and the Modern World* was based on the Lowell Lectures given at Harvard by a newly minted professor of philosophy, and perhaps that context is relevant to his assertion scientific modes of thought "are now broadly spread through the educated world." So perhaps we can conclude that there is now, just as there always has been, a big gulf between "the educated world" and the unwashed and unlettered. But Whitehead was quite aware that the Galilean-Newtonian "revolution" was the possession of only a very small number of people and that their beliefs bore slight relationship to those of the peasantry in Sussex, much less in Serbia or Siam. Although a number of scholars loosely referred (and refer) to science-induced tectonic and decisive shifts in "our" ways of thinking, or to those of "the West," Whitehead, addressing his Harvard audience, confined himself to "the educated world." So it must, then, be relevant that the 84 percent of contemporary Americans who profess belief in miracles does indeed drop when the responses of only those with *postgraduate degrees* are considered—that's to say, not just who are college educated but those with master's or doctoral degrees. The percentage of *these* elites who say they believe in miracles is *only* 72 percent and the percentage of college graduates who agree with the Gallup Poll's version of Darwinian evolution is 16.5 percent. So the possibility remains that we can still make some distinction of the general sort that Whitehead intended: suppose that "science" is what's believed at Harvard and Berkeley that's not believed at, say, Oral Roberts. Maybe that's right, but that's not *quite* what Whitehead said.

Perhaps, then, we should find some statistics about what *scientists* believe. A survey conducted in 1916 found that 40 percent of randomly selected American scientists professed belief in a personal God. This was a surprise to the author of the report and he expressed his confidence that the figure would surely drop as education spread.[19] But it has not. In a survey published in *Nature* in 1997 it was found that an identical 40 percent of American scientists counted themselves as believers in God, with only 45 percent willing to say they did not believe.[20] Those wanting to get the figure of scientists believing in a personal God or human immortality under 10 percent will have to accept a 1998 survey confined to members of the National Academy of Sciences, while the mathematicians among this elite were the most likely to believe, at about 15 percent.[21] Scientists, of course, are leading the charge in the recent American defense of Darwinism in the classroom, but, according to the Gallup Poll, only a bare majority of *them*—55 percent—actually assent to the Poll's version of Darwinian evolution.[22] Physician, cure thyself.

There is no reason to fetishize a Harris, Gallup, or any other systematic attitude survey. We do not know with any great specificity what people might *mean* when they say they believe in miracles (or, indeed, astrology), and the inadequacy of any simple-minded juxtaposition of "scientific" versus "fundamentalist" beliefs is indicated by the solid popularity of stem-cell research, even among evangelical Christians who are widely supposed to be against tampering with God-given human life. Religiosity seems to bear on embryo-destruction in abortion in a way it does not in stem-cell research.[23] And, if it were thought that religiosity translates into a "don't mess with God's Nature" attitude, then Americans again are much more favorably disposed toward genetically modified foods than Western Europeans or Japanese.[24] The legal scholar Ronald Dworkin has recently pointed out—without evidence, but plausibly enough—that not a lot should be inferred about overall attitudes to scientific expertise from evangelicals' doubts about Darwinism: "Almost all religious conservatives accept that the methods of empirical science are in general well designed for the discovery of truth . . . They would not countenance requiring or permitting teachers to teach, even as an alternate theory, what science has established as unquestionably and beyond challenge false: that the sun orbits the earth or that radioactivity is harmless, for example."[25] But it still seems safe to say that the great majority of the people professing belief in things like miracles have been presented with multiple articulations of what it might mean to "think scientifically" and that thinking that miracles happen is understood not to be part of the scientific game.[26] Quite a lot of people saying they believe in miracles, like quite a lot of the people saying that human beings were specially created by a divine agency, must be well aware that they are, in so saying, poking one in the eye of scientific authority. And so one thing we cannot sensibly mean when we say that we live in a Scientific Age or that Science Made the Modern World is that scientific beliefs have got much grip on the modern mind writ large. That just isn't the case. Maybe, if we mean anything legitimate at all by saying such things, we mean that the *Idea of Science* is widely held in respect. *That* seems plausible enough. Consider the litany of complaints from high scientific places about "public ignorance of science"—complaints that often are inspired by such statistics as those just cited. Complaints can actually help to establish the esteem in which science is held in our culture. It's been some time since I heard anyone gain a public platform for complaining about "public ignorance of sociological theory" or "public ignorance of the novels of Mrs. Gaskell." Nor do official worries about the proliferation of pseudo-science or junk science necessarily bear on the authority of science. Consider present-day concerns over "Intelligent Design" and "Creation Science," but note that these represent themselves as forms of science, not as non-science or as antiscience. Advocates of Intelligent Design want it taught in *science* classrooms. From a pertinent perspective, the problem today is not antiscience, but a contest for the proper winner of the designation "science." That's a sign that the designation "science" is a prize very much worth having. A writer in *The New York Times,* referring to the apparent upsurge in evangelical Christianity, recently announced that "Americans on the whole do not seem to care greatly for science," but such conclusions do not appear at all well grounded.[27] American faith in the power of science—or, more accurately, of science and technology—has long been enormous and continues to be. In the late 1950s, surveys found that a remarkable 83 percent of the U.S. public reckoned that world was "better off" because of science and only a negligible 2 percent thought it was "worse off."[28] Amid anxieties about "increasing public skepticism toward science," various surveys conducted in the 1970s—phrasing their questions somewhat differently—purported to find a decline in approval (to between 71 and 75 percent, with a negative assessment rising to between 5 and 7 percent)—though few other modern American institutions could hope to come close to that level of public favor.[29] In the most recent survey, Americans expressed a "great deal" of confidence (42 percent) in the scientific community, and significantly less in the banking system (29 percent), the presidency (22 percent), and, tellingly, organized religion (24 percent).[30] The Pew Research Center's Global Attitudes Project discovered that 19 percent of Americans surveyed recently accounted "Science/Technology" to be the "greatest achievement" of the U.S. government during the course of the twentieth century— more than twice as many as those who pointed to civil rights and more than three times as many as those giving the prize to the social security system. In the public mind, science and technology are endowed with colossal power: about 80 percent of Americans think that within the next fifty years science will ("probably/definitely") deliver cures for cancer and AIDS and would "improve [the] environment," compared to just 44 percent who believe that Jesus Christ will reappear on Earth during that period.[31]

Suppose we concede that scientific *beliefs*—or at least beliefs of the sort approved at Harvard—are not very widely distributed in modern culture. That means that the authority of science-the sense that we live in a scientific age-has to reside in something other than the widespread *understanding* of particular scientific facts or theories, no matter how important, foundational, or elementary they may be. This would be quite a concession in itself and we should reflect a lot more on what it means. But can't we nevertheless say that the authority and influence of science resides in something other than shared beliefs, something that nevertheless "belongs to" science? Consider, again, the notion of the Idea of Science. I've given some reasons to think that the Idea of Science possesses authority, even if a range of specific scientific beliefs does not. What might be meant by the Idea of Science? But there are problems saying much about such an Idea. If we want to talk about the Idea of Science apart from specific beliefs, then we probably are pointing at some notion of Scientific Method. Scientists—and, more important, philosophers of science—have been identifying, celebrating, and propagating the Scientific Method for a long time—arguably at least as far back as the time of Descartes, Newton, and Boyle. It's that universal, rational, and effective Method that has been said to account for the power of science and to mark it out from other modes of inquiry lacking such a Method. As the recent *New Yorker* piece announced, "The scientific method has come to shape our notion of progress and of modern life."[32]

The problem is that there is not now, and never has been, a consensus about what such a Method is.[33] The first two entries for "Scientific Method" that Google gave me opted for observation before the formulation of an explanatory hypothesis, followed by experimental tests of the hypothesis, though that account excludes all those sciences that are not experimental, for example, geology, meteorology, and many forms of evolutionary biology.[34] The current Wikipedia entry currently makes reference to the views of Thomas Kuhn, who, like Karl Popper, Imre Lakatos, and Paul Feyerabend, famously doubted whether theory-free observation ever occurred. *Science* magazine has usefully addressed the question by annotating a number of scientific papers to show the Scientific Method at work.[35] A "pragmatical scheme" of that seven-step Method is provided, starting with "Define the question," going through "Analyze the data," and concluding with "Publish results," but it's hard to look at this list without concluding that—"Perform experiment" apart—its directions can be found in any kind of systematic inquiry pretending to rigor, and not just in science.[36] Other entries early in the Google list give deductive, rather than inductive, inference pride of place and omit references to experiment.[37] Some make reference to "proof" or "confirmation" of a hypothesis; others point out—following Popper—that one can never prove but only disprove the validity of a hypothesis. Few bother to cite T. H. Huxley's view that science is "nothing but trained and organised common sense" or that of the Nobel-winning immunologist Peter Medawar that "The scientific method does not exist."[38]

In fact, if the authority of science—the way in which it is supposed to mark modernity—resides in some idea of Scientific Method, that would be as much as saying not just that academic *philosophy of science* rules the roost, but that some specific *version* of philosophy of science was the most authoritative form of modern culture. And somehow that doesn't seem right. The authority of philosophers in our culture doesn't come close to the authority of scientists. Much the same sort of argument, I think, applies to any Idea of Science that flows from identifying shared *conceptual* content. The Unity of Science Movement of the early and middle part of the twentieth century arose out of a worry that, while science *must,* of course, be conceptually unified, no one had yet definitively shown what the basis of that unity was. That situation has not changed, and, while scientists these days seem not to be much worried about "unity," leading-edge philosophers of science are now increasingly writing books and papers taking the "*disunity*" of science as their subject.[39]

I doubt that searching for some stable and plausible Idea of Science is going to get us very far in trying to describe the authority of science in the modern world, or in showing that science *does* have such authority. But, if I'm right, we're beginning to see the shape of a real problem: science, we say, marks modernity—it enjoys unique authority—but that authority does not seem to consist either in lay possession of any specific set of scientific beliefs— no matter how elementary or fundamental—nor in any stable sense of the Method scientists supposedly used to guarantee the power of their knowledge. Should we just agree that science has very little to do with Modern Culture—bizarre as that might sound—*or* that the authority of science resides in something besides knowledge of its beliefs or methods?

And so it seems that if we want to talk about the authority of science in the Modern World, we can't sensibly talk about our culture's knowledge of scientific beliefs or our grasp of some notion of Method. What seems to be essential is not knowing *science* but knowing where to look for it, knowing who are the relevant authorities, knowing that we can and should assent to what they said, *that* we can and should trust them in their proper domains. Pragmatically, there's a lot to recommend this state of affairs: it's unfortunate that the ideas of both Darwinian evolution and the heliocentric system have not taken better root in our culture, but, in general, *no one* can know very much of science, and so knowing who the relevant experts are is sufficient in the great majority of cases. This applies to scientists as well as the laity: even plant physiologists are likely to have a deficient knowledge of astrophysics and a cardiologist is going to go to a neurologist if she has persistent headaches. Expertise isn't considered to be fungible: it comes in various specialized flavors.

And so knowing where to look for the relevant experts has to involve some notion of relevant expertise, of relevant authority. When we say that our task is recognizing the experts in their proper domains, what *are* those domains? Putting the question that way identifies a sense in which scientific authority is now not greater but clearly much *less* than it once was. Consider what philosophers—following G. E. Moore in the first years of the twentieth century—call "the Naturalistic Fallacy." That fallacy is believing something that is impossible, moving logically from an "is-statement"—a description of how things are in the world-to an "ought-statement"—a prescription of how things ought to be. Put another way, science is one thing, morality another; and you should not think of deducing what's good from what is. But the Naturalistic Fallacy is not just a philosopher's boundary; during the course of the twentieth century, very many scientists publicly insisted that they possessed no special moral authority, and that questions of what ought to be done—for example, about the consequences of their own work—were not their preserve. As Edward Teller put it, it was the scientist's job to discover the laws of nature, not to pronounce on whether the laws permitted nuclear fusion ought to be mobilized for the construction of a hydrogen bomb.[40] You would think that Oppenheimer would have disagreed with such a sentiment, but on this point he was at one with Teller.[41]

Scientists—it was widely insisted by modern scientists themselves—possessed no particular moral authority. It was once assumed they did; now it was not. If moral authority is what you want, you should go to some other sort of person, and that's why the late Stephen Jay Gould referred to science and religion as "non-overlapping magisteria."[42] That division of labor between natural experts and ethical experts is now institutionalized, accepted almost as a matter of course. And yet it leads to a pervasive awkwardness in contemporary culture. Just as so many social and political decisions increasingly come to draw on massive amounts of specialized expertise—even to understand what they're *about*—so it is accepted that those who know most should accept radical restrictions on having consequential opinions about *what ought to be done*. Here, the up-curve of the reach of science in our social and political life meets the down-curve of scientists' acknowledged moral authority. Who are they, such that we can trust them—not just to know more about their specialized bits of the world, but to do the right thing?

"The scientist is not a priest." That's another way of identifying the limited authority of the modern scientist, and the non-priestly status of the scientist was much insisted on throughout the twentieth century by scientists themselves. At the same time, and perhaps responding to what was seen as the *increasing* cultural authority of science during the course of the century, the scientific community was accused of becoming "the new priesthood" and scientists as "the new brahmins."[43] An essay in the *Bulletin of the Atomic Scientists* about immediate postwar congressional engagements with science noted that, after Hiroshima, "scientists became charismatic figures of a new era, if not a new world, in which science was the new religion and scientists the new prophets . . . Scientists appeared to [politicians] as superior beings who had gone far ahead of the rest of the human race in knowledge and power . . . Congressmen perceived scientists as being in touch with a supernatural world of mysterious and awesome forces whose terrible power they alone could control. Their exclusive knowledge set scientists apart and made them tower far above other men."[44] It's a tension that remains unresolved: science is our most powerful form of knowledge; it's scientists—or at least those pretending to be scientists—that are turned to when we want an account of how matters stand in the natural world. But, however esoteric their knowledge is, it is not scientists who decide what ought to be done. For those decisions—and there are an increasing number of them that are potentially world-changing—it's politics as usual.

Knowing where to look for the relevant experts also involves some notion of what it is they know. In the early modern period, a common cultural distinction was made between mathematics and natural philosophy. Philosophy was understood as the search for Truth, for the realities behind appearances, for the real causal structure of the world. Mathematics, by contrast, was taken as the quest for regular patterns of natural relationships, such that you could use the resulting knowledge to predict and control, without necessarily taking a bet on what the world was really like. Copernicus was acting as a mathematician when he stipulated that the heliocentric system was to be regarded as a predictive tool, and Galileo was blurring the boundaries between mathematics and philosophy when he defied the Vatican in asserting the physical reality of Copernicanism.[45] I mention this old chestnut, just because it may have significance for our current problem of identifying who the relevant experts are.

At least from the early twentieth century, very many scientists—physicists, of course, but not just physicists—publicly asserted that they were not, so to speak, in the Truth Business.[46] Their task, it was insisted, was not metaphysics; it was not discovering ultimate realities. It was, rather, finding out what "works": what picture of nature was maximally coherent, with existing theories and evidence, and what picture of nature would allow scientists most powerfully to predict and control. Pragmatism was one version of such a sensibility, but so were those positions called operationalism, conventionalism, and phenomenalism. In 1899, the Johns Hopkins physicist Henry Rowland, making no allusions to pragmatism or to any other formal philosophy of science, explicitly contrasted the scientific with the "vulgar" or "ordinary crude" mind: the scientist alone properly appreciated that "There is no such thing as absolute truth and absolute falsehood."[47] By the 1920s, Albert Einstein was reminding the general reader that "It is difficult even to attach a precise meaning to the term 'scientific truth,'" its semantics varying radically according to context of use.[48] And C. P. Snow surely spoke for most scientists when he bumptiously stipulated that "By *truth*, I don't intend anything complicated . . . I am using the word as a scientist uses it. We all know that the philosophical examination of the concept of empirical truth gets us into some curious complexities, but most scientists really don't care."[49] The scientist was properly to be understood not on the model of the philosopher but on the model of the engineer and technician. Our culture used to insist on massive differences between science and technology and between the role of the scientist and that of the engineer. It's a distinction that now makes less and less sense: we're all engineers now, and the authority of science is increasingly based not on what scientists know but on what they can help make happen. It's a distinction that increasingly resonates in the public culture: a National Science Foundation survey in 1976 discovered that government funding of science was overwhelmingly popular but that only 9 percent of the respondents wanted any of their tax dollars used to support basic research.[50]

What difference does it make to the public authority of science if scientific knowledge is just *what works* and if the scientist is understood as an aid to the technologist? First, at one time it was believed that a world saturated with technology would not only be a modernized world but a secularized world. That turned out to be spectacularly untrue. The mere presence of advanced technology in a society seems to have little or nothing to do with how people think and what they value: some of the world's web wizards are jihadis and there seems to be no conflict between computer skill and religious fundamentalism. And we should be clear about another thing: engineers seem to include as many morally admirable people as any other group of professionals; some are more admirable than some scientists I know. But it's the institutions we're talking about here, and what virtues and authority are associated with the institutions. The technologist supplies what society wants; the scientist used to give society what it didn't know it wanted. That's a simplification, but, I think, a useful one: corporations, governments, and the military enlist experts in the natural world overwhelmingly on the condition that they can assist them in achieving useful goals—wealth and power. During the course of the twentieth century, the enterprise called science was effectively enfolded in the institutions dedicated to the production of wealth and the projection of power. That's where we started, and that's one way of describing the success of science in modernity. But one of the conditions of that success is, at the same time, a problem for the authority of science in the modern world.

So modern scientists are not priests. Their expertises are not fungible—either one form of technical expertise into another or technical expertise into moral authority. What the modern scientist *may* have left as a basis of authority is a kind of independence and a resulting notion of integrity. Yet the enfolding of science into the institutions of wealth-making and power-projecting makes that independence harder to recognize and acknowledge. And when scientific knowledge becomes patentable property, then the independence of science from civic institutions becomes finally invisible. We've gone some way in these directions—but not yet all the way. And so it's not a bad moment to reflect on where we've come from and where we might be going.

I started by recalling how easy it once was to talk about science as an independent cause of modernity, as modernity's characteristic form of culture, and as its distinct master authority. It's not so easy now. And one reason it's not so easy is that our ability to recognize relevant experts, and to recognize their independent authority, is harder and harder to do. The success of science has created its successor problem. That problem—the problem of the independent authority of science in our modern world—may be a problem *for* science, but, more important, it's a problem *in* our modern order of things. The place of science in the modern world is just the problem of *describing* the way we live now: what to believe, who to trust, what to do.

NOTES

Preface

1. The Pennsylvania department was founded with the then-fairly-standard name of "history and philosophy of science." At the time I entered as a graduate student, there was almost no interest in the *philosophy* of science, and the renaming did not mean there was any systematic engagement with *sociological* literature on science: it merely meant that the department saw itself as concerned with what were then loosely called "social aspects" of science.

2. Steven Shapin, "The Pottery Philosophical Society, 1819–1835: An Examination of the Cultural Uses of Provincial Science," *Science Studies* 2 (1972): 311–336; idem, "Property, Patronage, and the Politics of Science: The Founding of the Royal Society of Edinburgh," *The British Journal for the History of Science* 7 (1974): 1–41; idem, "The Audience for Science in Eighteenth-Century Edinburgh," *History of Science* 12 (1974): 95–121; idem, "Phrenological Knowledge and the Social Structure of Nineteenth-Century Edinburgh," *Annals of Science* 32 (1975): 219–243; idem, "The Politics of Observation: Cerebral Anatomy and Social Interests in the Edinburgh Phrenology Disputes," in *On the Margins of Science: The Social Construction of Rejected Knowledge,* ed. Roy Wallis, Sociological Review Monographs, vol. 27 (Keele: Keele University Press, 1979), pp. 139–178; idem, "*Homo phrenologicus:* Anthropological Perspectives on an Historical Problem," in *Natural Order: Historical Studies of Scientific Culture,* ed. Barry Barnes and Steven Shapin (Beverly Hills, CA, and London: Sage, 1979), pp. 41–71; Steven Shapin and Barry Barnes, "Head and Hand: Rhetorical Resources in British Pedagogical Writing, 1770–1850," *Oxford Review of Education* 2 (1976): 231–254; Steven Shapin and Barry Barnes, "Science, Nature and Control: Interpreting Mechanics' Institutes," *Social Studies of Science* 7 (1977): 31–74; Steven Shapin, "'Nibbling at the Teats of Science': Edinburgh and the Diffusion of Science in the 1830s," in *Metropolis and Province: Science in British Culture, 1780–1850,* ed. Ian Inkster and Jack Morrell (London: Hutchinson; Philadelphia: University of Pennsylvania Press, 1983), pp. 151–178.

3. Steven Shapin, "Social Uses of Science," in *The Ferment of Knowledge: Studies in the Historiography of Eighteenth-Century Science,* ed. G. S. Rousseau and Roy Porter (Cambridge: Cambridge University Press, 1980), pp. 93–139; idem, "Of Gods and Kings: Natural Philosophy and Politics in the Leibniz–Clarke Disputes," *Isis* 72 (1981): 187–215; Steven Shapin and Simon Schaffer, *Leviathan and the Air-Pump: Hobbes, Boyle, and the Experimental Life* (Princeton, NJ: Princeton University Press, 1985); Steven Shapin, *A Social History of Truth: Civility and Science in Seventeenth-Century England* (Chicago: University of Chicago Press, 1994); idem, *The Scientific Revolution* (Chicago: University of Chicago Press, 1996); idem, "Pump and Circumstance" (1984) (chapter 6 in this volume); idem, "The House of Experiment in Seventeenth-century England" (1988) (chapter 5 in this volume); idem, "Robert Boyle and Mathematics: Reality, Representation, and Experimental Practice," *Science in Context* 2 (1991): 23–58; idem, "Who Was Robert Hooke?" (1989) (chapter 9 in this volume).

4. Shapin, "Who is the Industrial Scientist?" (chapter 10 in this volume); idem, "The Scientist in 2008," *Seed Magazine,* no. 19 (December 2008), pp. 58–62; idem, *The Scientific Life: A Moral History of a Late Modern Vocation* (Chicago: University of Chicago Press, 2008).

5. Shapin, "The Philosopher and the Chicken" (1998) (chapter 11 in this volume); idem, "Descartes the Doctor: Rationalism and Its Therapies" (2000) (chapter 15 in this volume); idem, "How to Eat Like a Gentleman" (2003) (chapter 12 in this volume); idem, "Trusting George Cheyne" (2003) (chapter 13 in this volume); idem, "Expertise, Common Sense, and the Atkins Diet," in *Public Science in Liberal Democracy,* ed. Peter W. B. Phillips (Toronto: University of Toronto Press, 2007), pp. 174–193; idem, "Feeding, Feeling, Thinking," in *Gefühle zeigen: Manifestationsformen emotionaler Prozesse,* ed. Johannes Fehr and Gerd Folkers, Edition Collegium Helveticum, 5 (Zürich: Chronos Verlag, 2009), pp. 445–466.

6. One token of that type of writing is my little book *The Scientific Revolution* (1996), while others are represented by some contributions to *The New Yorker* and by a larger number of essays in *The London Review of Books.*

Chapter 1. Lowering the Tone in the History of Science

1. Woody Allen, "Thus Ate Zarathustra," in idem, *Mere Anarchy* (New York: Random House, 2008), pp. 141–146; orig. publ. in *The New Yorker* (3 July 2006).

2. Lesley Chamberlain, "A Spoonful of Dr Liebig's Beef Extract," *The Times Literary Supplement* (9 August 1996), p. 15: accessed 21 March 2009 at http://tls.timesonline.co.uk/article/0,,25332-1993806,00.html.

3. Shapin, "The Philosopher and the Chicken" (chapter 11 in this volume).

4. Friedrich Nietzsche to Franz Overbeck, Turin, Christmas 1888, in *Selected Letters of Friedrich Nietzsche,* ed. and trans. Christopher Middleton (Indianapolis, IN: Hackett Publishing, 1996), pp. 337–338.

5. http://www.mwscomp.com/movies/brian/brian-02.htm and http://www.youtube.com/watch?v=XiDmMBIyfsU (accessed 1 November 2008).

6. Ralf Dahrendorf, "The Intellectual and Society: The Social Function of the 'Fool' in the Twentieth Century," in *On Intellectuals: Theoretical Studies, Case Studies,* ed. Philip Rieff (Garden City, NY: Anchor Books, 1969), pp. 53–56.

7. Anthony Standen, *Science Is a Sacred Cow* (New York: E. P. Dutton, 1950); Spencer Klaw, *The New Brahmins: Scientific Life in America* (New York: William Morrow & Co., 1968).

8. John William Draper, *History of the Conflict between Science and Religion,* 4th ed. (New York: D. Appleton, 1875; orig. publ. 1874), pp. vi, 364. It was not Draper but White who preferred more restrictively to define the enemy as "Dogmatic Theology": Andrew Dickson White, *A History of the Warfare of Science with Theology in Christendom* (New York: D. Appleton, 1897), pp. ix, xii, 158. See also Shapin, "Science and the Modern World" (chapter 16 in this volume).

9. George Sarton, *The Life of Science: Essays in the History of Civilization* (Bloomington: Indiana University Press, 1948), pp. 61–63.

10. Or at least apocryphal: Jesus Ben Sirach, Ecclesiasticus 44:1: "Let us now praise famous men, and our fathers that begat us."

11. George Sarton, *The Study of the History of Science* (Cambridge, MA: Harvard University Press, 1936), p. 35.

12. Sarton, *The Study of the History of Science,* p. 45.

13. George Sarton, *The History of Science and the New Humanism,* new ed. (New Brunswick, NJ: Transaction Books, 1988; orig. publ. 1962), p. 47.

14. George Sarton, "Knowledge and Charity," *Isis* 5 (1923): 5–19, on p. 10.

15. Quoted by Lynn White, Jr., "Natural Science and Naturalistic Art in the Middle Ages," *American Historical Review* 52 (1947): 421–435, on p. 422.

16. Richard R. Yeo, "Genius, Method and Morality: Images of Newton in Britain, 1760–1860," *Science in Context* 2 (1988): 257–284.

17. For an interpretation of the cultural and social significance of this opposition in the context of early pedagogical thought, see Steven Shapin and Barry Barnes, "Head and Hand: Rhetorical Resources in British Pedagogical Writing, 1770–1850," *Oxford Review of Education* 2 (1976): 231–254.

18. See, for example, David Bloor, "Rationalism, Supernaturalism, and the Sociology of Knowledge," in *Scientific Knowledge Socialized,* ed. Imre Hronsky, Márta Fehér, and Balázs Dajka (Budapest: Akadémiai Kiadó, 1988), pp. 59–74; idem, "Epistemic Grace: Antirelativism as Theology in Disguise," *Common Knowledge* 13 (2007): 250–280.

19. Thomas S. Kuhn, "What Are Scientific Revolutions?" in idem, *The Road Since Structure: Philosophical Essays, 1970–1993, with an Autobiographical Interview,* ed. James Conant and John Haugeland (Chicago: University of Chicago Press, 2000), pp. 13–32, esp. pp. 16–17. For an emblematic instance of Koyré's stress on

historical specificity and coherence, see Alexandre Koyré, "Galileo and Plato," *Journal of the History of Ideas* 4 (1943): 400–428.

20. See Steven Shapin, "Discipline and Bounding: The History and Sociology of Science as Seen through the Externalism-Internalism Debate," *History of Science* 30 (1992): 333–369.

21. Among many examples of this genre of work, see J. E. McGuire and P. M. Rattansi, "Newton and the 'Pipes of Pan,'" *Notes and Records of the Royal Society* 21 (1966): 108–143; McGuire, "Force, Active Principles, and Newton's Invisible Realm," *Ambix* 15 (1968): 154–208; idem, "Boyle's Conception of Nature," *Journal of the History of Ideas* 33 (1972): 523–542; Charles Webster, *The Great Instauration: Science, Medicine, and Reform 1626–1660* (London: Duckworth, 1975). Arnold Thackray, who migrated from Leeds to Cambridge to Harvard, and then to the University of Pennsylvania when I was a graduate student, produced a body of intellectually challenging work in the history of early modern chemistry: see, for example, "'Matter in a Nutshell': Newton's *Opticks* and Eighteenth-Century Chemistry," *Ambix* 15 (1968): 29–53; idem, *Atoms and Powers: An Essay on Newtonian Matter-Theory and the Development of Chemistry* (Cambridge, MA: Harvard University Press, 1970).

22. See George Reisch, "Disunity in the *International Encyclopedia of Unified Science*," in *Logical Empiricism in North America*, ed. Gary L. Hardcastle and Alan W. Richardson (Minneapolis: University of Minnesota Press, 2003), pp. 197–215.

23. For example, John Dupré, *The Disorder of Things: Metaphysical Foundations of the Disunity of Science* (Cambridge, MA: Harvard University Press, 1993); Nancy Cartwright, *The Dappled World: A Study of the Boundaries of Science* (Cambridge: Cambridge University Press, 1999); Alexander Rosenberg, *Instrumental Biology, or the Disunity of Science* (Chicago: University of Chicago Press, 1994); Jerry A. Fodor, "Special Sciences (or: The Disunity of Science as a Working Hypothesis," *Synthese* 28 (1974): 97–115; Peter Galison and David J. Stump, eds., *The Disunity of Science: Boundaries, Contexts, and Power* (Stanford, CA: Stanford University Press, 1996); also Alison Wylie, "Rethinking Unity as a 'Working Hypothesis': How Archaeologists Exploit the Disunities of Science," *Perspectives on Science* 7 (1999): 293–317; Philip Kitcher, "Unification as a Regulative Ideal," *Perspectives on Science* 7 (1999): 337–348; Peter Galison, "The Americanization of Unity," *Dædalus* 127, no. 1 (Winter 1998): 45–71.

24. For Scientific Method, see, for example, John A. Schuster and Richard R. Yeo, eds., *The Politics and Rhetoric of Scientific Method: Historical Studies* (Dordrecht: D. Reidel, 1986); for natural philosophy and mathematics in the early modern period, see Robert S. Westman, "The Astronomer's Role in the Sixteenth Century: A Preliminary Study," *History of Science* 18 (1980): 105–147, and Mario Biagioli, *Galileo, Courtier: The Practice of Science in the Culture of Absolutism* (Chicago: University of Chicago Press, 1993); for contests over experimental and mechanical

philosophy in the seventeenth century, see, among many examples, Steven Shapin and Simon Schaffer, *Leviathan and the Air-Pump: Hobbes, Boyle, and the Experimental Life* (Princeton, NJ: Princeton University Press, 1985), and Peter Dear, *The Intelligibility of Nature: How Science Makes Sense of the World* (Chicago: University of Chicago Press, 2006), chapter 1.

25. See, for example, Peter J. Bowler and Iwan Rhys Morus, *Making Modern Science: A Historical Survey* (Chicago: University of Chicago Press, 2005); Andrew Ede and Lesley B. Cormack, *A History of Science in Society: From Philosophy to Utility* (Peterborough, ON: Broadview Press, 2004); Richard Olson, *Science Deified and Science Defied: The Historical Significance of Science in Western Culture*, 2 vols. (Berkeley: University of California Press, 1982, 1990); and Patricia Fara, *Science: A Four Thousand Year History* (Oxford: Oxford University Press, 2009).

26. See Steven Shapin, "Science," in *New Keywords: A Vocabulary of Culture and Society,* ed. Tony Bennett, Larry Grossberg, and Meaghan Morris (Oxford: Blackwell, 2005), pp. 314–317; Andrew Cunningham, "Getting the Game Right: Some Plain Words on the Identity and Invention of Science," *Studies in History and Philosophy of Science* 19 (1988): 365–389; Peter Dear, "What Is History of Science the History *Of?* Early Modern Roots of the Ideology of Modern Science," *Isis* 96 (2005): 390–406.

27. Isaac Newton, *The Mathematical Principles of Natural Philosophy,* trans. Andrew Motte (New York: Daniel Adee, 1848; trans. orig. publ. 1729), p. 506.

28. Quoted in Richard S. Westfall, *Never at Rest: A Biography of Isaac Newton* (Cambridge: Cambridge University Press, 1980), p. 473.

29. Westphal, *Never at Rest,* p. x.

30. Henry Brougham, *Lives of the Men of Letters and Science, Who Flourished in the Time of George III,* 2 vols. (London: Charles Knight, 1845–1846), vol. 1, p. 510. The material in this section draws upon Steven Shapin, *The Scientific Life: A Moral History of a Late Modern Vocation* (Chicago: University of Chicago Press, 2008), chapter 1.

31. Thomas Henry Huxley, "On the Method of Zadig [1880]," in idem, *Collected Essays,* vol. 4: *Science and Hebrew Tradition* (New York: D. Appleton, 1900), pp. 1–23, on p. 2.

32. Claude Bernard, *Introduction to the Study of Experimental Medicine,* trans. Henry Copley Greene (New York: Dover, 1957; orig. publ. 1865), pp. 40, 42–43.

33. Thorstein Veblen, *The Higher Learning in America: A Memorandum on the Conduct of Universities by Business Men* (New York: Sagamore Press, 1957; orig. publ. 1918), p. 5.

34. John Brooke, "Namier and Namierism," *History and Theory* 3 (1964): 331–347, on p. 333.

35. In my own case, that fascination with quotidian knowledge-making practice

was probably first expressed in Shapin and Schaffer, *Leviathan and the Air-Pump*, but see also Steven Shapin, "The Invisible Technician," *American Scientist* 77 (November–December 1989): 554–563; also idem, "The Invisible Technician," in *A Social History of Truth: Civility and Science in Seventeenth-Century England* (Chicago: University of Chicago Press, 1994), chapter 8. This interest was partly inspired by early ethnographic studies of laboratory work, for example, H. M. Collins, "The TEA Set: Tacit Knowledge and Scientific Networks," *Science Studies* 4 (1974): 165–186; idem, "The Seven Sexes: A Study in the Sociology of a Phenomenon, or the Replication of an Experiment in Physics," *Sociology* 9 (1975): 205–224.

36. Janet Browne, "I Could Have Retched All Night: Charles Darwin and His Body," in *Science Incarnate: Historical Embodiments of Natural Knowledge*, ed. Christopher Lawrence and Steven Shapin (Chicago: University of Chicago Press, 1998), pp. 240–287, on p. 243.

37. Shapin, "The Philosopher and the Chicken" (chapter 11 in this volume).

38. Rebecca M. Herzig, *Suffering for Science: Reason and Sacrifice in Modern America* (New Brunswick, NJ: Rutgers University Press, 2005).

39. Max Weber, "Science as a Vocation," in idem, *From Max Weber: Essays in Sociology*, ed. H. H. Gerth and C. Wright Mills (London: Routledge, 1991; essay orig. publ. 1919), pp. 129–156, on pp. 142, 149. This and the following paragraph draw on Shapin, *The Scientific Life*, chapters 1–2.

40. C. P. Snow, "Address by Charles P. Snow [to Annual Meeting of American Association for the Advancement of Science, 27 December 1960]," *Science* n.s. 133, no. 3448 (27 January 1961): 256–259; see also Shapin, *The Scientific Life*, chapter 2.

41. See Shapin, "How to Be Antiscientific" (chapter 3 in this volume).

42. Robert K. Merton, "The Normative Structure of Science," in idem, *The Sociology of Science: Theoretical and Empirical Investigations*, ed. Norman W. Storer (Chicago: University of Chicago Press, 1973; art. orig. publ. 1942), pp. 267–278, esp. pp. 275–276.

43. Shapin, *The Scientific Life*, chapter 3.

44. David A. Hollinger, "The Defence of Democracy and Robert K. Merton's Formulation of the Scientific Ethos," in *Knowledge and Society*, ed. Robert Alun Jones and Henrika Kuklick (Greenwich CT: JAI Press, 1983), vol. 4, pp. 1–15; idem, "Science as a Weapon in *Kulturkämpfe* in the United States During and After World War II," *Isis* 86 (1995): 440–454.

45. See Shapin, "Science and the Modern World."

46. See Shapin, *The Scientific Life*, chapters 4–5; idem, "Who Is the Industrial Scientist?" (chapter 10 in this volume).

47. W. B. Yeats, "The Second Coming," in idem, *The Poems*, ed. Daniel Albright (London: Dent, 1990), p. 235.

48. Edward Hallett Carr, *What Is History?* (New York: Vintage Books, 1961), p. 23.

Chapter 2. Cordelia's Love

1. Robert Boyle, *A Proëmial Essay . . . With Some Considerations touching Experimental Essays in General*, in *The Works of the Honourable Robert Boyle*, ed. Thomas Birch, 2nd ed., 6 vols. (London, J. & F. Rivington, 1772 [1661]), vol. 1, pp. 298–318, on p. 304.

2. Francis Bacon, *Preparative Towards a Natural and Experimental History* [= *Parasceve*], in *The Philosophical Works of Francis Bacon*, ed. James Spedding, Robert Leslie Ellis, and Douglas Denon Heath, 5 vols. (London: Longman and Co., 1858 [1620]), vol. 4, pp. 249–263, on p. 254.

3. Francis Bacon, *The Advancement of Learning*, in *The Philosophical Works of Francis Bacon*, vol. 3, pp. 253–491, on p. 281.

4. David Bloor, *Knowledge and Social Imagery*, 2nd ed. (Chicago: University of Chicago Press, 1991 [1976], chapters 1–2). "Our equivalence postulate is that all beliefs are on a par with one another with respect to the causes of their credibility. It is not that all beliefs are equally true or equally false, but that regardless of truth and falsity the fact of their credibility is to be seen as equally problematic" (Barry Barnes and David Bloor, "Relativism, Rationalism, and the Sociology of Knowledge," in *Rationality and Relativism*, ed. Martin Hollis and Steven Lukes [Oxford: Basil Blackwell, 1982], pp. 21–47, on p. 23).

5. Sociologists of scientific knowledge have long stressed the process of socialization as, so to speak, a "default explanation" of actors' beliefs (for example, Barry Barnes, "On the Reception of Scientific Beliefs," in *Sociology of Science: Selected Readings*, ed. Barry Barnes [Harmondsworth: Penguin, 1972], pp. 269–291, on p. 272): "The human actor adopts a way of life largely determined by his culture and the position he occupies within it; many of his beliefs and most of those crucial to the acquisition of further beliefs, will be found empirically to have been received in socialization processes. Theories of the socialization process will eventually provide the answer to most problems in the adoption of beliefs."

6. Shapin, "The House of Experiment in Seventeenth-century England" (chapter 5 in this volume); idem, "Placing the View from Nowhere: Historical and Sociological Problems in the Location of Science," *Transactions of the Institute of British Geographers*, n.s. 23 (1998): 5–12; Adi Ophir and Steven Shapin, "The Place of Knowledge: A Methodological Survey," *Science in Context* 4 (1991): 3–21; Steven Shapin, *A Social History of Truth: Civility and Science in Seventeenth-Century England* (Chicago: University of Chicago Press, 1994), chapter 4; idem, "The Philosopher and the Chicken" (chapter 11 in this volume).

7. For relevant history of credit-economies, see Simon Schaffer, "Defoe's Natural Philosophy and the Worlds of Credit," in *Nature Transformed: Science and Literature, 1700–1900*, ed. John R. R. Christie and Sally Shuttleworth (Manchester: Manchester University Press, 1989), pp. 13–44; idem, "A Social History of Plausibility: Country,

City and Calculation in Augustan Britain," in *Rethinking Social History: English Society 1570–1920 and Its Interpretation,* ed. Adrian Wilson (Manchester: Manchester University Press, 1993), pp. 128–157; for limits to skepticism, see Mary Douglas, "The Social Preconditions of Radical Scepticism," in *Power, Action and Belief: A New Sociology of Knowledge?* Sociological Review Monograph No. 32, ed. John Law (London: Routledge & Kegan Paul, 1986), pp. 68–87, and Shapin, *Social History of Truth,* chapter 1; for treatment of fraud and dishonesty in relation to science as a credit economy, see Steven Shapin, "Trust, Honesty, and the Authority of Science," in *Society's Choices: Social and Ethical Decision Making in Biomedicine,* ed. Ruth Ellen Bulger, Elizabeth Meyer Bobby, and Harvey V. Fineberg (for Committee on the Social and Ethical Impacts of Developments in Biomedicine, Institute of Medicine, National Academy of Sciences) (Washington, DC: National Academy Press, 1995), pp. 388–408.

8. See, for example, James A. Secord, ed., "The Big Picture: A Special Issue," *British Journal for the History of Science* 26 (1993): 387–483.

9. Augustine Brannigan and Michael Lynch, "On Bearing False Witness: Credibility as an Interactional Accomplishment," *Journal of Contemporary Ethnography* 16 (1987): 115–146.

10. For historical studies of the link between legal and scientific fact-making practices in early modern England, see, for example, Julian Martin, *Francis Bacon, the State, and the Reform of Natural Philosophy* (Cambridge: Cambridge University Press, 1992); Rose-Mary Sargent, "Scientific Expertise and Legal Expertise: The Way of Experience in Seventeenth-Century England," *Studies in History and Philosophy of Science* 20 (1989): 19–45; and Barbara J. Shapiro, "Law and Science in Seventeenth-Century England," *Stanford Law Review* 21 (1969): 727–766.

11. Michael Lynch and David Bogen, *The Spectacle of History: Speech, Text, and Memory in the Iran-Contra Hearings* (Durham, NC: Duke University Press, 1996), chapter 1.

12. See, in this connection, Barbara Herrnstein Smith's powerful argument ("The Unquiet Judge: Activism without Objectivism in Law and Politics," in *Rethinking Objectivity,* ed. Allan Megill [Durham, NC: Duke University Press, 1994 (1992)], pp. 289–311) against both the effective existence and the desirability of transcendental ("objectivist") standards of judgment in the law; and, for the relevance of her arguments to science studies debates over relativism, see Thomas F. Gieryn, "Objectivity for These Times," *Perspectives on Science* 2 (1994): 324–349, on pp. 342–345.

13. Brannigan and Lynch, "On Bearing False Witness," p. 116.

14. Mary Douglas, *Implicit Meanings: Essays in Anthropology* (London: Routledge & Kegan Paul, 1975), p. 238.

15. I do not here suppose that cultures are necessarily homogeneous with respect to their credibility-judgments or with respect to their sense of how credibility ought to be secured. (Nor, of course, did Mary Douglas.) On the contrary, it seems a pru-

dent maxim of method to presume that all cultures recognizable as such contain *conflicting* credibility-managing schemes.

16. Shapin, *Social History of Truth,* chapter 5.

17. I have strongly argued (for example, Shapin, *Social History of Truth,* chapter 1) that "trust in people" is an *ineliminable* feature of the credibility of factual claims. Yet I also make a distinction between necessary and sufficient conditions of credibility and I note that the identification of which people will count as trustworthy is scenically variable.

18. For recent accounts of these, and similar, passages of early modern pneumatics, see, for example, Steven Shapin and Simon Schaffer, *Leviathan and the Air-Pump: Hobbes, Boyle, and the Experimental Life* (Princeton, NJ: Princeton University Press, 1985), and, esp. Peter Dear, "Miracles, Experiments, and the Ordinary Course of Nature," *Isis* 81 (1990): 663–683; idem, *Discipline and Experience: The Mathematical Way in the Scientific Revolution* (Chicago: University of Chicago Press, 1995), chapter 7.

19. Trevor J. Pinch, "Towards an Analysis of Scientific Observation: The Externality and Evidential Significance of Observational Reports in Physics," *Social Studies of Science* 15 (1985): 3–36; Bruno Latour, *The Pasteurization of France,* trans. Alan Sheridan and John Law (Cambridge, MA: Harvard University Press, 1988), pp. 87–93; cf. Gerald Geison, *The Private Science of Louis Pasteur* (Princeton, NJ: Princeton University Press, 1995), chapter 6.

20. Here I can only hint at some differences in sensibility between my account and Latour's important model of the stabilization of claims through enrolling, controlling, and the constitution of "obligatory points of passage." Since I insist on the potential open-endedness of the resources for managing credibility, Latour's presumption of pragmatically maximizing actors strikes me as a bit too schematic and lacking in detail—too much like a global theory. Moreover, the Latourian model accounts for stability without the apparent invocation of a *normative order.* As John Law nicely noted some time ago, economic (and militaristic) models tend to be hollow because of their reluctance to acknowledge actors' pervasive concerns for maintaining the interactional order and the role of such concerns in economies of credibility. See Rob Williams and John Law, "Beyond the Bounds of Credibility," *Fundamenta Scientiae* 1 (1980): 295–315, on pp. 312–314; Steven Shapin, "Here and Everywhere: Sociology of Scientific Knowledge," *Annual Review of Sociology* 21 (1995): 289–321, on pp. 307–309.

21. Several of the general points about testing developed here have been concisely noted by Donald MacKenzie ("From Kwajalein to Armageddon? Testing and the Social Construction of Missile Accuracy," in *The Uses of Experiment: Studies in the Natural Sciences,* ed. David Gooding, Trevor Pinch, and Simon Schaffer [Cambridge: Cambridge University Press, 1989], pp. 409–435, on pp. 411–414) and Trevor J. Pinch ("'Testing—One, Two, Three . . . Testing!' Towards a Sociology of

Testing," *Science, Technology & Human Values* 18 [1993]: 25–41), to whose work I am obviously indebted.

22. MacKenzie, "From Kwajalein to Armageddon?"; idem, *Inventing Accuracy: A Historical Sociology of Nuclear Missile Guidance* (Cambridge, MA: MIT Press, 1990).

23. Steven Epstein, *Impure Science: AIDS, Activism, and the Politics of Knowledge* (Berkeley: University of California Press, 1996), p. 255.

24. I refer above to "formal discussions" because laboratory-bench informal conversation sometimes consequentially contrasted "scientific" to "political" (or "economic") considerations. That state of affairs only tended to change when discussions acquired a more "official" character.

25. Pinch, "Towards an Analysis of Scientific Observation."

26. Sheila Jasanoff, "Science, Politics, and the Renegotiation of Expertise at EPA," *Osiris* (2nd Series) 7 (1992): 195–217, on p. 202.

27. Zygmunt Bauman, *Intimations of Postmodernity* (London: Routledge, 1992), p. 71. There are obvious resonances here with Gaston Bachelard's (*The New Scientific Spirit,* trans. Arthur Goldhammer [Boston: Beacon Press, 1984 (1934)], p. 13) account of the "phenomeno-technology" of modern science: "The first achievement of the scientific spirit was to create reason in the image of the world; modern science has moved on to the project of constructing a world in the image of reason. Scientific work makes rational entities real."

28. Thomas Hobbes, *Leviathan,* ed. C. B. Macpherson (Harmondsworth: Penguin, 1968 [1651]), p. 166.

29. Graham Richards, *On Psychological Language and the Physiomorphic Basis of Human Nature* (London: Routledge, 1989), pp. 85–90.

30. Bauman, *Intimations of Postmodernity,* p. 73.

31. The scientific "boundary objects" described by Susan Leigh Star and James R. Griesemer ("'Translations' and Boundary Objects: Amateurs and Professionals in Berkeley's Museum of Vertebrate Zoology, 1907–39," *Social Studies of Science* 19 [1989]: 387–420)—objects whose properties *are* decided through transactions between actors in several "social worlds"—might possibly be treated as complex intermediaries between my "authorized" and "conversational objects." Also, it should not be concluded that "conversational objects" only populate the *Geisteswissenschaften:* recent scholarship seems to establish their pertinence to the career of medical ontology (for example, N. D. Jewson, "Medical Knowledge and the Patronage System in Eighteenth-Century England," *Sociology* 8 [1974]: 369–385; Cecil G. Helman, "'Feed a Cold, Starve a Fever'—Folk Models of Infection in an English Suburban Community and Their Relation to Medical Treatment," *Culture, Medicine and Psychiatry* 2 [1978]: 107–137; Charles E. Rosenberg, "The Therapeutic Revolution: Medicine, Meaning, and Social Change in Nineteenth-Century America," in idem, *Explaining Epidemics and Other Studies in the History of Medicine*

[Cambridge: Cambridge University Press, 1992 (1977)], pp. 9–31), while Dear's work on early modern natural philosophy (*Discipline and Experience*) draws attention to changing conceptions of "experience"—from that which is commonly available to that which requires authoritative testimony.

32. Barry Barnes and David Edge, "Science as Expertise," in *Science in Context: Readings in the Sociology of Science,* ed. Barry Barnes and David Edge (Milton Keynes: Open University Press, 1982), pp. 233–249, on p. 233.

33. Yaron Ezrahi, "The Political Resources of American Science," *Science Studies* 1 (1971): 117–134; Barnes and Edge, "Science as Expertise," p. 238.

34. Theodore M. Porter, *Trust in Numbers: The Pursuit of Objectivity in Science and Public Life* (Princeton, NJ: Princeton University Press, 1995).

35. Cf. Shapin, *Social History of Truth,* chapters 5–6 and pp. 409–417.

36. Niklas Luhmann, *Trust and Power: Two Works,* ed. Tom Burns and Gianfranco Poggi, trans. Howard Davis, John Raffan, and Kathryn Rooney (Chichester: John Wiley, 1979); Anthony Giddens, *The Consequences of Modernity* (Stanford, CA: Stanford University Press, 1989).

37. Porter, *Trust in Numbers,* p. ix.

38. Shapin, "Pump and Circumstance" (chapter 6 in this volume); idem, "The Politics of Observation: Cerebral Anatomy and Social Interests in the Edinburgh Phrenology Disputes," in *On the Margins of Science: The Social Construction of Rejected Knowledge,* Sociological Review Monograph No. 27, ed. Roy Wallis (Keele: Keele University Press, 1979), pp. 139–178.

39. The fact that I now happen to accept certain quantitatively expressed claims about the relative safety of air travel, or the yield of investments, may indeed have something to do with my sense that statistical propositions are more reliable than anecdote or simple assertion. After all, I spent many years acquiring that sense, in educational institutions, coached (and occasionally coerced) by the embodied authority of teachers, training me to accept the "just-so-ness" of numerical manipulations. The credibility of such claims also proceeds from the contingent fact that I am currently unable—lacking the necessary expertise—effectively to discern bias in specific modes of statistical inference or in the presuppositions on which specific data sets were assembled. Yet an item in tomorrow's newspapers may convince me that *these* quantitative claims were biased and misleading, thus potentially feeding a general sense that numerically expressed claims may be no more reliable, disinterested, or objective than anecdote.

Chapter 3. How to Be Antiscientific

1. Some of my work in this area includes Steven Shapin and Simon Schaffer, *Leviathan and the Air-Pump: Hobbes, Boyle, and the Experimental Life* (Princeton, NJ: Princeton University Press, 1985); Steven Shapin, *A Social History of Truth: Civility and Science in Seventeenth-Century England* (Chicago: University of Chicago

Press, 1994); idem, "Cordelia's Love" (chapter 2 in this volume); idem, "Here and Everywhere: Sociology of Scientific Knowledge," *Annual Review of* Sociology 21 (1985): 289–321; idem, "Rarely Pure and Never Simple: Talking about Truth," *Configurations* 7 (1999): 1–14.

2. Some well-known recent tracts by scientists expressing such sentiments are Lewis Wolpert, *The Unnatural Nature of Science: Why Science does not Make (Common) Sense* (London: Faber and Faber, 1992); Paul R. Gross and Norman Levitt, *Higher Superstition: The Academic Left and Its Quarrels with Science* (Baltimore: Johns Hopkins University Press, 1994); Paul R. Gross, Norman Levitt, and Martin W. Lewis, eds., *The Flight from Science and Reason* (New York: New York Academy of Sciences, 1996); Alan Sokal and Jean Bricmont, *Fashionable Nonsense: Postmodern Intellectuals' Abuse of Science* (New York: Picador USA, 1998); Steven Weinberg, "Night Thoughts of a Quantum Physicist," *Bulletin of the American Academy of Arts and Sciences* 49 (1995): 51–64; idem, "The Revolution that Didn't Happen," *New York Review of Books* 45, no. 15 (8 October 1998): 48–52.

3. After I had written this essay, I came across a broadly similar observation powerfully made by Israeli historian of physics Mara Beller ("The Sokal Hoax: At Whom are We Laughing?" *Physics Today* 51, no. 9 [September 1998]: 29–34; see also idem, "Criticism and Revolutions," *Science in Context* 10 [1997]: 13–37), though she focuses attention exclusively on the views of twentieth-century quantum physicists.

4. Sociologists of science, notably those "Edinburgh School" writers criticized by Steven Weinberg and others, have repeatedly stressed that the social component of scientific knowledge is *not* to be set against the causal role of unverbalized natural reality: the social component is seen as a condition for having experience of a recognized kind and for representing that experience in linguistic form. See, for example, David Bloor, *Knowledge and Social Imagery,* 2nd ed. (Chicago: University of Chicago Press, 1991 [1976]): "No consistent sociology could ever present knowledge as a fantasy unconnected with our experience of the material world around us" (p. 33), and Barry Barnes, *Interests and the Growth of Knowledge* (London: Routledge and Kegan Paul, 1977): "There is indeed one world, one reality, 'out there,' the source of all our perceptions" (pp. 25–26); see also idem, "Realism, Relativism, and Finitism," in *Cognitive Relativism and Social Science,* ed. Diederick Raven, Lieteke van Vucht Tijssen, and Jan de Wolf (New Brunswick, NJ: Transaction, 1992), pp. 131–147. I have no very satisfactory ideas why Defenders of Science should miss the facts right in front of their eyes.

5. Max Planck, *Scientific Autobiography and Other Papers,* trans. Frank Gaynor (New York: Philosophical Library, 1949), p. 106.

6. J. Robert Oppenheimer, *Science and the Common Understanding: The BBC Reith Lectures 1953* (New York: Oxford University Press, 1954), p. 4.

7. Weinberg, "The Revolution that Didn't Happen," p. 52. Only after this piece was drafted did I become aware of Richard Rorty's similar, but more vigorously

expressed, puzzlement about Weinberg's claim ("Thomas Kuhn, Rocks, and the Laws of Physics," *Common Knowledge* 6 [1997]: 6–16).

8. The controversy among scientists about whether or not science is about to be completed now even claims space in *The New York Times:* see the debate between John Horgan and John Maddox ("Resolved: Science is at an End. Or is It?" *The New York Times,* 10 November 1998, sec. F, p. 5). For pertinent claims, see Steven Weinberg, *Dreams of a Final Theory* (New York: Pantheon, 1992); John Horgan, *The End of Science: Facing the Limits of Knowledge in the Twilight of the Scientific Age* (Reading, MA: Helix Books, 1996); and Gunther Stent, *The Coming of the Golden Age: A View of the End of Progress* (Garden City, NY: Natural History Press, 1969). For historical commentary on recurrent announcements of the End of Science, see Simon Schaffer, "Utopia Unlimited: On the End of Science," *Strategies* 4/5 (1991): 151–181. For my own engagements with what scientists have meant by truth, see chapter 8 in this volume.

9. Albert Einstein, *Ideas and Opinions* (New York: Crown Publishers, 1954), p. 319. For Huxley, see Thomas Henry Huxley, "On the Educational Value of the Natural History Sciences [1854]," in *Science and Education: Essays,* vol. 3 of *Collected Essays* (New York: D. Appleton, 1900), pp. 38–65: "Science is, I believe, nothing but *trained and organised common sense*" (p. 45); for Planck, see *Scientific Autobiography and Other Papers,* p. 88.

10. Wolpert, *The Unnatural Nature of Science.*

11. For an interesting exploration of what scientists' professions of Popperianism might mean, see Michael J. Mulkay and G. Nigel Gilbert, "Putting Philosophy to Work: Karl Popper's Influence on Scientific Practice," *Philosophy of the Social Sciences* 11 (1981): 389–407; for psychological assessments of scientists' grasp of formal logic, see Michael J. Mahoney, "Psychology of the Scientist: An Evaluative Review," *Social Studies of Science* 9 (1979): 349–375, and idem and B. G. DeMonbreun, "Psychology of the Scientist: An Analysis of Problem-Solving Bias," *Cognitive Therapy and Research* 1 (1977): 229–238.

12. P. W. Bridgman, *Reflections of a Physicist,* 2nd ed. (New York: Philosophical Library, 1955), p. 81.

13. For the most aggressive recent affirmation of reductionist unity, see Edward O. Wilson, *Consilience: The Unity of Knowledge* (New York: Knopf, 1998), though Wilson now seems to have forgotten the complaints against rampant molecular reductionism he so eloquently expressed in his autobiographical *Naturalist* (New York: Warner Books, 1995), esp. chapter 12. Violently antireductionist statements by biologists are not, of course, hard to find: see, among many examples, Robert G. Shulman, "Hard Days in the Trenches," *FASEB Journal* 12 (1998): 255–258; Ernst Mayr, *This is Biology* (Cambridge, MA: Harvard University Press, 1997); Erwin Chargaff, *Essays on Nucleic Acids* (Amsterdam: Elsevier, 1963); idem, *Heraclitean Fire: Sketches from a Life before Nature* (New York: Rockefeller University Press,

1978); and Richard C. Lewontin, *Biology as Ideology: The Doctrine of DNA* (New York: HarperPerennial, 1993). For what it is worth, Wilson's vision of reductionist unity is devastatingly taken apart by the philosopher Jerry Fodor: "[Wilson] suspects that if we resist consilience, that's because we're suffering from pluralism, nihilism, solipsism, relativism, idealism, deconstructionism, and other symptoms of the French disease" ("Look! Review of *Consilience: The Unity of Knowledge,* by Edward O. Wilson," *London Review of Books* 20, no. 21 [29 October 1998], on pp. 3, 6).

14. Einstein, *Ideas and Opinions,* pp. 296 and 318. Quoted more fully, Einstein said: "If you want to find out anything from the theoretical physicists about the methods they use, I advise you to stick closely to one principle: don't listen to their words, fix your attention on their deeds" (ibid., p. 296).

15. I believe I owe this formulation to a conversation with Harry Collins many years ago.

16. Einstein, *Ideas and Opinions,* p. 224. For increasing pluralist sensibilities about science among philosophers, see, for example, John Dupré, *The Disorder of Things: Metaphysical Foundations of the Disunity of Science* (Cambridge, MA: Harvard University Press, 1993).

17. On this, see the classic essay by Isaiah Berlin, "The Divorce Between the Sciences and the Humanities," in *The Proper Study of Mankind,* ed. Henry Hardy and Roger Hausheer (New York: Farrar, Straus, and Giroux, 1998), pp. 326–358.

18. For example, Steven Weinberg's judgment that much philosophy of science "has nothing to do with science": "The fact that we scientists [which ones, please?] do not know how to state in a way that philosophers [which ones, please?] would approve what it is that we are doing in searching for scientific explanations does not mean that we are not doing something worthwhile. We could use help from professional philosophers in understanding what it is we are doing, but with or without help we shall keep at it" (*Dreams of a Final Theory,* p. 167, also p. 29). (I am delighted to hear that. I would be very disturbed, indeed, if I thought that natural scientists were taking marching orders from philosophers!)

Chapter 4. Science and Prejudice in Historical Perspective

1. The historical usages of the "Republic of Science" and the "Republic of Letters" are largely interchangeable. For their genealogies, and, esp., their eighteenth-century references, see Dena Goodman, *The Republic of Letters: A Cultural History of the French Enlightenment* (Ithaca, NY: Cornell University Press, 1994); Anne Goldgar, *Impolite Learning: Conduct and Community in the Republic of Letters, 1680–1750* (New Haven, CT: Yale University Press, 1995); Geoffrey V. Sutton, *Science for a Polite Society: Gender, Culture, and the Demonstration of Enlightenment* (Boulder, CO: Westview Press, 1995); Hans Bots and Françoise Waquet, *La république des lettres* (Paris: Belin, 1997); Zygmunt Bauman, *Legislators and Interpreters: On Modernity, Post-Modernity and Intellectuals* (Oxford: Polity Press, 1987), esp.

pp. 25–26. For seventeenth-century views on the relationships between virtuous social order and proper knowledge, see, for example, Steven Shapin, *The Scientific Revolution* (Chicago: University of Chicago Press, 1996), chapter 3.

2. Francis Bacon, "The New Organon [1623]," in *The Philosophical Works of Francis Bacon*, ed. James Spedding, Robert Leslie Ellis, and Douglas Denon Heath, 5 vols. (London: Longman and Co., 1857–1858), vol. 9, pp. 39–248, on pp. 58–64.

3. For this point, and the material in the following three paragraphs, see Shapin, *The Scientific Revolution*, chapters 2–3.

4. For the character of the experimental natural philosopher, see Shapin, "'A Scholar and a Gentleman'" (chapter 8 in this volume); also Mario Biagioli, "Etiquette, Interdependence, and Sociability in Seventeenth-Century Science," *Critical Inquiry* 22 (1996): 193–238; Lorraine J. Daston, "Baconian Facts, Academic Civility, and the Prehistory of Objectivity," *Annals of Scholarship* 8 (1991): 337–363.

5. For Boyle, see Steven Shapin, *A Social History of Truth: Civility and Science in Seventeenth-Century England* (Chicago: University of Chicago Press, 1994), chapter 4. Descartes's iconic anti-authoritarian and anti-traditionalistic gesture is in the *Discourse on Method*, part I.

6. Charles Taylor, *Sources of the Self: The Making of Modern Identity* (Cambridge, MA: Harvard University Press, 1989), chapters 8–9.

7. John Locke, *Essay Concerning Human Understanding* (1690), book I, chapter 3, sec. 24; Book IV, chapter 15, sec. 6.

8. For Leibniz, see Rudolph W. Meyer, *Leibnitz and the Seventeenth-Century Revolution*, trans. J. P. Stern (Chicago: Henry Regnery Co., 1952), and Ayval Ramati, "Harmony at a Distance: Leibniz's Scientific Academies," *Isis* 87 (1996): 430–452.

9. Lorraine J. Daston, "The Ideal and Reality of the Republic of Letters in the Enlightenment," *Science in Context* 4 (1991): 367–386; see also Gavin De Beer, *The Sciences Were Never at War* (London: Nelson, 1960).

10. For Gibbon, Montesquieu, and Hume, see Thomas J. Schlereth, *The Cosmopolitan Ideal in Enlightenment Thought: Its Form and Function in the Ideas of Franklin, Hume, and Voltaire, 1694–1790* (Notre Dame, IN: University of Notre Dame Press, 1977), pp. 47–48. For cosmopolitanism, solitude, and asceticism in images of the Western intellectual, see, for example, Shapin, "'The Mind Is Its Own Place'" (chapter 7 in this volume); idem, "The Philosopher and the Chicken" (chapter 11 in this volume).

11. The journalist was Jacques-Pierre Brissot de Warville, writing in his 1782 book *De la Vérité*: quoted in Roger Hahn, *The Anatomy of a Scientific Institution: The Paris Academy of Sciences, 1666–1803* (Berkeley: University of California Press, 1971), p. 153. For the idea of scientific genius during that period, see, for example, Simon Schaffer, "Genius in Romantic Natural Philosophy," in *Romanticism and the Sciences*, ed. Andrew Cunningham and Nicholas Jardine (Cambridge: Cambridge University Press, 1990), pp. 82–98, and Richard R. Yeo, "Genius, Method and Mo-

rality: Images of Newton in Britain, 1760–1860," *Science in Context* 2 (1988): 257–284. For the growing significance of "public opinion" in science, see Thomas H. Broman, "The Habermasian Public Sphere and 'Science in the Enlightenment,'" *History of Science* 36 (1998): 123–149. In these connections, see Steven Shapin, "The Image of the Man of Science," in *The Cambridge History of Science,* vol. 4, *Eighteenth-Century Science,* ed. Roy Porter (Cambridge: Cambridge University Press, 2003), pp. 159–183.

12. Quoted in Charles B. Paul, *Science and Immortality: The 'Éloges' of the Paris Academy of Sciences (1699–1791)* (Berkeley: University of California Press, 1980), p. 67; see also Hahn, *Anatomy of a Scientific Institution,* p. 165, and Keith M. Baker, *Condorcet: From Natural Philosophy to Social Mathematics* (Chicago: University of Chicago Press, 1975), esp. pp. 293–299.

13. Joseph Priestley, *An Examination of Dr Reid's Enquiry into the Human Mind on the Principles of Common Sense* (London, 1774), p. 74, and idem, *Experiments and Observations on Different Kinds of Air,* 3 vols. (London, 1774–1777), vol. 1, p. xiv (both quoted in Dorinda Outram, "Science and Political Ideology, 1790–1848," in *Companion to the History of Modern Science,* ed. R. C. Olby et al. [London: Routledge, 1990], pp. 1008–1023, on p. 1017).

14. Robert K. Merton, "Science and the Social Order [1938]," in idem, *The Sociology of Science,* ed. Norman W. Storer (Chicago: University of Chicago Press, 1973), pp. 254–266; idem, "The Normative Structure of Science [1942]," in ibid., pp. 267–278.

15. David A. Hollinger, "The Defense of Democracy and Robert K. Merton's Formulation of the Scientific Ethos," in *Knowledge and Society,* ed. Robert Alun Jones and Henrika Kuklick (Greenwich CT: JAI Press, 1983), vol. 4, pp. 1–15; idem, *Science, Jews, and Secular Culture: Studies in Mid-Twentieth Century American Intellectual History* (Princeton, NJ: Princeton University Press, 1996).

16. This is the main argument of Shapin, *A Social History of Truth,* esp. chapters 1, 5–6; Stephen P. Turner, "Merton's 'Norms' in Political and Intellectual Context," *Journal of Classical Sociology* 7 (2007): 161–178.

17. Shapin, *A Social History of Truth,* pp. 306–307 (for Leeuwenhoek) and chapters 3 and 8 (for trustworthiness and social standing in general).

18. For membership of the early Royal Society of London, see Michael Hunter, *The Royal Society and Its Fellows 1660–1700: The Morphology of an Early Scientific Institution* (Chalfont St. Giles: British Society for the History of Science, 1982); for the Paris Academy, see Hahn, *Anatomy of a Scientific Institution;* for women in French Enlightenment salons, see Dena Goodman, "Enlightenment Salons: The Convergence of Female and Philosophic Ambitions," *Eighteenth-Century Studies* 22 (1989): 329–350; for women in early modern science generally, see Londa Schiebinger, *The Mind Has No Sex? Women in the Origins of Modern Science* (Cambridge,

MA: Harvard University Press, 1989); for the Huguenot scholarly diaspora, see Goldgar, *Impolite Learning.*

19. Margaret 'Espinasse, "The Decline and Fall of Restoration Science," *Past and Present* 14 (1958): 71–89, on p. 86.

20. This is the intellectual argument for a scientific community representative of the society that supports it and that depends upon its epistemic products. For an entry to debates over whether the modern scientific community is genuinely open to all and meritocratic, see, for example, Anne Sayre, *Rosalind Franklin and DNA* (New York: W. W. Norton, 1975); Jonathan R. Cole, *Fair Science: Women in the Scientific Community* (New York: The Free Press, 1979); Harriet Zuckerman, Jonathan R. Cole, and John T. Bruer, eds., *The Outer Circle: Women in the Scientific Community* (New York: W. W. Norton, 1991); Barbara Laslett, Sally Gregory Kohlstedt, Helen Longino, and Evelynn Hammonds, eds., *Gender and Scientific Authority* (Chicago: University of Chicago Press, 1996).

21. See Daston, "Ideal and Reality of the Republic of Letters."

22. See Françoise Waquet, *Le Latin ou l'empire d'un signe* (Paris: Albin Michel, 1999).

23. Isaiah Berlin, "The Divorce between the Sciences and the Humanities [1974]," in idem, *The Proper Study of Mankind: An Anthology of Essays,* ed. Henry Hardy and Roger Hausheer (New York: Farrar, Straus and Giroux, 1998), pp. 326–358, on p. 327; also idem, "The Counter-Enlightenment [1973]," in ibid., pp. 243–268.

24. Taylor, *Sources of the Self,* esp. pp. 167ff.

25. Edmund Burke, *Reflections on the Revolution in France,* ed. Conor Cruise O'Brien (Harmondsworth: Penguin, 1986; orig. publ. 1790), pp. 183–184. For endorsement of the Englishness of Reason-skepticism, see Stephen Toulmin, *Return to Reason* (Cambridge, MA: Harvard University Press, 2001), as discussed in Steven Shapin, "Dear Prudence," *London Review of Books* 24, no. 2 (24 January 2002), pp. 25–27.

26. For a fine attempt at reviving a more modest, "case-by-case" approach to ethical decision-making, see Albert R. Jonsen and Stephen Toulmin, *The Abuse of Casuistry: A History of Moral Reasoning* (Berkeley: University of California Press, 1988).

Chapter 5. The House of Experiment in Seventeenth-century England

1. An outstanding exception is Owen Hannaway, "Laboratory Design and the Aim of Science: Andreas Libavius versus Tyco Brahe," *Isis* 77 (1986): 585–610. See also Peter Galison, "Bubble Chambers and the Experimental Workplace," in *Observation, Experiment, and Hypothesis in Modern Physical Science,* ed. Peter Achinstein and Owen Hannaway (Cambridge, MA: MIT Press, 1985), pp. 309–373; Larry Owens, "Pure and Sound Government: Laboratories, Playing Fields, and Gymnasia

in the Nineteenth-Century Search for Order," *Isis* 76 (1985): 182–194; and, although not concerned with knowledge-making processes, Sophie Forgan, "Context, Image and Function: A Preliminary Enquiry into the Architecture of Scientific Societies," *The British Journal for the History of Science* 19 (1986): 89–113.

2. See, for example, Erving Goffman, *The Presentation of Self in Everyday Life* (London: Allen Lane, Penguin Press, 1969), chapter 3; Richard Sennett, *The Fall of Public Man* (Cambridge: Cambridge University Press, 1974), esp. chapter 5; Shirley Ardener, "Ground Rules and Social Maps for Women: An Introduction," in *Women and Space: Ground Rules and Social Maps,* ed. Shirley Ardener (London: Croom Helm, 1981), chapter 1; Clark E. Cunningham, "Order in the Atoni House," in *Right and Left: Essays on Dual Symbolic Classification,* ed. Rodney Needham (Chicago: University of Chicago Press, 1973), pp. 204–238.

3. Anthony Giddens, *The Constitution of Society: Outline of the Theory of Structuration* (Cambridge: Polity Press, 1984), chapter 3 (esp. his discussion of the work of the Swedish geographer Torsten Hägerstrand); Bill Hillier and Julienne Hanson, *The Social Logic of Space* (Cambridge: Cambridge University Press, 1984), esp. pp. ix–xi, 4–5, 8–9, 19. For Foucauldian perspectives, see Michel Foucault, "Questions on Geography," in idem, *Power/Knowledge: Selected Interviews and Other Writings, 1972–1977,* ed. Colin Gordon (Brighton: Harvester Press, 1980), pp. 63–77; idem, *Discipline and Punish: The Birth of the Prison,* trans. Alan Sheridan (New York: Vintage, 1979); and Adi Ophir, "The City and the Space of Discourse: Plato's Republic—Textual Acts and Their Political Significance" (Ph.D. diss., Boston University, 1984).

4. For a survey of the evaluations of evidence in this setting, see Barbara J. Shapiro, *Probability and Certainty in Seventeenth-Century England: A Study of the Relationships between Natural Science, Religion, History, Law, and Literature* (Princeton, NJ: Princeton University Press, 1983), esp. chapter 2; for treatment of experimental practice in these connections, see Steven Shapin and Simon Schaffer, *Leviathan and the Air-Pump: Hobbes, Boyle, and the Experimental Life* (Princeton, NJ: Princeton University Press, 1985), esp. chapter 2.

5. John Locke, *Essay Concerning Human Understanding,* in Locke, *Works,* 10 vols. (London: Thomas Tegg, 1823), vol. 3, on pp. 97–100; and John Dunn, "The Concept of 'Trust' in the Politics of John Locke," in *Philosophy in History: Essays on the Historiography of Philosophy,* ed. Richard Rorty, J. B. Schneewind, and Quentin Skinner (Cambridge: Cambridge University Press, 1984), pp. 279–301.

6. This is an observational version of what Harry Collins has called "the experimenter's regress": H. M. Collins, *Changing Order: Replication and Induction in Scientific Practice* (London: Sage, 1985), chapter 4.

7. Even in legal writings centrally concerned with the evaluation of testimony, the need to spell out the grounds of persons' differential credibility was apparently rarely felt; see, for example, Shapiro, *Probability and Certainty,* pp. 179–188; Julian

Martin, "'Knowledge is Power': Francis Bacon, the State and the Reform of Natural Philosophy" (Ph.D. diss., University of Cambridge, 1988) esp. chapter 3; cf. Peter Dear, "*Totius in verba:* Rhetoric and Authority in the Early Royal Society," *Isis* 76 (1985): 145–161, on pp. 153–157. These problems are dealt with extensively in Steven Shapin, *A Social History of Truth: Civility and Science in Seventeenth-Century England* (Chicago: University of Chicago Press, 1994), esp. chapters 5–6.

8. Robert Boyle, "An Hydrostatical Discourse, occasioned by the Objections of the Learned Dr. Henry More" (1672), in Boyle, *Works,* ed. Thomas Birch, 6 vols. (London, 1772), vol. 3, pp. 596–628, on p. 626. In other circumstances Boyle elected to credit the testimony of divers: see Boyle, "Of the Temperature of the Submarine Regions," ibid., pp. 342–349, on p. 342; and Shapin and Schaffer, *Leviathan and the Air-Pump,* pp. 217–218.

9. For the Coga episode, see *The Correspondence of Henry Oldenburg,* ed. A. Rupert Hall and Marie Boas Hall, 13 vols. (Madison: University Wisconsin Press; London, Mansell/Taylor & Francis, 1965–1986), vol. 3, pp. 611, 616–617; vol. 4, pp. xx–xxi, 6, 59, 77; "An Account of the Experiment of *Transfusion,* practiced upon a *Man* in *London,*" *Philosophical Transactions,* 9 December 1667, no. 30, pp. 557–559; Henry Stubbe, *Legends no Histories* (London, 1670), p. 179.

10. See Hannaway, "Laboratory Design," pp. 585–586; and (on alchemical usage) Shapin and Schaffer, *Leviathan and the Air-Pump,* p. 57 and note. For references to the "laboratory" as an intensely private space, see Gabriel Plattes, "Caveat for Alchymists," in Samuel Hartlib, comp., *Chymical, Medicinal, and Chyrurgical Addresses* (London, 1665), p. 87.

11. I will use the term more loosely, although it should be clear from the context what sort of place is being referred to. Like Hannaway ("Laboratory Design," p. 585), I accept that an intensive investigation of scientific sites would be obliged to take in such places as the anatomical theater, the astronomical observatory, the curiosity cabinet, and the botanic garden.

12. For futile planning by the Royal Society in the late 1660s to construct experimental facilities in the grounds of Arundel House, see Michael Hunter, "A 'College' for the Royal Society: The Abortive Plan of 1667–1668," *Notes and Records of the Royal Society* 38 (1983–1984): 159–186. Wren's plan (p. 173) called for "a fair Elaboratory" in the basement of the proposed house. Before the society was founded there were several proposals for experimental colleges that included laboratories: see [William Petty], *The Advice of W.P. to Mr. Samuel Hartlib . . .* (London, 1648; rpt. in *The Harleian Miscellany* [London, 1745]), vol. 6, pp. 1–13 (pp. 5, 7 for the "Chymical Laboratory"); John Evelyn to Robert Boyle, 3 September 1659, in *The Diary and Correspondence of John Evelyn,* ed. W. Bray, 3 vols. (London, 1852), vol. 3, pp. 116–120. A year after the society's foundation a plan emerged for a "Philosophical Colledge," again with "great laboratories for Chymical Operations": Abraham Cowley, *A Proposition for the Advancement of Experimental Philosophy*

(London, 1661), p. 25. For the Ashmolean laboratory, see R. F. Ovenell, *The Ashmolean Museum 1683–1894* (Oxford: Clarendon Press, 1986), pp. 16–17, 22; and Edward Lhuyd to John Aubrey, 12 February 1686, quoted in Stanley Mendyk, "Robert Plot: Britain's 'Genial Father of County Natural Histories,'" *Notes and Records of the Royal Society* 39 (1985): 159–177, on p. 174 n. 28 (for confusion about what space was designated by "ye Labradory").

13. Robert Boyle, "The Sceptical Chymist," in Boyle, *Works,* vol. 1, pp. 458–586, on p. 461, intended to draw "the chymists' doctrine out of their dark and smokey laboratories" and bring "it into the open light." The contemporary Dutch-Flemish pictorial genre of "the alchemist in his laboratory" generally depicted alchemical workplaces in this way, without the *necessary* implication of criticism: see C. R. Hill, "The Iconography of the Laboratory," *Ambix* 22 (1975): 102–110; Jane P. Davidson, *David Teniers the Younger* (London: Thames & Hudson, 1980), pp. 38–43; and Davidson, "'I Am the Poison Dripping Dragon': Iguanas and Their Significance in the Alchemical and Occult Paintings of David Teniers the Younger," *Ambix* 34 (1987): 62–80 (who diverges from Hill in her acceptance that Teniers' many paintings of the laboratory genre are probably accurate and informed representations of such sites).

14. So accusations that the Royal Society's meeting places were *not* public might count as particularly devastating: see, for example, Thomas Hobbes, "Dialogus physicus de natura aeris" (1661), in Hobbes, *Latin Works,* ed. Sir William Molesworth, 5 vols. (London: J. Bohn, 1839–1845), vol. 4, pp. 233–296, on p. 240; Shapin and Schaffer, *Leviathan and the Air-Pump,* pp. 112–115, 350; and Stubbe, *Legends no Histories,* "Preface," sig. *3.

15. Among secondary sources relatively rich in material relating to these sites, see Robert G. Frank, Jr., *Harvey and the Oxford Physiologists: A Study of Scientific Ideas* (Berkeley: University of California Press, 1980), chapter 3; Charles Webster, *The Great Instauration: Science, Medicine and Religion 1626–1660* (London: Duckworth, 1975), esp. pp. 47–63, 89–98, 130–157; R. T. Gunther, *Early Science in Oxford,* 15 vols. (Oxford: privately printed, 1923–1967), vol. 1, pp. 7–51; Betty Jo Dobbs, "Studies in the Natural Philosophy of Sir Kenelm Digby," parts I–III, *Ambix* 18 (1971): 1–20; 20 (1973): 144–163; 21 (1974): 1–28; Ronald Sterne Wilkinson, "The Hartlib Papers and Seventeenth-Century Chemistry, Part II," *Ambix* 17 (1970): 85–110; Lesley Murdin, *Under Newton's Shadow: Astronomical Practices in the Seventeenth Century* (Bristol: Adam Hilger, 1985); and Michael Hunter, *Science and Society in Restoration England* (Cambridge: Cambridge University Press, 1981).

16. Robert Boyle to Lady Ranelagh, 6 May 1647, in Boyle, *Works,* vol. 1, p. xxxvi, and vol. 4, pp. 39–40; G. Agricola to Boyle, 6 April 1668, ibid., vol. 6, pp. 650–651; and James Randall Jacob, "Robert Boyle, Young Theodicean" (Ph.D. diss., Cornell University, 1969), pp. 129–138. For the Stalbridge house, see R. E. W. Maddison, "Studies in the Life of Robert Boyle, F.R.S., Part VI: The Stalbridge Period, 1645–

1655, and the Invisible College," *Notes and Records of the Royal Society* 18 (1963): 104–124; Maddison, *The Life of the Honourable Robert Boyle F.R.S.* (London: Taylor & Francis, 1969), chapter 2; Nicholas Canny, *The Upstart Earl: A Study of the Social and Mental World of Richard Boyle, First Earl of Cork 1566–1643* (Cambridge: Cambridge University Press, 1982), pp. 68, 73, 98–99.

17. Lady Ranelagh to Boyle, 12 October [1655], in Boyle, *Works,* vol. 6, pp. 523–524; Maddison, *Life of Boyle,* chapter 3.

18. For Chelsea, see *The Diary of John Evelyn,* ed. E. S. de Beer (London: Oxford University Press, 1959), pp. 410, 417; Henry Oldenburg to Robert Boyle, 10 September 1666, in Oldenburg, *Correspondence,* vol. 3, pp. 226, 227n.; and Maddison, *Life of Boyle,* p. 94. For Leese, see Boyle to Oldenburg, 13 June 1666, in Oldenburg, *Correspondence,* vol. 3, p. 160; Mary Rich, *Memoir of Lady Warwick: Also her Diary, from A.D. 1666 to 1672* (London: Religious Tract Society, 1847?), esp. pp. 51, 161–163, 242–243; and Maddison, *Life of Boyle,* pp. 74, 132, 142.

19. Lady Ranelagh to Boyle, 13 November [1666] (as given, but more likely to be 1667), in Boyle, *Works,* vol. 6, pp. 530–531 (where Katherine offers her "backhouse" to be converted into a laboratory); Thomas Birch, "Life of Boyle," ibid., vol. 1, pp. vi–clxxi, on pp. cxlv, cxxix; John Aubrey, *Brief Lives,* ed. Oliver Lawson Dick (Harmondsworth: Penguin, 1972), p. 198; and Maddison, *Life of Boyle,* pp. 128–129, 133–137, 177–178.

20. Gilbert Burnet, *Select Sermons . . . and a Sermon at the Funeral of the Honourable Robert Boyle* (Glasgow, 1742), pp. 204–205; Maddison, *Life of Boyle,* pp. 134–134; and Webster, *Great Instauration,* pp. 61–63.

21. D. C. Martin, "Former Homes of the Royal Society," *Notes and Records of the Royal Society* 22 (1967): 12–19; I. R. Adamson, "The Royal Society and Gresham College 1660–1711," *Notes and Records of the Royal Society* 33 (1978–1979): 1–21; and Charles Richard Weld, *A History of the Royal Society,* 2 vols. (London: J.W. Parker, 1848), vol. 1, pp. 80–85, 192–198. For a satirical account of the Royal Society at Gresham, and esp. of the Repository and "the *Elaboratory-keepers* Apartment," see [Edward Ward], *The London Spy,* 4th ed. (London, 1709), pp. 59–60.

22. Hunter, "A 'College' for the Royal Society," and Adamson, "The Royal Society and Gresham College," pp. 5–6. For Arundel House, see David Howarth, *Lord Arundel and His Circle* (New Haven, CT: Yale University Press, 1985), and Graham Perry, *The Golden Age Restor'd: The Culture of the Stuart Court, 1603–42* (Manchester: Manchester University Press, 1981), chapter 5. For Peacham, see Henry Peacham, *The Complete Gentleman . . . ,* ed. Virgil B. Heltzel (1622, 1634, 1661; Ithaca, NY: Cornell University Press, 1962), pp. ix–xx.

23. Thomas Birch, *The History of the Royal Society of London,* 4 vols. (London, 1756–1757), vol. 1, pp. 123–124, 250; and Oldenburg to Boyle, 10 June 1663, in Boyle, *Works,* vol. 6, p. 147.

24. Margaret 'Espinasse, *Robert Hooke* (London: Heinemann, 1956), pp. 4–5;

John Ward, *Lives of the Professors of Gresham College* (London, 1740), pp. 91, 178; and Adamson, "The Royal Society and Gresham College," p. 4. The contents of Hooke's rooms at his death in 1703 are detailed in Public Record Office (London) MS PROB 5/1324. For further treatment of Hooke's status, see Shapin, "Who Was Robert Hooke?" (chapter 9 in this volume).

25. 'Espinasse, *Robert Hooke,* pp. 106–107, 113–127, 131–138, 141–147. The major source for Hooke's domestic life and activities for periods from the early 1670s: Henry W. Robinson and Walter Adams, eds., *The Diary of Robert Hooke, M.A., M.D., F.R.S., 1672–1680* (London: Taylor & Francis, 1935), and *The Diary of Robert Hooke, November 1688 to March 1690; December 1692 to August 1693,* in Gunther, *Early Science in Oxford,* vol. 10, pp. 69–265. An unreflectively Freudian account of Hooke's sex life is found in Lawrence Stone, *The Family, Sex and Marriage in England 1500–1800* (London: Weidenfeld & Nicolson, 1977), pp. 561–563.

26. Boyle to Lady Ranelagh, 31 August 1648, in Boyle, *Works,* vol. 6, pp. 49–50, and vol. 1, p. xlv; cf. Boyle to [Benjamin Worsley?], n.d. (probably late 1640s), ibid., vol. 6, pp. 39–40.

27. Boyle Papers, Royal Society, vol. 8, fol. 128, quoted in Jacob, "Boyle, Young Theodicean," p. 158 (quotation). See also Harold Fisch, "The Scientist as Priest: A Note on Robert Boyle's Natural Theology," *Isis* 44 (1953): 252–265; Shapin and Schaffer, *Leviathan and the Air-Pump,* p. 319; and, on Sunday experiments, Jacob, "Boyle, Young Theodicean," pp. 153–154. In one of Boyle's later notebooks he recorded how many experiments he had performed day by day for periods between 1684 and 1688. By then, the rule had clearly become "never on Sunday": Boyle Papers, Commonplace Book, 190, fols. 167–171. See also Maddison, *Life of Boyle,* p. 187; and idem, "Studies in the Life of Robert Boyle, F.R.S., Part IV: Robert Boyle and Some of His Foreign Visitors," *Notes and Records of the Royal Society* 11 (1954): 38–53, on p. 38.

28. Boyle to Benjamin Worsley, n.d., in Boyle, *Works,* vol. 6, pp. 39–41; and Boyle to Lady Ranelagh, 13 November ?, ibid., pp. 43–44. (Both letters probably date from the late 1640s.) See also Marie Boas, *Robert Boyle and Seventeenth-Century Chemistry* (Cambridge: Cambridge University Press, 1958), pp. 15–16, 19, 21.

29. Note the monastic flavor of the Cowley, Evelyn, and Petty plans for philosophical colleges. Evelyn referred explicitly to a Carthusian model: Evelyn, *Diary and Correspondence,* vol. 3, p. 118. See also Shapin, "'The Mind Is Its Own Place'" (chapter 7 in this volume).

30. Gilbert Burnet, *History of His Own Time,* 6 vols. (Oxford: Oxford University Press, 1833), vol. 1, p. 351; cf. Burnet, *Select Sermons,* pp. 202, 210: Boyle "had neither designs nor passions."

31. [Robert Boyle], "An Epistolical Discourse of Philaretus to Empiricus . . . inviting All True Lovers of Vertue and Mankind, to a Free and Generous Communication of Their Secrets and Receits in Physick" (prob. written 1647), in Hartlib,

comp., *Chymical, Medicinal, and Chyrugical Addresses,* pp. 113–150, rpt. in Margaret E. Rowbottom, "The Earliest Published Writing of Robert Boyle," *Annals of Science* 6 (1948–1950): 376–389, on pp. 380–385.

32. On foreigners: Aubrey, *Brief Lives,* p. 198; Eustace Budgell, *Memoirs of the Lives and Characters of the Illustrious Family of the Boyles,* 3rd ed. (London, 1737), p. 144; R. E. W. Maddison, "Studies in the Life of Robert Boyle, F.R.S., Part I: Robert Boyle and Some of His Foreign Visitors," *Notes and Records of the Royal Society* 9 (1951): 1–35, on p. 3; Birch, "Life of Boyle," p. cxlv. For Magalotti: W. E. Knowles Middleton, "Some Italian Visitors to the Early Royal Society," *Notes and Records of the Royal Society* 33 (1978–1979): 157–173, on p. 163; *Lorenzo Magalotti at the Court of Charles II: His "Relazione d'Inghilterra" of 1668,* ed. and trans. W. E. Knowles Middleton (Waterloo, ON: Wilfrid Laurier University Press, 1980), p. 8. Similar hospitality was extended in 1669, when Magalotti escorted the Grand Duke of Tuscany to Boyle's Pall Mall laboratory: [Lorenzo Magalotti], *Travels of Cosmo the Third, Grand Duke of Tuscany* (London: J. Mawman, 1821), pp. 291–293, and R. W. Waller, "Lorenzo Magalotti in England 1668–1669," *Italian Studies* 1 (1937): 49–66. For Evelyn's comments: John Evelyn to William Wotton, 29 March 1696, quoted in Maddison, "Studies in the Life of Boyle, Part IV," p. 38.

33. Boyle Papers, vol. 37, fol. 166 (no date but probably mid- to late 1640s). For background to this manuscript ("The Gentleman"), see J. R. Jacob, *Robert Boyle and the English Revolution: A Study in Society and Intellectual Change* (New York: Burt Franklin, 1977), pp. 48–49.

34. Boyle to Oldenburg, 8, ca. 16 and 30 September and 9 December 1665; Oldenburg to Spinoza, 12 October 1665; Sir Robert Moray to Oldenburg, 4 December 1665, in Oldenburg, *Correspondence,* vol. 2, pp. 502, 509, 537, 568, 627, 639; Evelyn to Wotton, 29 March 1696, quoted in Maddison, "Studies in the Life of Boyle, Part IV," p. 38; and Maddison, *Life of Boyle,* pp. 186–188.

35. On the sign: Weld, *History of the Royal Society,* vol. 1, p. 136n; Maddison, *Life of Boyle,* pp. 177–178; cf. *Diary of Hooke,* in Gunther, *Early Science in Oxford,* vol. 10, p. 139 (entry for 29 July 1689): "To Mr. Boyle. Not to be spoken wth." For the advertisement: Boyle Papers, vol. 35, fol. 194. There are a number of other drafts of this advertisement in the Boyle Papers; one is printed in Maddison, *Life of Boyle,* p. 177.

36. Burnet, *Selected Sermons,* p. 201; Maddison, "Studies in the Life of Boyle, Part I," pp. 2–3.

37. Richard Brathwait, *The English Gentleman* (London, 1630), pp. 65–66, sig. Nnn2r; Lawrence Stone and Jeanne C. Fawtier Stone, *An Open Elite? England 1540–1880* (Oxford: Clarendon Press, 1984), pp. 307–310; and Boyle Papers, vol. 37, fol. 160v. For Boyle and gentility, see Shapin, *A Social History of Truth,* esp. chapter 4.

38. Boas, *Boyle and Seventeenth-Century Chemistry,* p. 207. On the role of the

virtuoso and the expectation that his collections would be accessible to the visits of others, see Walter E. Houghton, Jr., "The English Virtuoso in the Seventeenth Century," *Journal of the History of Ideas* 3 (1942): 51–73, 190–219, and Oliver Impey and Arthur MacGregor, eds., *The Origins of Museums: The Cabinet of Curiosities in Sixteenth- and Seventeenth-Century Europe* (Oxford: Clarendon Press, 1985).

39. There was, of course, another model available in principle to identify the conditions of privacy. This was the alchemist's laboratory, but Boyle worked hard to discredit that model, even as he spent time in the relatively open laboratories of the Hartlib circle. See, for example, Samuel Hartlib to Boyle, 28 February 1654 and 14 September 1658, in Boyle, *Works,* vol. 6, pp. 78–83, 114–115.

40. Weld, *History of the Royal Society,* vol. 1, pp. 224–225.

41. Birch, *History of the Royal Society,* vol. 1, pp. 264–265. On the society's increasing concern for secrecy and the limitation of public access in the 1670s, see Michael Hunter and Paul B. Wood, "Towards Solomon's House: Rival Strategies for Reforming the Early Royal Society," *History of Science* 24 (1986): 49–108, on pp. 74–75. The test for admittance to fellowship of a special claim to scientific knowledge was not formalized until 1730: Maurice Crosland, "Explicit Qualifications as a Criterion for Membership of the Royal Society: A Historical Review," *Notes and Records of the Royal Society* 37 (1983): 167–187.

42. Joseph Hill to Boyle, 20 April 1685, in Boyle, *Works,* vol. 6, p. 661. At about the same time, an eminent cleric wrote announcing his visit but assured the weary Boyle that it would be sufficient to have "some servant of yours" delegated "to shew me your laboratory": Bishop of Cork to Boyle, 12 June 1683, ibid., p. 615.

43. One would like to know much more about the specific sites within the house where experimental work was done and where experimental discourses were held. In addition to rooms designated for public use, the *closet,* where many virtuosi kept their curiosities (including scientific instruments) and where intimate conversations often took place, should be of particular interest. The closet was a room variously located and variously employed in the seventeenth century, but many examples were situated off the bedroom, meaning that this was a private space, access to which acknowledged or accorded intimacy. For the closet and the divisions of domestic space generally, see Mark Girouard, *Life in the English Country House: A Social and Architectural History* (Harmondsworth: Penguin, 1980), esp. pp. 129–130, and Gervase Jackson-Stops and James Pipkin, *The English Country House: A Grand Tour* (London: Weidenfeld & Nicolson, 1985), chapter 9. For an analysis of the internal layout of the house in relation to social structure, see Norbert Elias, *The Court Society,* trans. Edmund Jephcott (Oxford: Basil Blackwell, 1983), chapter 3.

44. Robert Boyle, "An Account of Philaretus [i.e., Mr. R. Boyle] during his Minority," in Boyle, *Works,* vol. 1, pp. xii–xxvi, on p. xiii.

45. Lorenzo Magalotti to Cardinal Leopold, 14 February 1668, in Middleton,

"Some Italian Visitors to the Royal Society," p. 160; Waller, "Magalotti in England," pp. 52–53; *Magalotti at the Court of Charles II*, ed. and trans. Middleton, p. 8.

46. For divergences of opinion among the fellowship about whether or not mere spectators ought to be encouraged, see Hunter and Wood, "Towards Solomon's House," pp. 71, 87–92. Hooke took a particularly strong line that *participants* only were wanted; see also *Philosophical Experiments and Observations of the Late Eminent Dr. Robert Hooke*, ed. W. Derham (London, 1726), pp. 26–27.

47. Margaret 'Espinasse, "The Decline and Fall of Restoration Science," *Past and Present* 14 (1958): 71–89, on p. 86, and Quentin Skinner, "Thomas Hobbes and the Nature of the Early Royal Society," *Historical Journal* 12 (1969): 217–239, on p. 238. See also Michael Hunter, *The Royal Society and Its Fellows 1660–1700: The Morphology of an Early Scientific Institution* (Chalfont St. Giles, Bucks: British Society for the History of Science, 1982), p. 8. On Hobbes and the Royal Society, see Shapin and Schaffer, *Leviathan and the Air-Pump*, pp. 131–139.

48. Samuel Sorbière, *A Voyage to England, containing Many Things relating to the State of Learning, Religion, and Other Curiosities of that Kingdom* (London, 1709; trans. of 1664 French original), pp. 35–38; *Journal des voyages de Monsieur de Monconys* (Lyons, 1666), separately paginated "Seconde Partie: Voyage d'Angleterre," p. 26. See also Thomas Molyneux to William Molyneux, 26 May 1683, in K. Theodore Hoppen, "The Royal Society and Ireland: William Molyneux, F.R.S. (1656–1698)," *Notes and Records of the Royal Society* 18 (1963): 125–135, on p. 126, and Maddison, "Studies in the Life of Boyle, Part I," pp. 14–21. There is a brief "official" account of the society's rooms in Thomas Sprat, *History of the Royal Society* (London, 1667), p. 93.

49. *Account of the Proceedings in the Council of the Royal Society, in Order to Remove from Gresham College* (London, 1707?), partly rpt. in Weld, *History of the Royal Society*, vol. 1, pp. 82–83; Martin, "Former Homes of the Royal Society," p. 13. A brief account of the society's rooms in the last year of its occupancy of Gresham College is in *London in 1710: From the Travels of Zacharias Conrad von Uffenbach*, trans. and ed. W. H. Quarrell and Margaret Mare (London: Faber & Faber, 1934), pp. 97–102. Uffenbach described the "wretchedly ordered" repository and library, and the "very small and wretched" room where the society usually met. At that time it was sparsely decorated with portraits of its members (including, Uffenbach claimed, a picture of Hooke, about which he was mistaken or which was subsequently lost), two globes, a model of a contrivance for rowing, and a large pendulum clock. There is no pictorial record of an internal space occupied by the Royal Society in the seventeenth century. For an engraving (of doubtful date) recording a Crane Court meeting, see T. E. Allibone, *The Royal Society and Its Dining Clubs* (Oxford: Pergamon Press, 1976), frontispiece.

50. Magalotti to Cardinal Leopold, 21 February 1668, in Middleton, "Some

Italian Visitors to the Royal Society," pp. 160–161; Magalotti, *Travels of Cosmo,* pp. 185–186. Too much weight should not, perhaps, be given to Magalotti's evidence. He seems to have been confused about where exactly he was: in the 1669 account he said that he went with the Grand Duke "to Arundel House, in the interior of Gresham College." Moreover, he seems to have derived portions of his version of the society from Sorbière's earlier account.

51. *Journal des voyages de Monconys,* pp. 26–28, 47, 55–57; Sorbière, *Voyage to England,* p. 37; Maddison, "Studies in the Life of Boyle, Part I," pp. 16–19 (Monconys), 21 (Sorbière); Magalotti to Cardinal Leopold, 21 February 1668, in Middleton, "Some Italian Visitors to the Royal Society," pp. 161–162.

52. *Journal des voyages de Monconys,* p. 26 (quoted); Magalotti, *Travels of Cosmo,* pp. 186–187; Sorbière, *Voyage to England,* pp. 36–37. For Commons practice, see Sir Thomas Smith, *De republica anglorum* (London, ca. 1600), pp. 51–52. (This practice differs from that of the House of Lords, where speakers address "My lords.")

53. On set speeches: Edward Chamberlayne, *Angliae notitia: or the Present State of England,* 7th ed. (London, 1673), p. 345. Cf. Sir Thomas Erskine May, *A Treatise on the Law, Privileges, Proceedings and Usage of Parliament,* ed. T. Lonsdale Webster and William Edward Grey, 11th ed. (London: William Clowes & Sons, 1906), pp. 310, 314–315, 344–345, and Lord Campion, *An Introduction to the Procedure of the House of Commons* (London: Macmillan, 1958), pp. 190, 192. On "opposite opinions": Magalotti, *Travels of Cosmo,* pp. 187–188. Cf. J. E. Neale, *The Elizabethan House of Commons* (London: Jonathan Cape, 1949), pp. 404–407, and Smith, *De republica anglorum,* p. 52. (Needless to say, these were stipulations of ideal behavior: violations of the norms in Commons were frequent.)

54. Sorbière, *Voyage to England,* p. 38, Magalotti, *Travels of Cosmo,* p. 187. Cf. George Henry Jennings, comp., *An Anecdotal History of the British Parliament . . .* (London: Horace, Cox, 1880), p. 433; Vernon F. Snow, *Parliament in Elizabethan England: John Hooker's Order and Usage* (New Haven, CT: Yale University Press, 1977), p. 164; and Neale, *Elizabethan House of Commons,* p. 364.

55. *Journal des voyages de Monconys,* p. 26; Sorbière, *Voyage to England,* p. 36; Magalotti, *Travels of Cosmo,* p. 186; Thomas Sprat, *Observations on Mons. de Sorbière's Voyage into England* (1665; London, 1708), pp. 164–165; idém, *History of the Royal Society,* p. 94. For the Commons mace: Erskine May, *Usage of Parliament,* p. 155; Campion, *Procedure of Commons,* pp. 54, 73; and for the society's mace: Margery Purver, *The Royal Society: Concept and Creation* (London: Routledge & Kegan Paul, 1967), p. 140.

56. Quoted in Hunter and Wood, "Towards Solomon's House," p. 81. The same author described the council as "the Societys Parliament" (pp. 68, 83).

57. The directed coordination of the society's labors prompted Robert Hooke to compare it to "a Cortesian army well Disciplined and regulated though their number

be but small" (quoted in Hunter and Wood, "Towards Solomon's House," p. 87). On coffeehouses see, for example, Aytoun Ellis, *The Penny Universities: A History of the Coffee-Houses* (London: Secker & Warburg, 1956), esp. pp. 46–47 (on rules of order). The Royal Society "club" of the 1670s held much of its conversation in city coffeehouses like Garraway's and Jonathan's. On occasion, experimental performances were even staged at coffeehouses. For the connections between the late-Interregnum Harringtonian Rota club, meeting at Miles's coffeehouse, and the early Royal Society, see Anna M. Strumia, "Vita istituzionale della Royal Society seicentesca in alcuni studi recenti," *Rivista Storica Italiana* 98 (1986): 500–523, on pp. 520–523.

58. Sprat, *History of the Royal Society,* pp. 73, 100; and, on attendance, Hunter, *The Royal Society and Its Fellows,* pp. 16–19, and J. L. Heilbron, *Physics at the Royal Society during Newton's Presidency* (Los Angeles: William Andrews Clark Memorial Library, 1983), p. 4.

59. Hunter, *Science and Society in Restoration England,* p. 46.

60. The only record of a crowd scene in one of Boyle's laboratories is an account of the visit in 1677 of the German chemist Johann Daniel Kraft, when the display of phosphorus attracted "the confused curiosity of many spectators in a narrow compass." However, "no strangers were present" when the secret of the phosphorus was later revealed: Robert Hooke, *Lectures and Collections made by Robert Hooke . . .* (London, 1678), pp. 273–282; see also J. V. Golinski, "A Noble Spectacle: Research on Phosphorus and the Public Cultures of Science in the Early Royal Society," *Isis* 80 (1989): 11–39.

61. Shapin, "Invisible Technicians," in *A Social History of Truth,* chapter 8; see also R. E. W. Maddison, "Studies in the Life of Robert Boyle, F.R.S., Part V: Boyle's Operator: Ambrose Godfrey Hanckwitz, F.R.S.," *Notes and Records of the Royal Society* 11 (1955): 159–188 (see p. 159 for a partial list of Boyle's technicians).

62. Robert Boyle, "A Continuation of New Experiments Physico-Mechanical, touching the Spring and Weight of the Air . . . The Second Part" (1680), in Boyle, *Works,* vol. 4, pp. 505–593, on pp. 506–507.

63. Among very many examples, see esp. a note on the ineptitude of John Mayow (identified only by his initials): Robert Boyle, "A Continuation of New Experiments Physico-Mechanical touching the Spring and Weight of the Air" (1669), in Boyle, *Works,* vol. 3, pp. 175–276, on p. 187.

64. Burnet, *Select Sermons,* p. 200; Birch, "Life of Boyle," p. lx: "The irreligious fortified themselves against all that was said by the clergy, with this, that *it was their trade,* and that *they were paid for it.*"

65. Sprat, *History of the Royal Society,* pp. 63–67, 76, 407, 427, 431, 435; also Robert Hooke, *Micrographia* (London, 1665), "Preface," sig. g1v. For visitors' accounts of social diversity in the society, see, for example, Magalotti, *Travels of Cosmo,* pp. 186–188, and Sorbière, *Voyage to England,* p. 37.

66. Joseph Glanvill, *Scepsis scientifica* (London, 1665), "Preface," sig. c1r; also sig. b4v for "a Society *of persons that can command both* Wit *and* Fortune." Peter Dear rightly notes evidence that the testimony of "lowly folk" might be credible because their accounts were less likely than those of the educated to be colored by theoretical commitments. This view was, however, rarely expressed and, as Dear says, was more than counterbalanced by the consideration that "gentlemen were trustworthy just because they *were* gentlemen": Dear, *"Totius in verba,"* pp. 156–157.

67. Sprat, *History of the Royal Society,* pp. 65–67. On seventeenth-century English thought about the master-servant relationship and the political significance of servitude, see C. B. Macpherson, *The Political Theory of Possessive Individualism: Hobbes to Locke* (Oxford: Oxford University Press, 1970), esp. chapter 3; Christopher Hill, "Pottage for Freeborn Englishmen: Attitudes to Wage-Labour," in idem, *Change and Continuity in Seventeenth-Century England* (Cambridge, MA: Harvard University Press, 1975), chapter 10.

68. Sprat, *History of the Royal Society,* pp. 67–69. Cf. John Webster, *Academiarum examen* (London, 1654), p. 106, where it was recommended that youth be educated "so they may not be sayers, but doers, not idle spectators, but painful operators . . . which can never come to pass, unless they have Laboratories as well as Libraries, and work in the fire, better than build Castles in the air."

69. Sprat, *History of the Royal Society,* p. 70. On fellows' freedom of judgment: Lotte Mulligan and Glenn Mulligan, "Reconstructing Restoration Science: Styles of Leadership and Social Composition of the Early Royal Society," *Social Studies of Science* 11 (1981): 327–364, on p. 330 (quoting William Croone); on the moral economy of the experimental community generally: Shapin and Schaffer, *Leviathan and the Air-Pump,* pp. 310–319, 332–344; and for Hobbes's suggestion that the Royal Society did indeed have "masters": ibid., pp. 112–115. On the presumed equality of all gentlemen: J. C. D. Clark, *English Society 1688–1832: Ideology, Social Structure and Political Practice during the Ancien Regime* (Cambridge: Cambridge University Press, 1985), p. 103.

70. See, for example, Peacham, *The Complete Gentleman,* p. 24, and Brathwait, *The English Gentleman,* pp. 83–84.

71. Ismael Boulliau to Christiaan Huygens, 6 December 1658, in Harcourt Brown, *Scientific Organizations in Seventeenth Century France (1620–1680)* (Baltimore: Williams & Wilkins, 1934), pp. 87–89; see also pp. 108, 119, 126–127 (on civility), and p. 96 (for Oldenburg's familiarity with proceedings at Montmor's house).

72. Boyle, "Hydrostatical Discourse," p. 615. For this episode, and for Boyle–More relations generally, see Shapin and Schaffer, *Leviathan and the Air-Pump,* pp. 207–224.

73. The episode concerned reports by physicians in Danzig regarding the trans-

fusion of animal blood into humans: see Oldenburg to Boyle, 10 December 1667, in Boyle, *Works,* vol. 6, pp. 254–255, and Oldenburg, *Correspondence,* vol. 4, pp. 26–28.

74. Sprat, *History of the Royal Society,* pp. 97–102; Glanvill, *Scepsis scientifica,* "Preface," sig. c1r.

75. There are judicial resonances here, but they should not be overemphasized. What was most often being "tried" in experiment was some hypothesis or other explanatory item. In law the trial is of matter of fact, and the jury's judgment is of what counts as fact. However, the best parallel is between the experimental show and the "show trial," where the matter of fact is known (or decided upon) in advance. Both bear the same relation to their respective genuine trials. For the judicial process of "bringing matters to a trial," see Martin, "'Knowledge Is Power,'" chapter 3.

76. In this connection see David Gooding's excellent work on Faraday at the Royal Institution. Gooding studied the passage from the basement laboratory to the ground floor lecture theater as movement in the epistemological status of experimental phenomena: Gooding, "'In Nature's School': Faraday as an Experimentalist," in *Faraday Rediscovered: Essays on the Life and Work of Michael Faraday, 1791–1867,* ed. David Gooding and Frank A. L. James (Basingstoke: Macmillan, 1985), pp. 106–135. See also H. M. Collins, "Public Experiments and Displays of Virtuosity: The Core-Set Revisited," *Social Studies of Science* 18 (1988): 725–748.

77. 'Espinasse, *Robert Hooke,* pp. 106–147; Shapin, "Who Was Robert Hooke?" (chapter 9 in this volume). The quotations, instances of which could be multiplied indefinitely, are from Hooke's 1672–1680 *Diary,* pp. 15, 37.

78. Thomas S. Kuhn, "Mathematical versus Experimental Traditions in the Development of Physical Science," in idem, *The Essential Tension: Selected Studies in Scientific Tradition and Change* (Chicago: University of Chicago Press, 1977), pp. 31–65, on p. 43.

79. Quotations are from Hooke's 1672–1680 *Diary,* pp. 27–29, 33. For Boyle's views on what counted as an experimental failure, see Shapin and Schaffer, *Leviathan and the Air-Pump,* pp. 185–201.

80. For uses of Foucauldian notions of "discipline" and "docile bodies" in the sociology of scientific knowledge: Michael Lynch, "Discipline and the Material Form of Images: An Analysis of Scientific Visibility," *Social Studies of Science* 15 (1985): 37–66, and Bruno Latour, *Science in Action: How to Follow Scientists and Engineers through Society* (Milton Keynes: Open University Press, 1987), chapter 3.

81. Historians concerned with other issues have known this for some time, for example, Hunter, *Science and Society in Restoration England,* p. 46; Hunter and Wood, "Towards Solomon's House," p. 76; and Penelope M. Gouk, "Acoustics in the Early Royal Society 1660–1680," *Notes and Records of the Royal Society* 36 (1981–1982): 155–175, on p. 170.

82. *Philosophical Experiments of Hooke,* ed. Derham, pp. 26–28, and Magalotti

to Cardinal Leopold, 20 February 1668, in Maddison, "Some Italian Visitors to the Royal Society," p. 161.

83. Birch, *History of the Royal Society,* vol. 3, p. 124 (entry for 12 February 1674), and vol. 2, p. 115 (entry for 26 September 1666).

84. Richard Waller, "The Life of Dr. Robert Hooke," in *The Posthumous Works of Robert Hooke . . . ,* ed. Waller (London, 1705), pp. i–xxviii, on p. iii (also quoted in Weld, *History of the Royal Society,* vol. 1, p. 438). Weld was one of the first historians to note this characteristic of Royal Society experiments: "These experiments were generally repetitions of experiments already made in private and exhibited afterwards for the satisfaction and information of the Society" (p. 136n).

85. Birch, *History of the Royal Society,* vol. 1, pp. 177, 194, 260 (entries for 14 January, 11 February, and 17 June 1663).

86. Birch, *History of the Royal Society,* vol. 1, pp. 139, 212, 218, 220, 238, 248, 254–255, 268, 274–275, 286–287, 295, 299–301, 305, 310, 386.

87. The story of anomalous suspension is told in Shapin and Schaffer, *Leviathan and the Air-Pump,* chapter 6. The society ultimately came round to the view that anomalous suspension authentically existed, and, therefore, that experiments *not* revealing it were incompetent. This shift crucially involved Boyle's personal experience of the phenomenon and Huygens's visit to London to produce anomalous suspension before witnesses. For similar doubt of Hooke's experimental testimony, see "An Account of the Experiment made by Mr. *Hook,* of Preserving Animals Alive by Blowing through Their Lungs with Bellows," *Philosophical Transactions,* 21 October 1667, No. 28, pp. 539–540.

88. Birch, *History of the Royal Society,* vol. 2, p. 189 (entry for 25 July 1667).

Chapter 6. Pump and Circumstance

1. Robert Boyle, "New Experiments Physico-Mechanical, touching the Spring of the Air . . . ," in Boyle, *Works,* ed. Thomas Birch, 6 vols. (London, 1772), vol. 1, pp. 1–117. (All subsequent references to Boyle's writings are to this edition and are cited as *RBW.*)

2. Ian Hacking, *The Emergence of Probability: A Philosophical Study of Early Ideas about Probability, Induction, and Statistical Inference* (Cambridge: Cambridge University Press, 1975), esp. chapters 3–5; Barbara J. Shapiro, *Probability and Certainty in Seventeenth-Century England: A Study of the Relationships between Natural Science, Religion, History, Law and Literature* (Princeton, NJ: Princeton University Press, 1983), esp. chapter 2.

3. Newton's place in the development of a probabilist view of physical science is ambiguous. Certain of his critics thought that he aimed at the necessary assent that most English natural philosophers had agreed to eschew; see Zev Bechler, "Newton's 1672 Optical Controversies: A Study in the Grammar of Scientific Dissent," in *The*

Interaction between Science and Philosophy, ed. Yehuda Elkana (Atlantic Highlands, NJ: Humanities Press, 1974), pp. 115–142.

4. The usual form in which Boyle phrased this was the statement that God might produce the same effects in nature through very different causes; therefore, "it is a very easy mistake for men to conclude that because an effect may be produced by such determinate causes, it must be so, or actually is so." Boyle, "Some Considerations touching the Usefulness of Experimental Natural Philosophy," *RBW,* vol. 2, pp. 1–201, on p. 45 (orig. publ. 1663). See also Laurens Laudan, "The Clock Metaphor and Probabilism: The Impact of Descartes on English Methodological Thought, 1650–65," *Annals of Science* 22 (1965): 73–104; G. A. J. Rogers, "Descartes and the Method of English Science," *Annals of Science* 29 (1972): 237–255; Henry G. van Leeuwen, *The Problem of Certainty in English Thought 1630–1690* (The Hague: M. Nijhoff, 1963), pp. 95–96; Shapiro, *Probability and Certainty,* pp. 44–61.

5. This is especially evident in historians' treatment (or lack thereof) of criticisms of seventeenth-century experimentalism by philosophers who denied both the central role of experimental procedures and the foundational status of the matter of fact. For example, insofar as Thomas Hobbes's criticisms of Boyle's experimental program have been discussed, historians have preferred to conclude that he "misunderstood" Boyle, or that he "failed to appreciate" the power of experimental methods: see, among others, Frithiof Brandt, *Thomas Hobbes' Mechanical Conception of Nature* (Copenhagen: Levin & Munksgaard, 1928), pp. 377–378; Marie Boas Hall, "Boyle, Robert," in *Dictionary of Scientific Biography* (New York: Charles Scribner's Sons, 1970), vol. 2, p. 379; Louis Trenchard More, *The Life and Works of the Honourable Robert Boyle* (London: Oxford University Press, 1944), pp. 97, 239. Hobbes's anti-experimentalism is treated in Steven Shapin and Simon Schaffer, *Leviathan and the Air-Pump: Hobbes, Boyle, and the Experimental Life* (Princeton, NJ: Princeton University Press, 1985).

6. The use of the word *technology* in reference to the "software" of literary practices and social relations may appear jarring, but it is in fact etymologically justified, as Carl Mitcham nicely shows: Carl Mitcham, "Philosophy and the History of Technology," in *The History and Philosophy of Technology,* ed. George Bugliarello and Dean B. Doner (Urbana: University of Illinois Press, 1979), pp. 163–201, esp. pp. 172ff. The Greek *techne* has behind it the Indo-European stem *tekhn,* probably meaning "woodwork" or "carpentry." By using "technology" to refer to social and literary practices, as well as to hardware, I stress that all three are *knowledge-producing tools.*

7. See, for example, Boyle, "An Examen of Mr. T. Hobbes his Dialogus Physicus de Natura Aeris . . . ," in *RBW,* vol. 1, pp. 186–242, on p. 241 (orig. publ. 1662); Boyle, "Animadversions upon Mr. Hobbes's Problemata de Vacuo," in ibid., vol. 4, pp. 104–128, on p. 105 (orig. publ. 1674). The explication of the behavior of liquids

in the gardener's pot was a set-piece in the mid-seventeenth-century contest between rival physical systems; see Thomas Hobbes, "Concerning Body," in *The English Works of Thomas Hobbes,* ed. Sir William Molesworth, 11 vols. (London, 1839–1845), vol. 1, pp. 414–415 (orig. publ. 1656); compare Boyle, "Examen of Hobbes," pp. 191–193.

8. Boyle described his pump in "New Experiments," pp. 6–11. One of the best accounts of the original pump and subsequent designs is still George Wilson, "On the Early History of the Air-Pump in England," *Edinburgh New Philosophical Journal* 46 (1849): 330–354; see also Robert G. Frank, Jr., *Harvey and the Oxford Physiologists: A Study of Scientific Ideas* (Berkeley: University of California Press, 1980), pp. 128–130.

9. The only information we have concerning the cost of the Boyle pump indicates that a version of the *receiver* ran to £5: Thomas Birch, *The History of the Royal Society of London,* 4 vols. (London, 1756–1757), vol. 2, p. 184. Given the expense of machining the actual pumping apparatus, an estimate of £25 for the entire engine might be conservative. So, an air-pump would have cost more than the annual salary of the curator of the Royal Society, Robert Hooke, who was the London pump's chief operator. Christiaan Huygens's elder brother Constantijn pulled out of a pump-building project, "being afraid of the cost": Christiaan Huygens, *Oeuvres complètes,* 22 vols. (The Hague: M. Nijhoff, 1888–1950), vol. 3, p. 389. The Accademia del Cimento in Florence did not even try to build a *Machina Boyleana,* even though they had the necessary texts at hand: W. E. Knowles Middleton, *The Experimenters: A Study of the Accademia del Cimento* (Baltimore: Johns Hopkins University Press, 1971), pp. 263–265. Full details of the career of the air-pump in the 1660s are in Shapin and Schaffer, *Leviathan and the Air-Pump,* chapter 6.

10. Boyle, "Examen of Hobbes," p. 193. Both "elaborate" and systematic experimentation were also recommended as the bases for constructing well-framed theories. Those theories "that are grounded but upon few and obvious experiments, are subject to be contradicted" by new findings; see Boyle, "A Proëmial Essay . . . with Some Considerations touching Experimental Essays in General," in *RBW,* vol. 1, pp. 299–318, on p. 302 (orig. publ. 1661).

11. See, for example, Boyle, "The Sceptical Chymist," in *RBW,* vol. 1, pp. 458–586, on p. 460 (orig. publ. 1661): here Boyle suggests that many "experiments" reported by the alchemists "questionless they never tried." For an insinuation that Henry More may not actually have performed experiments adduced against Boyle's findings, see idem, "An Hydrostatical Discourse, Occasioned by the Objections of the Learned Dr. Henry More," in *RBW,* vol. 3, pp. 596–628, on pp. 607–608 (orig. publ. 1672). Compare the response of Boyle to Pascal's trials and their reporting. Boyle described the replication of the Puy-de-Dôme experiment in "New Experiments," pp. 14, 43; and by Power, Towneley, and himself in "A Defence of the Doctrine touching the Spring and Weight of the Air . . . against the Objections of

Franciscus Linus," in *RBW*, vol. 1, pp. 118–185, on pp. 151–155 (orig. publ. 1662). Yet Boyle doubted the reality of Pascal's other reports of underwater trials; see "Hydrostatical Paradoxes, made out by New Experiments . . . ," in *RBW*, vol. 2, pp. 738–797, on pp. 745–746 (orig. publ. 1666): "though the experiments [Pascal] mentions be delivered in such a manner, as is usual in mentioning matters of fact; yet I remember not, that he expressly says, that he actually tried them, and therefore he might possibly have set them down, as things that *must* happen, upon a just confidence, that he was not mistaken in his ratiocinations . . . Whether or not Monsieur Pascal ever made these experiments himself, he does not seem to have been very desirous, that others should make them after him." For the role of thought experiments in the history of science, see Alexandre Koyré, *Galileo Studies* (Atlantic Highlands, NJ: Humanities Press, 1978), p. 97; Thomas S. Kuhn, "A Function for Thought Experiments," in idem, *The Essential Tension* (Chicago: University of Chicago Press, 1977), pp. 240–265; Charles B. Schmitt, "Experience and Experiment: A Comparison of Zabarella's View with Galilee's in *De motu,*" *Studies in the Renaissance* 16 (1969): 80–137.

12. Boyle, "Two Essays, Concerning the Unsuccessfulness of Experiments," in *RBW*, vol. 1, pp. 318–353, on p. 343 (orig. publ. 1661); Boyle, "Sceptical Chymist," p. 486. Cf. Boyle, "Animadversions on Hobbes," p. 110: here Boyle rejected Hobbes's claim to have observed a phenomenon that Boyle regarded as implausible; Hobbes "does not here affirm, that he, or any he can trust, has seen the thing done . . . Wherefore, till I be better informed of the matter of fact, I can scarce look upon what Mr. Hobbes says . . . as other than his conjecture."

13. Boyle, "Some Considerations about the Reconcileableness of Reason and Religion," in *RBW*, vol. 4, pp. 151–191, on p. 182 (orig. publ. 1675); see also Lorraine J. Daston, "The Reasonable Calculus: Classical Probability Theory, 1650–1840" (unpublished Ph.D. diss., Harvard University, 1979), pp. 90–91; on testimony: Hacking, *Emergence of Probability,* chapter 3; on evidence in seventeenth-century English law, see Shapiro, *Probability and Certainty,* chapter 5; Simon Schaffer, "Making Certain (essay review of Shapiro)," *Social Studies of Science* 14 (1984): 137–152, esp. pp. 146–147 (for the legal analogy of scientific witnessing).

14. Thomas Sprat, *History of the Royal Society* (London, 1667), p. 100.

15. One of the ways by which Hobbes attacked the experimental program was to insinuate that the Royal Society was *not* a public place: not everyone could come to witness experimental displays; see Thomas Hobbes, "Dialogus physicus de natura aeris . . . ," in Hobbes, *Opera philosophica,* ed. Sir William Molesworth, 5 vols. (London, 1839–1845), vol. 4, pp. 233–296, on p. 240 (orig. publ. 1661): "Cannot anyone who wishes come, since as I suppose they meet in a public place, and give his opinion on the experiments which are seen as well as they? Not at all . . . the place where they meet is not public." Thomas Birch praised Boyle because "his laboratory was constantly open to the curious"; see *RBW*, vol. 1, p. cxlv.

16. Boyle, "New Experiments," p. 1; idem, "The History of Fluidity and Firmness," in *RBW*, vol. 1, pp. 377–442, on p. 410 (orig. publ. 1661); idem, "Defence against Linus," p. 173; the place of experiment is dealt with in Shapin, "The House of Experiment in Seventeenth-century England" (chapter 5 in this volume).

17. Robert Hooke, *Philosophical Experiments and Observations* (London, 1726), pp. 27–28.

18. Sprat, *History of the Royal Society*, pp. 98–99; see also Shapiro, *Probability and Certainty*, pp. 21–22.

19. Boyle, "New Experiments," pp. 33–34; idem, "A Discovery of the Admirable Rarefaction of Air . . . ," in *RBW*, vol. 3, pp. 496–500, on p. 498 (orig. publ. 1671); idem, "Sceptical Chymist," p. 460.

20. Boyle, "The Christian Virtuoso," in *RBW*, vol. 5, pp. 508–540, on p. 529 (orig. publ. 1690); see also Shapiro, *Probability and Certainty*, chapter 5 (esp. p. 179). For a study of social accounting systems in the evaluation of observation reports, see Ron Westrum, "Science and Social Intelligence about Anomalies: The Case of Meteorites," *Social Studies of Science* 8 (1978): 461–493. Explicit concern for the quality of testimony was much more intense in natural history than it was in experimental philosophy. In the latter, access to experimental devices was disciplined by their cost and location; so, not everyone could *in practice* offer experimental testimony, while those that did were of known character, reliability and probity. By contrast, the offering of observation reports was almost completely undisciplined, and the reliability of such testimony was a matter of fundamental concern. The relationship between social standing and credibility is dealt with in my *A Social History of Truth: Civility and Science in Seventeenth-Century England* (Chicago: University of Chicago Press, 1994).

21. Marie Boas [Hall], *Robert Boyle and Seventeenth-Century Chemistry* (Cambridge: Cambridge University Press, 1958), pp. 40–41; Boyle, "New Experiments," p. 2; idem, "The Experimental History of Colours," in *RBW*, vol. 1, pp. 662–778, on p. 633 (orig. publ. 1663). Cf. p. 664, where certain "easy and recreative experiments, which require but little time, or charge, or trouble in the making" were recommended to be tried by ladies. Richard Jones was the "Pyrophilus" to whom other essays were addressed.

22. Boyle, "A Continuation of New Experiments Physico-Mechanical, touching the Spring and Weight of the Air," in *RBW*, vol. 3, pp. 175–276, on p. 176. This was written in 1668 and printed a year later. Boyle was not being entirely straightforward here: Huygens's air-pump in the Netherlands had in 1662 produced a matter of fact—the so-called anomalous suspension of water—that seriously troubled Boyle's explanatory schema. Boyle never referred to this finding in print; see Shapin and Schaffer, *Leviathan and the Air-Pump*, chapter 6.

23. Boyle, "A Continuation of New Experiments, Physico-Mechanical . . . The Second Part," in *RBW*, vol. 4, pp. 505–93, on pp. 505, 507 (orig. publ. 1680).

24. This practice can be contrasted with the iconography of the anti-experimentalist Hobbes, whose natural philosophy texts included only a few images of experimental systems, and these very simple and highly stylized. In giving his account of the air-pump and how it worked, Hobbes deliberately scorned the use of pictures; see Hobbes, "Dialogus physicus," pp. 235, 242. For studies of engraving and printmaking in scientific texts, see William M. Ivins, Jr., *Prints and Visual Communication* (Cambridge, MA: MIT Press, 1969), esp. pp. 33–36, and Elizabeth L. Eisenstein, *The Printing Press as an Agent of Change* (Cambridge: Cambridge University Press, 1979), esp. pp. 262–270, 468–471.

25. Hooke to Boyle, 25 August and 8 September 1664, in *RBW*, vol. 6, pp. 487–490, and R. E. W. Maddison, "The Portraiture of the Honourable Robert Boyle, FRS," *Annals of Science* 15 (1959): 141–214.

26. Boyle, "Continuation of New Experiments," p. 178.

27. This essay was completed before I was able to read Svetlana Alpers's brilliant *The Art of Describing: Dutch Art in the Seventeenth Century* (Chicago: University of Chicago Press, 1983). Alpers analyzes the purposes and the conventions of realistic pictures in seventeenth-century Holland, demonstrating substantial links between English empiricist theories of knowledge and Dutch picturing. Her chapter on "The Craft of Representation" is a superb examination of the pictorial conventions for generating realist responses. Evidently, the Dutch were trying to achieve by way of picturing what the English were attempting by way of the reform of prose.

28. Boyle, "New Experiments," pp. 1–2 (emphases added). The role of circumstantial detail in Boyle's prose and in that of other early Fellows of the Royal Society is treated in Shapiro, *Probability and Certainty*, chapter 7. See also Peter Dear, "*Totius in verba*: Rhetoric and Authority in the Early Royal Society," *Isis* 76 (1985): 145–161, and Jan V. Golinski, "Robert Boyle: Scepticism and Authority in Seventeenth-Century Chemical Discourse," in *The Figural and the Literal: Problems of Language in the History of Science and Philosophy*, ed. Andrew E. Benjamin, Geoffrey N. Cantor, and John R. R. Christie (Manchester: Manchester University Press, 1987), pp. 58–82.

29. There is probably a connection between Boyle's justification for circumstantial reporting and Bacon's argument in favor of "initiative" (as opposed to "magistral") methods of communication in science: see, for example, Devon Leigh Hodges, "Anatomy as Science," *Assays* 1 (1981): 73–89, esp. pp. 83–84; Lisa Jardine, *Francis Bacon: Discovery and the Art of Discourse* (Cambridge: Cambridge University Press, 1974), pp. 174–178; K. R. Wallace, *Francis Bacon on Communication and Rhetoric* (Chapel Hill: University of North Carolina Press, 1943), pp. 18–19. The magistral method, Bacon said, "requires that what is told should be believed; the initiative that it should be examined." Initiative methods display the processes by which conclusions were reached; magistral methods mask those processes. Although Boyle's inspiration may, plausibly, have been Baconian, the "influence" of Bacon is

sometimes much exaggerated. It is useful to remember that it was Boyle, not Bacon, who actually developed the literary forms of experimental communication; it is hard to imagine two more different forms than Bacon's aphorisms and Boyle's experimental narratives.

30. Boyle, "Proëmial Essay," pp. 305–306; cf. idem, "New Experiments," p. 1; Richard S. Westfall, "Unpublished Boyle Papers relating to Scientific Method," *Annals of Science* 12 (1956): 63–73, 103–117.

31. Boyle, "Unsuccessfulness of Experiments," pp. 339–340, 353. Recognizing that contingencies might affect experimental outcomes was also a way of tempering inclinations to reject good testimony too readily. If an otherwise reliable authority stipulated an outcome not immediately obtained, one was advised to persevere; see ibid., pp. 344–345; idem, "Continuation of New Experiments," pp. 275–276; idem, "Hydrostatical Paradoxes," p. 743; Westfall, "Unpublished Boyle Papers," pp. 72–73.

32. Boyle, "New Experiments," p. 26. For an example of Boyle reporting an experimental failure, see ibid., pp. 69–70. A critic like Hobbes could capitalize upon Boyle's reported failures, or, more interestingly, deconstruct Boyle's reported successes by identifying further contingencies that affected experimental outcomes; see, for instance, Hobbes, "Dialogus physicus," pp. 245–246.

33. Boyle, "Proëmial Essay," pp. 300–301, 307; cf. "Sceptical Chymist," pp. 469–470, 486, 584. Several of the less modest personalities of seventeenth-century English science were individuals who lacked the gentle birth that routinely enhanced the credibility of testimony: for example, Hobbes, Hooke, Wallis, and Newton.

34. The best source for Boyle's social situation and temperament is J. R. Jacob, *Robert Boyle and the English Revolution: A Study in Social and Intellectual Change* (New York: Burt Franklin, 1977), chapters 1–2. For the "characters" of seventeenth-century scholars and gentlemen, see Shapin, "'A Scholar and a Gentleman'" (chapter 8 in this volume).

35. Boyle, "Proëmial Essay," pp. 318, 304. For the importance of the lens and the perceptual model of knowledge in seventeenth-century epistemology, see Alpers, *Art of Describing,* chapter 3. The goal for Boyle, as for many other philosophers concerned with linguistic reform, was *plain-speaking.* For the linguistic program of the early Royal Society and its connections with experimental philosophy, see Francis Christensen, "John Wilkins and the Royal Society's Reform of Prose Style," *Modern Language Quarterly* 7 (1946): 179–187, 279–290; Richard F. Jones, "Science and Language in England of the Mid-Seventeenth Century," *Journal of English and Germanic Philology* 31 (1932): 315–331; idem, "Science and English Prose Style in the Third Quarter of the Seventeenth Century," *Publications of the Modern Language Association of America* 45 (1930): 977–1009; Vivian Salmon, "John Wilkins' Essay (1668): Critics and Continuators," *Historiographica Linguistica* 1 (1974): 147–163; M. M. Slaughter, *Universal Languages and Scientific Taxonomy in the*

Seventeenth Century (Cambridge: Cambridge University Press, 1982), esp. pp. 104–186; Hans Aarsleff, *From Locke to Saussure: Essays on the Study of Language and Intellectual History* (London: Athlone Press, 1982), pp. 225–277; Michael Hunter, *Science and Society in Restoration England* (Cambridge: Cambridge University Press, 1981), pp. 118–119; Shapiro, *Probability and Certainty*, pp. 227–246; and the sources cited in notes 28 and 29. For Boyle's attack on the "confused," "equivocal," and "cloudy" language of the alchemists, see "Sceptical Chymist," pp. 460, 520–522, 537–539, and, for his criticisms of Hobbes's expository "obscurity," see "Examen of Hobbes," p. 227.

36. Boyle, "Proëmial Essay," p. 307. On "wary and diffident expressions," see also "New Experiments," p. 2; and compare Sprat, *History of the Royal Society*, pp. 100–101; Joseph Glanvill, *Scepsis scientifica* (London: Kegan, Paul, Trench, 1885; orig. publ. 1665), pp. 200–201. For discussions of Boyle's remarks in the context of probabilist and fallibilist models of knowledge, see Shapiro, *Probability and Certainty*, pp. 26–27; van Leeuwen, *Problem of Certainty*, p. 103; Daston, *The Reasonable Calculus*, pp. 164–165.

37. Boyle, "Hydrostatical Discourse," p. 596.

38. Boyle, "New Experiments," p. 2.

39. Boyle, "Proëmial Essay," pp. 313, 317.

40. On the "Idols" and fallibilism, see Shapiro, *Probability and Certainty*, pp. 61–62.

41. Boyle, "Some Specimens of an Attempt to Make Chymical Experiments Useful to Illustrate the Notions of the Corpuscular Philosophy. The Preface," in *RBW*, vol. 1, pp. 354–359, on p. 355 (orig. publ. 1661); idem, "Proëmial Essay," p. 302. On the corrupting effects of "preconceived hypothesis or conjecture," see idem, "New Experiments," p. 47; and for doubts about the correctness of Boyle's professed unfamiliarity with the writings of Descartes and other systematists, see Westfall, "Unpublished Boyle Papers," p. 63; Laudan, "Clock Metaphor," p. 82n.; Marie Boas [Hall], "Boyle as a Theoretical Scientist," *Isis* 41 (1950): 261–268; idem, "The Establishment of the Mechanical Philosophy," *Osiris* 10 (1952): 412–541, on pp. 460–461; Frank, *Harvey and the Oxford Physiologists*, pp. 93–97. My concern here is not with the veracity of Boyle's professions but with the reasons why he made them.

42. Shadwell's play was performed in 1676. There is some evidence that Hooke believed *he* was the model for Gimcrack; see Richard S. Westfall, "Hooke, Robert," in *Dictionary of Scientific Biography*, vol. 6, pp. 481–488, on p. 483. Charles II, the Royal Society's patron, was also said to have found the weighing of the air rather funny.

43. For the extent to which experimental philosophy was, in fact, popular, see Hunter, *Science and Society*, chapters 3, 6.

44. This is not intended as an exhaustive catalogue of the measures necessary for institutionalization. Obviously, patronage was required and alliances had to be forged with existing powerful institutions.

45. Boyle, "Sceptical Chymist," esp. pp. 468, 513, 550, 584.

46. Boyle, "Proëmial Essay," p. 303.

47. Boyle's way of dealing with the hermetics drew on the views of the Hartlib group of the 1640s and 1650s. By contrast, there were those who rejected the findings of late alchemy (for example, Hobbes) and those who rejected the process of assimilation (for example, Newton).

48. Boyle, "Examen of Hobbes," pp. 233, 197; idem, "Defence against Linus," p. 122.

49. Boyle, "New Experiments," p. 10.

50. Boyle, "Continuation of New Experiments," pp. 250–258. Note that in other contexts Boyle encouraged speech of immaterial entities such as spirits; what he said was that such items ought to be purged from the routine discourse of experimental philosophy; see, for example, "Hydrostatical Discourse," p. 608.

51. Boyle, "New Experiments," pp. 11–12; cf. idem, "The General History of the Air . . . ," in *RBW*, vol. 5, pp. 609–743, on pp. 614–615 (orig. publ. 1692).

52. These problems were structurally similar to those afflicting Newton later in the century. Newton said that he wished to speak of gravitation as a mathematical regularity, without venturing an account of its physical cause. Newton's allies and enemies alike found it difficult to accept such mathematical statements as the end-product of philosophical inquiry; see Alexandre Koyré, *Newtonian Studies* (Chicago: University of Chicago Press, 1968), pp. 115–163, 273–282.

53. Boyle, "Defence against Linus," pp. 135, 137; idem, "Examen of Hobbes," pp. 191, 207; idem, "Animadversions on Hobbes," p. 112.

54. Hobbes, "Dialogus physicus," pp. 271, 273, 278; for Boyle's reply, see "Examen of Hobbes," pp. 193–194.

55. Cf. Boas [Hall], "Boyle"; idem, "Establishment," pp. 475–477.

56. Boyle, "Proëmial Essay," p. 312.

57. Boyle, "Proëmial Essay," p. 311.

58. See Robert P. Multhauf, "Some Nonexistent Chemists of the Seventeenth Century: Remarks on the Use of the Dialogue in Scientific Writing," in Allen G. Debus and Robert P. Multhauf, *Alchemy and Chemistry in the Seventeenth Century* (Los Angeles: William Andrews Clark Memorial Library, 1966), pp. 31–50.

59. Boyle, "Sceptical Chymist," p. 486. In the preface Boyle says that he will not "declare my own opinion"; he wishes to be "a silent auditor of their discourses" (pp. 460, 466–467).

60. The consensus that emerges is very like the position from which Carneades starts, but the plot of *The Sceptical Chymist* involves disguising that fact. Interestingly, the consensus is not total (a point nicely made by Golinski, "Robert Boyle"):

Eleutherius indicates reservations about Carneades's arguments; and Philoponus (a more "hard-line" alchemist who is absent for the bulk of the symposium) might not, in Eleutherius's opinion, have been persuaded. The obvious contrast is with the form and function of the dialogue in the writings of Boyle's anti-experimentalist adversary Hobbes, esp. the *Dialogus physicus, Problemata physica,* and *Decameron physiologicum.* Boyle strongly disapproved of Hobbes's dialogues, in which the "Hobbes" character demanded, and secured, absolute assent from his interlocutor; see Boyle, "Animadversions on Hobbes," p. 105.

61. Boyle, "Sceptical Chymist," p. 462.

62. Actually, the great bulk of the talk is between Carneades and Eleutherius. The other two participants inexplicably absent themselves from most of the proceedings. This is possibly an accident due to Boyle's self-confessed sloppiness with his manuscripts; he was continually apologizing for losing pages of his drafts.

63. Boyle, "Sceptical Chymist," p. 469.

64. Boyle's responses to his adversaries are examined in Shapin and Schaffer, *Leviathan and the Air-Pump,* chapter 5.

65. Boyle, "Examen of Hobbes," p. 190; cf. idem, "Hydrostatical Discourse," pp. 596–597; idem, "Animadversions on Hobbes," pp. 104–105; idem, "Defence against Linus," pp. 118–121.

66. Boyle, "Defence against Linus," p. 122; cf. idem, "Examen of Hobbes," pp. 197, 208, 233.

67. Boyle, "Defence against Linus," pp. 163, 120.

68. Boyle, "Examen of Hobbes," p. 186.

69. Boyle, "Examen of Hobbes," p. 188; cf. p. 190.

70. Hobbes, "Six Lessons to the Professors of the Mathematics . . . in the University of Oxford," in Hobbes, *The English Works,* vol. 7, pp. 181–356, on p. 356 (orig. publ. 1656).

71. Hobbes, "Considerations upon the Reputation, Loyalty, Manners, and Religion of Thomas Hobbes," in idem, *The English Works,* vol. 4, pp. 409–440, on p. 440 (orig. publ. 1662).

72. Boyle, "New Experiments," p. 3. The phrase "wonderful pacifick year" is from Sprat, *History of the Royal Society,* p. 58.

73. Sprat, *History of the Royal Society,* pp. 53, 56.

74. The focus of the essay has been upon an individual, yet its purpose is not *individualistic.* Boyle was a major innovator and practitioner of the new linguistic technologies. Nevertheless, he proposed them as routine practices for a *community,* and it is clear that Boyle's proposals were widely applauded and implemented, especially in the early Royal Society. His sentiments on this subject were echoed by Sprat, Glanvill, and Hooke, among others. For further details on the relevant English community of language users, see Shapiro, *Probability and Certainty,* chapter 7.

75. Bruno Latour and Steve Woolgar, *Laboratory Life: The Social Construction*

of Scientific Facts (Beverly Hills, CA: Sage, 1979), chapter 2; and, for a fine study of the role of instruments in scientific observation reports, see T. J. Pinch, "Towards an Analysis of Scientific Observation: The Externality and Evidential Significance of Observational Reports in Physics," *Social Studies of Science* 15 (1985): 3–36.

76. Boyle to Robert Moray, July 1662, in Huygens, *Oeuvres complètes*, vol. 4, pp. 217–220; cf. Boyle, "Defence against Linus," pp. 152–153.

77. Sprat, *History of the Royal Society*, pp. 85 (for "eyes and hands"), 98–99 (for the individual and the collective), 28–32 (for "tyrants" in philosophy). For the disciplining of the Royal Society's public, see Jacob, *Robert Boyle and the English Revolution*, esp. p. 156, and idem, *Henry Stubbe, Radical Protestantism and the Early Enlightenment* (Cambridge: Cambridge University Press, 1983), pp. 59–63; also some perceptive remarks in Yaron Ezrahi, "Science and the Problem of Authority in Democracy," in *Science and Social Structure: A Festschrift for Robert K. Merton, Transactions of the New York Academy of Sciences,* ed. Thomas F. Gieryn, series 11, vol. 39 (New York: New York Academy of Sciences, 1980), pp. 43–60, esp. 46–53; and Shapin, "The House of Experiment in Seventeenth-century England" (chapter 5 in this volume).

78. See esp. the work of H. M. Collins, whose metaphor of completed and consensual scientific knowledge as "the ship in the bottle" nicely crystallizes this point: for example, idem, "The Seven Sexes: A Study in the Sociology of a Phenomenon, or the Replication of Experiments in Physics," *Sociology* 9 (1975): 205–224; idem, "Son of Seven Sexes: The Social Destruction of a Physical Phenomenon," *Social Studies of Science* 11 (1981): 33–62. Cf. Latour and Woolgar, *Laboratory Life,* p. 176: "Our argument is not just that facts are socially constructed. We also wish to show that the process involves the use of certain devices whereby all traces of production are made extremely difficult to detect." For an historical study making a similar point, see Steven Shapin, "The Politics of Observation: Cerebral Anatomy and Social Interests in the Edinburgh Phrenology Disputes," in *On the Margins of Science: The Social Construction of Rejected Knowledge,* ed. Roy Wallis, *Sociological Review Monograph,* no. 27 (Keele: Keele University Press, 1979), pp. 139–178.

79. Ludwik Fleck, *Genesis and Development of a Scientific Fact,* trans. F. Bradley and Thaddeus J. Trenn, ed. Thaddeus J. Trenn and Robert K. Merton (Chicago: University of Chicago Press, 1979), pp. 103, 105 (orig. publ. in German, 1935).

Chapter 7. "The Mind Is Its Own Place"

1. Eleanor M. Sickels, *The Gloomy Egoist* (New York: Columbia University Press, 1932); Maynard Mack, *The Garden and the City: Retirement and Politics in the Later Poetry of Pope, 1731–1743* (Toronto: University of Toronto Press, 1969); John Sitter, *Literary Loneliness in Mid-Eighteenth-Century England* (Ithaca, NY: Cornell University Press, 1982), esp. pp. 92–94; P. M. Pasinetti, *Life for Art's Sake: Studies in the Literary Myth of the Romantic Artist* (New York: Garland Publishing, 1985).

2. Howard S. Becker, *Art Worlds* (Berkeley: University of California Press, 1982), pp. 22–24, 114–115; Svetlana Alpers, *Rembrandt's Enterprise: The Studio and the Market* (London: Thames and Hudson, 1988), esp. pp. 2–5, chapter 3; Janet Wolff, *The Social Production of Art* (London: Macmillan, 1981), chapter 2; cf. Caroline A. Jones, "Andy Warhol's 'Factory': The Production Site, Its Context, and Impact on the Work of Art," *Science in Context* 4 (1991): 101–131.

3. Samuel Johnson, *The Rambler*, vols. 3–5 of *The Works of Samuel Johnson*, ed. W. J. Bate and Albrecht B. Strauss, 15 vols. (New Haven, CT: Yale University Press, 1969), vol. 5, pp. 261–263. Johnson also held medical and physical theories of why intellectuals and artists flourished in garrets. On the one hand, purer air assisted the operation of genius; on the other, the swifter motion of the elevated philosopher relative to ground level acted as a stimulant (ibid., pp. 262–263). Seventeenth-century advocates of private worship recommended that the room set apart for this purpose be as high in the house as possible: "some secret Property there is in such high and eminent places, whence we may behold the heavens and over-look the earth, which . . . much raiseth the soul and elevates the affections, as if we derived or partaked more from Heaven, by how much nearer we come to it" (Edward Wetenhall, *Enter into Thy Closet: or, A Method and Order for Private Devotion*, 5th ed. [London, 1684 (1660)]).

4. Edward Gibbon, *The History of the Decline and Fall of the Roman Empire*, ed. J. B. Bury, 2nd ed., 7 vols. (London: Methuen, 1901), vol. 5, p. 337; see also W. B. Carnochan, *Gibbon's Solitude: The Inward World of the Historian* (Stanford, CA: Stanford University Press, 1987), pp. 7–8.

5. William Penn, *The Fruits of Solitude*, ed. Joseph Besse (London: J. M. Dent, Everyman's Library, 1915 [1648]), p. 25.

6. Anthony Storr, *The School of Genius* (London: André Deutsch, 1988).

7. Jean-Jacques Rousseau, *The Reveries of the Solitary Walker*, trans. and ed. Charles C. Butterworth (New York: New York University Press, 1979 [1782]), esp. pp. 5–6, 30–31, 62–70.

8. Daniel Defoe, *The Life and Strange Surprizing Adventures of Robinson Crusoe* (London: Oxford University Press, 1972 [1719]), pp. 128, 135–136.

9. Maurice Gonnaud, *An Uneasy Solitude: Individual and Society in the Work of Ralph Waldo Emerson*, trans. Lawrence Rosenwald (Princeton, NJ: Princeton University Press, 1987 [1964]), pp. 30, 34.

10. Henry David Thoreau, *Walden*, in *Walden and Civil Disobedience* (New York: Penguin, 1983 [1854]), p. 372.

11. Alexis de Tocqueville, *Democracy in America*, trans. Henry Reeve, 4 vols. (London: Saunders and Otley, 1835–1840), vol. 3, p. 206.

12. Henry Pemberton, *A View of Sir Isaac Newton's Philosophy* (London, 1728), preface; Frank E. Manuel, *A Portrait of Isaac Newton* (Cambridge, MA: Harvard University Press, 1968), pp. 82–83.

13. Richard S. Westfall, *Never at Rest: A Biography of Sir Isaac Newton* (Cambridge: Cambridge University Press, 1980), chapter 3, pp. 191–192.

14. René Descartes, "Discourse on Method," in *The Philosophical Works of Descartes,* ed. and trans. Elizabeth S. Haldane and G. R. T. Ross, 2 vols. (New York: Dover, 1955 [1637]), vol. 1, pp. 79–130, on pp. 88, 91, 100; cf. John A. Schuster, "Cartesian Method as Mythic Speech: A Diachronic and Structural Analysis," in *The Politics and Rhetoric of Scientific Method,* ed. John A. Schuster and R. R. Yeo (Dordrecht: Reidel, 1986), pp. 33–95.

15. Dorinda Outram, *Georges Cuvier: Vocation, Science and Authority in Post-Revolutionary France* (Manchester: Manchester University Press, 1984). For the significance of Charles Darwin's solitude—on the *Beagle* and at Down House—see Martin J. S. Rudwick, "Charles Darwin in London: The Integration of Public and Private Science," *Isis* 73 (1982): 186–206, esp. pp. 188–189, and James R. Moore, "Darwin of Down: The Evolutionist as Squarson-Naturalist," in *The Darwinian Heritage,* ed. David Kohn (Princeton, NJ: Princeton University Press, 1985), pp. 435–481. But see Janet Browne, *Charles Darwin: A Biography,* 2 vols. (New York: Alfred A. Knopf, 1995–2002).

16. Brian Vickers, "Introduction," in *Arbeit, Musse, Meditation: Betrachtungen zur Vita Activa und Vita Contemplativa,* ed. Brian Vickers (Zürich: Verlag der Fachvereine Zürich, 1985), pp. 1–19, on p. 2.

17. Aristotle, *Ethica Nichomachea,* ed. W. D. Ross (Oxford: Clarendon Press, 1925), 1169–1170; idem, *Politica,* trans. Benjamin Jowett (Oxford: Clarendon Press, 1921), 1325a; Jill Kraye, "Moral Philosophy," in *The Cambridge History of Renaissance Philosophy,* ed. Charles Schmitt et al. (Cambridge: Cambridge University Press, 1988), pp. 303–386, on pp. 335–339; Alasdair MacIntyre, *A Short History of Ethics* (New York: Collier, 1966), p. 83; idem, *After Virtue: A Study in Moral Theory,* 2nd ed. (Notre Dame, IN: University of Notre Dame Press, 1984), chapter 11.

18. Aristotle, *Ethica Nichomachea,* 1178b.

19. Aristotle, *Politica,* 1324a.

20. Aristotle, *Ethica Nichomachea,* 1177a–b.

21. Aristotle, *Ethica Nichomachea,* 1179a.

22. Cicero, *Offices: De Officiis, Laelius, Cato Major and Select Letters* (London: J. M. Dent, Everyman's Library, 1909), pp. 13, 68–70.

23. Hannah Arendt, *The Human Condition* (Chicago: University of Chicago Press, 1958), pp. 12–17.

24. George Duby, "Solitude: Eleventh to Thirteenth Century," in *A History of Private Life,* vol. 2, *Revelations of the Medieval World,* ed. George Duby, trans. Arthur Goldhammer (Cambridge, MA: Harvard University Press, 1988), pp. 509–533.

25. Peter Brown, *The Body and Society: Men, Women and Sexual Renunciation in Early Christianity* (London: Faber & Faber, 1989), chapters 11, 18.

26. Ann K. Warren, *Anchorites and Their Patrons in Medieval Europe* (Berkeley: University of California Press, 1985), esp. pp. 2–17, 94, 100–101.

27. Cary J. Nederman, "Nature, Sin, and the Origins of Society: The Ciceronian Tradition in Medieval Political Thought," *Journal of the History of Ideas* 49 (1988): 3–26; Letizia A. Panizza, "Active and Contemplative in Lorenzo Valla: The Fusion of Opposites," in *Arbeit, Musse, Meditation,* ed. Vickers, pp. 181–223; Quentin Skinner, "Political Philosophy," in *Cambridge History of Renaissance Philosophy,* ed. Schmitt et al., pp. 389–452, esp. pp. 428–429, 449; Paul Oskar Kristeller, "The Active and Contemplative Life in Renaissance Humanism," in *Arbeit, Musse, Meditation,* ed. Vickers, pp. 133–152, on pp. 137–139.

28. Aristotle, *Politica,* 1253a; Thomas Stanley, *The History of Philosophy,* 3 vols. (London, 1655–1660), vol. 1, part 1, p. 169; Stefano Guazzo, *The Civile Conversation of M. Steeven Guazzo,* ed. Sir Edward Sullivan, trans. George Pettie and Bartholomew Young, 2 vols. (London: Constable, 1925 [1581]), vol. 1, p. 30; Sir Thomas Smith, *De Republica Anglorum* (London, [1583] [ca. 1600]), p. 13; Robert Burton, *The Anatomy of Melancholy,* ed. Floyd Dell and Paul Jordan-Smith (New York: Tudor Publishing, 1927 [1628]), p. 216; Francis Bacon, *The Moral and Historical Works of Lord Bacon, Including his Essays . . . ,* ed. Joseph Devey (London: Henry G. Bohn, 1852 [1597]), p. 73.

29. Fritz Caspari, *Humanism and the Social Order in England* (Chicago: University of Chicago Press, 1954), p. 80; Kristeller, "Active and Contemplative Life."

30. Owen Hannaway, "Laboratory Design and the Aim of Science: Andreas Libavius Versus Tycho Brahe," *Isis* 77 (1986): 585–610; Mario Biagioli, "The Social Status of Italian Mathematicians, 1450–1600," *History of Science* 27 (1989): 41–95, on pp. 51–52; Shapin, "The House of Experiment in Seventeenth-century England" (chapter 5 in this volume); and idem, "'A Scholar and a Gentleman'" (chapter 8 in this volume).

31. Caspari, *Humanism and the Social Order,* pp. 79–80.

32. The term *courtesy text,* and the appreciation of a loosely linked "courtesy" genre, dates from the work of Victorian scholars in the Early English Text Society. Courtesy texts are generally understood as works dealing with manners, civility, and practical daily conduct. For discussion of this literature, see Steven Shapin, *A Social History of Truth: Civility and Science in Seventeenth-Century England* (Chicago: University of Chicago Press, 1994), esp. chapters 2–3.

33. Burton, *Anatomy of Melancholy,* p. 260; Guazzo, *Civile Conversation,* vol. 1, p. 24; cf. Richard Brathwait, *The English Gentleman* (London, 1630), p. 234.

34. Burton, *Anatomy of Melancholy,* p. 260; cf. Guazzo, *Civile Conversation,* vol. 1, pp. 18–19.

35. Samuel Butler, "A Melancholy Man," in *The Genuine Remains in Verse and Prose of Mr. Samuel Butler,* ed. Robert Thyer, 2 vols. (London, 1759), vol. 2, pp. 134–136, on p. 134.

36. Guazzo, *Civile Conversation,* vol. 1, p. 33.

37. Michel Eyquem de Montaigne, *Essays*, ed. L. C. Harmer, trans. John Florio, 3 vols. (London: J. M. Dent, Everyman's Library, 1965 [1580]), vol. 1: "Of Pedantisme," pp. 135–136.

38. Sir Henry Wotton, *A Philosophical Survey of Education or Moral Architecture and the Aphorisms of Education*, ed. H. S. Kermode (London: Hodder & Stoughton, for the University Press of Liverpool, 1938 [orig. published in Wotton, *Reliquiæ Wottonianæ* . . . , 3rd ed., 1651]), p. 25; cf. Philip Dormer Stanhope, Earl of Chesterfield, *Letters to His Son and Others* (London: J. M. Dent, Everyman's Library, 1984 [1774]), p. 48.

39. George C. Brauer, *Education of a Gentleman: Theories of Gentlemanly Education in England, 1660–1775* (New York: Bookman Associates, 1959), pp. 60–62; Fenela Ann Childs, "Prescriptions for Manners in English Courtesy Literature, 1690–1760, and Their Social Implications" (D.Phil. diss., Oxford University, 1984), pp. 215–221.

40. Francis Bacon, "The Advancement of Learning [1605]," in idem, *The Philosophical Works of Francis Bacon*, ed. James Spedding, Robert Leslie Ellis, and Douglas Denon Heath, 5 vols. (London, 1857–1858), vol. 3, pp. 253–491.

41. Bacon, "The Advancement of Learning [1605]," p. 268.

42. Bacon, "The Advancement of Learning [1605]," p. 327.

43. Julian Martin, "'Knowledge is Power': Francis Bacon, the State, and the Reform of Natural Philosophy" (Ph.D. diss., Cambridge University, 1988), pp. 86–91, 107–108, 228–229.

44. Peter Dear, "*Totius in Verba:* Rhetoric and Authority in the Early Royal Society," *Isis* 76 (1985): 145–161; Shapin, "Pump and Circumstance" (chapter 6 in this volume); Steven Shapin and Simon Schaffer, *Leviathan and the Air-Pump: Hobbes, Boyle, and the Experimental Life* (Princeton, NJ: Princeton University Press, 1985), chapter 2; J. V. Golinski, "Robert Boyle: Scepticism and Authority in Seventeenth-Century Chemical Discourse," in *The Figural and the Literal: Problems of Language in the History of Science and Philosophy 1630–1800*, ed. Andrew F. Benjamin, Geoffrey N. Cantor, and John R. R. Christie (Manchester: Manchester University Press, 1987), pp. 58–82; and idem, "A Noble Spectacle: Research on Phosphorus and the Public Cultures of Science in the Early Royal Society," *Isis* 80 (1989): 11–39.

45. Thomas Sprat, *The History of the Royal-Society of London* (London, 1667), pp. 13–14, 19.

46. Sprat, *History,* pp. 96–97.

47. William Petty, "The Advice of W.P. to Mr. Samuel Hartlib, for the Advance of Some Particular Parts of Learning," in *The Harleian Miscellany* (London, 1745 [1648]), vol. 6, pp. 1–13, on pp. 5–7.

48. Evelyn to Boyle, 3 September 1659, in John Evelyn, *The Diary and Correspondence of John Evelyn*, ed. W. Bray, 3 vols. (London, 1852), vol. 3, pp. 116–120.

49. Abraham Cowley, *A Proposition for the Advancement of Experimental Philosophy* (London, 1661), esp. sig. A5, pp. 14–15, 22, 35, 40.

50. Sprat, *History of the Royal-Society,* esp. pp. 53, 56. The political significance of the rhetoric of rural retirement in the civil wars and early Restoration merits extended treatment in this connection. It would, for example, be apposite to compare the debates over solitude in the poetry of Andrew Marvell ("Appleton House," "The Garden") and Cowley ("The Wish," "Of Solitude") with Sprat's portrayal of Interregnum Oxford as a place where philosophers might breathe "a freer air" and converse "in quiet." Michael Hunter ("A 'College' for the Royal Society: The Abortive Plan of 1667–1668," *Notes and Records of the Royal Society* 38 [1983–1984]: 159–186) describes the Royal Society's plans in the late 1660s for purpose-designed quarters and their relationship with earlier thinking about the proper venue for experimental natural philosophy.

51. Thomas Birch, "The Life of the Honourable Robert Boyle [1744]," in Robert Boyle, *Works,* ed. Thomas Birch, 6 vols. (London, 1772), vol. 1, pp. vi–clxxi, on p. cxlv; Gilbert Burnet, "Character of a Christian Philosopher, in a Sermon Preached January 7, 1691–1692, at the Funeral of the Hon. Robert Boyle," in idem, *Lives, Characters, and an Address to Posterity,* ed. John Jebb (London: James Duncan, 1833), pp. 325–376, on p. 361; R. E. W. Maddison, "Studies in the Life of Robert Boyle, F.R.S. Part I, Robert Boyle and Some of His Foreign Visitors," *Notes and Records of the Royal Society* 9 (1951): 1–35; idem, "Studies in the Life of Robert Boyle, F.R.S. Part IV. Robert Boyle and Some of His Foreign Visitors," *Notes and Records of the Royal Society* 11 (1954): 38–53; J. R. Jacob, *Robert Boyle and the English Revolution: A Study in Social and Intellectual Change* (New York: Burt Franklin, 1977), chapter 4; Robert Boyle, "An Account of the Two Sorts of the Helmontian Laudanum . . . ," in idem, *Works,* vol. 4, pp. 149–150, on p. 149; Shapin, "The House of Experiment in Seventeenth-century England" (chapter 5 in this volume).

52. Burnet, "Character of a Christian Philosopher," pp. 358, 368; Shapin, *A Social History of Truth,* chapter 4.

53. Robert Boyle, "An Account of Philaretus [= R. Boyle] during his Minority [1648–1649]," in R. E. W. Maddison, *The Life of the Honourable Robert Boyle F.R.S.* (London: Taylor & Francis, 1969), pp. 2–45, on pp. 32–35; Jacob, *Robert Boyle and the English Revolution,* pp. 38–40.

54. Boyle Papers (Royal Society, London), 37, f. 166.

55. For example, Henry Oldenburg, *Correspondence,* ed. A. Rupert Hall and Marie Boas Hall, 13 vols. (Madison: University of Wisconsin Press; London: Mansell; Taylor & Francis, 1965–1986), vol. 2, pp. 509, 613, 639.

56. Robert Boyle, "The Excellency of Theology, Compared with Natural Philosophy [1674]," in idem, *Works,* vol. 4, pp. 1–66, on pp. 53–54; Shapin, "The House of Experiment in Seventeenth-century England" (chapter 5 in this volume).

57. Robert Boyle, "Some Motives and Incentives to the Love of God, Pathetically Discoursed of, in a Letter to a Friend [= 'Seraphick Love'] [1659]," in idem, *Works,* vol. 1, pp. 243–298, on pp. 246–249, 259, 281, 290, 293.

58. Harold Fisch, "The Scientist as Priest: A Note on Robert Boyle's Natural Theology," *Isis* 44 (1953): 252–265; Jacob, *Robert Boyle and the English Revolution,* pp. 96–118.

59. For example, John Beale to Boyle, 17 October 1663, in Boyle, *Works,* vol. 6, pp. 341–342; Richard Lower, *Diatribæ Thomæ Willisii M.D. et Prof. Oxon. de Febribus Vindicatio adversus Edmundum de Meara, Ormoniensem Hiberum M.D.,* ed. and trans. Kenneth Dewhurst as *Richard Lower's "Vindicatio": A Defense of the Experimental Method* (Oxford: Sandford Publications, 1983 [1665]), p. 198; Joseph Glanvill, *Plus Ultra; or, the Progress and Advancement of Learning, since the Days of Aristotle* (London, 1668), p. 93.

60. For an account of Boyle's laboratory as densely inhabited—by support personnel—see Steven Shapin, "The Invisible Technician," *American Scientist* 77 (1989): 554–563; idem, *A Social History of Truth,* chapter 8.

61. Robert Boyle, "Occasional Reflections upon Several Subjects [1665]," in idem, *Works,* vol. 2, pp. 323–460, on pp. 334–340; idem, "Some Considerations Touching the Usefulness of Experimental Natural Philosophy [1663]," in idem, *Works,* vol. 2, pp. 1–201, and vol. 3, pp. 393–457, on pp. 8, 32, 57, 61–62.

62. Thomas S. Kuhn, "Mathematical versus Experimental Traditions in the Development of the Physical Sciences," in idem, *The Essential Tensions: Selected Studies of Scientific Tradition and Change* (Chicago: University of Chicago Press, 1977), pp. 31–65; Steven Shapin, "Science and the Public," in *Companion to the History of Modern Science,* ed. R. C. Olby et al. (London: Routledge, 1990), pp. 990–1007.

63. Steven Shapin, "Robert Boyle and Mathematics: Reality, Representation, and Experimental Practice," *Science in Context* 2 (1988): 23–58; idem, *A Social History of Truth,* chapter 7.

64. Shapin and Schaffer, *Leviathan and the Air-Pump,* pp. 100–102, 143–154, 328–329.

65. 18 August 1676, in Sir Isaac Newton, *Correspondence,* ed. H. W. Turnbull, J. D. Scott, A. R. Hall, and L. Tilling, 7 vols. (Cambridge: Cambridge University Press, 1959–1977), vol. 2, p. 79.

66. Robert Charles Iliffe, "'The Idols of the Temple': Isaac Newton and the Private Life of Anti-Idolatry" (Ph.D. diss., Cambridge University, 1989), esp. chapters 5–6.

67. Westfall, *Never at Rest,* pp. 74–75, 78–79, 191–196. J. V. Golinski ("The Secret Life of an Alchemist," in *Let Newton Be!* ed. John Fauvel, Raymond Flood, Michael Shortland, and Robin Wilson [Oxford: Oxford University Press, 1988], pp. 147–167, esp. pp. 151–155) discusses the distinction between private and public knowledge in Newton's chemistry. Newton distinguished "vulgar chymistry"—laboratory imitation of mechanical processes—from "a more subtle secret & noble way of working"—laboratory reenaction of the processes of growth and life. It was knowledge of the latter that Newton wanted kept private. Golinski contrasts the relatively open laboratory maintained by Boyle with Newton's private, and largely

self-manned, laboratory attached to his Trinity College rooms. He also usefully points out the extent to which Boyle and Newton were agreed that certain items of chemical knowledge not be made public, though it still seems likely that Boyle—far more than Newton—wanted chemistry moved into the public domain. See also Westfall, *Never at Rest,* pp. 361, 371–377.

68. Westfall, *Never at Rest,* pp. 98–99.

69. Newton to Governors of Christ's Hospital, 3 April 1682, in Newton, *Correspondence,* vol. 2, p. 375.

70. Both quoted in Richard Yeo, "Genius, Method, and Morality: Images of Newton in Britain, 1760–1860," *Science in Context* 2 (1988): 257–284, on pp. 266, 276.

71. Newton to Henry Oldenburg, 6 January and 6 February 1671–1672, in Newton, *Correspondence,* vol. 1, pp. 80, 92, 100; Simon Schaffer, "Glass Works: Newton's Prisms and the Uses of Experiment," in *The Uses of Experiment: Studies in the Natural Sciences,* ed. David Gooding, Trevor Pinch, and Simon Schaffer (Cambridge: Cambridge University Press, 1989), pp. 67–104; Lord Adrian, "Newton's Rooms at Trinity," *Notes and Records of the Royal Society* 18 (1963): 17–24.

72. Zev Bechler, "Newton's 1672 Optical Controversies: A Study in the Grammar of Scientific Dissent," in *The Interaction between Science and Philosophy,* ed. Yehuda Elkana (Atlantic Highlands, NJ: Humanities Press, 1974), pp. 115–142.

73. Newton to Oldenburg, 20 February 1671–1672, 11 June 1672, 13 November 1675, 29 February 1675–1676, 18 August and 28 November 1676, in Newton, *Correspondence,* vol. 1, pp. 116–117, 186–187, 356, 422; vol. 2, pp. 81, 184–185; Iliffe, "Idols of the Temple," pp. 202–207.

74. Newton to Oldenburg, 10 February 1671–1672, in Newton, *Correspondence,* vol. 1, pp. 108–119.

75. Newton to Hooke, 5 February 1675–1676, in Newton, *Correspondence,* vol. 1, pp. 416–417.

76. Simon Schaffer, "Godly Men and Mechanical Philosophers: Souls and Spirits in Restoration Natural Philosophy," *Science in Context* 1 (1987): 55–85.

77. Shapin, "The House of Experiment in Seventeenth-century England" (chapter 5 in this volume); idem, "Who Was Robert Hooke?" (chapter 9 in this volume).

78. David Gooding, "'In Nature's School': Faraday as an Experimentalist," in *Faraday Rediscovered: Essays on the Life and Work of Michael Faraday,* ed. David Gooding and Frank A. J. L. James (London: Macmillan, 1985), pp. 106–135, esp. p. 108.

79. H. M. Collins, "Public Experiments and Displays of Virtuosity: The Core-Set Revisited," *Social Studies of Science* 18 (1988): 725–748, on p. 728.

80. Erving Goffman, *The Presentation of Self in Everyday Life* (London: Allen Lane, Penguin Press, 1969; orig. publ. 1959), chapter 3.

81. Mary Douglas, *Cultural Bias,* Occasional Paper No. 34 of the Royal Anthropological Institute of Great Britain and Ireland (London: RAI, 1978), p. 44. Pasinetti (*Life for Art's Sake,* p. 29) comments on this feature of the voice of the supposedly

isolated poet: "The role of solitariness implies the presence of an audience; the performance implies acknowledgment of a social context which has, if nothing else, the function of being the target, for instance, of the hero's anger or rebellion . . . The solitary voice does not exist except communally." And Arendt (*The Human Condition,* p. 22) notes that "No human life, not even the life of the hermit in nature's wilderness, is possible without a world which directly or indirectly testifies to the presence of other human beings."

82. Jonathan Swift, "Thoughts on Various Subjects," in idem, *The Prose Works,* ed. H. Davis, 14 vols. (Oxford: Blackwell, 1939–1968), vol. 1, pp. 241–245, on p. 242.

83. Bruno Latour's work is a powerful attack on the notion that "society" happens outside of "science" (for example, *Science in Action: How to Follow Scientists and Engineers Through Society* [Milton Keynes: Open University Press, 1987], esp. chapter 1; idem, *The Pasteurization of France,* trans. Alan Sheridan and John Law [Cambridge, MA: Harvard University Press, 1988], esp. pp. 3–7, 13–16).

84. Of course, knowledge that applies everywhere is likely not to apply very accurately anywhere in particular. Most writers who insist on both the global character of mathematical and scientific knowledge and its universal application tend to overlook the immense amount of work that is done to create and sustain the artificial and formal environments in which "application" happens. See, for example, Eric Livingston (*The Ethnomethodological Foundations of Mathematics* [London: Routledge & Kegan Paul, 1986]) on mathematical transcendence as a "locally produced" phenomenon and David Bloor's critique ("The Living Foundations of Mathematics," *Social Studies of Science* 17 [1987]: 337–358).

85. Jean Lave's work ("The Values of Quantification," in *Power, Action and Belief: A New Sociology of Knowledge?* ed. John Law, Sociological Review Monograph, no. 32 [London: Routledge & Kegan Paul, 1986], pp. 88–111; *Cognition in Practice: Mind, Mathematics, and Culture in Everyday Life* [New York: Cambridge University Press, 1988]) marvellously displays both the situatedness of lay arithmetical skills and the resources available, even in lay society, to distinguish context-dependent knowledge from what is taken to be the genuine commodity.

86. Diogenes Laërtius, *The Lives and Opinions of Eminent Philosophers,* trans. C. D. Yonge (London: Henry G. Bohn, 1853), pp. 59, 234, 240, 255; Stanley, *History of Philosophy,* vol. 3, part IV, p. 20.

87. Epictetus, *The Discourses and the Manual,* ed. P. E. Matheson, 2 vols. (Oxford: Clarendon Press, 1916), vol. 2, p. 235.

88. Arendt, *The Human Condition,* pp. 75–76; also pp. 16, 20.

89. Cicero, *Offices,* p. 114; Guazzo, *Civile Conversation,* vol. 1, pp. 49–52.

90. Montaigne, *Essays,* vol. 1, pp. 253–255; also Adi Ophir, "A Place of Knowledge Re-created: The Library of Michel de Montaigne," *Science in Context* 4 (1991): 163–189.

91. John Milton, *Paradise Lost,* in *The Complete Poetry and Selected Prose*

(New York: Modern Library, 1950 [1667]), pp. 98–99; see also Christopher Hill, *The World Turned Upside Down: Radical Ideas During the French Revolution* (Harmondsworth: Penguin, 1975), p. 175; idem, *Milton and the English Revolution* (Harmondsworth: Penguin, 1979), pp. 308–311.

92. Sir Thomas Browne, *The Religio Medici and Other Writings,* ed. C. H. Herford (London: J. M. Dent, Everyman's Library, 1940 [1643]), p. 82.

Chapter 8. *"A Scholar and a Gentleman"*

1. It is not clear from the context in *The Prelude* whether Wordsworth pointed to "scholars" and "gentlemen" as two categories of individuals (as well he might have done in relation to contemporary Cambridge) or as two aspects of all the individuals concerned.

2. Additional late eighteenth-century uses appear in an editorial comment in an edition of Izaak Walton's *Lives* (Thomas Zouch, ed., *Walton's Lives,* 2 vols. [York: Wilson, Spence and Mawman, 1796], vol. 2, p. 265). Another appears in a review of Della Crusca's *Poetry of the World* in the 1788 volume of the *Monthly Review* (as quoted in Derek Roper, *Reviewing before the Edinburgh, 1788–1802* [London: Methuen, 1978], pp. 77–78).

3. See, for example, John Henry Newman, *Discourses on the Scope and Nature of University Education . . .* (Dublin: James Duffy, 1852), pp. 175–183. Newman here identified "gentleman's knowledge" and the role of the university in supplying it, though I have not located his use of the relevant commonplace.

4. William Winstanley, *The Lives of the Most Famous English Poets* (London, 1687), p. 191. Antonia Fraser (*Cromwell: The Lord Protector* [New York: Alfred A. Knopf, 1973], p. 659) quotes (without specifying a source) an almost identical form of words applied to the regicide clergyman Dr. John Hewett in the Interregnum.

5. Clearly, both the gentle and the learned worlds in the sixteenth and seventeenth centuries interacted internationally and many of their attributed characteristics were shared between national settings. My concern is with the situation in England. I use sources deriving from non-English contexts when such sources were important for Englishmen, whether in translation or not. When there is evidence that English patterns importantly diverged from those elsewhere or that the English accounted themselves different in relevant respects from other nationalities, I draw attention to such evidence. I make no claims of the sort that the English were unique or that every culture was the same.

6. My use of this notion bears only partial resemblance to that of sociologists, for example, Michael Mulkay, *Science and the Sociology of Knowledge* (London: George Allen & Unwin, 1979), pp. 71–72, 93–95, and G. N. Gilbert and Michael Mulkay, *Opening Pandora's Box: A Sociological Analysis of Scientists' Discourse* (Cambridge: Cambridge University Press, 1984), pp. 40, 56–62, and, following Mulkay, Thomas F. Gieryn, "Boundary-Work and the Demarcation of Science from

Non-Science: Strains and Interests in Professional Ideologies of Scientists," *American Sociological Review* 48 (1983): 781–795, on p. 783.

7. Robert K. Merton, *Science, Technology and Society in Seventeenth-Century England* (New York: Harper Torchbook, 1970; orig. publ. 1938); Max Weber, *The Protestant Ethic and the Spirit of Capitalism,* trans. Talcott Parsons (New York: Charles Scribner's, 1958; orig. publ. 1904–1905), esp. pp. 72, 181–182; also Thomas F. Gieryn, "Distancing Science from Religion in Seventeenth-Century England," *Isis* 79 (1988): 582–593, on pp. 583–587 (for a claim about Durkheimian inspiration). For culture as the stock of legitimations, see Mary Douglas in, for example, *How Institutions Think* (Syracuse, NY: Syracuse University Press, 1986), esp. pp. 45–47, 63, 112; idem, *Risk Acceptability According to the Social Sciences,* Russell Sage Foundation, Social Research Perspectives, Occasional Reports on Current Topics, No. 11 (New York: Russell Sage Foundation, 1986), esp. pp. 67–68; idem, *Cultural Bias* (London: RAI, 1978); and see an overview by several of her colleagues: Michael Thompson, Richard Ellis, and Aaron Wildavsky, *Cultural Theory* (Boulder, CO: Westview Press, 1990), esp. chapter 1.

8. Quentin Skinner, "Some Problems in the Analysis of Political Thought and Action," in *Meaning and Context: Quentin Skinner and His Critics,* ed. James Tully (Cambridge: Polity Press, 1988), pp. 79–118, on p. 117 (art. orig. publ. 1974); also idem, "Language and Social Change," in ibid., pp. 119–132, esp. pp. 131–132, and idem, "The Idea of a Cultural Lexicon," *Essays in Criticism* 29 (1979): 205–224, on p. 215. For an independent analysis of *motives* as both situated repertoires for describing and evaluating action *and* as determinants of action, see C. Wright Mills, "Situated Actions and Vocabularies of Motive," *American Sociological Review* 5 (1940): 904–913.

9. For the relationship between some of this literature and macro-social change, see Norbert Elias, *The Civilizing Process,* vol. 1, *The History of Manners,* and vol. 2, *Power & Civility,* trans. Edmund Jephcott (New York: Pantheon, 1982; orig. publ. 1939), esp. vol. 1, chapter 2; and, for further use of this literature, see Steven Shapin, *A Social History of Truth: Civility and Science in Seventeenth-Century England* (Chicago: University of Chicago Press, 1994), esp. chapters 2–3.

10. *The Institucion of a Gentleman* (London, 1555), sig. A.iii.

11. Epictetus, *Epictetus his Manuell,* trans. John Healey (London, 1616), p. 91 (quoted in W. Lee Ustick, "Changing Ideals of Aristocratic Character and Conduct in Seventeenth-Century England," *Modern Philology* 30 [1933]: 147–166, on p. 149). Cf. Epictetus, *The Discourses and Manual . . . ,* ed. P. E. Matheson, 2 vols. (Oxford: Clarendon Press, 1916), vol. 2, p. 235. The place of neo-Stoicism in early modern English social thought needs fuller treatment, especially in relation to ethical theory and its practical use in justifying social roles; see, for example, Gerhard Oestreich, *Neostoicism and the Early Modern State,* ed. Brigitta Oestreich and H. G. Koenigsberger, trans. David McLintock (Cambridge: Cambridge University

Press, 1982), and Jill Kraye, "Moral Philosophy," in *The Cambridge History of Renaissance Philosophy*, ed. Charles B. Schmitt et al. (Cambridge: Cambridge University Press, 1988), pp. 303–386, on pp. 360–374; also Shapin, "'The Mind Is Its Own Place'" (chapter 7 in this volume); idem, *A Social History of Truth: Civility and Science in Seventeenth-Century England* (Chicago: University of Chicago Press, 1994), chapters 2, 4.

12. Lewis Einstein, *The Italian Renaissance in England* (New York: Columbia University Press, 1902), chapters 1–2; Fritz Caspari, *Humanism and the Social Order in Tudor England* (Chicago: University of Chicago Press, 1954), esp. pp. 10–14, 80, 136–152; Ruth Kelso, *The Doctrine of the English Gentleman in the Sixteenth Century*, University of Illinois Studies in Language and Literature (Urbana: University of Illinois Press, 1929), vol. 14, pp. 1–288, esp. pp. 111–148.

13. George Pettie, "Preface" to Stefano Guazzo, *The Civile Conversation of M. Steeven Guazzo*, trans. George Pettie and Bartholomew Young, ed. Sir Edward Sullivan, 2 vols. (London: Constable and Co., 1925; orig. publ. in Italian in 1574 and in English in 1581, 1586), vol. 1, pp. 8–9; see also Lawrence Stone, *The Crisis of the Aristocracy 1558–1641* (Oxford: Clarendon Press, 1965), p. 672, and, for the general humanist stress on education for the nobility, see Quentin Skinner, *The Foundations of Modern Political Thought*, 2 vols. (Cambridge: Cambridge University Press, 1978), vol. 1, pp. 241–243.

14. Baldessare Castiglione, *The Book of the Courtier*, trans. Charles S. Singleton (Garden City, NY: Anchor Doubleday, 1959; orig. publ. 1528; orig. English trans. 1561), p. 67.

15. Shakespeare, *Hamlet*, III, i.

16. Translator's preface to Pierre Gassendi, *The Mirrour of True Nobility and Gentility*, trans. W. Rand (London, 1657), sig. A4 (quoted in Joseph M. Levine, *Dr. Woodward's Shield: History, Science, and Satire in Augustan England* [Berkeley: University of California Press, 1977], p. 120), and see Levine, *Humanism and History: Origins of Modern English Historiography* (Ithaca, NY: Cornell University Press, 1987), p. 86 (for Gassendi) and pp. 126–131 (for education and English humanism generally).

17. James Cleland, *The Instruction of a Young Noble-man* (Oxford, 1612), pp. 134–139.

18. *The Rich Cabinet furnished with varieties of Excellent Discriptions, Exquisite Characters, Witty Discourses, and Delightfull Histories, Devine and Morrall . . . Whereunto is Annexed the Epitome of Good Manners, extracted from Mr. Iohn de la Casa* (London, 1616), p. 76r.

19. Richard Braithwait, *The English Gentleman* (London, 1630), sig. Nnnr. For Puritan views of bookish learning, see John Morgan, *Godly Learning: Puritan Attitudes towards Reason, Learning, and Education, 1560–1640* (Cambridge: Cambridge University Press, 1986), esp. p. 71: "The greatest danger of learning . . . was

that it would become an end in itself, the focus of the scholar's endeavour, not subordinated to godly utility."

20. Henry Peacham, *The Complete Gentleman . . .* , ed. Virgil B. Heltzel (Ithaca, NY: Cornell University Press, 1962; orig. publ. 1622), pp. 28–29, 145. On self-fashioning, see Stephen Greenblatt, *Renaissance Self-Fashioning: From More to Shakespeare* (Chicago: University of Chicago Press, 1980). On Peacham's place in the English culture of virtuosity, see Walter E. Houghton, Jr., "The English Virtuoso in the Seventeenth Century," *Journal of the History of Ideas* 3 (1942): 51–73, 190–219, esp. p. 67. On the Arundel circle of virtuosi, see Levine, *Humanism and History*, pp. 83–87, and David Howarth, *Lord Arundel and His Circle* (New Haven, CT: Yale University Press, 1985).

21. Stone, *Crisis*, pp. 672–677; Caspari, *Humanism*, pp. 5–13; Kelso, *English Gentleman*, pp. 113–114; W. Gordon Zeeveld, *Foundations of Tudor Policy* (Westport, CT: Greenwood Press, 1981); and see J. H. Hexter, "The Education of the Aristocracy in the Renaissance," in idem, *Reappraisals in History: New Views on History and Society in Early Modern Europe,* 2nd ed. (Chicago: University of Chicago Press, 1979), pp. 45–70, on pp. 61–65.

22. See, for example, Mark H. Curtis, *Oxford and Cambridge in Transition 1558–1642: An Essay on the Changing Relations between the English Universities and English Society* (Oxford: Clarendon Press, 1959); Lawrence Stone, "The Educational Revolution in England, 1560–1640," *Past and Present* 28 (1964): 41–80; idem, "The Size and Composition of the Oxford Student Body 1580–1910," in *The University in Society,* 2 vols., ed. Lawrence Stone (Princeton, NJ: Princeton University Press, 1974), vol. 1, pp. 3–110; James McConica, "Scholars and Commoners in Renaissance Oxford," in ibid., vol. 1, pp. 151–181; Hugh Kearney, *Scholars and Gentlemen: Universities and Society in Pre-Industrial Britain 1500–1700* (Ithaca, NY: Cornell University Press, 1970); Morgan, *Godly Learning,* chapter 11.

23. The phrase about "seminaries" is from J. Bass Mullinger, *The University of Cambridge,* 3 vols. (Cambridge: Cambridge University Press, 1873–1911), vol. 2, p. 250; the latter phrase is from Curtis, *Oxford and Cambridge in Transition,* p. 56 (see also pp. 61, 82). For statistics, see Lawrence Stone, "The Size and Composition of the Oxford Student Body 1580–1910," in *The University in Society,* vol. 1, pp. 3–110, and James McConica, "Scholars and Commoners in Renaissance Oxford," in ibid., pp. 151–181.

24. Hexter, "Education of the Aristocracy," p. 50.

25. Kearney, *Scholars and Gentlemen,* pp. 23–25; Stone, "Educational Revolution," p. 70.

26. Quoted in Stone, "Educational Revolution," p. 76.

27. Quoted in Hexter, "Education of the Aristocracy," p. 50.

28. Curtis, *Oxford and Cambridge,* pp. 128–144; see also Kearney, *Scholars and Gentlemen,* pp. 38–39; Stone, "Educational Revolution," p. 70.

29. John Earle, *Microcosmographie,* facsimile ed. of autograph manuscript (Leeds: Scolar Press, 1966; orig. 1628), f. 81.

30. Seth Ward, *Vindiciae academiarum* (Oxford, 1654), p. 50; see also Mordechai Feingold, *The Mathematicians' Apprenticeship: Science, Universities and Society in England, 1560–1640* (Cambridge: Cambridge University Press, 1984), p. 30.

31. Stone, *Crisis,* p. 674.

32. Hexter, "Education of the Aristocracy," pp. 46–49.

33. Quoted in Stone, "Educational Revolution," p. 74. For relevant statistics, see McConica, "Scholars and Commoners"; Stone, "Size and Composition"; Hexter, "Education of the Aristocracy," p. 54.

34. For the "explosion," see Stone, *Crisis,* p. 676; Cecil, quoted in Kearney, *Scholars and Gentlemen,* p. 25; see also pp. 26–30; Curtis, *Oxford and Cambridge,* pp. 129–131, 135, 143–144.

35. For the careers of university graduates, see Hexter, "Education of the Aristocracy," pp. 55–56; for universities as vehicles of social mobility, see Kearney, *Scholars and Gentlemen,* p. 33, and Curtis, *Oxford and Cambridge,* p. 271; for remarks on the particular effects of university residence on commitment to science, see R. G. Frank, Jr., "Science, Medicine, and the Universities of Early Modern England: Background and Sources," *History of Science* 11 (1973): 194–216, 239–269, on pp. 198–199. A more fine-grained account than can be offered here would discuss the learned professions (and trade) as occasional repositories of the gentry's younger sons, the role of the universities in facilitating such moves, and the consequences of a professional career for gentle status; see, for example, Peter Earle, *The Making of the English Middle Class: Business, Society and Family Life in London, 1600–1730* (London: Methuen, 1989), esp. pp. 60–75, 88–89.

36. McConica, "Scholars and Commoners"; Stone, "Size and Composition." The distinction between gentlemen and scholars was (and to varying extents remains) visible in college costume.

37. For career patterns of university fellows in the seventeenth and eighteenth centuries, see Geoffrey Holmes, *Augustan England: Professions, State and Society, 1680–1730* (London: George Allen & Unwin, 1982), pp. 36–40. Fellowships were normally held for ten years after taking the B.A., while the graduate was awaiting a college living, at which point they would be resigned for a clerical career. But fellows holding professorships or readerships were in a separate category of career academics, especially those content to remain celibate.

38. For a fuller account, see Shapin, "'The Mind is Its Own Place'" (chapter 7 in this volume).

39. The opposed attributes of "scholar" and "gentleman" were frequently inscribed within the standard literary form of the collection of "characters"—brief sketches of the virtues and typical behaviors of different social types ("the merchant," "the mathematician," "the pedant," "the zealot," "the citizen," etc.). The

genre was inherited from Antiquity (esp. Theophrastus's *Characters*), and early modern practitioners included *inter alia* Jean de la Bruyère, John Selden, Samuel Butler, John Earle, and the anonymous author of *The Rich Cabinet*. So, the literate classes had readily available to them standardized portrayals of the attributes of gentlemen and men of learning.

40. For example, Bruce T. Moran, "Princes, Machines and the Valuation of Precision in the Sixteenth Century," *Sudhoffs Archiv* 61 (1977): 209–228; idem, "Science at the Court of Hesse-Kassel: Informal Communication, Collaboration and the Role of the Prince-Practitioner in the Sixteenth Century" (Ph.D. thesis, University of California at Los Angeles, 1978); Robert S. Westman, "The Astronomer's Role in the Sixteenth Century: A Preliminary Study," *History of Science* 18 (1980): 105–147; Mario Biagioli, "The Social Status of Italian Mathematicians, 1450–1600," *History of Science* 27 (1989): 41–95; idem, "Galileo's System of Patronage," *History of Science* 28 (1990): 1–62; idem, "Galileo the Emblem Maker," *Isis* 81 (1990): 230–258.

41. For antagonism between the knight and the scholar, see Kelso, *English Gentleman*, pp. 111–113. On the seventeenth-century English universities perpetuating the image of the scholar as cleric, see Kearney, *Scholars and Gentlemen*, pp. 168–169.

42. Recent work on Galileo as court-philosopher shows how he used the social ambiguity of his position in an ambitious attempt to establish himself as a person of substance and honor, a courtier whose work flattered the prince-patron in a way exactly analogous to the role of other courtiers; see esp. Biagioli, "Galileo's System of Patronage," and idem, "Galileo the Emblem Maker." Nevertheless, it remains unclear how that attempt was perceived by his princely patrons and even more obscure how plausible was the attempted association of the life of honor and the philosophical role. For example, much Italian and French courtesy literature specifying who could and could not instigate a duel put scholars, professionals, and philosophers outside the well-born community of honor and resentment. Dueling did, however, become more prevalent among the rising classes of seventeenth-century England as a way of publicly displaying entitlement to gentle status. See Frederick Robertson Bryson, *The Point of Honor in Sixteenth-Century Italy: An Aspect of the Life of the Gentleman,* Publications of the Institute of French Studies (New York: Columbia University [Press], 1935), p. 33; V. G. Kiernan, *The Duel in European History: Honour and the Reign of Aristocracy* (Oxford: Oxford University Press, 1989), esp. pp. 102–103, 133–134; and François Billacois, *The Duel: Its Rise and Fall in Early Modern France,* trans. Trista Selous (New Haven, CT: Yale University Press, 1990); also Shapin, *A Social History of Truth,* pp. 107–114.

43. Richard Pace, *De fructu* (Basel, 1517), quoted in Hexter, "Education of the Aristocracy," p. 46 (cf. Kelso, *English Gentleman*, p. 113); *Rich Cabinet*, p. 134v; Cleland, *Instruction of a Young Noble-man*, pp. 134–135; Robert Burton, *The Anatomy of Melancholy*, ed. Floyd Dell and Paul Jordan-Smith (New York: Tudor,

1927; orig. publ. 1628), p. 273 (see also pp. 266–267). Castiglione (*Book of the Courtier*, p. 67) alleged that the French "not only do not esteem, but they abhor letters, and consider all men of letters to be very base; and they think it is a great insult to call anyone a clerk."

44. For discussions of the bases of gentle standing in the English setting, see, for example, Peter Laslett, *The World We Have Lost* (London: Methuen, 1965), chapter 2; Keith Wrightson, *English Society 1580–1680* (London: Hutchinson, 1982), chapter 1.

45. Castiglione, *Book of the Courtier*, pp. 109, 139 (for decorum); pp. 43–44, 98, 154 (for genteel *sprezzatura*, or nonchalance, and the avoidance of affectation); p. 111 (for criticism of displays of melancholia); Elias, *The Civilizing Process*, vol. 1, pp. 80–84 (for courtesy); cf. Giovanni della Casa, *Galateo or the Book of Manners*, trans. R. S. Pine-Coffin (Harmondsworth: Penguin; orig. publ. 1558), pp. 93–97. See also Kelso, *English Gentleman*, pp. 83–85 (for greater English stress on civic usefulness compared to Italian emphasis on personal perfection) and pp. 94–96 (on magnanimity). For enthusiasm as "vulgar," see Ustick, "Changing Ideals," p. 150.

46. Guazzo, *Civile Conversation*, vol. 1, pp. 56, 114, 164. (In sixteenth- and seventeenth-century usage, "conversation" meant not only face-to-face verbal exchange, but all forms of relations involved in living in society.) Among many seventeenth-century injunctions to decorum in the courtesy literature, see Cleland, *Instruction of a Young Noble-man*, pp. 171–173, 188; [Antoine de Courtin], *The Rules of Civility; or, Certain Ways of Deportment observed amongst All Persons of Quality* . . . (London, 1685; orig. publ. in French 1671; orig. English trans. 1673), esp. pp. 4–5. For a survey of the idea of decorum, see Fenela Ann Childs, *Prescriptions for Manners in English Courtesy Literature, 1690–1760, and Their Social Implications* (D.Phil. thesis, Oxford University, 1984), esp. pp. 46–59, 65–67, 95, 193–201; Kelso, *English Gentleman*, pp. 83–84; Greenblatt, *Renaissance Self-Fashioning*, esp. pp. 162–163; George C. Brauer, Jr., *The Education of a Gentleman: Theories of Gentlemanly Education in England, 1660–1775* (New York: Bookman Associates, 1959), chapter 5.

47. Mariateresa Fumagalli Beonio Brocchieri, "The Intellectual," in *Medieval Callings*, ed. Jacques Le Goff, trans. Lydia G. Cochrane (Chicago: University of Chicago Press, 1988), pp. 180–209, on pp. 205–206; Hugh Trevor-Roper, "Robert Burton and *The Anatomy of Melancholy*," in idem, *Renaissance Essays* (Chicago: University of Chicago Press, 1985), pp. 239–274, on pp. 257–258. For Aristotle on melancholy, see *Problems*, ed. and trans. W. S. Hett, 2 vols. (London: Heinemann, 1957), vol. 2, p. 155.

48. For example Burton, *Anatomy of Melancholy*, pp. 259, 458–460; Guazzo, *Civile Conversation*, vol. 1, pp. 16–52. For the iconology of scholarly solitude, see, for example, various portrayals of Sts. Anthony, Jerome, and Augustine (for example, Fra Filippo Lippi, Vincenzo di Catena, Sandro Botticelli, Vittore Carpaccio, Titian, Ma-

rinus van Reymerswaele); Johannes Vermeer's Louvre *L'astronome;* Gerrit Dou's *Astronomer by Candlelight* (Getty Museum); Rembrandt's Louvre *Philosophe en meditation;* and Albrecht Dürer's *Melencholia I.* Further general comments on the representation of intellectual solitude are in Shapin, "'The Mind Is Its Own Place'" (chapter 7 in this volume).

49. See, for example, Shakespeare's usage in *Love's Labour's Lost,* III, i ("A domineering pedant o'er the boy") and *Twelfth Night,* III, ii ("like a pedant that keeps a school").

50. Michel Eyquem de Montaigne, "Of Pedantry," in *The Complete Essays of Montaigne,* trans. Donald M. Frame (Stanford, CA: Stanford University Press, 1965; orig. English trans. 1580), pp. 97–106, on pp. 98, 100–101.

51. Guazzo, *Civile Conversation,* vol. 1, p. 33; Montaigne, *Essays,* p. 98; Burton, *Anatomy of Melancholy,* p. 262; Richard Lingard, *A Letter of Advice to a Young Gentleman Leaving the University Concerning His Behaviour and Conversation in the World,* ed. Frank C. Erb (New York: McAuliffe & Booth, 1907; orig. publ. 1670), p. 3; Lodowick Bryskett, *A Discourse of Civill Life,* ed. Thomas E. Wright (Northridge, CA: San Fernando Valley State College Press, 1970; orig. publ. 1606), p. 13. (The Bryskett text is basically a free and synthetic translation of sixteenth-century Italian texts by Giraldi, Piccolomini, and Guazzo.)

52. Roger Ascham, *The Scholemaster* (London, 1570), p. 5v. On the prevalence of injunctions against too deep a study of mathematics for gentlemen, see Feingold, *Mathematicians' Apprenticeship,* esp. pp. 29–30, 190–193.

53. Francis Bacon, "Of Studies," in *The Moral and Historical Works of Lord Bacon,* ed. Joseph Devey (London: Bohn, 1852), pp. 136–137.

54. Montaigne, *Essays,* p. 106.

55. Bryskett, *Discourse of Civill Life,* p. 13.

56. John Evelyn, *Publick Employment and an Active Life prefer'd to Solitude, and all its Appanages . . .* (London, 1667), pp. 83, 93. For this text as part of a Restoration argument over the active versus contemplative life, see Brian Vickers, "Public and Private Life in Seventeenth-Century England: The Mackenzie-Evelyn Debate," in idem, ed., *Arbeit, Musse, Meditation: Betrachtungen zur 'Vita activa' und 'Vita contemplativa'* (Zürich: Verlag der Fachvereine Zürich, 1985), pp. 257–278.

57. Samuel Butler, "The Abuse of Learning. Fragments of a Second Part," in idem, *Satires and Miscellaneous Poetry and Prose,* ed. René Lamar (Cambridge: Cambridge University Press, 1928), p. 80.

58. Burton, *Anatomy of Melancholy,* p. 261.

59. John Milton, "The Seventh Prolusion: A Speech in Defense of Learning Delivered in the [Christ's] College Chapel," trans. Thomas R. Hartmann, in *The Prose of John Milton,* ed. J. Max Patrick (London: University of London Press; New York: New York University Press, 1968; comp. 1632), pp. 15–28, quoting pp. 19–20; for criticism of schoolmen's pedantry, see pp. 23–24.

60. Sir Henry Wotton, *A Philosophical Survey of Education or Moral Architecture and Aphorisms of Education,* ed. H. S. Kermode (London: Hodder & Stoughton, for the University Press of Liverpool, 1938; orig. publ. 1651), p. 25.

61. Clement Ellis, *The Gentile Sinner, or England's Brave Gentleman . . . Both As he is, and as he should be,* 5th ed. (Oxford, 1672; orig. publ. 1660), p. 26.

62. Edward, Lord Herbert of Cherbury, *The Autobiography,* 2nd rev. ed., ed. Sidney Lee (London: George Routledge, 1906), pp. 26, 35.

63. John Milton, "Of Education, to Master Samuel Hartlib . . . ," in idem, *Complete Poetry & Selected Prose* (New York: Modern Library, 1950; orig. publ. 1644), pp. 663–676, on pp. 666–667. Note the contemporary resonance of "generous" as pertaining to nobility, the defining characteristic of the "gentle."

64. For Ramism as pragmatic humanism, see Anthony Grafton and Lisa Jardine, *From Humanism to the Humanities: Education and the Liberal Arts in Fifteenth- and Sixteenth-Century Europe* (Cambridge, MA: Harvard University Press, 1986), chapter 7; for the English Ramist reform program, see Kearney, *Scholars and Gentlemen,* chapter 3; for Paracelsian-sectarian attacks on the universities, see ibid., chapter 7, and Allen G. Debus, *Science and Education in the Seventeenth Century: The Webster–Ward Debate* (London: Macdonald, 1970); and, for identification of *alchemical* discourse as "pedantic," see, for example, William Gilbert, *De magnete,* trans. P. Fleury Mottelay (New York: Dover, 1958; orig. publ. 1600), p. l.

65. Ward, *Academiarum examen,* p. 33.

66. Quoted in R. W. Meyer, *Leibnitz and the Seventeenth-Century Revolution,* trans. J. P. Stern (Chicago: Henry Regnery, 1952), p. 96. On the urban siting of new chemical practice, see Owen Hannaway, "Laboratory Design and the Aim of Science: Andreas Libavius versus Tycho Brahe," *Isis* 77 (1986): 585–610; cf. Shapin, "The House of Experiment in Seventeenth-century England" (chapter 5 in this volume).

67. Thomas Hobbes, *Behemoth: The History of the Causes of the Civil Wars of England,* ed. William Molesworth (New York: Burt Franklin, 1963; orig. publ. 1668), pp. 51, 53, 75.

68. Thomas Hobbes, *Leviathan* (Harmondsworth: Penguin, 1968; orig. publ. 1651), pp. 115, 117.

69. A fuller version of this general argument, and the evidence in the following several paragraphs, is Shapin, "'The Mind Is Its Own Place'" (chapter 7 in this volume). For an appreciation of the humanist contribution to seventeenth-century scientific movement, see Barbara J. Shapiro, "Early Modern Intellectual Life: Humanism, Religion and Science in Seventeenth-Century England," *History of Science* 29 (1991): 45–71.

70. Francis Bacon, "The Advancement of Learning," in *The Philosophical Works of Francis Bacon,* ed. James Spedding, Robert Leslie Ellis, and Douglas Denon Heath, 5 vols. (London: Longman, 1857–1858), vol. 3, pp. 253–491, on pp. 268–271, 313. Bacon here rhetorically played with the etymologically pure meaning of *pedants* as

"teachers and preceptors." For a parallel citation of philosopher-princes, see Milton, "The Seventh Prolusion," p. 21, and, for the general seventeenth-century tendency to posit a Golden Age of gentleman-scholars, see Brauer, *Education of a Gentleman*, p. 59.

71. Thomas Elyot, *The Boke named The Governour* (London: J. M. Dent, 1907; orig. publ. 1531), pp. 51–52, 207; see also Skinner, *Foundations of Modern Political Thought*, vol. 1, p. 243.

72. Bacon, "Advancement of Learning," pp. 273, 277, 280, 314; see also idem, "Of Studies," p. 136.

73. Bacon, "Advancement of Learning," pp. 282–295. On the Baconian reforms of natural philosophy as civic humanism, see Julian Martin, "'Knowledge is Power': Francis Bacon, the State, and the Reform of Natural Philosophy" (Ph.D. thesis, Cambridge University, 1988), esp. chapters 4–5; also Moody E. Prior, "Bacon's Man of Science," in *Roots of Scientific Thought,* ed. Philip P. Wiener and Aaron Noland (New York: Basic Books, 1957), pp. 382–389.

74. The study of gentlemanly norms and conduct in the early Royal Society has developed rapidly in recent years. In the paragraphs below I offer new material, as well as drawing upon my own previously published work, and acknowledging excellent studies by Peter Dear, Simon Schaffer, J. V. Golinski, Robert Iliffe, and others.

75. Detailed analyses of the society's social composition indicate the predominance of the gentry and aristocracy, and, despite much utilitarian rhetoric, the paucity of merchants and tradesmen: Michael Hunter, *The Royal Society and Its Fellows 1660–1700: The Morphology of an Early Scientific Institution,* British Society for the History of Science Monographs, 4 (Chalfont St Giles: BSHS, 1982), p. 5 (quoting letter from Francis Vernon to Henry Oldenburg, 1 May 1669). Of course, my argument about the advertised suitability of the new practice for gentlemen does not necessarily depend upon any given statistical state of affairs concerning the society's social composition, though a striking exemplar of the aristocratic experimental philosopher provided important public legitimation.

76. Thomas Sprat, *The History of the Royal Society of London* (London, 1667), pp. 404–409, 417.

77. Sprat, *History,* pp. 331–332. For an account of attitudes toward traditional learning among the early Royal Society, see Michael Hunter, *Science and Society in Restoration England* (Cambridge: Cambridge University Press, 1981), chapter 6.

78. Quoted in Michael Hunter, *John Aubrey and the Realm of Learning* (New York: Science History Publications, 1975), p. 41.

79. Joseph Glanvill, *Scepsis scientifica: or, Confest Ignorance, the Way to Science* (London: Kegan Paul, Trench, 1885; orig. publ. 1665), pp. liii–liv, lix–lx.

80. Sprat, *History,* pp. 26–27; for melancholy see also pp. 16–17, 21, 336, 345, 417.

81. Shapin, "Pump and Circumstance" (chapter 6 in this volume); idem, "The

House of Experiment in Seventeenth-century England" (chapter 5 in this volume); idem, *A Social History of Truth,* esp. chapters 5–6.

82. Glanvill, *Scepsis scientifica,* pp. 201–202; for Scholasticism as litigious and as the philosophy of base and servile men, see also ibid., pp. 136–142, 196–198. Cf. idem, *Essays on Several Important Subjects in Philosophy and Religion* (London, 1676), pp. 31–32; idem, *Plus ultra: or, The Progress and Advancement of Knowledge, since the Days of Aristotle . . .* (London, 1668), pp. 148–149.

83. Glanvill, *Scepsis scientifica,* pp. lix–lx. Glanvill seems to have been unusually attuned to the ideal conjunction between knowledge, piety, and manners. In his disputes with the Aristotelian cleric Richard Crosse, Glanvill took his adversary to task for conduct "so little agreeing with the *Discretion* of a *Wiseman,* the *Charity* of a *Christian,* or the *Civility* of a *Gentleman*" (Glanvill, *Plus ultra,* "Preface," sig. [A7]). For the Glanvill–Crosse controversy, see James R. Jacob, *Henry Stubbe, Radical Protestantism and the Early Enlightenment* (Cambridge: Cambridge University Press, 1983), pp. 78–84.

84. On Boyle's social standing and predicament during the Civil War years, see J. R. Jacob, *Robert Boyle and the English Revolution: A Study in Social and Intellectual Change* (New York: Burt Franklin, 1977), chapter 2, though the stress there on Boyle's sharp break with the main lines of Renaissance ethical theory seems overstated. The major source for Boyle early moral and social views is now *The Early 'Essays' and 'Ethics' of Robert Boyle,* ed. John T. Harwood (Carbondale: Southern Illinois University Press, 1991); see also Shapin, *A Social History of Truth,* chapter 4 ("Who Was Robert Boyle?").

85. The claim that Christianity underwrote commitment to the common good was widespread in English Puritan circles, and was forcefully expressed by Bacon: "never in any age has there been any philosophy, sect, religion, law, or other discipline, which did so highly exalt the good which is communicative, and depress the good which is private and particular, as the Holy Christian Faith . . . It decides the question touching the preferment of the contemplative or active life" (Francis Bacon, "De augmentis scientarum," in *Philosophical Works,* vol. 5, pp. 7–8). For Christianity and civic responsibility among the English gentry, see Kelso, *English Gentleman,* p. 57; Brauer, *Education of a Gentleman,* pp. 15–16; Michael Walzer, *The Revolution of the Saints: A Study of the Origins of Radical Politics* (Cambridge, MA: Harvard University Press, 1965), chapter 7.

86. Here the *Englishness* of this pattern can be stressed, since in no other country was Christian piety so strongly associated with an idea of gentility; see, for example, Ustick, "Changing Ideals," pp. 154–163. Moreover, English commentators themselves widely insisted upon the moral integrity of the English gentleman compared to his French and Italian counterparts; see, for example, Einstein, *Italian Renaissance in England,* chapter 4; idem, *Tudor Ideals* (London: G. Bell, 1921), p. 69; Childs, "Prescriptions for Manners," pp. 89–93; Lingard, *Letter of Advice,* pp. 22–23.

87. Glanvill, *Plus ultra*, p. 93.

88. Gilbert Burnet, "Character of a Christian Philosopher, in a Sermon Preached January 7, 1691–2, at the Funeral of the Hon. Robert Boyle," in idem, *Lives, Characters, and an Address to Posterity*, ed. John Jebb (London: James Duncan, 1833), pp. 325–376, quoting pp. 333, 348, 352, 360–361, 367.

89. [William Ramesey], *The Gentlemans Companion: or, A Character of True Nobility, and Gentility . . . by a Person of Quality . . .* (London, 1676; comp. 1669), p. 15. For arguments against "dogmatism" and "contentious disputes" in general, see pp. 71–74.

90. Gilbert Burnet, *Thoughts on Education*, ed. John Clarke (Aberdeen: Aberdeen University Press, 1914; comp. ca. 1668), pp. 61, 70.

91. William Wotton, *Reflections upon Ancient and Modern Learning* (London, 1694), pp. 353–355.

92. Ramesey, *The Gentlemans Companion*, pp. 131–132. Also on the list were works of Copernicus, Galileo, Descartes, Henry More, William Harvey, Walter Charleton, Thomas Willis, Kenelm Digby, Thomas Browne, Joseph Glanvill, and Robert Hooke—"his *Micrography; and the rest of our New Experiments."*

93. Daniel Defoe, *The Compleat English Gentleman*, ed. Karl D. Bülbring (London: David Nott, 1890; orig. comp. ca. 1728–1729), pp. 228, 191–192, 65, 69. Defoe also exempted some few other English gentlemen from the general indictment: England has "some of the greatest men that ever the world produced, exquisite in science, compleat in the politest learning, bright in witts . . . ; no science, no commendable study, no experimentall knowledge, no humane attainment, but they excell in" (ibid., pp. 89–90).

94. Anon. editor's note in *The Original Works of William King, LL.D.*, 3 vols. (London, 1776), vol. 2, p. 98.

95. Zouch, ed., *Walton's Lives*, vol. 2, p. 265.

96. See, for example, Roy Porter, *English Society in the Eighteenth Century* (Harmondsworth: Penguin, 1982), esp. chapter 6. In this connection it is relevant that the English universities in the period from the Restoration to the mid-eighteenth century were experiencing a decline in their student numbers and fortunes. As Porter notes, "It was grinding tutors and slumbrous pedants who made an academic career in the colleges: the great scholars of Georgian England . . . were not dons" (ibid., p. 179). On the conservatism of the universities and on their continuing significance as clerical seminaries ca. 1700, see Kearney, *Scholars and Gentlemen*, chapter 10; Holmes, *Augustan England*, pp. 34–42; and John Gascoigne, *Cambridge in the Age of the Enlightenment: Science, Religion and Politics from the Restoration to the French Revolution* (Cambridge: Cambridge University Press, 1989), esp. pp. 17–19.

97. Defoe, *English Gentleman*, p. 203; see also Brauer, *Education of a Gentleman*, p. 67.

98. Lingard, *Advice to a Young Gentleman*, p. 9; De Courtin, *The Rules of Civil-*

ity, pp. 28–29; Adam Petrie, *Rules of Good Deportment, or of Good Breeding* (Edinburgh, 1720), pp. 46, 58. For surveys of attitudes toward learning and pedantry in the late seventeenth-century courtesy literature, see Childs, "Prescriptions for Manners," pp. 78, 215–218; Brauer, *Education of a Gentleman,* chapter 3.

99. Samuel Butler, *Characters and Passages from Note-Books,* ed. A. R. Waller (Cambridge: Cambridge University Press, 1908), pp. 136–137.

100. Samuel Butler, "Satyr upon the Imperfection and Abuse of Human Learning Part 1st," in idem, *Satires,* p. 68.

101. Obadiah Walker, *Of Education especially of Young Gentlemen* (Oxford, 1673), pp. 112, 249–251; see also Brauer, *Education of a Gentleman,* pp. 66–68.

102. William Temple, "An Essay upon the Ancient and Modern Learning," in idem, *The Works of Sir William Temple, Bart.,* 4 vols. (London: Rivington, 1814; orig. publ. 1690), vol. 3, pp. 444–486, on pp. 484–486. And for late seventeenth-century criticism of humanistic arrogance and humanists' contribution to pedantry, see Jean LeClerc, *Parrhasiana, or, Thoughts upon Several Subjects,* trans. Anon. (London, 1700), pp. 181–183 (quoted in B. C. Southgate, "'No Other Wisdom'? Humanist Reactions to Science and Scientism in the Seventeenth Century," *The Seventeenth Century* 5 [1990]: 71–92, on p. 87).

103. Anthony Ashley Cooper, Third Earl of Shaftesbury, *Characteristics of Men, Manners, Opinions, Times,* ed. John M. Robertson, 2 vols. in 1 (Indianapolis: Bobbs-Merrill, 1964; orig. publ. 1711), vol. 1, pp. 53, 81, 88, 196, 214.

104. *The Spectator,* no. 105, 30 June 1711, in *The Spectator,* ed. Donald F. Bond, 5 vols. (Oxford: Oxford University Press, 1965), vol. 1, pp. 436–438.

105. Joseph Addison and Richard Steele, *The Tatler,* ed. George A. Aitken, 4 vols. (London: Duckworth, 1898), vol. 1, p. 148 (17–19 May 1709); p. 176 (26–28 May 1709); vol. 4, pp. 245–247 (28–31 October 1710); see also (on pedants) vol. 3, p. 235 (11–13 April 1710); vol. 4, p. 49 (25–27 July 1710).

106. Samuel Johnson, *The Rambler,* ed. W. J. Bate and Albrecht B. Strauss, vols. 3–5 of 15 vols. *Works* (New Haven, CT: Yale University Press, 1958–1985), vol. 5, pp. 36–37 (10 April 1750); pp. 151–153 (12 November 1751); see also (on scholarly seclusion and incivility) pp. 70–75 (17 September 1751) and pp. 176–181 (3 December 1751).

107. Philip Dormer Stanhope, Earl of Chesterfield, *Letters to His Son and Others* (London: J. M. Dent, Everyman's Library; comp. ca. 1746–1753), pp. 16, 71, 136, 145, 184, 191, 263, 270–271.

108. John Locke, *Some Thoughts Concerning Education* (Cambridge: Cambridge University Press, 1899; orig. publ. 1690), pp. 74, 129, 153; see also Brauer, *Education of a Gentleman,* p. 59.

109. Brauer, *Education of a Gentleman,* pp. 166, 169–170; see also Lawrence Stone, "Practical Wisdom for Élite Boys," *Times Literary Supplement* (2–8 March 1990), pp. 229–230.

110. For example, R. H. Syfret, "Some Early Reactions to the Royal Society," *Notes and Records of the Royal Society* 7 (1950): 207–258; idem, "Some Early Critics of the Royal Society," *Notes and Records of the Royal Society* 8 (1950): 20–64; Marjorie Hope Nicolson, *Science and Imagination* (Ithaca, NY: Cornell University Press, 1956); and the synoptic treatment in Hunter, *Science and Society,* esp. chapter 6.

111. Wotton, *Ancient and Modern Learning,* "Conclusion," as quoted in Syfret, "Early Critics," p. 44; for sensitivity to the power of the "wits" ridicule, see also Sprat, *History,* pp. 417–419. And, from the point of view of an "ancient" scholar, see Temple, "Ancient and Modern Learning," pp. 485–486.

112. Samuel Butler, "Pædants," in idem, *Satires,* p. 166.

113. Samuel Butler, "The Elephant in the Moon," in idem, *Satires,* pp. 3–30; see also Nicolson, *Science and Imagination,* pp. 170–171.

114. Butler, *Satires,* p. 33. Of course, according to Pepys, Charles II also wanted to know what his philosophers had done apart from weighing the air, and Thomas Shadwell repeated what must by then have been a very stale joke in *The Virtuoso* of 1676. For the king's view of the Royal Society, see, for example, W. E. Knowles Middleton, "What did Charles II Call the Fellows of the Royal Society?" *Notes and Records of the Royal Society* 32 (1977): 13–17, and Hunter, *Science and Society,* pp. 130–131; for Shadwell, see *The Virtuoso,* ed. Marjorie Hope Nicolson and David Stuart Rodes (Lincoln: University of Nebraska Press, 1966), p. 110 (V.ii); Syfret, "Early Critics," pp. 54–55, 58–59.

115. Shadwell, *The Virtuoso,* pp. 22 (I.i), 68 (III.iii). For the background to this play, see Nicolson and Rodes, "Introduction," ibid., pp. xi–xxvi. Historians have claimed that Gimcrack was modeled on Robert Hooke. Most plausibly, he is a pastiche of a number of well-known Royal Society figures. There are certainly large dollops of Boyle in the character: his title (Sir Nicholas) points in that direction, as does his Puritanical piety. There are also quite specific references to Boyle's published work.

116. William King, "Useful Transactions in Philosophy, and Other Sorts of Learning . . . ," in *Original Works,* vol. 2, pp. 57–178 (orig. publ. 1709), on pp. 98–99 (for Boyle's phosphorus); 103–114, 121–125 (for Leeuwenhoek); 135 (for dregs); idem, "A Journey to London, In the Year 1698. After the Ingenious Method of that made by Dr. Martin Lister to Paris . . . ," in ibid. (orig. publ. 1699), vol. 1, p. 198 (for cats).

117. John Wilkins, as quoted in Syfret, "Early Critics," p. 55; see also Nicolson, *Science and Imagination,* chapter 6; John T. Harwood, "Rhetoric and Graphics in *Micrographia,*" in *Robert Hooke: New Studies,* ed. Michael Hunter and Simon Schaffer (Woodbridge: Boydell, 1989), pp. 119–147, esp. pp. 139, 142; and Shapin, "Who Was Robert Hooke?" (chapter 9 in this volume).

118. *The Tatler*, 10–12 January 1709–1710, 24–26 August 1710, vol. 3, p. 31; vol. 4, p. 110.

119. *The Tatler*, 10–12 October 1710, vol. 14, pp. 209–210. For satire of "Sir Nicholas Gimcrack," see ibid., pp. 112–113, 133–138, and for additional criticism of Royal Society science as trivial and useless, see ibid., pp. 171–172, 320–326.

120. [Mary Astell or Judith Drake?], *An Essay in Defense of the Female Sex in which are inserted the Character of a Pedant, A Squire, A Beau, A Virtuoso* (London, 1696), quoted in Levine, *Dr. Woodward's Shield*, p. 125. Here, as elsewhere, critics often had specific individuals in view, whose actual behavior at times was more than equal to any caricature. William Woodward was very widely pointed to as the pattern of a vain and pompous virtuoso; William King reserved special odium for Martin Lister and Hans Sloane; and Butler, Swift, and others had merry sport with Boyle's pious philosophy. For Woodward, see ibid., chapters 5–7; for female *virtuosa*, see Nicolson, *Science and Imagination*, pp. 186–193. Scientific culture in specifically *female* society has, arguably, been treated more fully in recent years than in a specifically masculine context; see, among many examples, Patricia Phillips, *The Scientific Lady: A Social History of Woman's Scientific Interests 1520–1918* (New York: St. Martin's Press, 1990); Ann B. Shteir, "Linneaus's Daughters: Women and British Botany," in *Women and the Structure of Society: Selected Research from the Fifth Berkshire Conference on the History of Women*, ed. Barbara J. Harris and Jo Ann K. McNamara (Durham, NC: Duke University Press, 1984), pp. 67–73.

121. Shaftesbury, *Characteristics*, vol. 2, p. 253.

122. Levine, *Dr. Woodward's Shield*, p. 125 (emphases added).

123. Shaftesbury, *Characteristics*, vol. 1, pp. 193–196; cf. Samuel Butler, *Characters*, p. 276: "The end of all Knowledge, is to understand what is Fit to be don; For to know what has been, and what is, and what may be, dos but tend to that."

124. Syfret, "Early Critics," pp. 56–58; Houghton, "English Virtuoso."

125. Shadwell, *The Virtuoso*, p. 47 (II.ii).

126. Shadwell, *The Virtuoso*, pp. 114, 119 (V.ii, V.iii).

127. Temple, "Some Thoughts upon Reviewing the Essay of Ancient and Modern Learning," in idem, *Works*, vol. 3 (essay comp. ca. 1695), pp. 487–518, on pp. 516–517. For an appreciation of the seventeenth-century literary and moral case against scientistic ambition, see Southgate, "'No Other Wisdom'?"

128. Jonathan Swift, *Gulliver's Travels*, ed. Miriam Kosh Starkman (New York: Bantam, 1962; comp. 1721–1725; orig. publ. 1735), p. 175; see also Nicolson, *Science and Imagination*, pp. 110–154, 193–199.

129. *The Tatler*, 5–7 September 1710, vol. 4, p. 134.

130. William King, "The Transactioneer," in idem, *Original Works*, vol. 7, pp. 1–56 (orig. publ. 1700), esp. pp. 16, 32, 49; idem, "Useful Transactions," passim; Temple, "Some Thoughts upon Reviewing Ancient and Modern Learning," pp. 516–517.

131. Samuel Butler, "A Virtuoso," in *Characters,* p. 81; cf. Levine, *Dr. Woodward's Shield,* p. 124; Syfret, "Early Critics," p. 60.

132. *The Tatler,* 7–9 December 1710, vol. 4, p. 326.

133. Temple, "Some Thoughts upon Reviewing Ancient and Modern Learning," p. 517.

134. King, "The Transactioneer," p. 17.

135. Shadwell, *The Virtuoso,* p. 111 (V.ii). Cf. Robert Boyle, "Some Observations about Shining Flesh, Both of Veal and of Pullet . . . ," in idem, *Works,* ed. Thomas Birch, 6 vols. (London: Rivington, 1772; orig. publ. 1672), vol. 3, pp. 651–655.

136. Shadwell, *The Virtuoso,* p. 19 (I.i).

137. Samuel Butler, "An Occasional Reflection on Dr. Charleton's Feeling a Dog's Pulse at Gresham-College. By R. B. Esq.," in idem, *Satires,* pp. 341–343; Jonathan Swift, "A Meditation upon a Broom-Stick: According to the Style and Manner of the Honourable Robert Boyle's *Meditations.* Written in the Year 1703," in idem, *The Prose Works,* ed. Herbert Davis, 14 vols. (Oxford: Basil Blackwell, 1939–1969), vol. 1, pp. 239–240; letter from Duke of Buckingham to Earl of Rochester, July 1677, in *The Letters of John Wilmot, Earl of Rochester,* ed. Jeremy Treglown (Chicago: University of Chicago Press, 1980), pp. 146–147. See also Shadwell, *The Virtuoso,* p. 18 (I.i) for Boyle's *meletetiques* (or meditational techniques).

138. Shaftesbury, *Characteristics,* vol. 1, pp. 50–52; cf. John Redwood, *Reason, Ridicule and Religion: The Age of Enlightenment in England 1660–1750* (Cambridge, MA: Harvard University Press, 1976), esp. pp. 39, 84–86.

139. Samuel Butler, "A Philosopher," in idem, *Characters,* pp. 57–58. The reference is to Walter Charleton's text of 1654.

140. Swift, *Gulliver's Travels,* pp. 161–162.

141. For structural conditions bearing on the expectation of "militant combativeness" in the world of learning, see, for example, A. Rupert Hall, *Philosophers at War: The Quarrel between Newton and Leibniz* (Cambridge: Cambridge University Press, 1980), esp. pp. 2–7.

142. Levine, *Dr. Woodward's Shield,* esp. pp. 81–91.

143. Both quoted in Levine, *Dr. Woodward's Shield,* pp. 123–124.

144. Swift, *Gulliver's Travels,* p. 161.

145. [William Darrell], *The Gentleman Instructed, in the Conduct of a Virtuous and Happy Life . . . Written for the Instruction of a Young Nobleman . . . ,* 8th ed. (London: E. Smith, 1723; orig. publ. 1704–1712), p. 15. This general form of injunction was evidently quite standard in the contemporary courtesy literature.

146. Samuel Johnson, *The Rambler,* 17 September 1751, vol. 5, pp. 71–75.

147. David Elliston Allen, *The Naturalist in Britain: A Social History* (London: Allen Lane, 1976), esp. chapters 2–3; Roy S. Porter, "Science, Provincial Culture and Public Opinion in Enlightenment England," *British Journal for Eighteenth-Century Studies* 3 (1980): 20–46; David Philip Miller, "The Royal Society of London 1800–

1835: A Study of the Cultural Politics of Scientific Organization" (Ph.D. thesis, University of Pennsylvania, 1981), pp. 42–46; Levine, *Humanism and History,* chapters 3–4; Stan A. E. Mendyk, *'Speculum Britanniae': Regional Study, Antiquarianism, and Science in Britain to 1700* (Toronto: University of Toronto Press, 1989), esp. pp. 239–244.

148. Examples are Robert E. Schofield, *The Lunar Society of Birmingham: A Social History of Provincial Science and Industry in Eighteenth-Century England* (Oxford: Clarendon Press, 1963); Arnold Thackray, "Natural Knowledge in Cultural Context: The Manchester Model," *American Historical Review* 79 (1974): 672–709; *Metropolis and Province: Science in British Culture, 1780–1850,* ed. Ian Inkster and Jack Morrell (London: Hutchinson, 1983).

149. By the late nineteenth century, scientific naturalists like John Tyndall were once more urging support of science "Not as a servant of Mammon" but "as the strengthener and enlightener of man."

150. For historically changing social locations of moral authority and technical knowledge, see T. W. Heyck, *The Transformation of Intellectual Life in Victorian England* (London: Croom Helm, 1982), chapter 3.

151. Simon Schaffer, "Genius in Romantic Natural Philosophy," in *Romanticism and the Sciences,* ed. Andrew Cunningham and Nicholas Jardine (Cambridge: Cambridge University Press, 1990), pp. 82–98.

152. Richard Yeo, "Genius, Method and Morality: Images of Newton in Britain, 1760–1860," *Science in Context* 2 (1988): 257–284; Shapin, "'The Mind Is Its Own Place'" (chapter 7 in this volume).

153. Newman, *Discourses,* pp. 180–181; Thomas Carlyle, "Signs of the Times [1829]," in idem, *Selected Writings,* ed. Alan Shelston (Harmondsworth: Penguin, 1971), pp. 61–85, esp. pp. 64–71.

154. James Moore persuasively argues that Charles Darwin in the years after his removal to Down in 1842 was experimenting with his entitlement to the recognized role as "squarson-naturalist": J. R. Moore, "Darwin of Down: The Evolutionist as Squarson-Naturalist," in *The Darwinian Heritage,* ed. David Kohn (Princeton, NJ: Princeton University Press, 1985), pp, 435–481, esp. pp. 440–443.

155. Few modern commentators on these matters have been as maliciously perceptive as Simon Raven, for example, *The Decline of the Gentleman* (New York: Simon and Schuster, 1962), pp. 10–11, 180–181.

Chapter 9. Who Was Robert Hooke?

1. John Aubrey, *Brief Lives,* ed. Andrew Clark, 2 vols. (Oxford: Clarendon Press, 1898), vol. 1, pp. 410–411; Anthony à Wood, *Fasti Oxoniensis II,* ed. Philip Bliss (London: Rivington, 1820), pp. 628–631.

2. A good source for these quotidian patterns is Margaret 'Espinasse, *Robert Hooke* (London: Heinemann, 1956), esp. chapter 6. No one working in this area can

fail to owe a large debt to the writings of 'Espinasse, whether or not they share her impulse to repair Hooke's reputation. The significance of ill-health for Hooke's daily activities cannot be overestimated: see Lucinda McCray Beier, *Sufferers and Healers: The Experience of Illness in Seventeenth-Century England* (London: Routledge & Kegan Paul, 1987), esp. pp. 151, 65–70. Since this piece was written, further biographies of Hooke have appeared: Stephen Inwood, *The Man Who Knew Too Much: The Strange and Inventive Life of Robert Hooke, 1635–1703* (London: Macmillan, 2002); Lisa Jardine, *The Curious Life of Robert Hooke: The Man Who Measured London* (London: HarperCollins, 2003); Allan Chapman, *England's Leonardo: Robert Hooke and the Seventeenth-Century Scientific Revolution* (Philadelphia: Institute of Physics, 2005); see also the contributions in *Robert Hooke: New Studies*, ed. Michael Hunter and Simon Schaffer (Woodbridge: The Boydell Press, 1989), and Stephen Pumfrey, "Ideas Above His Station: A Social Study of Hooke's Curatorship of Experiments," *History of Science* 29 (1991): 1–44.

3. Of course, toward the end of his life, and particularly after the mid-1690s, Hooke's deteriorating physical condition dictated a very considerable slowing down of the pace of his activities. And, apart from the effects of age and ill-health, it also appears that he became increasingly reclusive and temperamental, especially after the death of his niece Grace Hooke in 1687: Richard Waller, "The Life of Dr. Robert Hooke," in Hooke, *The Posthumous Works of Robert Hooke*, ed. Waller (London, 1705), pp. i–xxviii, esp. pp. xxiv–xxvii: Hooke became over time "to a Crime close and reserv'd."

4. On Hooke's mechanical and architectural activities: J. A. Bennett, "Robert Hooke as Mechanic and Natural Philosopher," *Notes and Records of the Royal Society* 35 (1980–1981): 33–48, and contributions to Hunter and Schaffer, eds., *Robert Hooke: New Studies.*

5. From 1678, Hooke may have had access to the laboratory of John Mapletoft, the Gresham professor of physic. Hooke was also landlord of living-spaces occupied by others in the stables of Gresham College. In 1688, the Royal Society rented rooms from him, though it is not clear for what purpose: C. R. Weld, *A History of the Royal Society,* 2 vols. (London: J. W. Parker, 1848), vol. 1, pp. 318–319.

6. The lodgings of the geometry professor opened behind the "reading hall" of Gresham College, where Hooke lectured and where the precursor group to the Royal Society met; see John Ward, *The Lives of the Professors of Gresham College* (London: J. Moore, 1740), p. 91.

7. On the "opratory" and "turret," see Hooke to Robert Boyle, 3 February 1666, in Robert Boyle, *Works,* ed. Thomas Birch, 6 vols. (London: Rivington, 1772), vol. 6, p. 505; Robert Hooke, *The Diary of Robert Hooke, 1672–1680,* ed. Henry W. Robinson and Walter Adams (4 November 1675), p. 191; Ward, *Lives of the Professors of Gresham College,* pp. 91, 178; Ian Adamson, "The Royal Society and Gresham

College, 1661–1711," *Notes and Records of the Royal Society* 33 (1978): 1–21, on p. 4.

8. Thomas Birch, *The History of the Royal Society of London,* 4 vols. (London: A. Millar, 1756–1757), vol. 2, p. 189 (25 July 1667); vol. 3, p. 100 (6 November 1673). The circumstances in which the Royal Society originally paid Hooke to reside at Gresham College (before he became geometry professor) are discussed in Simon Schaffer, "Wallification: Thomas Hobbes on School Divinity and Experimental Pneumatics," *Studies in History and Philosophy of Science* 19 (1988): 275–298. Hooke was commanded by the society in the autumn of 1663 to live at Gresham for at least four days a week specifically to get the new pump for the compressing of air ready for the king's projected entertainment at the Royal Society (p. 294).

9. For places of residence as workplaces, see Peter Laslett, *The World We Have Lost* (London: Methuen, 1965), pp. 1–10; also Shapin, "The House of Experiment in Seventeenth-century England" (chapter 5 in this volume), with which this chapter partly overlaps.

10. Adamson, "The Royal Society and Gresham College," pp. 5–6.

11. For the role of technicians in this setting, see Steven Shapin, *A Social History of Truth: Civility and Science in Seventeenth-Century England* (Chicago: University of Chicago Press, 1994), chapter 8.

12. Aubrey, *Brief Lives,* vol. 1, pp. 411, 415; cf. vol. 1, p. 43, where Hooke was listed as one of Aubrey's "amici." The suggestion that Hooke and Wren may have been related by marriage is in 'Espinasse, *Robert Hooke,* p. 114.

13. On meetings in Hooke's lodgings, see, for example, Hooke to Robert Boyle, 5 September 1667, in Boyle, *Works,* vol. 6, pp. 508–509 (where Hooke mentioned the presence of "about half a score of the Society . . . at my chamber this afternoon, where we had some discourse of philosophical matters"). For the society's presence in Hooke's quarters, see, for example, Robert Hooke, *Diary, 1672–1680,* pp. 108 (18 June 1674) ("Councell in my Dining Room.") and 129 (9 November 1674) ("Councell at my chamber."). The council met occasionally at Hooke's lodgings through the 1680s: Birch, *History,* vol. 4, pp. 226, 228 (21 and 24 November 1683). For possible allusions to the society being entertained in Hooke's rooms, see Hooke, *Diary, 1672–1680,* p. 141 (14 January 1674–1675): "Society Dind here," though whether "here" meant his rooms or elsewhere in Gresham or in nearby eating-houses is unclear in most cases. See also pp. 132 (23 November 1674) and 149 (25 February 1674–1675), where "Society Dind Here, I not" may support the case that "here" did not necessarily mean Hooke was playing host.

14. Hooke, *Diary, 1672–80,* p. 148 (18 February 1674–1675): "Mr. Newton, Cambridge, here."

15. Waller, "Life of Hooke," p. xxvii. The sense of Waller's description of Hooke as "Cynick" is partly lost to the modern ear: it may have resonated with its Greek

derivation (*kynikos,* doglike, surly, snarling, disinclined to recognize or believe in goodness or selflessness—following *Chambers Dictionary*), or it may have referred to the Athenian sect of philosophers, the Cynics or followers of the dog, who "deliberately flouted convention, 'doing in public what is generally considered should be done in private'": John Silverlight, "Words," *The Observer,* 26 July 1987.

16. Waller, "Life of Hooke," p. xxvii.

17. On Newton's solitude, see J. V. Golinski, "The Secret Life of an Alchemist," in *Let Newton Be!* ed. John Fauvel et al. (Oxford: Oxford University Press, 1988), pp. 147–168, and R. C. Iliffe, "'The Idols of the Temple': Isaac Newton and the Private Life of Anti-Idolatry" (Ph.D. thesis, Cambridge University, 1989), esp. chapter 5.

18. See, in these connections, Shapin, "'The Mind Is Its Own Place'" and idem, "The Philosopher and the Chicken" (chapters 7 and 11 in this volume).

19. R. E. W. Maddison, "Studies in the Life of Robert Boyle, F.R.S. Part I. Robert Boyle and Some of His Foreign Visitors," *Notes and Records of the Royal Society* 9 (1951): 1–35, esp. p. 3; idem, "Studies in the Life of Robert Boyle, F.R.S. Part IV. Robert Boyle and Some of His Foreign Visitors," *Notes and Records of the Royal Society* 11 (1954): 38–53, esp. p. 38. For the significance of public and private life in Boyle's work, see Shapin, "The House of Experiment in Seventeenth-century England" (chapter 5 in this volume, pp. 69–70).

20. Gilbert Burnet, *Select Sermons . . . and a Sermon at the Funeral of the Honourable Robert Boyle* (Edinburgh: Foulis, 1742), p. 201.

21. Others who entertained Hooke frequently at their homes include Viscount Brouncker, Jonas Moore, Christopher Wren, and Seth Ward.

22. Henry Peacham, *The Complete Gentleman,* ed. Virgil B. Heltzel, from 1622, 1634, and 1661 editions (Ithaca, NY: Cornell University Press, 1962), p. 24.

23. On the coffeehouse and science, see Michael Hunter, *Science and Society in Restoration England* (Cambridge: Cambridge University Press, 1981), pp. 33–34, 76–77; Aytoun Ellis, *The Penny Universities: A History of the Coffee-Houses* (London: Secker & Warburg, 1956), pp. 37–52, 73–88, 255–263; Steven Shapin and Simon Schaffer, *Leviathan and the Air-Pump: Hobbes, Boyle, and the Experimental Life* (Princeton, NJ: Princeton University Press, 1985), pp. 292–293. For experimental trials and displays at coffeehouses and taverns, see, for example, Hooke, *Diary, 1672–1680,* pp. 276–277 (3 March 1676–1677), 279 (15 March 1676–1677), 431 (15 November 1679).

24. Brian Cowan, *The Social Life of Coffee: The Emergence of the British Coffee House* (New Haven, CT: Yale University Press, 2005).

25. Ellis, *The Penny Universities,* pp. 43–44, 73.

26. Hooke, *Diary, 1672–1680,* p. 289 (11 May 1677): "Saw Mr. Boyle at Garways." Given Hooke's economical and erratic way with punctuation, even this reference could mean that he saw Boyle and *then* went to the coffeehouse. A further

possible reference—"Read Newtons letters to Boyle at Garways" (p. 434 [24 December 1679])—probably does not bear the reading that Boyle was in attendance.

27. The distinction between trials and shows, and their distribution in private and public space, are discussed in Shapin, "The House of Experiment in Seventeenth-century England" (chapter 5 in this volume).

28. Hooke, *Diary, 1672–1680,* pp. 191 (3 November 1675), 343 (5 February 1677–1678), 364 (20 June 1678). In Hooke's published texts deference to Boyle was, of course, magnified according to recognizably standard formulas governing client-patron relations. In Hooke's *Micrographia,* Boyle was "the most illustrious Mr. Boyle," "the truly honourable Mr. Boyle," "the Incomparable Mr. Boyle," and "the most illustrious and incomparable Mr. Boyle" (Hooke, *Micrographia* [London, 1665], "Preface," sig. dv, and pp. 54–55, 69, 227).

29. There is some evidence in the *Diary* of a falling-off of honorifics over time. This may be a function of the rather more stenographic style of later *Diary* entries, or it may, indeed, testify to Hooke's alleged growing cynicism.

30. Hooke, *Diary, 1672–1680,* pp. 69 (14 November 1673), 75 (16 December 1673), 215 (30 January 1675–1676), 450 (2 August 1680). See also Aubrey, *Brief Lives,* vol. 2, p. 312, for Hooke telling Aubrey immediately after the event that Wren had been knighted.

31. Hooke, *Diary, 1672–1680,* pp. 81 (20 January 1673–1674) and 364 (20 June 1678).

32. Geoffrey Keynes, *A Bibliography of Dr. Robert Hooke* (Oxford: Oxford University Press, 1966), p. ix; also Pumfrey, "Ideas Above His Station."

33. Hooke, *Diary, 1672–1680,* pp. 333 (13 December 1677), 340 (16 January 1677–1678).

34. Aubrey, *Brief Lives,* vol. 1, pp. 410–411.

35. Robert Hooke, *An Attempt for the Explication of the Phaenomena, observable in an Experiment published by the Honourable Robert Boyle, Esq.* . . . (London, 1661), "The Epistle Dedicatory," sig. A2-A3. And see the panegyric to John Wilkins in *Micrographia,* "Preface," sig. dv, for Boyle as Hooke's "particular patron."

36. Hooke, *Micrographia,* "Preface," pp. d1v. Michael Hunter cites an interesting Hooke manuscript of 1683 which, in connection with the Cutlerian lecturer, makes very clear the distinction Hooke recognized between his command of the "speculative & rational part" and "tradesmen's" knowledge of the "operative part" of mechanical knowledge: Bowood House William Petty MSS H[8]15, cited in Michael Hunter, "Science, Technology and Patronage: Robert Hooke and the Cutlerian Lectureship," in idem, *Establishing the New Science: The Experience of the Early Royal Society* (Woodbridge: The Boydell Press, 1989), p. 313.

37. Between 1662 and 1664, Hooke continued to refer to himself as Boyle's creature. In 1663, he wrote to Boyle about his encounter with Hobbes at Richard Reeve's instrument shop, speculating whether Hobbes realized "to whom I belonged,"

and on 5 June of the same year he begged Boyle to dispense with his service "from attending on you for two or three days . . . having wholly resigned myself to your disposal": Boyle, *Works,* vol. 6, pp. 486–487, 482.

38. See, for example, Hooke, *Micrographia,* "Preface," sig. dv: "the wonderful progress made by the *Noble Engine* of the *most illustrious Mr. Boyle,* whom it becomes me to mention with all honour, not only as my particular Patron, but as the *Patron* of *Philosophy* it self; which he every day *increases* by his *Labours,* and *adorns* by his *Example.*" And see also ibid., p. 227, for Hooke's attribution to Boyle of the law relating pressures and volumes of air. Cf. Waller, "Life of Hooke," p. iii, where Hooke's manuscript autobiography claims responsibility for *making* the air-pump while identifying it as *Boyle's:* "in 1658 or 9, I contriv'd and perfected the Air-pump for Mr *Boyle,*" though the device Hooke constructed remained "Mr. *Boyle's Pneumatick Engine.*"

39. For Hooke's work on Boyle's laboratory, see, for example, Hooke, *Diary, 1672–1680,* pp. 247 (28 August 1676), 257 (18 November 1676), 260 (1 December 1676), 279 (17 March 1676–1677), 307 (18 August 1677), 308 (24 August 1677), 460 (28 December 1680). And see ibid., p. 362 (13 June 1678): "to Sir Ch. Wrens, then Mr. Boyles who had sent for me."

40. Hooke, *Diary, 1672–1680,* p. 265 (31 December 1676).

41. Bennett, "Robert Hooke as Mechanic," esp. p. 34.

42. Sir Robert Moray to Henry Oldenburg, 12 and 16 November 1665, in *The Correspondence of Henry Oldenburg,* ed. A. R. and M. B. Hall, 13 vols. (Madison: University of Wisconsin Press, 1965–1986), vol. 2, pp. 605–607, 608–611 (cf. Moray to Oldenburg, 10 October 1665, ibid., p. 560). The context of these remarks was the dispute over comets between Auzout and Hevelius, for which see Shapin, *A Social History of Truth,* chapter 6. An informal group of Fellows at Oxford urgently required Hooke's observations and interpretations in order to help resolve the controversy between the two foreign astronomers.

43. It is possible, but not certain, that this shift in usage corresponds to Hooke's change in status from non-remunerated to remunerated curator. Hooke's original appointment in November 1662 was without "recompense"; formal arrangements for paying him were not made until the end of July 1664. (Birch, *History,* vol. 1, pp. 123–124, 453).

44. Hooke, *Diary, 1672–1680,* p. 197 (3 December 1675).

45. For tensions between the often conflicting aims of Hooke, the Royal Society, and Sir John Cutler, see Hunter, "Science, Technology, and Patronage."

46. Birch, *History,* vol. 3, p. 364 (13 December 1677).

47. The ways in which Oldenburg dealt with his dependent status vis-à-vis Boyle and the Royal Society are briefly treated in Steven Shapin, "O Henry [essay-review of Oldenburg, Correspondence]," *Isis* 78 (1987): 417–424. Unlike Hooke, Olden-

burg adapted to his dependency by assuming a cloak of invisibility. Cf. Michael Hunter, "Promoting the New Science: Henry Oldenburg and the Early Royal Society," *History of Science* 26 (1988): 165–181, esp. pp. 170–171. While most modern historians unreservedly take Oldenburg's side, Hooke was not without contemporary supporters. Hooke's friend Christopher Wren shared his opinion of Oldenburg's duplicity; see Stephen Wren, publ., *Parentalia: or, Memoirs of the Family of the Wrens* (London: T. Osborn, 1750), pp. 199, 247.

48. Hooke, *Diary, 1672–1680*, pp. 188 (15 October 1675), 193 (11 November 1675), 253 (8–14 October 1676).

49. For Hooke's dealings with Newton, see A. R. and M. B. Hall, "Why Blame Oldenburg?" *Isis* 53 (1962): 482–491; A. R. Hall and R. S. Westfall, "Did Hooke Concede to Newton?" *Isis* 58 (1967): 403–405.

50. It is not clear that Mayow was engaged on a remunerated basis, and it would be wrong to assume that Boyle did not recognize the partial integrity of such remunerated assistants as Papin and Hooke. Yet no other assistants were ever named by Boyle. For some information on Boyle's technicians, see R. E. W. Maddison, "Studies in the Life of Robert Boyle, F.R.S. Part V. Boyle's Operator: Ambrose Godfrey Hanckwitz, F.R.S.," *Notes and Records of the Royal Society* 11 (1955): 159–188, see p. 159 for a partial list.

51. See Shapin, *A Social History of Truth*, chapter 8.

52. Aubrey, referring to Boyle's laboratory at Lady Ranelagh's house, said that Boyle had "severall servants (prentices to him) to looke to it," but there is no evidence whatever that Boyle offered training to his technicians: Aubrey, *Brief Lives*, vol. 1, p. 121. For guild patterns of master-apprenticeship relationships in the London instrument-making trade, see M. A. Crawforth, "Instrument Makers in the London Guilds," *Annals of Science* 44 (1987): 319–377. It was expected that masters make themselves responsible for their apprentices' moral conduct. Hooke was not a guild member, though some of his behavior is consistent with a strong guild influence. Certainly, Hooke's father considered a formal apprenticeship for him. Waller ("Life of Hooke," p. li) said that Hooke's father "thought to put him Apprentice to some easy Trade (as a Watchmakers or Limners)."

53. Hooke to Aubrey, 24 August 1675, in R. T. Gunther, "Life and Work of Robert Hooke" (*Early Science in Oxford*, 15 vols. [Oxford: privately printed, 1923–1967]), vol. 7, pp. 434–435). For Gresham College lecture audiences, see, for example, Hooke, *Diary, 1672–1680*, pp. 430 (13 November 1679), 431 (20 November 1679). It was, of course, standard for seventeenth-century bourgeois and aristocratic households to include residential servants, typically engaged on an annual basis, though apprentices' term was commonly seven years. And modern historians have noted the intimacy and lack of concern for privacy of such arrangements.

54. Birch, *History*, vol. 3, pp. 343, 369, 409, 417, 419, 421, 427, 429. For Wicks

and Hunt, see H. W. Robinson, "The Administrative Staff of the Royal Society 1663–1861," *Notes and Records of the Royal Society* 4 (1946): 193–205, on pp. 194–197.

55. 'Espinasse, *Robert Hooke,* pp. 131–138; Hooke, *Diary, 1672–1680,* p. 338 (3 January 1677–1678); cf. p. 302 (20 July 1677).

56. On Boyle and the moral constitution of the experimental philosopher, see Shapin and Schaffer, *Leviathan and the Air-Pump,* esp. chapters 2, 7; Simon Schaffer, "Godly Men and Mechanical Philosophers: Souls and Spirits in Restoration Natural Philosophy," *Science in Context* 1 (1987): 55–85, esp. pp. 75–77.

57. For treatment of this theme, see Shapin, "'A Scholar and a Gentleman'" (chapter 8 in this volume); idem, *A Social History of Truth,* chapter 4.

58. Robert Boyle, "The Christian Virtuoso . . . The First Part," in Boyle, *Works,* vol. 5, pp. 508–540, on p. 522; idem, "The Christian Virtuoso . . . The Second Part," ibid., vol. 6, pp. 673–796, on p. 717.

59. Boyle, "The Christian Virtuoso . . . The First Part," pp. 522–523, 536.

60. Boyle, "The Christian Virtuoso . . . The First Part," pp. 525–526, 528; Shapin, "Pump and Circumstance" (chapter 6 in this volume).

61. The management of trust was a practical problem for the experimental community in the mid- to late seventeenth century. In the main, the problems potentially posed by reliance upon natural historical and experimental testimony were dealt with by mobilizing taken-for-granted identifications of "honorable men" or "credible witnesses." The importance of this practical solution was, however, periodically indicated by the reaction of responsible agents to its breakdown. See, for example, the handling of experimental testimony from a group of physicians in Danzig, *Oldenburg,* vol. 3, pp. 548–549; vol. 4, pp. 26–28; and for detailed treatment, Shapin, *A Social History of Truth,* esp. chapters 1, 5–6.

62. Shapin and Schaffer, *Leviathan and the Air-Pump,* p. 319; Shapin, "The House of Experiment in Seventeenth-century England" (chapter 5 in this volume); Harold Fisch, "The Scientist as Priest: A Note on Robert Boyle's Natural Theology," *Isis* 44 (1953): 252–265.

63. Boyle, "The Christian Virtuoso . . . The Second Part," p. 757.

64. Burnet, *Select Sermons,* p. 195.

65. Thomas Sprat, *History of the Royal Society* (London, 1667), pp. 67–69; Shapin, "The House of Experiment in Seventeenth-century England" (chapter 5 in this volume).

66. See, among many examples, Gilbert Burnet, *History of His Own Time,* 6 vols. (Oxford: University Press, 1833), vol. 1, p. 351 ("[Boyle] was looked on by all who knew him as a very perfect pattern"); Maddison, "Studies in the Life of a Boyle. Part IV," esp. p. 38; Daniel Defoe, *The Compleat English Gentleman,* ed. Karl D. Bülbring (London: D. Nutt, 1890), p. 69.

67. Waller, "Life of Hooke," pp. iv–vi.

68. Hunter, "Science, Technology and Patronage."

69. Hooke, *Diary, 1672–1680*, pp. 116 (3 August 1674), 123 (26 September 1674).

70. Hooke, *Diary, 1672–1680*, p. 62 (29 September 1673).

71. Hooke, *Diary, 1672–1680*, p. 370 (3 August 1678); cf. p. 364 (25 June 1678); "deliverd Mr. Boyle 12 Journall de Scavans, he owes me 6sh."

72. Hooke, *Diary, 1672–1680*, p. 103 (16 May 1674).

73. Hooke, *Diary, 1672–1680*, pp. 245 (31 July–10 August 1676), 136 (17 December 1674).

74. For example, Hooke, *Diary, 1672–1680*, p. 309 (26 August 1677).

75. Waller, "Life of Hooke," pp. xiii, xxvli; Aubrey, *Brief Lives*, vol. 1, p. 411. The contents of Hooke's trunk are listed in the manuscript inventory. One of Hooke's contemporaries reckoned his total wealth at his death at £1200: letter from Sir. Godfrey Copley to ?, ca. 1703, quoted in [art.] "Hooke (Robert)," in Alexander Chalmers, *The General Biographical Dictionary*, new ed. 32 vols. (London: J. Nichols, 1812–1817), vol. 18, pp. 28–35, on pp. 132n.–133n. A passionate attempt to gloss over Hooke's alleged meanness is 'Espinasse, *Robert Hooke*, pp. 141–142.

76. Aubrey, *Brief Lives*, vol. 1, p. 411.

77. 'Espinasse, *Robert Hooke*, p. 74; Bennett, "Robert Hooke as Mechanic," p. 41.

78. On the rewards of Hooke's architectural and surveying work: H. W. Robinson, "Robert Hooke as Surveyor and Architect," *Notes and Records of the Royal Society of London* 6 (1949): 48–55; on the 1657 contract: Waller, "Life of Hooke," pp. iv–v.

79. 'Espinasse, *Robert Hooke*, p. 149.

80. For example, Hooke, *Diary, 1672–1680*, p. 395 (3 February 1678–1679). The glass-making project of 1691 also involved a patent. See Robert Hooke, *Lectures De Potentia Restitutiva, or of Spring . . .* (London, 1678), in Gunther, *Cutlerian Lectures*, p. 338 (for reference to Hooke securing a patent).

81. Hooke, *Diary, 1672–1680*, p. 236 (8 June 1676).

82. See, for example, Hooke, *Diary, 1672–1680*, pp. 131 (17 November 1674) ["Drew up proposals about Secresy and Secretary"] and 205–206 (1 January 1675–1676); Birch, *History*, vol. 3, pp. 137–138 (15 October 1674).

83. Peacham, *The Complete Gentleman*, p. 24.

84. Henry Oldenburg to Robert Boyle, 10 December 1667, in *Oldenburg*, vol. 4, pp. 26–28.

85. Hooke, *Diary, 1672–1680*, pp. 229 (1 May 1676) and 276 (25 February 1676–1677). This text was almost certainly [Antoine de Courtin], *The Rules of Civility or Certain Ways of Deportment observed amongst all Persons of Quality* (Lon-

don, 1671), translated from the French *Nouveau Traité de la Civilité.* This was a popular guidebook to the exact ritual and ceremonial forms of manners to be observed when dealing with one's equals, inferiors, and superiors.

86. Edward Chamberlayne, *Angliae Notitia; or the Present State of England,* 7th ed. (London, 1673), pp. 320–321, 328. See also Ruth Kelso, *The Doctrine of the English Gentleman in the Sixteenth Century,* University of Illinois Studies in Language and Literature, vol. 14 (Urbana: University of Illinois Press, 1929), pp. 1–288, esp. p. 78. For warnings about merchants' secrecy and search for present profit, see Sprat, *History,* pp. 65–67; for the Royal Society's freedom from "sordid interests," see Joseph Glanvill, *Scepsis Scientifica* (London: Kegan Paul, Trench, 1885; orig. publ. 1665), p. lxiv; also Shapin, *A Social History of Truth,* chapter 3.

87. For Copley: Chalmers, *Biographical Dictionary,* pp. 130 and 132n, and 'Espinasse, *Robert Hooke,* p. 142; for Molyneux, Thomas Molyneux to William Molyneux, 9 June 1683, quoted in K. Theodore Hoppen, "The Royal Society and Ireland: William Molyneux, F.R.S. (1656–1698)," *Notes and Records of the Royal Society of London* 18 (1963): 125–135, on p. 127; for Leibniz: Leibniz to Oldenburg, 26 February 1672–1673, in *Oldenburg,* vol. 9, p. 494; for Moray: Moray to Oldenburg, 8 January 1665–1666, in ibid., vol. 3, p. 9; for Oldenburg: Oldenburg to Boyle, 6 October 1664, in ibid., vol. 2, p. 248 (the editors suggest [p. 250 note] that Hooke was the person referred to, and other sources make this seem highly probable).

88. Lawrence Stone, *The Family, Sex and Marriage in England 1500–1800* (New York: Harper & Row, 1977), pp. 561–563. Stone does not appear to recognize any problems in equating the number of orgasms recorded in Hooke's *Diary* with the number Hooke actually experienced, nor in making judgments on this basis about Hooke's "sexual drive."

89. Hooke, *Diary, 1672–1680,* pp. 176 (21 August 1675) and 177 (27 August 1675). The manuscript inventory of Hooke's possessions in 1703 recorded "a picture of a Naked woman without a frame" in the cellar.

90. Hooke, *Micrographia,* p. 198. Other references to God are on pp. 2, 8, 95, 105, 124–125, 133–134, 154, 171–172, 179, 189–190, 193–195, 207, 242. See also Michael Aaron Dennis, "Graphic Understanding: Instrument and Interpretation in Robert Hooke's *Micrographia,*" *Science in Context* 3 (1989): 309–364.

91. References to the Deluge in Robert Hooke, "Lectures and Discourse of Earthquakes . . . ," in *Posthumous Works,* pp. 210–450, are on pp. 319–320, 322, 324, 328, 341, 408, 412, 414–416, 422–424. On Hooke's attitude to biblical authority in his geological work, see esp. Martin Rudwick, *The Meaning of Fossils: Episodes in the History of Paleontology* (London: Macdonald, 1972), chapter 2.

92. Waller, "Life of Hooke," p. xxviii.

93. Hooke, *Diary, 1672–1680,* pp. 235 (2 June 1676), 226 (13 April 1676), 232 (13 May 1676), 201 (15 December 1675).

94. Waller, "Life of Hooke," p. xxviii. For a reference to Hooke's purchase of a Welsh Bible: Hooke, *Diary, 1672–1680,* p. 411 (8 and 10 May 1679). For Hooke's library: H. A. Feisenberger, "The Libraries of Newton, Hooke and Boyle," *Notes and Records of the Royal Society* 21 (1966): 42–55, esp. p. 50.

95. Hooke, *Diary, 1672–1680,* pp. 335 (21 December 1677), 447 (27 June 1680 [Sunday], when Hooke also recorded "Not at church in the afternoon"), 387 (8 December 1678 [Sunday]), 241 (9 July 1676 [Sunday]), for the St. Helen's parson. For Burnet's sermons: pp. 354 (21 April 1678 [Sunday]), 387 (8 December 1678 [Sunday]), 439 (22 February 1679–80 [Sunday]). See also ibid., p. 353 (14 April 1678 [Sunday]): "at Temple church with Mr. Godfrey." Sunday entries in Hooke's *Diary 1688–1693* (*Early Science in Oxford,* ed. Gunther, vol. 10, pp. 69–265) periodically record parish names (St. Peter's, St. Helen's), which may indicate church attendance. (St. Peter's and St. Helen's were both parish churches within short walks of Gresham College, and Hooke was buried in the latter.) These references are not common before 1692. See, for example, Hooke, *Diary, 1688–1693,* pp. 196, 222, 229, 242, 244, 253.

96. For theological discourse, see, for example, Hooke, *Diary, 1672–1680,* pp. 163 (8 June 1675) [for theological conversation with Sydenham], 250 (20 September 1676) [for conversation with Hoskins regarding Tillotson's theology], 387 (5 December 1678) [again with Hoskins "about Creeds, Spirits, Antichrist, &c."], 368 (24 July 1678) ["much discourse about Spinosa quakers"], 376 (9 September 1678) [visited Tillotson; "Discoursd much of Criticall Learning of the French Bible"], 382 (27 October 1678 [Sunday]) [at Jonathan's coffeehouse: "Chaff about Religion"]; for Mosaic philosophy and Caballa: p. 292 (24 and 26 May 1677); for the "Rosicrucian" Society: p. 242 (14 July 1676); for *The Virtuoso:* p. 166 (25 June 1675); for Oliver Hill as "quaker": p. 338 (3 January 1677–1678). In *Micrographia,* "The Preface," sig. g2r, Hooke wrote that in the Reverend John Wilkins "we have an evident instance, what the true and the *primitive unpassionate Religion was,* before it was *sowred* by particular Factions," though the allusion was a fairly routine Latitudinarian formula in the early Restoration. For Waller's suggestion of unorthodoxy, see "Life of Hooke," p. xxviii.

97. C. B. Macpherson, *The Political Theory of Possessive Individualism: Hobbes to Locke* (Oxford: Oxford University Press, 1970), esp. chapter 3; Christopher Hill, "Pottage for Freeborn Englishmen: Attitudes towards Wage-Labour," in idem, *Change and Continuity in Seventeenth-Century England* (Cambridge: Cambridge University Press, 1975), pp. 219–238.

98. Chamberlayne, *Angliae Notitia,* pp. 320–321; [Daniel Defoe], *Complete English Tradesman* (London, 1726), pp. 275–292. For Defoe "trading lies," for example, asking more for an item than one knew one would actually accept, were normal and permissible departures from literal truth-telling for the tradesmen: "the

Tradesmen's promises, similarly, ought to be taken "with a contingent dependence upon the circumstances of trade" (pp. 276, 281); also Shapin, *A Social History of Truth*, chapters 3, 8.

99. Richard Brathwait, *The English Gentleman* (London, 1630), p. 84.

100. Francis Bacon, "Of Truth," in idem, *The Moral and Historical Works of Lord Bacon*, ed. Joseph Devey (London: Henry G. Bohn, 1852), pp. 1–4.

101. See, for example, Robert Hooke, "A General Scheme, or Idea of the Present State of Natural Philosophy," in *Posthumous Works*, pp. 1–70, on p. 63; see also Shapin, *A Social History of Truth*, chapter 5.

102. I will not treat the handling of Hooke's mechanic testimony, though there is evidence that portions of this were disbelieved by his colleagues. It is possible that this behavior had a bearing upon the evaluation of his experimental testimony. For example, Copley discussed Hooke's persistent claim that "he knew a certain and infallible method of discovering the longitude at sea; yet it is evident that his friends distrusted his asseveration of this discovery; and . . . little credit was then given to it in general" (Copley letter, quoted in Chalmers, *Biographical Dictionary*, p. 133n). It is not clear what Hooke's associates thought of his continued claims that he could "fly," nor, indeed, what Hooke was thinking when he made such claims.

103. There are many references to these "failures" in the *Journal-Book* for the period between ca. December 1662 and ca. October 1663; see, for example, Birch, *History*, vol. 1, pp. 139, 212, 268. See also Hooke to Boyle, [ca. July 1663], in Boyle, *Works*, vol. 6, pp. 484–485, on p. 484.

104. The career of anomalous suspension in the 1660s and 1670s is described in Shapin and Schaffer, *Leviathan and the Air-Pump*, esp. pp. 248–250. Hooke continued to theorize about the cause of anomalous suspension into the mid-1680s; see Hooke, "Lectures and Discourses of Earthquakes," pp. 365–370.

105. Birch, *History*, vol. 1, p. 177 (14 January 1662–1663).

106. Birch, *History*, vol. 3, pp. 61, 77–78 (20 November 1672, 5 and 19 March 1672–1673).

107. Birch, *History*, vol. 4, pp. 261–262 (27 February 1683–1684); Stephen Pumfrey, "Mechanizing Magnetism in Restoration England—The Decline of Magnetic Philosophy," *Annals of Science* 44 (1987): 1–22, on p. 13; cf. Michael Hunter, "Reconstructing Restoration Science," pp. 458–459, for the Royal Society's discontent with Hooke's performance of his duties in the late 1670s and early 1680s; and Shapin, "The House of Experiment in Seventeenth-century England" (chapter 5 in this volume).

108. "An Account of an Experiment made by Mr. Hooke, of Preserving Animals alive by Blowing through their Lungs with Bellows," *Philosophical Transactions* 3 (1667): 539–540.

109. This evidence involves Henry More's controversies with Boyle that largely concerned the proper interpretation of pneumatic experiments and the relationships

between experimental natural philosophy and theology. Although Boyle wrote in 1672 that More "did indeed deny the matter of fact [which Boyle narrated] to be true," he noted that his adversary was "too civil, to give me *in terminus* the lye," and it is possible to see the disputed point as interpretative in nature: Robert Boyle, "Hydrostatical Discourse," in Boyle, *Works,* vol. 3, pp. 596–628, on p. 615; also Shapin and Schaffer, *Leviathan and the Air-Pump,* pp. 217–218. (Henry More was an inactive member of the Royal Society.) On the literary and social techniques for securing assent to experimental testimony, see Shapin, "Pump and Circumstance" (chapter 6 in this volume).

Chapter 10. Who Is the Industrial Scientist?

1. Polemical writing against the obstruction or perversion of pure science by the interests and norms of business appeared in the late nineteenth and early twentieth centuries—see notably Henry Rowland, "A Plea for Pure Science," in *The Physical Papers of Henry Augustus Rowland* (Baltimore: Johns Hopkins Press, 1902; art. orig. publ. 1883), pp. 593–613; Thorstein Veblen, *The Higher Learning in America: A Memorandum on the Conduct of Universities by Business Men* (New York: Sagamore Press, 1957; orig. publ. 1918).

2. Robert K. Merton, "The Normative Structure of Science," in idem, *The Sociology of Science,* ed. Norman W. Storer (Chicago: University of Chicago Press, 1973; art. orig. publ. 1942), pp. 267–278, on pp. 273, 275, 278. See also Bernard Barber, *Science and the Social Order* (New York: Collier Books, 1952); Warren O. Hagstrom, *The Scientific Community* (Carbondale: Southern Illinois University Press, 1975; orig. publ. 1965); Norman W. Storer, *The Social System of Science* (New York: Holt, Rinehart & Winston, 1966); idem, "Science and Scientists in an Agricultural Research Organization: A Sociological Study" (Ph.D. thesis, Cornell University, 1961); Walter Hirsch, *Scientists in American Society* (New York: Random House, 1968).

3. Merton, "Normative Structure," p. 260.

4. Here I acknowledge the continuing interest of my former colleague Barry Barnes's early publications, identifying in the Mertonian story empirical and theoretical awkwardnesses that flowed from a specific theory of socialization, and pointing to an alternative such theory in the work of Howard Becker: Barry Barnes, "Making Out in Industrial Research," *Science Studies* 1 (1971): 157–175; cf. David Bloor and Celia Bloor, "Twenty Industrial Scientists: A Preliminary Exercise," in *Essays in the Sociology of Perception,* ed. Mary Douglas (London: Routledge & Kegan Paul, 1982), pp. 83–102; Michael Aaron Dennis, "Accounting for Research: New Histories of Corporate Laboratories and the Social History of American Science," *Social Studies of Science* 17 (1987): 479–518. However, my purpose here is not the same as Barnes's: though there is, indeed, criticism of the Mertonian account in this essay, my major impulse here is to show the interest of interpreting it as a historical

product. And this is done at greater length in Steven Shapin, *The Scientific Life: A Moral History of a Late Modern Vocation* (Chicago: University of Chicago Press, 2008), chapters 4–6.

5. William Kornhauser, *Scientists in Industry: Conflict and Accommodation* (Berkeley: University of California Press, 1962); Simon Marcson, *The Scientist in American Industry: Some Organizational Determinants in Manpower Utilization* (New York: Harper & Row, 1960).

6. For example, Edward H. Litchfield, "Notes on a General Theory of Administration," *Administrative Science Quarterly* 1 (1956): 3–29; see also James D. Thompson, "On Building an Administrative Science," *Administrative Science Quarterly* (1956): 102–111; Talcott Parsons, "Suggestions for a Sociological Approach to the Theory of Organizations (I) and (II)," *Administrative Science Quarterly* (1956): 63–85, 225–239.

7. Alvin W. Gouldner, "Cosmopolitans and Locals: Toward an Analysis of Latent Social Roles," *Administrative Science Quarterly* 2 (1957–1958): 281–306, 444–480.

8. For representative practical expressions of such anxieties, see, for example, John R. Steelman, *Science and Public Policy: A Report to the President by John R. Steelman, Chairman, The President's Scientific Research Board,* 5 vols. (Washington, DC: Government Printing Office, 1947), vol. 3, p. 28 (also p. 141); Phillip N. Powers, "Industrial Research Workers and Defense," in *Selection, Training, and Use of Personnel in Industrial Research,* Proceedings of the Second Annual Conference on Industrial Research, June 1951, ed. David B. Hertz and Albert H. Rubenstein (New York: King's Crown Press, 1952), pp. 94–112, esp. p. 98; Donald M. Laughlin, "The Engineer Shortage: Challenge to Management," *Personnel* 29 (1952): 418–421; Charles D. Orth, III, "More Productivity from Engineers," *Harvard Business Review* 35, no. 2 (March–April 1957): 54–62, esp. p. 58; also David Kaiser, "Cold War Requisitions, Scientific Manpower, and the Production of American Physicists after World War II," *Historical Studies in the Physical Sciences* 33 (2002): 131–159.

9. *Scientific Creativity: Its Recognition and Development: Selected Papers from the Proceedings of the First, Second, and Third University of Utah Conferences: "The Identification of Creative Scientific Talent,"* ed. Calvin W. Taylor and Frank Barron (New York: John Wiley & Sons, 1963), see esp. contributions by Maury H. Chorness of the Air Force Personnel Research Laboratory; J. H. McPherson of Dow; and N. E. Golovin of ARPA, who made his concern with the United States–USSR creativity-gap quite explicit. The research of one of this volume's editors (Calvin Taylor) was funded by the Air Force Personnel Laboratory and that of other contributors by the ONR.

10. Steelman, *Science and Public Policy,* vol. 3, pp. 28, 141, 205–252.

11. Charles D. Orth, III, "The Optimum Climate for Industrial Research," *Harvard Business Review* 37, no. 2 (March–April 1959): 55–64, on pp. 55–56, 64; re-

printed in *Science and Society,* ed. Norman Kaplan (Chicago: Rand-McNally, 1965), pp. 194–210. For insistence that Progressive-era industrial scientists identified very *strongly* with the companies that employed them and embraced their values, see Patrick J. McGrath, *Scientists, Business, and the State, 1890–1960* (Chapel Hill: University of North Carolina Press, 2002), esp. pp. 19–21.

12. William H. Whyte, Jr., *The Organization Man* (New York: Simon and Schuster, 1956), pp. 205–206.

13. Whyte, *The Organization Man,* pp. 206–207, 211, 214. (No such "studies" were actually cited.) And see in these connections fine work by David Kaiser, esp. his "The Postwar Suburbanization of American Physics," *American Quarterly* 56 (2004): 851–888.

14. Elmer W. Engstrom, "What Industry Requires of the Research Worker," in *Human Relations in Industrial Research, including Papers from the Sixth and Seventh Annual Conferences on Industrial Research: Columbia University, 1955 and 1956,* ed. Robert Teviot Livingstone and Stanley H. Milberg (New York: Columbia University Press, 1957), pp. 69–79, on p. 69. (Engstrom was a radio engineer who had been a director of research at RCA since the 1940s, and in 1951—when his talk was delivered—was senior executive vice president of that company.)

15. Carleton R. Ball, "Personnel, Personalities and Research," *Scientific Monthly* 23 (1926): 33–45, on p. 35.

16. The closest thing I can find to an expression of role-conflict-through-academic-socialization from an industrial source is Lowell W. Steele, "Rewarding the Industrial Scientist: A Problem of Conflicting Values," in *Human Relations in Industrial Research,* ed. Livingstone and Milberg, pp. 163–175. But while Steele was indeed an executive at General Electric, it is relevant that he was not a natural scientist or engineer by training and had no direct shop-floor experience. Steele was an industrial personnel manager at the GE Research Laboratory, in charge of salary practices. He took a Harvard M.B.A. in 1948 (where he may well have encountered the views of the previously mentioned Charles Orth) and, later, an MIT Ph.D. in economics and social science under the supervision of Herbert A. Shepard. Whyte's *Organization Man* (pp. 211–212) used Steele's earlier views prominently as an authority on the crushing demands for loyalty and conformity rightly made on scientists by commercial organizations: Lowell W. Steele, "Personnel Practices in Industrial Laboratories," *Personnel* 29 (1953): 469–476, on p. 471. Steele evidently had a change of heart on this issue between 1953 and 1957.

17. I. Gorog, "Successful Adjustment to Industrial Research: How Can University and Industry Assist," *Research Management* 9 (1966): 5–13, on pp. 7, 10–11. Gorog was a 1964 Berkeley Ph.D. in electrical engineering, then working at RCA. See also J. J. Tietjen, "The Transition from Graduate School to Industrial Research," *Research Management* 9 (1966): 109–113.

18. Horace A. Secrist, director of research at The Kendall Company, noted that

industrial management "often does not fully realize how great an adaptation the scientist does succeed in making when he moves from the university to the industrial laboratory," drawing attention here not to programmatic norms but to some mundane aspects of adjusting to *teamwork:* Secrist, "Motivating the Industrial Research Scientist," *Research Management* 3 (1960): 57–64, on pp. 58–59. And a research manager at Sun Oil noted the importance of showing recruits—through organization charts—where they fit in the corporate scheme of things: E. M. Kipp, "Introduction of the Newly Graduated Recruit to Industrial Research," *Research Management* 3 (1960): 39–47. On the pragmatics of adjustment see, for example, Warner Eustis (of The Kendall Company), "Personnel Policies and Personality Problems," in *Research in Industry: Its Organization and Management,* ed. C. C. Furnas (New York: D. Van Nostrand, 1948), pp. 277–294; John A. Van Raalte, "Reflections on the Ph.D. Interview," *Research Management* 9 (1966): 307–317, esp. pp. 307, 315.

19. See Tom Burns, "Research, Development and Production: Problems of Conflict and Cooperation," in *Administering Research and Development: The Behavior of Scientists and Engineers in Organizations,* ed. Charles D. Orth, III, Joseph C. Bailey, and Francis W. Wolek (Homewood, IL: Richard D. Irwin, Inc. and the Dorsey Press, 1964), pp. 112–129 (orig. publ. *IRE Transactions on Engineering Management* [March 1961], pp. 15–23), for a study of tensions between the research facility and production functions in British companies, and p. 123 for a participant's idyllic account of relations within the laboratory: "In the lab, we're very happy—sort of happy family relationships. The lab chief must select people on the grounds of getting on with others; they certainly do get on with everybody"; see also Shapin, *The Scientific Life,* chapter 6 (for teamwork).

20. Steele, "Personnel Practices in Industrial Laboratories," p. 471.

21. There are countless examples of such sentiments among industrial research administrators. For a representative instance, see C. E. Kenneth Mees, *The Organization of Industrial Scientific Research* (New York: McGraw-Hill, 1920), pp. 20–21; also John J. Carty (of AT&T), "Science and Business: An Address to the Chamber of Commerce of the United States, May 8, 1924," *Reprint and Circular Series of the National Research Council* (Washington, DC: National Research Council, 1929), pp. 1–2; David F. Noble, *America by Design: Science, Technology, and the Rise of Corporate Capitalism* (New York: Alfred A. Knopf, 1977), pp. 129–131; Willis Jackson (of Metropolitan-Vlckers [UK]), "Discussion," in *The Direction of Research Establishments: Proceedings of a Symposium held at the National Physical Laboratory [Teddington] on 26th, 27th & 28th September 1956* (New York: Philosophical Library, 1957), p. A.3: insisting that "the primary function of university research was in its relation to teaching," he urged that the goals of academic research should never be defined by any outside sponsor, and "a clear distinction should be made between universities and national as well as industrial laboratories"; John A. Leermakers, "Basic Research in Industry," *Industrial Laboratories* 2, no. 3 (March 1951):

2–3: "It is in the universities that true intellectual freedom, so essential to the development of the scientific attitude is found." Note, however, that Leermakers was then an associate of Eastman Kodak's Kenneth Mees, whose views on the freedom required by *industrial* research are quoted below.

22. Leonard S. Reich, *The Making of American Industrial Research: Science and Business at GE and Bell, 1876–1926* (Cambridge: Cambridge University Press, 1985), pp. 69–70, 75, 82–83, 92, 111, 126–127; also Daniel J. Kevles, *The Physicists: The History of a Scientific Community in Modern America* (New York: Vintage Books, 1979), pp. 99–101.

23. David A. Hounshell and John Kenly Smith, Jr., *Science and Corporate Strategy: Du Pont R&D, 1902–1980* (Cambridge: Cambridge University Press, 1988), pp. 230–231.

24. Charles P. Steinmetz, "Scientific Research in Relation to the Industries," *Journal of the Franklin Institute* 182 (1916): 711–718, on pp. 711–712.

25. Frank B. Jewett, "Industrial Research," *Reprint and Circular Series of the National Research Council,* #4 (Washington, DC: National Research Council, 1919), p. 7 (quoted in Kevles, *The Physicists,* p. 100).

26. Carty, "Science and Business," quoted in Noble, *America by Design,* p. 115.

27. Waldo H. Kliever, "Design of Research Projects and Programs," *Industrial Laboratories* 3, no. 10 (October 1952): 6–13, on p. 6; see also McGrath, *Scientists, Business, and the State,* pp. 12–14.

28. James W. Hackett and B. L. Steierman, "The Organization of a Fundamental Research Effort at Owens-Illinois Glass Company," *Research Management* 6 (1963): 81–92, on pp. 85–86.

29. C. George Evans, *Supervising R&D Personnel* (New York: American Management Association, 1969), p. 24.

30. Hackett and Steierman, "The Organization of a Fundamental Research Effort at Owens-Illinois Glass Company," pp. 85–86; see also James B. Fisk (Bell Labs), "Basic Research in Industrial Laboratories," in *Symposium on Basic Research, Sponsored by the National Academy of Sciences, the American Association for the Advancement of Science, and the Alfred P. Sloan Foundation,* ed. Dael Wolfle (Washington, DC: American Association for the Advancement of Science, 1959), pp. 159–167, on p. 163.

31. Quoted in Francis Bello, "The World's Greatest Industrial Laboratory," *Fortune* (November 1958): 148–157, on p. 163.

32. C. E. Kenneth Mees and John A. Leermakers, *The Organization of Industrial Scientific Research,* 2nd ed. (New York: McGraw-Hill, 1950), p. 223; Malcolm H. Hebb, with Miles J. Martin, "Free Inquiry in Industrial Research," *Research Management* 1 (1958): 67–83, on p. 67. Both Hebb and Martin were physicists who moved to the GE Research Laboratory from academic appointments sometime after 1955.

33. For example, Merritt L. Kastens, "Research—A Corporate Function," *Industrial Laboratories* 8, no. 10 (October 1957): 92–101; D. L. Williams, *Planning of Research and Development* (New York: Wallace Clark, 1947), p. 5; David B. Hertz, *The Theory and Practice of Industrial Research* (New York: McGraw-Hill, 1950), p. 202; Harvey Brooks, "Can Science Be Planned?" in idem, *The Government of Science* (Cambridge, MA: MIT Press, 1968), chapter 3, esp. pp. 59–72.

34. Noble, *America by Design,* chapter 7, esp. p. 118: "As the industrial research laboratories grew in size, the role of the scientists within them came more and more to resemble that of workmen on the production line and science became essentially a management problem."

35. Arthur W. Baum, "Doctor of the Darkroom," *The Saturday Evening Post* (25 October 1947): 15–17, 47, 50, 52, on p. 17. (I thank Ray Curtin of the Eastman Kodak Company for this reference.) See also Mees, *The Organization of Industrial Scientific Research,* p. 102; idem, *From Dry Plates to Ektachrome Film: A Story of Photographic Research* (New York: Ziff-Davis, 1961), p. 50; Darrell H. Voorhies, *The Co-ordination of Motive, Men and Money in Industrial Research, A Survey of Organization and Business Practice Conducted by the Department on Organization of the Standard Oil Company of California* (San Francisco: Standard Oil Company of California, 1946), pp. 53–54; Arthur Gerstenfeld, *Effective Management of Research and Development* (Reading, MA: Addison-Wesley, 1970), pp. 18–23 (for four-year payback and large- versus small-company figures); C. Wilson Randle, "Problems of Research and Development Management," *Harvard Business Review* 37, no. 1 (January–February 1959): 128–136, on p. 131 (for lack of formal evaluating criteria).

36. C. E. Kenneth Mees, "The Organization of Industrial Scientific Research," *Science* 43, no. 1118 (2 June 1916): 763–773, on p. 768.

37. Quoted in T. A. Boyd, *Professional Amateur: The Biography of Charles Franklin Kettering* (New York: E. P. Dutton, 1957), p. 118.

38. See, notably, Richard R. Nelson, "The Economics of Invention: A Survey of the Literature," *The Journal of Business* 32 (1959): 101–127; cf. David A. Hounshell, "The Medium is *the* Message, or How Context Matters: The RAND Corporation Builds an Economics of Innovation, 1946–1962," in *Systems, Experts, and Computers: The Systems Approach in Management and Engineering, World War II and After,* ed. Thomas P. Hughes and Agatha C. Hughes (Cambridge, MA: The MIT Press, 2000), pp. 255–310.

39. Mees and Leermakers, *The Organization of Industrial Scientific Research,* pp. 235–237; see also F. Russell Bichowsky, *Industrial Research* (Brooklyn, NY: Chemical Publishing Co., 1942), p. 106 (for the advisability of keeping a "slush fund"—say, 5 percent of the research budget—for individual researchers, with the assent of group leaders, to do with pretty much as they pleased).

40. Norman A. Shepard (American Cyanamid), "How Can We Build Better

Teamwork within Our Research Organizations?" *Chemical and Engineering News* 23, no. 9 (10 May 1945): 804–807, on p. 805 (quoting Willard Dow).

41. Reich, *The Making of American Industrial Research,* p. 75 (for one-third); Kendall Birr, *Pioneering in Industrial Research: The Story of the General Electric Research Laboratory* (Washington, DC: Public Affairs Press, 1957), p. 37 (for one-half).

42. Letter from Fieser to James B. Conant, 26 November 1927, quoted in David A. Hounshell, "The Evolution of Industrial Research in the United States," in *Engines of Innovation: U.S. Industrial Research at the End of an Era,* ed. Richard S. Rosenbloom and William J. Spencer (Cambridge, MA: Harvard University Press, 1996), pp. 13–85, on p. 27; see also Hounshell and Smith, *Science and Corporate Strategy,* pp. 228, 299. (Fieser ultimately turned down the Du Pont position.) But see McGrath, *Scientists, Business, and the State,* pp. 35 and 208 n. 6, for vigorous dissent from Hounshell's sketch of Conant as a typical academic "snob" about industrial research. Conant did indeed maintain that "first-rate" students should wind up in universities rather than industrial laboratories, but he was himself heavily involved in industrial research and saw nothing awkward in these relationships.

43. Robert N. Anthony, *Management Controls in Industrial Research Organizations* (Boston: Graduate School of Business Administration, Harvard University, 1952), pp. 134–136.

44. Hounshell, "Evolution of Industrial Research," pp. 26–27. Evidently the then-common practice of allowing industrial research workers free time could be unknown to social scientists commenting on the condition of scientists in industry during the 1950s. A Columbia professor of education, for example, wrote of his impression that such an idea was regarded by industrial research administrators as "fantastic" and even laughable: Lyman Bryson, "Researchers in Industry," *Human Relations in Industrial Research,* ed. Livingstone and Milberg, pp. 129–137, on p. 130.

45. Quoted again in Mees and Leermakers, *The Organization of Industrial Scientific Research,* p. 244. (This so-called "second edition" of 1950 was in fact an almost totally different book than the 1920 original.) I have not yet been able to locate the source of this quotation in Little's publications. Suspicion of the descriptive relevance of organization charts continued to be vigorously expressed by A. D. Little management consultants well after World War II: see, for example, Sherman Kingsbury, in association with Lawrence W. Bass and Warren C. Lothrop, "Organizing for Research," in *Handbook of Industrial Research Management,* ed. Carl Heyel (New York: Reinhold Publishing Corporation, 1959), pp. 65–99, on pp. 65, 77–78.

46. Mees, *The Organization of Industrial Scientific Research,* pp. 67–68.

47. C. E. Kenneth Mees, "Discussion [of Michael Polanyi, 'The Foundations of Freedom in Science']," in *Physical Science and Human Values,* ed. E. P. Wigner (Princeton, NJ: Princeton University Press, 1947), pp. 140–141. In these connections, it is interesting to note that, around this time, Mees was collaborating with the

English zoologist John R. Baker on a semi-popular history of science: Baker had been a leading light, and associate of Polyani, in the Society for Freedom in Science: see C. E. Kenneth Mees, with the co-operation of John R. Baker, *The Path of Science* (London: Chapman & Hall, 1946).

48. Mees's discussion of a piece by Thomas Midgley, in Standard Oil Development Company, *The Future of Industrial Research: Papers and Discussion* (New York: Standard Oil Development Company, 1943), p. 48; cf. Mees and Leermakers, *The Organization of Industrial Scientific Research*, p. 233. Many research directors felt obliged to take a position—for, against, or qualifying—Mees's aphorisms: see, for example, Shepard, "How Can We Build Better Teamwork?" p. 804.

49. This has been quoted many times, including Kliever, "Design of Research Projects," p. 11; Nelson, "The Economics of Invention," p. 125; Rob Kaplan, *Science Says: A Collection of Quotations on the History, Meaning, and Practice of Science* (New York: Stonesong Press, 2001), pp. 105–106; John Jewkes, David Sawers, and Richard Stillerman, *The Sources of Invention* (London: Macmillan, 1958), p. 138.

50. O. E. Buckley, quoted in Jewkes, Sawers, and Stillerman, *The Sources of Invention*, pp. 137–138.

51. Kliever, "Design of Research Projects and Programs," p. 13.

52. Mees, "Discussion of Midgley," p. 49; idem, *From Dry Plates to Ektachrome Film*, pp. 291–292; idem and Leermakers, *The Organization of Industrial Scientific Research*, p. 149; Baum, "Doctor of the Darkroom," p. 47. The scientist concerned was K. C. D. Hickman, and the new company founded to manufacture vitamins A and E using this technology was Distillation Products, Inc.

53. Francis Bello, "Industrial Research: Geniuses Now Welcome," *Fortune* (January 1956): 96–99, 142, 144, 149–150. Bello, like Whyte, exempted the industrial research "stars" (Bell, GE, Eastman Kodak) from his criticisms, and, while he reckoned that "There are still a good many laboratories where signs might still be posted: 'No geniuses wanted,'" he applauded big industry's willingness—perhaps, he thought, prompted by *Fortune*'s arguments over the years-to develop "more flexibility in accommodating unusual personalities": ibid., p. 96.

54. Noble, *America by Design*, chapter 6.

55. Hounshell and Smith, *Science and Corporate Strategy*, pp. 300–301.

56. Quoted in Ralph T. K. Cornwell, "Professional Growth of the Research Man," in *Research in Industry*, ed. Furnas, pp. 295–307, on p. 301.

57. See, for example, the vigorously pro-publication, and wholly pragmatic, views of a research manager at GE: R. W. Schmitt, "Why Publish Scientific Research from Industry?" *Research Management* 4 (1961): 31–41.

58. Cornwell (Sylvania), "Professional Growth of the Research Man," p. 301; Robert E. Wilson (Standard Oil, Indiana), "The Attitude of Management toward Research," *Chemical and Engineering News* 27, no. 5 (31 January 1949): 274–277,

on p. 277; H. B. McClure (Union Carbide), "External Communication of Research Results," in *Selection, Training, and Use of Personnel in Industrial Research*, ed. Hertz and Rubenstein, pp. 161–176, on pp. 163, 166; Schmitt (General Electric), "Why Publish Scientific Research from Industry?" p. 32. (Uniquely, in my experience, this research manager offered evidence that he had read some works of Robert Merton, citing Merton's claims about the essential openness of genuine science while completely setting aside Merton's characterization of what industrial research was like: ibid., pp. 36, 39–40.) See also Mees and Leermakers, *The Organization of Industrial Scientific Research*, pp. 277–280, 313; Hertz, *The Theory and Practice of Industrial Research*, pp. 347–350; Fisk, "Basic Research in Industrial Laboratories," p. 164. For a "must" policy on open publication, see Norman Shepard, "How Can We Build Better Teamwork?" p. 805.

59. Hounshell and Smith, *Science and Corporate Strategy*, pp. 223, 301.

60. Van Raalte, "Reflections on the Ph.D. Interview," p. 310.

61. Cornwell, "Professional Growth of the Research Man," pp. 301–302.

62. Reich, *The Making of American Industrial Research*, p. 110. A 1961 survey of publication practices for basic research findings among 174 U.S. companies showed that 14 percent of them published "substantially all" basic research findings; 26 percent published "most"; 45 percent published "some"; and 16 percent published "none": *Publication of Basic Research Findings in Industry* (Washington, DC: National Science Foundation, 1961), pp. 11ff., quoted in Hirsch, *Scientists in American Society*, p. 64.

63. Mees, "The Organization of Industrial Scientific Research," p. 771; idem, *The Organization of Industrial Scientific Research*, p. 100; see also D. E. H. Edgerton, "Industrial Research in the British Photographic Industry, 1879–1939," in *The Challenge of New Technology: Innovation in British Business Since 1850*, ed. Jonathan Liebenau (Aldershot: Gower, 1988), pp. 106–134, on p. 122.

64. An industrial research director interviewed by Lowell Steele in the early 1950s remarked that "he was constantly amazed at the speed at which his scientists became interested in the welfare of the company." Steele was then aware—though his later change of mind has been noted—that many commentators thought it impossible to get scientists to become "company conscious," but cited such evidence to argue that "this criticism is unsound": Steele, "Personnel Practices in Industrial Laboratories," p. 471.

65. Alfred H. Nissan, "Similarities and Differences between Industrial and Academic Research," *Research Management* 9 (1966): 211–219, on pp. 211–212.

66. As early as the late 1920s, James Bryant Conant was noting the change: among chemists awarded the Ph.D. at Harvard between 1907 and 1917, practically all intended an academic career; a decade later, "the majority of those who received the advanced degree would enter industrial employment either at once or after a few years": James B. Conant, *My Several Lives: Memoirs of a Social Inventor* (New

York: Harper & Row, 1970), p. 25. Government statistics showed that in the late 1940s 54 percent of all U.S. Ph.D. chemists worked in industrial laboratories, 10 percent in government laboratories, and only 33 percent in academia, without specifying the percentage of the latter active in research: cited in Hagstrom, *The Scientific Community*, p. 38. The 1950 Census was the first to specify personnel in scientific fields other than chemistry, and it showed that of about 150,000 total "natural and physical scientists"—including those without higher degrees—only 33,000 (or a little over a fifth) were working as teachers in institutions of higher education, the only notable exception being mathematicians, 80 percent of whom were in higher education: figures reproduced from National Manpower Council, *A Policy for Scientific and Professional Manpower: A Statement by the Council with Facts and Issues Prepared by the Research Staff* (New York: Columbia University Press, 1953), p. 47. See also, for example, Hirsch, *Scientists in American Society*, pp. 4–5, 60, and Albert E. Hickey, Jr., "Basic Research: Should Industry Do More of It?" *Harvard Business Review* 36, no. 4 (July–August 1958): 115–122, figures on pp. 116–117.

67. Writing in 1914, Kenneth Mees gave reasons why the universities were unlikely to become natural homes for scientific research: "For the last fifty years it has been assumed that the proper home for scientific research is the university, and that scientific discovery is one of the most important—if not the most important—functions which a university can fulfill. In spite of this only a few of the American universities, which are admittedly among the best equipped and most energetic of the world, devote a very large portion of their energies to research work, while quite a number prefer to divert as little energy as possible from the business of teaching, which they regard as the primary function of the university": C. E. Kenneth Mees, "The Future of Scientific Research [Editorial]," *Journal of Industrial and Engineering Chemistry* 6 (1914): 618–619, on p. 618.

68. For example, Veblen, *The Higher Learning in America*; Robert E. Kohler, "The Ph.D. Machine: Building on the Collegiate Basis," *Isis* 81 (1990): 638–662.

69. John W. Servos, "The Industrial Relations of Science: Chemical Engineering at MIT, 1900–1939," *Isis* 71 (1980): 531–549, esp. p. 532, for poor research resources at MIT in the early twentieth century.

70. J. F. Downie Smith, "Academic and Industrial Research," *Research Management* 5 (1962): 257–275, on p. 261.

71. By the late 1960s, troubled by the changes accompanying the rise of Big Science, some academic sociologists nervously broached the idea that, even for the academic scientist, "autonomy is a relative matter: his 'role set' involves at the very least the judgment of his peers, not to speak of the constraints imposed on him by his need to have a continuous flow of funds for equipment and research assistance—or just plain time to think": Hirsch, *Scientists in American Society*, p. 67. And, for more full-blooded skepticism, see Norman Kaplan, "Organization: Will It Choke or Pro-

mote the Growth of Science?" in *The Management of Scientists,* ed. Karl Hill (Boston: Beacon Press, 1964), pp. 103–127, esp. p. 114.

72. See, notably, Kaplan, "Organization: Will it Choke or Promote the Growth of Science?"; also Stephen Cotgrove and Steven Box, *Science, Industry and Science: Studies in the Sociology of Science* (London: George Allen and Unwin, 1970), and Barnes, "Making Out in Industrial Research."

73. Steelman, *Science and Public Policy,* vol. 3, pp. 142, on p. 208 (emphasis added). Of all groups responding, 11 percent preferred government, 31 percent industry, and 48 percent academia. Note that when industrial scientists alone were asked to put money considerations aside, 58 percent thought industry gave the most satisfaction.

74. For such perceptions, see, for example, Hirsch, *Scientists in American Society,* p. 53.

75. Paul F. Lazarsfeld and Wagner Thielens, Jr. (with a field report by David Riesman), *The Academic Mind: Social Scientists in a Time of Crisis* (Glencoe, IL: The Free Press, 1958), pp. 148–149.

76. See, for example, David A. Hollinger, "Money and Academic Freedom in a Half-Century after McCarthyism: Universities and the Force Fields of Capital," in *Unfettered Expression: Freedom in American Intellectual Life,* ed. Peggie J. Hollingsworth (Ann Arbor: University of Michigan Press, 2000), pp. 161–184.

Chapter 11. The Philosopher and the Chicken

1. David Brewster, *The Life of Sir Isaac Newton* (London: John Murray, 1831), p. 341n; for a representative twentieth-century retelling of the chicken story, see Grove Wilson, *The Human Side of Science* (New York: Cosmopolitan Book Corp., 1929), p. 198.

2. Maurice O'C. Drury, "Conversations with Wittgenstein," in *Recollections of Wittgenstein,* ed. Rush Rhees (Oxford: Oxford University Press, 1984), pp. 97–171, on p. 125.

3. Norman Malcolm, *Ludwig Wittgenstein: A Memoir* (Oxford: Oxford University Press, 1962), p. 40.

4. Drury, "Conversations with Wittgenstein," p. 156.

5. Here I should say that stories about truth-lovers' stomachs are only one potential focus for thinking about disembodiment as a topic in general epistemology. One could imagine an extended study divided into chapters: the face, the eyes, the loins, the skin, the hands, gesture, costume, the body in solitude. (See, for example, the topical organization of Richard Broxton Onians, *The Origins of European Thought about the Body, the Mind, the Soul, the World, Time and Fate,* 2nd ed. [Cambridge: Cambridge University Press, 1988].) I concentrate here on the belly partly because of the strength of the opposition between it and the mind.

6. This was the same intellectual subject that Nietzsche recognized and opposed: "a 'pure, will-less, painless, timeless subject of knowledge,'" "knowledge-in-itself": Friedrich Nietzsche, "What Is the Meaning of Ascetic Ideals?" in *The Genealogy of Morals,* trans. Horace B. Samuel, in *The Philosophy of Nietzsche* (New York: Random House, 1940 [1887]), pp. 717–793, on p. 744.

7. For the iconography of intellectuals, see, for example, Paul Zanker, *The Mask of Socrates: The Image of the Intellectual in Antiquity,* trans. Alan Shapiro (Berkeley: University of California Press, 1995); Angus Fletcher, "Iconographies of Thought," *Representations* 28 (1989): 99–112; and Janet Browne, "I Could Have Retched All Night: Charles Darwin and His Body," in *Science Incarnate: Historical Embodiments of Natural Knowledge,* ed. Christopher Lawrence and Steven Shapin (Chicago: University of Chicago Press, 1998), pp. 240–287. I have treated the related topic of solitude as an epistemological resource in Shapin, "'The Mind Is Its Own Place'" (chapter 7 in this volume).

8. See also Peter Dear, "A Mechanical Microcosm: Bodily Passions, Good Manners, and Cartesian Mechanism," in *Science Incarnate,* ed. Lawrence and Shapin, pp. 51–82; Robert Iliffe, "Isaac Newton: Lucatello Professor of Mathematics," in ibid., pp. 121–155; and Simon Schaffer, "Regeneration: The Body of Natural Philosophers in Restoration England," in ibid., pp. 83–120.

9. For the significance of similar tropes in non-Western as well as European cultures, see, for example, Jack R. Goody, *Cooking, Cuisine, and Class: A Study in Comparative Sociology* (Cambridge: Cambridge University Press, 1982), chapter 4.

10. Plato, *Phaedrus* in *The Collected Dialogues,* ed. Edith Hamilton and Huntington Cairns (Princeton, NJ: Princeton University Press, 1961), 259 b–e.

11. Plato, *Phaedrus,* 64d–66d, 67e; cf. idem, *Gorgias,* 524–527. For treatment of the pervasive (but "very curious") association between death and philosophy, see Hannah Arendt, *The Life of the Mind,* 2 vols. in 1 (San Diego: Harcourt Brace, 1987), vol. 1, pp. 79–81.

12. Diogenes Laërtius, *Lives of Eminent Philosophers,* trans. R. D. Hicks, 2 vols. (Cambridge, MA: Harvard University Press, 1979), vol. 1, p. 41; cf. Thomas Stanley, *The History of Philosophy* (London, 1687), p. 410.

13. Whitney J. Oates, ed., *The Stoic and Epicurean Philosophers: The Complete Extent Writings of Epicurus, Epictetus, Lucretius, Marcus Aurelius* (New York: Random House, 1940), p. 48 (for Epicurus); for Stoic dietetics, see Epictetus, *The Discourses of Epictetus, with the Encheiridion and Fragments,* trans. George Long (New York: A. L. Burt, [1888?]), pp. 434, 439, 443; Seneca, *Moral Essays,* 3 vols., trans. John W. Basore (Cambridge, MA: Harvard University Press, 1958), vol. 1, pp. 128, 151; and see also Martha Nussbaum, *The Therapy of Desire: Theory and Practice in Hellenistic Ethics* (Princeton NJ: Princeton University Press, 1994), pp. 112–114, and Peter Brown, *The Body and Society: Men, Women and Sexual Renunciation in Early Christianity* (London: Faber and Faber, 1989), p. 27.

14. Stanley, *History of Philosophy*, pp. 493, 506–507, 511, 518 (quoted passage), 564; see also Mirko D. Grmek, *Diseases in the Ancient Greek World*, trans. Mireille Muellner (Baltimore: Johns Hopkins University Press, 1989), chapter 9 ("The Harm in Broad Beans"); E. R. Dodds, *The Greeks and the Irrational* (Boston: Beacon Press, 1957), pp. 143–154; Piero Camporesi, *The Magic Harvest: Food, Folklore, and Society*, trans. Joan Krakover Hall (Cambridge: Polity Press, 1993), pp. 11, 15.

15. For an introduction to the origins of this pun, see Marx W. Wartofsky, *Feuerbach* (Cambridge: Cambridge University Press, 1977), pp. 413–414, 451 n. 6.

16. Iamblichus, *On the Pythagorean Life*, trans. Gillian Clark (Liverpool: Liverpool University Press, 1989), esp. pp. 4–7, 14, 24, 43–44, 47–48 (on Pythagorean dietetics); Porphyry, *On Abstinence from Animal Food*, trans. Thomas Taylor, ed. Esme Wynne-Tyson (London: Centaur Press, 1965), for example, pp. 54–56, 64, 99–100; see also Caroline Walker Bynum, *The Resurrection of the Body in Western Christianity, 200–1336* (New York: Columbia University Press, 1995), pp. 33–41; Catherine Osborne, "Ancient Vegetarianism," in *Food in Antiquity*, ed. John Wilkins, David Harvey, and Mike Dobson (Exeter: University of Exeter Press, 1995), pp. 214–224, on pp. 218–223.

17. See also Caroline Walker Bynum, *Holy Feast and Holy Fast: The Religious Significance of Food to Medieval Women* (Berkeley: University of California Press, 1987), p. 3. Forty days and forty nights was also the period of Elijah's fast: 1 Kings 19:8. For Judaic and early Christian conceptions of food as embodying God's knowledge, see Gillian Feeley-Harnik, *The Lord's Table: Eucharist and Passover in Early Christianity* (Philadelphia: University of Pennsylvania Press, 1981), pp. 82–91.

18. Brown, *Body and Society*, pp. 218–221. Interviewed about the first volume of his *History of Sexuality*, Michel Foucault "confessed" that "sex is boring," and that it was so for the Greeks and early Christians as well: "[Sex] was not a great issue. Compare, for instance, what they say about the place of food and diet. I think it is very, very interesting to see the move, the very slow move, from the privileging of food, which was overwhelming in Greece, to interest in sex. Food was still much more important during the early Christian days than sex. For instance, in the rules for monks, the problem was food, food, food. Then you can see a very slow shift during the Middle Ages when they were in a kind of equilibrium . . . and after the seventeenth century it was sex" (Michel Foucault, "On the Genealogy of Ethics: An Overview of a Work in Progress," in *Michel Foucault: Beyond Structuralism and Hermeneutics*, ed. Hubert L. Dreyfus and Paul Rabinow, 2nd ed. [Chicago: University of Chicago Press, 1983], pp. 229–252, on p. 229).

19. Abba Daniel (ca. 450), in Desert Fathers, *The Sayings of the Desert Fathers (Apophthegmata Patrum)*, trans. Benedicta Ward (London: A. R. Mowbray and Co., 1975), pp. 43–44; cf. Herbert Musurillo, "The Problem of Ascetical Fasting in the Greek Patristic Writers," *Traditio* 12 (1956): 1–64. Note the typical gesture here at Max Weber (Max Weber, "Religious Rejections of the World and Their Directions

[1915]," in *From Max Weber: Essays in Sociology,* ed. H. H. Gerth and C. Wright Mills [London: Routledge, 1991], pp. 323–359, on p. 327) called the "Janus-face" of asceticism: the world and the flesh are denied, but in such a way as to attain mastery—if not of this world then of a greater world.

20. Desert Fathers, *The Lives of the Desert Fathers (The Historia Monachorum in Aegypto),* trans. Norman Russell (Oxford: Mowbray, 1981), p. 109; Bynum, *Holy Feast and Holy Fast,* p. 38; Graham Gould, *The Desert Fathers on Monastic Community* (Oxford: Clarendon Press, 1993), p. 143; Piero Camporesi, *The Anatomy of the Senses: Natural Symbols in Medieval and Early Modern Italy,* trans. Allan Cameron (Cambridge: Polity Press, 1994), esp. chapter 4.

21. Brown, *Body and Society,* p. 223; see also Veronika Grimm, "Fasting Women in Judaism and Christianity in Late Antiquity," in *Food in Antiquity,* ed. Wilkins, Harvey, and Dobson, pp. 225–240, on pp. 231–234. Ancient theories of "innate heat" and its relation to diet are treated in Everett Mendelsohn, *Heat and Life: The Development of the Theory of Animal Heat* (Cambridge, MA: Harvard University Press, 1964), chapter 2, and in Owsei Temkin, "Nutrition from Classical Antiquity to the Baroque," in *Human Nutrition: Historic and Scientific,* ed. Iago Galdston (New York: International University Press, 1960), pp. 78–97, on pp. 85–88. For continuing medical speculation on the natural dietetics of human beings before the Fall and in Antiquity, see, for example, George Cheyne, *An Essay of Health and Long Life* (London, 1724), pp. 91–92; James Mackenzie, *The History of Health, and the Art of Preserving It,* 3rd ed. (Edinburgh, 1760), pp. 17–53; and William Smith, *A Sure Guide in Sickness and Health . . .* (London, 1776), pp. 78–81. And for the causal influence of dietetics on the sexual appetite, see Bynum, *Holy Feast and Holy Fast,* p. 37; Camporesi, *The Anatomy of the Senses,* pp. 67–69; and Aline Rouselle, *Porneia: On Desire and the Body in Antiquity,* trans. Felicia Pheasant (Oxford: Blackwell, 1988), pp. 169–178. The dependence of lust on diet remained proverbial into the early modern period; see Erasmus's quotation (Desiderius Erasmus, *Proverbs or Adages,* ed. and trans. Richard Taverner [Gainesville, FL: Scholars' Facsimiles and Reprints, 1569 (reprinted 1956)], p. 34v) of the adages "Without meate and drinke the lust of the body is colde"; "The beste way to tame carnall lust, is to kepe abstinence of meates and drinkes"; and "A licourouse [licentious] mouth, a licourouse taile."

22. St. Augustine, *Confessions,* trans. Henry Chadwick (Oxford: Oxford University Press, 1991), pp. 171, 204–207. St. Gregory of Nyssa (d. 395) described taste as "the mother of all vice" (quoted in Bynum, *Holy Feast and Holy Fast,* p. 38). And Camporesi refers (*The Anatomy of the Senses,* p. 65; cf. p. 147) to a Christian "anti-cuisine," aiming at "an alienation of taste . . . a cuisine with a minus sign, a protest against the physiological game we are forced to play by the organic cycles of the flesh."

23. In the second century, Porphyry (*On Abstinence from Animal Food,* p. 54)

wrote specifically against taking a *variety* of foods, for such diversity only fed the "variety of pleasure . . . and in this respect resembles venereal enjoyments, and the drinking of foreign wines."

24. Bynum, *Resurrection of the Body,* p. 102; cf. pp. 124–128, 148. For an anthropological interpretation of the Eucharist, see Feeley-Harnik, *The Lord's Table,* esp. pp. 63–70.

25. Cf. Ecclesiastes 5:18; 8:15; 10:19; Luke 12:19.

26. Moses Maimonides, *The Eight Chapters of Maimonides on Ethics (Shemonah Perakim): A Psychological and Ethical Treatise,* trans. and ed. Joseph I. Gorfinkle (New York: Columbia University Press, 1912 [composed ca. 1160]), pp. 60–62; idem, *The Medical Aphorisms of Moses Maimonides,* trans. and ed. Fred Rosner and Suessman Muntner, 2 vols. (New York: Yeshiva Press, 1970–1971 [composed ca. 1187–1190]), vol. 1, pp. 122; vol. 2. pp. 41–46; cf. Bynum *Holy Feast and Holy Fast,* p. 36 (for Patristic citation of Old Testament examples of holy fasting). In the seventeenth century, Spinoza's dietetics substantially fell in with the dominant tradition of Jewish philosophical moderation: "It is the part of a wise man to refresh and recreate himself with moderate and pleasant food and drink . . . For the human body is composed of very numerous parts, of diverse nature, which continually stand in need of fresh and varied nourishment, so that the whole body may be equally capable of performing all the actions, which follow from the necessity of its own nature; and, consequently, so that the mind may also be equally capable of understanding many things simultaneously. This way of life, then, agrees best with our principles, and also with general practice" (Benedict de Spinoza, "The Ethics [1677]," in *On the Improvement of Understanding, The Ethics, Correspondence,* trans. R. H. M. Elwes [New York: Dover Books, 1955], pp. 43–271, on pp. 219–220; cf. p. 241).

27. For example, Diogenes Laërtius, *Lives of Eminent Philosophers,* vol. 1, p. 165 (of Socrates: "He would say that the rest of the world lives to eat, while he himself ate to live"); Galen, *On the Passions and Errors of the Soul,* trans. Paul W. Harkins (Columbus: Ohio State University Press, 1963), pp. 49–51; Seneca, *Moral Essays,* vol. 2, pp. 119, 137, 157; Epictetus, *Discourses,* pp. 458–459.

28. Wesley D. Smith ("The Development of Classical Dietetic Theory," in *Hippocratica: Actes du Colloque Hippocratique de Paris [4–9 Septembre 1978],* ed. M. D. Grmek [Paris: CNRS, 1980], pp. 439–436, on pp. 443–444) refers to such counsel, and the dietetic knowledge that underpinned it, as "the common property of the culture": "It is probably because the tradition belonged to everyone that it did not easily take the impress of a special point of view or group and persisted essentially unchanged through the centuries." In classical usage, *dietetics* included the study and regulation of food and drink, but the term more generally signified regimen or the management of ways of living, or, in medical terms of art, the "non-naturals." For civic moderation, see Shapin, "How to Eat Like a Gentleman" (chapter 12 in this volume).

29. For surveys of the dietetic literature of Early and Late Antiquity, see, for example, Ludwig Edelstein, "The Dietetics of Antiquity [1931]," in *Ancient Medicine: Selected Papers of Ludwig Edelstein,* ed. Owsei Temkin and C. Lilian Temkin (Baltimore: Johns Hopkins University Press, 1967), pp. 303–316, esp. pp. 308–316 (for recognition of the special dietetic requirements of the scholar and philosopher); Owsei Temkin, *Galenism: Rise and Decline of a Medical Philosophy* (Ithaca, NY: Cornell University Press, 1973), esp. pp. 26, 36–39, 85; Smith, "Development of Classical Dietetic Theory"; and, notably, Michel Foucault, *The History of Sexuality,* trans. Robert Hurley, 3 vols. (vol. 1 = *An Introduction;* vol. 2 = *The Use of Pleasure;* vol. 3 = *The Care of the Self*) (New York: Vintage Books, 1988, 1990), vol. 2, pp. 97–139; also vol. 3, pp. 140–141.

30. Caesar wanted "men about me that are fat": *Julius Caesar,* 1.2. And see also Shakespeare's association of thinness, diet, and intelligence: "Fat paunches have lean pates; and dainty bits / Make rich ribs, but bankrupt quite the wits" (*Love's Labour's Lost,* 1.1); and "Methinks sometimes I have no more wit than a Christian or an ordinary man has: but I am a great eater of beef and I believe that does harm to my wit" (*Twelfth Night,* 1.3). The link was proverbial. An early modern English saying pronounced that "The sparing diet is the spirit's feast"; another (attributed to Socrates) judged that "The belly is the head's grave"; an Italian proverb said "Capo, grasso, cervello magro"; and St. Jerome referred to an old Greek adage: "A gross belly does not produce a refined mind": see Morris Palmer Tilley, *A Dictionary of the Proverbs in England in the Sixteenth and Seventeenth Centuries* (Ann Arbor: University of Michigan Press, 1966), for example, pp. 44, 156, 526.

31. St. Benedict, *The Rule of St. Benedict,* trans. Anthony C. Meisel and M. L. del Mastro (Garden City, NY: Image Books, 1975), pp. 80–81. One rule was to take and to consume what one was given without complaint and even without speech. In another exercise it would be necessary to recover precise historical distinctions between the practices designated by *abstinence, temperance, fasting,* and related locutions, though, as Bynum points out (*Holy Feast and Holy Fast,* pp. 37–38), the one term *abstinence* came to refer to practices as diverse as refraining from certain types of food, taking only one meal a day, eating no cooked foods, and eating nothing at all for a period; see also Rouselle, *Porneia,* pp. 167–169.

32. Linda Georgianna, *The Solitary Self: Individuality in the "Ancrene Wisse"* (Cambridge, MA: Harvard University Press, 1981), pp. 25–37; Aviad M. Kleinberg, *Prophets in Their Own Country: Living Saints and the Making of Sainthood in the Later Middle Ages* (Chicago: University of Chicago Press, 1992), pp. 19, 135–141; Frank Bowman, "Of Food and the Sacred: Cellini, Teresa, Montaigne," *L'Esprit Créateur* 16 (1976): 111–133, esp. pp. 111–114. For Thomas Aquinas's debate with himself over whether extreme abstinence counted as virtue or vice, see *Summa Theologica,* trans. Fathers of the English Dominican Province, 3 vols. (New York: Benziger Brothers, 1947–1948), vol. 2, pp. 1783–1792.

33. Quoted in Kleinberg, *Prophets in Their Own Country,* p. 135 and n. 14.

34. Rudolph M. Bell, *Holy Anorexia* (Chicago: University of Chicago Press, 1985), pp. 25–27; see also Piero Camporesi, "The Consecrated Host: A Wondrous Excess," trans. Anna Cancogne, in *Fragments for a History of the Human Body,* ed. M. Feher et al. (New York: Zone Books, 1989), vol. 1, pp. 221–237.

35. Bell, *Holy Anorexia,* chapter 3.

36. Bynum, *Holy Feast and Holy Fast,* pp. 24–29 (quoting p. 29).

37. Bynum, *Resurrection of the Body,* p. 221, and Julia M. H. Smith, "The Problem of Female Sanctity in Carolingian Europe c. 780–920," *Past and Present* 146 (1995): 3–37, on pp. 18–20; see also Ian MacLean, *The Renaissance Notion of Women: A Study in the Fortunes of Scholasticism and Medical Science in European Intellectual Life* (Cambridge: Cambridge University Press, 1980), esp. pp. 41–46; Bell, *Holy Anorexia;* Walter Vandereycken and Ron van Deth, *From Fasting Saints to Anorexic Girls: The History of Self-Starvation* (New York: New York University Press, 1994); Grimm, "Fasting Women," pp. 229–230; and Steven Shapin, *A Social History of Truth: Civility and Science in Seventeenth-Century England* (Chicago: University of Chicago Press, 1994), pp. 86–91 (for women's physical and social natures in relation to their effective participation in knowledge-making).

38. Thomas Elyot, *The Castel of Helth* (London, 1541), pp. 11v, 15v–16r (quoted passage), 20r, 32r–33v, 42r–43r, 53v–54r. For St. Paul, see 1 Timothy 5:23.

39. Luigi Cornaro, *The Temperate Life* [1558], in *The Art of Living Long,* ed. William F. Butler (New York: Fowler and Wells, 1903), pp. 37–113, on pp. 59–60, 75, 87. Cornaro deplored (p. 112) the fact that so many men then in monastic orders no longer lived the temperate lives originally intended for them and were "for the greater part, unhealthy, melancholy, and dissatisfied."

40. Michel Eyquem de Montaigne, "Of Experience (comp. 1588)," in *The Complete Essays of Montaigne,* trans. Donald M. Frame (Stanford, CA: Stanford University Press, 1965), pp. 815–857, on p. 827.

41. Montaigne, "Of Experience," pp. 830–832, 843.

42. Francis Bacon, "The History of Life and Death," in *The Philosophical Works of Francis Bacon,* ed. James Spedding, Robert Leslie Ellis, and Douglas Denon Heath, 5 vols. (London: Longman and Co., 1857–1858), vol. 5, pp. 213–335, on pp. 217, 251, 261, 280. For eighteenth-century medical agreement about the longevity of ancient philosophers and its dietetic cause, see Mackenzie, *History of Health,* pp. 243–244.

43. Bacon, "History of Life and Death," p. 299. Bacon also disagreed (pp. 301–302) with dominant religious and philosophical recommendations of dietary simplicity: a variety of dishes was, he said, *better* for digestion, and daintily sauced foods likewise assisted the making of good bodily juices.

44. Bacon, "History of Life and Death," pp. 261, 277, 304. Eighteenth-century medical dietetics rounded on Bacon's advocacy of occasional excess; see, for exam-

ple, Mackenzie, *History of Health,* pp. 125–126 (cf. pp. 207–212): to be "warmed with wine" does indeed assist conversation, and even philosophizing, but a "chearful glass" is not to be confused with surfeit. It was popularly but falsely attributed to Hippocrates that "getting drunk once or twice every month [w]as conducive to health." See Shapin, "How to Eat Like a Gentleman" (chapter 12 in this volume).

45. Marcilio Fincino, *Three Books on Life* [1489], trans. and ed. Carol V. Kaske and John R. Clark (Binghamton, NY: Renaissance Society of America, 1989); see also Raymond Klibansky, Erwin Panofsky, and Fritz Saxl, *Saturn and Melancholy: Studies in the History of Natural Philosophy, Religion, and Art* (London: Thomas Nelson, 1964).

46. Aristotle, *Problems,* 953a.10–15; Seneca, *Moral Essays,* vol. 1, p. 29; see also Wolf Lepenies, *Melancholy and Society,* trans. Jeremy Gaines and Doris Jones (Cambridge, MA: Harvard University Press, 1992), pp. 13–16, 31–32; Bennett Simon, *Mind and Madness in Ancient Greece: The Classical Roots of Modern Psychiatry* (Ithaca, NY: Cornell University Press, 1978), pp. 228–237.

47. [Walter Charleton], *Two Discourses, I. Concerning the Different Wits of Men: II. Of the Mysterie of the Vintners* (London, 1669), pp. 104–105. I set aside the important and related question of the relationship between genius and *mental* illness. The physiological fragility of the learned continued to be described, explained, and dietetically managed through the eighteenth and nineteenth centuries and into recent times; see, among many examples, Bernardino Ramazzini, *Diseases of Tradesmen* [1700], ed. Herman Goodman (New York: Medical Lay Press, 1933), pp. 61–65; Cheyne, *Essay of Health and Long Life* (1724), pp. xiii–xiv, 33–38, 83–87; idem, *The English Malady: Or, A Treatise of Nervous Diseases of All Kinds, as Spleen, Vapours, Lowness of Spirits, Hypochondriacal, and Hysterical Distempers, &c* (London, 1733), p. 38; Mackenzie, *History of Health,* pp. 137–140, 155–162, 187–188, 197, 223; David Lindsay Watson, "Sick Scientists," in idem, *Scientists are Human* (London: Watts and Co., 1938), pp. 29–32.

48. Evelyn to William Wotton, 30 March 1696, in John Evelyn, *Diary and Correspondence of John Evelyn, F.R.S.,* 4 vols. (London: Henry Colburn, 1854), vol. 3, p. 351. For the widely distributed late medieval and early modern delusion that one's body was made of glass, see Gill Speak, "An Odd Kind of Melancholy: Reflections on the Glass Delusion in Europe (1440–1680)," *History of Psychiatry* 1 (1990): 191–206.

49. Gilbert Burnet, "Character of a Christian Philosopher, in a Sermon Preached January 7, 1691–1692 , at the Funeral of Hon. Robert Boyle," in *Lives, Characters, and An Address to Posterity,* ed. John Jebb (London: James Duncan, 1833), pp. 325–376, on pp. 351, 360–362, 366–367; and see Shapin, *A Social History of Truth,* pp. 151–156, 185–187.

50. Richard Ward, *The Life of the Learned and Pious Dr Henry More* [1710],

abridged and ed. by M. F. Howard (London: Theosophical Publishing Society, 1911), p. 83.

51. Henry More, "Preface General," in idem, *A Collection of Several Philosophical Writings of Dr Henry More,* 2nd ed., 2 vols. (London, 1662), vol. 1, p. viii. More was here specifically situating his views in the Pythagorean and early Christian traditions of such writers as Plotinus.

52. Ward, *Life of More,* pp. 82–83. Cf. Boethius's sixth-century claim that "The body of a holy man is formed of pure aether": Ancius Manlius Serinus Boethius, *The Consolation of Philosophy,* trans. I. T., ed. William Anderson (Carbondale: Southern Illinois University Press, 1963 [composed ca. 523]), p. 99.

53. Ward, *Life of More,* pp. 84–85, 123–124, 230. Apart from bouts of fasting, More's Cambridge diet was not said to be extraordinary: he sometimes did not refrain from meat during Lent (since that abstinence "quite altered the Tone of his Body"), and "His Drink was for the most part the College Small Beer: Which, in his pleasant way of speaking, he would say sometimes, was 'Seraphical, and the Best Liquor in the World'" (ibid., p. 122; see also More to Anne Conway, 5 April and 5 August 1662, in Anne Conway, *The Conway Letters: The Correspondence of Anne, Viscountess Conway, Henry More, and Their Friends 1642–1684,* ed. Marjorie Hope Nicolson, rev. ed. Sarah Hutton [Oxford: Clarendon Press, 1992], pp. 200, 205).

54. For More's extended treatment of the dietetics of "enthusiasm," its relation to melancholy and philosophizing, and its management, see "Enthusiasmus Triumphatus," in his *Collection of Philosophical Writings,* vol. 1, esp. pp. 14, 37, 47. For contemporary views of enthusiasm and its medical management, see Michael Heyd, *"Be Sober and Reasonable": The Critique of Enthusiasm in the Seventeenth and Early Eighteenth Centuries* (Leiden: E. J. Brill, 1995), esp. chapters 2–3, and see also Rob Iliffe, "'That Puzleing Problem': Isaac Newton and the Political Physiology of Self," *Medical History* 39 (1995): 433–458, on pp. 436–439.

55. Finch to Conway, 27 April 1652; More to Conway, 28 March 1653 and 3 September 1660, in Conway, *Letters,* pp. 63, 75, 164; see also Conway to her husband, 16 September 1664, in ibid., p. 230.

56. John Finch speculated that Anne's terrible headaches might arise from "the closeness of the sutures [or pores] in your head which may hinder the perspiring of vapours; but in regard few of your sex have that inconvenience," and instructed her not to cool herself excessively "when you are very hott or sweat in your bed": Finch to Conway, 27 April 1652 and 9 April 1653, in Conway, *Letters,* pp. 63, 79; and for treatment of attempts to cure her headaches, see Schaffer, "Regeneration."

57. More to Conway, 1 May 1654, 5 June, 4 and 27 December 1660, and 31 December 1663; Conway to More, 28 November 1660, in Conway, *Letters,* pp. 96, 164, 181, 184, 220; see also Ward, *Life of More,* p. 146.

58. George Cheyne, *Natural Method of Cureing Diseases of the Body and Dis-*

orders of the Mind (London, 1742), p. 81. During periods of intense concentration, Cheyne added, Newton took "a little sack and water, without any regulation . . . as he found a craving or failure of spirits."

59. Quoted in Louis Trenchard More, *Isaac Newton: A Biography* (New York: Scribner's, 1934), pp. 129, 132; see also ibid., p. 206 and esp. Iliffe, "Isaac Newton: Lucatello Professor."

60. More, *Isaac Newton,* pp. 247, 250; Richard S. Westfall, *Never at Rest: A Biography of Isaac Newton* (Cambridge: Cambridge University Press, 1980), pp. 103–104 (for the fat cat), 580, 850–851, 866; and see William Stukeley, *Memoirs of Sir Isaac Newton's Life,* ed. A. Hastings White (London: Taylor and Francis, 1752 [reprinted 1936]), pp. 48, 60–61, 66. Newton's niece contradicted reports of his vegetarianism, and, according to More (*Isaac Newton,* p. 135), "said that he followed the rule of St. Paul to take and eat what comes from the shambles without asking questions for conscience's sake." Andrew Combe's influential mid-nineteenth-century dietetic text also denied Newton's status as an icon of vegetarianism; there was, Combe said, much evidence (including the gout from which Newton suffered) that "he did not usually confine himself to a vegetable diet": *The Physiology of Digestion, considered with Relation to the Principles of Dietetics,* ed. James Coxe, 9th ed. (Edinburgh: Maclachlan and Stewart, 1849; orig. publ. 1841), p. 149.

61. More, *Isaac Newton,* pp. 131–132, 206–207, 247.

62. Westfall, *Never at Rest,* p. 103. For important treatment of images of Newton's person and mind in the eighteenth and nineteenth centuries, see Richard R. Yeo, "Genius, Method, and Morality: Images of Newton in Britain, 1760–1860," *Science in Context* 2 (1988): 257–284.

63. George Wilson, *The Life of the Honble Henry Cavendish* (London: Cavendish Society, 1851), pp. 154 (for mutton), 169–170, 185 (for head, eyes, and hands); see also A. J. Berry, *Henry Cavendish: His Life and Scientific Work* (London: Hutchinson, 1960), pp. 15, 22. The late twentieth-century circulation of the Cavendish mutton story is indicated by David Oldroyd, "Social and Historical Studies of Science in the Classroom?" *Social Studies of Science* 20 (1990): 747–756, on pp. 751, 756 n.1.

64. Sinclair Lewis, *Arrowsmith* [1925] (New York: New American Library, 1980), p. 267. Gottlieb was here giving voice to a continuing conception of the scientific vocation as a *calling* rather than as a *job.* This subject is treated at length in Steven Shapin, *The Scientific Life: A Moral History of a Late Modern Vocation* (Chicago: University of Chicago Press, 2008), chapters 1–3.

65. Albert Einstein, *Autobiographical Notes,* trans. and ed. Paul Arthur Schilpp (La Salle, IL: Open Court, 1979), p. 3. See also Yaron Ezrahi, "Einstein and the Light of Reason," in *Albert Einstein: Historical and Cultural Perspectives,* ed. Gerald Holton and Yehuda Elkana (Princeton, NJ: Princeton University Press, 1982),

pp. 253–278, esp. pp. 268–273, for Einstein's representation of the relativistic physicist as lonely saint and magus.

66. Maud Ellmann, *The Hunger Artists: Starving, Writing, and Imprisonment* (Cambridge, MA: Harvard University Press, 1993), esp. pp. 17, 27–32, 63–69.

67. The "carnivalesque" inversion of the "proper" relationship between mind and belly was hinted at by Mikhail Bakhtin (*Rabelais and His World,* trans. Hélène Iswolsky [Bloomington: Indiana University Press, 1984], p. 171): "Most of the epithets and comparisons applied by Rabelais to spiritual things have what one might call an edible character. The author boldly states that he writes only while eating and drinking, and adds: 'Is that not the proper time to commit to the page such sublime themes and such profound wisdom?'" Indeed, the prologue to *Gargantua* makes explicit reference to the carnal habits of Diogenes the Cynic.

68. A case could also be made for Galileo as a secularizing seventeenth-century natural philosopher associated with convivial connoisseurship (especially in wines), even though stories about him also picked out his "abstemiousness" and tendencies toward melancholy; see Camporesi, *The Magic Harvest,* pp. 51–59.

69. Westfall, *Never at Rest,* pp. 580, 866; More, *Isaac Newton,* p. 127.

70. Will Durant, *The Age of Faith: A History of Medieval Civilization—Christian, Islamic, and Judaic—From Constantine to Dante: A.D. 325–1300* (New York: Simon and Schuster, 1950), p. 786. For images of late medieval monks as gluttons, see, for example, Bakhtin, *Rabelais and His World,* chapters 4–5.

71. Shapin, *The Scientific Life,* chapter 3. You'll be hungry by now, so here's my recipe for *Fricasée du poulet épistémologique:* joint one free-range chicken; brown in olive oil; in same pan add chopped garlic and soaked dried ceps (porcini); add one cup dry vermouth; reduce a little, then slowly braise covered for forty-five minutes; remove chicken to warm plate, then add some soaking water from the mushrooms and a quarter cup of sherry vinegar to the pan; reduce on high heat, pour over chicken, and garnish with chervil or Italian parsley. Serve with "foreign wine." *Bon appetit!*

72. It is in this connection that I want to draw attention to the apparently systematic changes in the topical content of intellectual biographies from the period before ca. 1850–ca. 1930 to more recent treatments. Biographical accounts in the earlier period routinely contained sections entitled "Appearance and Manner of Living" or otherwise offered detailed accounts of what intellectuals looked like, how they conducted their personal and social lives, and, indeed, what and how they ate. (For a perspicuous late example, see J. H. W. Stuckenberg, *The Life of Immanuel Kant* [London: Macmillan, 1882], chapters 4, 6.) And earlier cultures worked with conceptions of knowledge and the knower in which such details were vitally important. In the space formally occupied by such conceptions, more modern intellectual biography now confronts a great "problem," that of the narrative and the causal

relationship between what is "personal" and what is "intellectual." Following Freud, there is a recognized (if controversial) idiom for speaking of the link between the gonads and the mind, but the very suggestion that significant stories may be told connecting belly and mind now has the character of a joke; see Shapin, "Lowering the Tone in the History of Science" (chapter 1 in this volume).

73. William Butler Yeats, "Per Amica Slientia Lunae," in *Mythologies* (New York: Collier Books, 1969), pp. 317–369, on p. 341.

74. Friedrich Nietzsche, *The Gay Science* [1887], trans. Walter Kaufmann (New York: Vintage Books, 1974), pp. 34–35; and see Geoffrey Galt Harpham, *The Ascetic Imperative in Culture and Criticism* (Chicago: University of Chicago Press, 1987), chapter 4 (for Nietzschean and Foucauldian topics). For Nietzsche's intense philosophic and personal interest in dietetics, see Lesley Chamberlain, "A Spoonful of Dr Liebig's Beef Extract," *Times Literary Supplement* no. 4874 (9 August 1996): 14–15, on p. 15: "No more greasy, stodgy, beer-washed idealistic Christian German food for me! I shall curl up with gut pain, vomit if you don't give me Italian vegetables."

Chapter 12. How to Eat Like a Gentleman

1. Some examples from the Tudor to the early Hanoverian period include Thomas Elyot, *The Castel of Helthe* (1539); John Goeurot, *The Regiment of Life* (1546); Andrewe Boorde, *Compendyous Regyment or a Dyetary of Healthe* (1547); Luigi Cornaro, *The Temperate Life* (1558–1566; common in English translations of the Italian original); Thomas Cogan, *The Haven of Health* (1589); Leonard Lessius, *Hygiasticon* (ca. 1600; again in translation); William Vaughan, *Naturall and Artificial Directions for Health* (1602); Sir John Harington's verse translation of the *Regimen Sanitatis Salernitanum* (as *The Englishmans Doctor*) (1607), and several other prior translations of the same; the anonymous *The Skilful Physician* (1656); Nicholas Culpeper, *A Physicall Directory* (1649); John Archer, *Every Man His Own Doctor* (1671); Thomas Tryon, *The Way to Health, Long Life, and Happiness* (1683); and George Cheyne, *An Essay of Health and Long Life* (1724). Studies of the early modern English popular medical literature notably include Paul Slack, "Mirrors of Health and Treasures of Poor Men: The Use of the Vernacular Medical Literature of Tudor England," in *Health, Medicine and Mortality in the Sixteenth Century,* ed. Charles Webster (Cambridge: Cambridge University Press, 1979), pp. 237–273; Andrew Wear, "The Popularization of Medicine in Early Modern England," in *The Popularization of Medicine 1650–1850,* ed. Roy Porter (London: Routledge, 1992), pp. 17–41; John Henry, "Doctors and Healers: Popular Culture and the Medical Profession," in *Science, Culture and Popular Belief in Renaissance Europe,* ed. Stephen Pumfrey, Paolo L. Rossi, and Maurice Slawinski (Manchester: Manchester University Press, 1991), pp. 191–221, esp. pp. 198–201; Henry E. Sigerist, "The *Regimen Sanitatis Salernitanum* and Some of Its Commentators," in idem, *Land-*

marks in the History of Hygiene (London: Oxford University Press, 1956), pp. 20–35; Charles Webster, *The Great Instauration: Science, Medicine and Reform 1626–1660* (London: Duckworth, 1975), chapter 4 (for the ideological and political context of popular medical writing); and, for a fine survey of British and Continental dietetics, Heikki Mikkeli, *Hygiene in the Early Modern Medical Tradition,* Annals of the Finnish Academy of Sciences and Letters, Humaniora, no. 305 (Helsinki: Academia Scientiarum Fennica, 1999).

2. For studies of the courtesy and related genres, see Anna Bryson, *From Courtesy to Civility: Changing Codes of Conduct in Early Modern England* (Oxford: Oxford University Press, 1998); Frank Whigham, *Ambition and Privilege: The Social Tropes of Elizabethan Courtesy Literature* (Berkeley: University of California Press, 1984); John E. Mason, *Gentlefolk in the Making: Studies in the History of Courtesy Literature and Related Topics from 1531 to 1774* (New York: Octagon Books 1971; orig. publ. 1935); George C. Brauer, Jr., *The Education of a Gentleman: Theories of Gentlemanly Education in England, 1660–1775* (New York: Bookman Associates, 1959); W. L. Ustick, "Advice to a Son: A Type of Seventeenth Century Conduct Book," *Studies in Philology* 29 (1932): 409–441; idem, "Changing Ideals of Aristocratic Character and Conduct in Seventeenth-Century England," *Modern Philology* 30 (1933): 147–166; Fenela Ann Childs, "Prescriptions for Manners in English Courtesy Literature, 1690–1760, and Their Social Implications" (D.Phil. diss., Oxford University, 1984); Gertrude Elizabeth Noyes, *Bibliography of Courtesy and Conduct Books in Seventeenth-Century England* (New Haven, CT: Tuttle, Morehouse & Taylor, 1937); and Peter Burke, *The Fortunes of the 'Courtier': The European Reception of Castiglione's 'Cortegiano'* (Oxford: Polity Press, 1995).

3. For example, J. T. Cliffe, *The World of the Country House in Seventeenth-Century England* (New Haven, CT: Yale University Press, 1999), pp. 166–167.

4. Of Francis Osborne's *Advice to His Son,* Pepys said that "I shall not never admire it enough for sense and language": *The Diary of Samuel Pepys,* ed. Robert C. Latham and William Matthews, 11 vols. (London: HarperCollins, 1995), vol. 4, p. 96 (5 April 1663); also vol. 2, p. 199 (19 October 1661); vol. 5, p. 10 (9 January 1663–1664) and p. 27 (27 January 1663–1664). The decidedly ungenteel experimentalist Robert Hooke possessed four or five courtesy books, though they did not notably affect either his deportment or his diet: Leona Rostenberg, *The Library of Robert Hooke: The Scientific Book Trade of Restoration England* (Santa Monica, CA: Modoc Press, 1989), pp. 198, 205–206; Shapin, "Who Was Robert Hooke?" (chapter 9 in this volume).

5. Aubrey's selection included works by Peacham, Osborne, Ralegh, de Courtin, and Shaftesbury, and the essays of Montaigne and Bacon: *Aubrey on Education: A Hitherto Unpublished Manuscript by the Author of 'Brief Lives,'* ed. J. E. Stephens (London: Routledge & Kegan Paul, 1972; comp. from 1669 to ca. 1694), pp. 131–132.

6. Ludwig Edelstein, "The Dietetics of Antiquity," in idem, *Ancient Medicine: Selected Papers of Ludwig Edelstein,* ed. Owsei Temkin and C. Lilian Temkin (Baltimore: Johns Hopkins Press, 1967; art. orig. publ. 1931), pp. 303–316, on p. 303.

7. Owsei Temkin, *Galenism: Rise and Decline of a Medical Philosophy* (Ithaca, NY: Cornell University Press, 1973), pp. 39–40 (quoting Galen, *De sanitate tuenda*).

8. Owsei Temkin, *Hippocrates in a World of Pagans and Christians* (Baltimore: Johns Hopkins University Press, 1991), pp. 15, 45–47; see also Sigerist, "Galen's Hygiene," in idem, *Landmarks in the History of Hygiene,* pp. 1–19, esp. p. 12.

9. Michel Foucault, *The History of Sexuality,* 3 vols., trans. Robert Hurley (New York: Vintage Books, 1988–1990), vol. 2, pp. 97–139; see also vol. 3, pp. 140–141. (In these connections, Foucault did not cite Edelstein et al., nor, in the approved French academic manner, did he acknowledge the work of any other living or recently deceased scholar. But it is hard to imagine that, as his references imply, Foucault relied solely on ancient primary sources for his knowledge of dietetics.) For present-day classicists' and moral philosophers' reflections on ethics and body management in Antiquity, see also Martha Nussbaum, *The Therapy of Desire: Theory and Practice in Hellenistic Ethics* (Princeton, NJ: Princeton University Press, 1994); John Cottingham, *Philosophy and the Good Life: Reason and the Passions in Greek, Cartesian and Psychoanalytic Ethics* (Cambridge: Cambridge University Press, 1998); Alexander Nehamus, *The Art of Living: Socratic Reflections from Plato to Foucault* (Berkeley: University of California Press, 1998); and esp. James Davidson, *Courtesans and Fishcakes: The Consuming Passions of Classical Athens* (London: HarperCollins, 1997).

10. Keith Thomas, "Health and Morality in Early Modern England," in *Morality and Health,* ed. Allan M. Brandt and Paul Rozin (New York: Routledge, 1997), pp. 15–34, on p. 20; see also idem, *Man and the Natural World: Changing Attitudes in England 1500–1800* (London: Allen Lane, 1983), pp. 289–300 (esp. for vegetarian dietetics and ethics); and Slack, "Mirrors of Health."

11. For example, William Coleman, "Health and Hygiene in the *Encyclopédie:* A Medical Doctrine for the Bourgeoisie," *Journal of the History of Medicine* 29 (1974): 399–421; idem, "The People's Health: Medical Themes in 18th-Century French Popular Literature," *Bulletin of the History of Medicine* 51 (1977): 55–74; Christopher J. Lawrence, "William Buchan: Medicine Laid Open," *Medical History* 19 (1975): 20–36; Charles E. Rosenberg, "Medical Text and Social Context: Explaining Buchan's *Domestic Medicine,*" *Bulletin of the History of Medicine* 57 (1983): 22–42; idem, "The Therapeutic Revolution: Medicine, Meaning, and Social Change in Nineteenth-Century America," *Perspectives in Biology and Medicine* 20 (1977): 485–506; idem, "Florence Nightingale on Contagion: The Hospital as Moral Universe," in idem, ed., *Healing and History: Essays for George Rosen* (New York: Science History Publications, 1982), pp. 116–136; Ginnie Smith, "Prescribing the Rules of Health: Self-Help and Advice in the Late Eighteenth Century," in *Patients*

and Practitioners: Lay Perceptions of Medicine in Pre-Industrial Society, ed. Roy Porter (Cambridge: Cambridge University Press, 1985), pp. 249–282.

12. Thomas Elyot, *The Book Named The Governor,* ed. S. E. Lehmberg (London: Dent, 1962; orig. publ. 1531), p. 214.

13. [Thomas Gainsford], *The Rich Cabinet furnished with varieties of Excellent Discriptions, Exquisite Characters, Witty Discourses, and Delightfull Histories, Devine and Morrall . . . Whereunto is Annexed the Epitome of Good Manners, extracted from Mr. John de la Casa . . .* (Amsterdam: Da Capo Press for Theatrum Orbis Terrarum, 1972; orig. publ. 1616), p. 143^v.

14. James VI, king of Scotland, later James I of England, *Basilicon Doron,* facsimile edition (Menston: The Scolar Press, 1969; orig. publ. 1599), p. 126.

15. James Cleland, *The Instruction of a Young Noble-man* (Oxford, 1612), p. 211. Large sections of Cleland's tract are just (unacknowledged) quotations or close paraphrases of the king's *Basilicon Doron.*

16. Henry Peacham, *The Complete Gentleman,* ed. Virgil B. Heltzel (Ithaca, NY: Cornell University Press, for The Folger Shakespeare Library, 1962; orig. publ. 1622, 1634), pp. 151–152.

17. Gilbert Burnet, *Thoughts on Education,* ed. John Clarke (Aberdeen: Aberdeen University Press, 1914; posth. publ. 1761; comp. ca. 1668), p. 13. Burnet concurred with common early modern hereditary sentiment in urging that care be taken in selecting children's wet-nurses for their dietary temperance and good moral character (ibid., pp. 14–15).

18. Clement Ellis, *The Gentile Sinner, or England's Brave Gentleman Character'd in a Letter to a Friend: Both As He Is, and As He Should Be,* 4th ed. (Oxford, 1668; orig. publ. 1660), p. 193.

19. Jean Gailhard, *The Compleat Gentleman: Or Directions for the Education of Youth,* 2 parts, separately paginated (London, 1678), part 1, p. 88.

20. John Locke, *Some Thoughts Concerning Education,* ed. R. H. Quick (Cambridge: Cambridge University Press, 1899; orig. publ. 1693), pp. 9–12.

21. Anthony Ashley Cooper, Third Earl of Shaftesbury, *An Inquiry Concerning Virtue, in Two Discourses . . .* (London, 1699), pp. 162–173; see also Henry Richard Fox Bourne, *The Life of John Locke,* 2 vols. (Darmstadt: Scientia Verlag Aalen, 1969; orig. publ. 1876), vol. 2, pp. 255–264.

22. Bryson, *From Courtesy to Civility,* p. 121; Lawrence Stone, *The Crisis of the Aristocracy 1558–1641* (Oxford: Clarendon Press, 1965), pp. 555–562; Roy Porter and G. S. Rousseau, *Gout: The Patrician Malady* (New Haven, CT: Yale University Press, 1998), esp. pp. 48–59; see also Roy Porter, *English Society in the Eighteenth Century* (Harmondsworth: Penguin, 1982), pp. 3–35, 233–235; Andrew B. Appleby, "Diet in Sixteenth-Century England: Sources, Problems, Possibilities," in *Health, Medicine and Mortality in the Sixteenth Century,* ed. Webster, pp. 97–116; and Peter Earle, *The Making of the English Middle Class: Business, Society and Family Life in*

London, 1660–1730 (Berkeley: University of California Press, 1989), pp. 272–281. Earle estimates that even those Londoners in "the middle station" of society consumed meat four or five days a week.

23. Peacham, *The Complete Gentleman,* pp. 153, 235–236, 238–239.

24. Richard Lingard, *A Letter of Advice to a Young Gentleman Leaving the University Concerning His Behaviour and Conversation in the World,* ed. Frank C. Erb (New York: McAuliffe & Booth, 1907; orig. publ. 1670), p. 41; see also Thomas, "Health and Morality," p. 21, and Burke, *The Fortunes of the 'Courtier,'* pp. 113–115.

25. In the 1540s, Sir Thomas Elyot celebrated the nutritional virtues of English roast beef, while recognizing the possible dangers of excess: "Biefe of Englande to Englysshemen, whiche are in helth, bringeth stronge nouryshynge, but it maketh grosse bloude, and ingendreth melancoly" (*The Castel of Helthe* [London, 1541], p. 16ʳ).

26. Porter, *English Society in the Eighteenth Century,* pp. 34–35.

27. James Boswell, "The Life of Samuel Johnson," in *The Portable Johnson & Boswell,* ed. Louis Kronenberger (New York: Viking, 1947), pp. 122–123.

28. On this, see the learned and entertaining Davidson, *Courtesans and Fishcakes.*

29. Desiderius Erasmus, *The Education of a Christian Prince,* trans. Lester K. Born (New York: Octagon, 1965; orig. publ. 1516), p. 209.

30. Baldesar Castiglione, *The Book of the Courtier,* trans. Charles S. Singleton (Garden City, NY: Doubleday Anchor, 1959; orig. publ. 1528), p. 135.

31. Stefano Guazzo, *The Civile Conversation of M. Steeven Guazzo,* trans. George Pettie and Barth[olomew] Young, ed. Sir Edward Sullivan, 2 vols. (London: Constable and Co., 1925; orig. publ. in Italian in 1574 and in English 1581, 1586), vol. 2, p. 137.

32. François, duc de La Rochefoucauld, *The Maxims of La Rochefoucauld,* trans. Louis Kronenberger (New York: Random House, 1959; orig. publ. 1665), p. 142.

33. See, among very many such ancient arguments for dietary moderation influential in early modern England, Plutarch, "Rules for the Preservation of Health," in *Plutarch's Lives and Miscellanies,* ed. A. H. Clough and William W. Goodwin, 5 vols. (New York: The Colonial Company, 1905), vol. 1, pp. 251–279, and "Plutarch's Symposiacs," in ibid., vol. 3, pp. 197–460, esp. pp. 290–295, 339, 394–398. Of course, the average educated Englishman—even equipped with decent school Latin—was more likely to encounter both the ethical and the medical knowledge of Antiquity via early modern English summaries and compendia.

34. On the body as divine temple in English Protestant thought, see, for example, Keith Thomas, "Cleanliness and Godliness in Early Modern England," in *Religion, Culture and Society in Early Modern Britain: Essays in Honour of Patrick Collinson,* ed. Anthony Fletcher and Peter Roberts (Cambridge: Cambridge University Press, 1994), pp. 56–83, esp. pp. 62–63; idem, "Health and Morality," pp. 16–18.

35. Giovanni Della Casa, *Galateo,* trans. Konrad Eisenbichler and Kenneth R. Bartlett (Toronto: Centre for Reformation and Renaissance Studies, 1986; orig. publ. 1558, and widely available in English translation from 1576), pp. 9–10, 57–59; Antoine de Courtin, *The Rules of Civility; or, Certain Ways of Deportment observed amongst all Persons of Quality upon Several Occasions, newly revised and much enlarged* (London, 1685; orig. publ. 1671), pp. 122–145; Peacham, *The Complete Gentleman,* p. 153; Ellis, *The Gentile Sinner,* p. 189; Gailhard, *The Compleat Gentleman,* part 1, p. 90; Josiah Dare, *Counsellor Manners: His Last Legacy to His Son* (New York: Coward-McCann. 1929; orig. publ. 1672), pp. 16–18; John Evelyn, *A Character of England,* in *Harleian Miscellany,* ed. T. Park (London, 1808–1813), vol. 10, pp. 189–198; Cleland, *The Instruction of a Young Noble-man,* pp. 211–212; Bryson, *From Courtesy to Civility,* pp. 83, 93, 121; idem, "The Rhetoric of Status: Gesture, Demeanour and the Image of the Gentleman in Sixteenth- and Seventeenth-Century England," in *Renaissance Bodies: The Human Figure in English Culture c. 1540–1660,* ed. Lucy Gent and Nigel Llewellyn (London: Reaktion Books, 1990), pp. 136–153, on pp. 145, 150–151.

36. Norbert Elias, *The Civilizing Process,* trans. Edmund Jephcott, 2 vols. [vol. 1 = *The History of Manners;* vol. 2 = *The Court Society*] (Oxford: Basil Blackwell, 1978, 1983), esp. vol. 1, chapter 2, part 4 ("On Behavior at Table"); also Stephen Mennell, *All Manners of Food: Eating and Taste in England and France from the Middle Ages to the Present* (Oxford: Basil Blackwell, 1985); idem, "On the Civilizing of Appetite," in *The Body: Social Process and Cultural Theory,* ed. Mike Featherstone, Mike Hepworth, and Bryan S. Turner (London: Sage, 1991), pp. 126–156; Janet Whatley, "Food and the Limits of Civility," *Sixteenth Century Journal* 15 (1984): 387–400; cf. the historical criticisms of Elias in Bryson, *From Courtesy to Civility,* p. 105. Cultural anthropologists, of course, have made a meal of food, and food-giving, symbolism; see, among very many examples, Mary Douglas and Jonathan L. Gross, "Food and Culture: Measuring the Intricacy of Rule Systems," *Social Science Information* 20 (1981): 1–35; Jack R. Goody, *Cooking, Cuisine and Class* (Cambridge: Cambridge University Press, 1982); Claude Lévi-Strauss, *The Raw and the Cooked,* trans. John and Doreen Weightman (New York: Harper & Row, 1969; orig. publ. 1964); idem, *From Honey to Ashes,* trans. John and Doreen Weightman (New York: Harper & Row, 1973; orig. publ. 1966); and idem, *The Origin of Table Manners,* trans. John and Doreen Weightman (New York: Harper & Row, 1978; orig. publ. 1968). So, too, have cultural historians, for example, Caroline Walker Bynum, *Holy Feast and Holy Fast: The Religious Significance of Food to Medieval Women* (Berkeley: University of California Press, 1987); Peter Brown, *The Body and Society: Men, Women and Sexual Renunciation in Early Christianity* (London: Faber and Faber, 1989); and Piero Camporesi, *Bread of Dreams: Food and Fantasy in Early Modern Europe,* trans. David Gentilcore (Cambridge: Polity Press, 1989).

37. King James, *Basilicon Doron,* p. 124.

38. Gailhard, *The Compleat Gentleman,* part 2, pp. 67–68; see also Cleland, *The Instruction of a Young Noble-man,* p. 210; Della Casa, *Galateo,* p. 57; De Courtin, *Rules of Civility,* pp. 131–133; Thomas, "Health and Morality," p. 27.

39. King James, *Basilicon Doron,* pp. 124, 126. Here, as elsewhere, Cleland closely followed royal advice: *The Instruction of a Young Noble-man,* pp. 207–213. See also Erasmus, *The Education of a Christian Prince,* p. 209, and Sir John Harington's popular verse translation of the Salernitan canon: *The School of Salernum: Regimen Sanitatis Salerni* (Salerno: Ente Provinciale per Il Turismo, 1957; orig. publ. 1607), p. 50: "A King that cannot rule him in his dyet, / Will hardly rule his Realme in peace and quiet." (Harington was a favorite of Henry, prince of Wales, and James Cleland dedicated his *Instruction* to Harington.)

40. Laurent Joubert, *The Second Part of the Popular Errors,* trans. Gregory David de Rocher (Tuscaloosa: University of Alabama Press, 1995; orig. publ. 1579), pp. 256–257. I here quote a physician on the subject of princely obligations because Joubert's account is particularly clear and extended, but his counsel is pervasively, if more diffusely, echoed in the practical ethical literature.

41. [William Cecil, Baron Burghley], *The Counsell of a Father to His Sonne, in Ten Severall Precepts, Left as a Legacy at His Death* (London, 1611), broadsheet; Dare, *Counsellor Manners,* pp. 60–61.

42. Gainsford, *The Rich Cabinet,* p. 36^v.

43. Bryson, *From Courtesy to Civility,* pp. 83–85; Brauer, *The Education of a Gentleman,* pp. 25–27; Thomas, "Health and Morality," p. 27; De Courtin, *Rules of Civility,* p. 123.

44. Burghley, *Counsell of a Father;* Peter Charron, *Of Wisdome,* trans. Sansom Lennard (London, 1612), p. 540.

45. Gainsford, *The Rich Cabinet,* p. 51^r.

46. Castiglione, *The Courtier,* pp. 299, 302.

47. Elyot, *The Governor,* p. 209.

48. Ellis, *The Gentile Sinner,* p. 131; see also the similarly Puritanical Richard Bra[i]thwait, *The English Gentleman. Containing Sundry Excellent Rules or Exquisite Observations, tending to Direction of Every Gentleman, of Selecter Ranke and Qualitie* (London, 1630), pp. 305–372, esp. pp. 306, 310.

49. Walter Ralegh, "Sir Walter Ralegh's Instructions to His Son and to Posterity," in idem, *The Works of Sir Walter Ralegh, Kt.,* 8 vols. (London, 1876), vol. 8, pp. 557–570, on p. 568. Ralegh was here quoting Anacharsis, reputed as one of the Seven Sages of Antiquity.

50. Lingard, *A Letter of Advice,* pp. 16–17.

51. William de Britaine, *Humane Prudence, or the Art by which a Man May Raise Himself & Fortune to Grandeur,* 3rd ed. (London, 1686), p. 31.

52. Locke, *Some Thoughts Concerning Education,* p. 25.

53. Castiglione, *The Courtier*, p. 302.

54. Elyot, *The Governor*, p. 209.

55. Lodowick Bryskett, *A Discourse of Civill Life*, ed. Thomas E. Wright (Northridge, CA: San Fernando Valley State College, 1970; orig. publ. 1606), pp. 48, 162. This was apparently a popular early modern interpretation of Plato's *Republic*, IV. 430e–432b.

56. Gainsford, *The Rich Cabinet*, p. 144^r; Brathwait, *The English Gentleman*, p. 311.

57. King James, *Basilicon Doron*, pp. 100–101; see also Brathwait, *The English Gentleman*, pp. 305, 311. The identification of temperance as master-virtue was not, of course, uncontested in the early modern period. For *prudence* "as the generall Queene, superintendent, and guide of all other vertues," see Charron, *Of Wisdome*, p. 350. Temperance was, for Charron, "not a speciall vertue, but generall and common, the seasoning sauce of all the rest" (ibid., p. 532). The translator dedicated the book to Henry, prince of Wales.

58. For the Galenic sources of the doctrine and phrase, see L. J. Rather, "The 'Six Things Non-Natural': A Note on the Origins and Fate of a Doctrine and a Phrase," *Clio Medica* 3 (1968): 337–347; Saul Jarcho, "Galen's Six Non-Naturals: A Bibliographic Note and Translation," *Bulletin of the History of Medicine* 44 (1970): 372–377; Peter Niebyl, "The Non-Naturals," *Bulletin of the History of Medicine* 45 (1971): 486–492; Jerome J. Bylebyl, "Galen on the Non-Natural Causes of Variation in the Pulse," *Bulletin of the History of Medicine* 45 (1971): 482–485; and Mikkeli, *Hygiene*, chapter 1. For a study of the non-naturals in French early modern popular medicine, see Antoinette Emch-Dériaz, "The Non-Naturals Made Easy," in *The Popularization of Medicine*, ed. Porter, pp. 134–159, and Coleman, "Health and Hygiene in the *Encyclopédie*."

59. Henry Peacham, *The Truth of Our Times* [1638] (bound together with idem, *The Complete Gentleman*), pp. 175–239, on p. 239.

60. Locke, *Some Thoughts Concerning Education*, pp. 9–24. Among other practical ethical tracts organizing their advice—to a greater or lesser extent—through a list of the non-naturals, see King James's *Basilicon Doron*.

61. For the medical context of Descartes's views on the passions, see, for example, Shapin, "Descartes the Doctor" (chapter 15 in this volume). For an entry into the early modern literature on the passions, see, for example, Susan James, *Passion and Action: The Emotions in Seventeenth-Century Philosophy* (Oxford: Clarendon Press, 1997); Stephen Gaukroger, ed., *The Soft Underbelly of Reason: The Passions in the Seventeenth Century* (London: Routledge, 1998); Cottingham, *Philosophy and the Good Life*, esp. chapter 3; and Jon Elster, *Alchemies of the Mind: Rationality and the Emotions* (Cambridge: Cambridge University Press, 1999).

62. George Cheyne, *An Essay of Health and Long Life* (London, 1724), p. 5.

This passage is also quoted by Thomas's fine "Health and Morality," p. 24 (and see ibid., pp. 20–24, for Thomas's appreciation of the ethical significance of the non-naturals).

63. Francis Bacon, "The History of Life and Death . . . ," in idem, *The Philosophical Works of Francis Bacon,* ed. James Spedding, Robert Leslie Ellis, and Douglas Denon Heath, 5 vols. (London: Longman and Co., 1857–1858; essay post. publ. 1636), vol. 5, pp. 213–335, on p. 29.

64. In a mid-sixteenth-century translation of Erasmus's *Proverbs* the Latin tag was given as "Without meate and drinke the lust of the body is colde"; alternatively, "The beste way to tame carnall lust, is to kepe abstinence of meates and drinkes"; and "A licorouse [licentious] mouth a licourouse taile." Desiderius Erasmus, *Proverbs or Adages,* ed. and trans. Richard Taverner (London, 1569), p. 34ᵛ. For medical proverbs generally, see Archer Taylor, *The Proverb* (Berlin: Peter Lang, 1985; orig. publ. 1931), pp. 121–129.

65. Cleland, *The Instruction of a Young Noble-man,* p. 209; see also Gainsford, *The Rich Cabinet,* p. 134ᵛ.

66. Peacham, *The Complete Gentleman,* p. 151; see also Gainsford, *The Rich Cabinet,* p. 36ᵛ; Castiglione, *The Courtier,* p. 302. The literature produced by religious ascetics from the early Christian period is especially rich in appreciations of the causal influence of diet on sexual desire, while such ancient secular thinkers as Seneca laid much stress on the diet-anger connection. For an entry to this material, see Shapin, "The Philosopher and the Chicken" (chapter 11 in this volume), and, esp., Bynum, *Holy Feast and Holy Fast,* pp. 35–37.

67. Charron, *Of Wisdome,* p. 540.

68. George Cheyne, *The English Malady: or, a Treatise of Nervous Diseases of All Kinds,* ed. Roy Porter (London: Tavistock/Routledge, 1991; orig. publ. 1733); see also Anita Guerrini, *Obesity and Depression in the Enlightenment: The Life and Times of George Cheyne* (Norman: University of Oklahoma Press, 2000), esp. chapter 6; Bryan S. Turner, "The Government of the Body: Medical Regimens and the Rationalization of Diet," *British Journal of Sociology* 33 (1982): 254–269; idem, "The Discourse of Diet," *Theory, Culture and Society* 1 (1982): 23–32; Shapin, "Trusting George Cheyne" (chapter 13 in this volume).

69. Shapin, "The Philosopher and the Chicken" (chapter 11 in this volume).

70. Robert Burton, *The Anatomy of Melancholy,* ed. Floyd Dell and Paul Jordan-Smith (New York: Tudor Publishing Co., 1927; orig. publ. 1628), p. 200. The seventeenth-century sense of "Cockney" (a hen's egg, and by analogy a coddled child) did indeed pick out town-dwellers in general and Londoners in particular, but more directly pointed to people who were effete and squeamish—milksops. See also Piero Camporesi, *The Anatomy of the Senses: Natural Symbols in Medieval and Early Modern Italy,* trans. Allan Cameron (Cambridge: Polity Press, 1994), p. 65

(for hunger as "the cheapest and most universal of drugs"), and chapter 4 generally (for the early modern cultural significance of dietary abstinence).

71. Peacham, *The Complete Gentleman,* p. 154. Peacham was here quoting Ecclesiasticus, a well-known compilation of maxims from the second century BCE. The full passage is: "What is life to a man derived of wine?/Was it not created to warm men's hearts?/Wine brings gaiety and high spirits,/if a man knows when to drink and when to stop;/but wine in excess makes for bitter feelings/and leads to offence and retaliation" (Jesus Ben Sira, *Ecclesiasticus or The Wisdom of Jesus Son of Sirach,* ed. John G. Snaith [Cambridge: Cambridge University Press, 1974], p. 154).

72. See notably Mikhail Bakhtin, *Rabelais and His World,* trans. Hélène Iswolsky (Bloomington: Indiana University Press, 1984; orig. publ. 1965). As Peter Burke nicely pointed out, "It was meat which put the *carne* in Carnival" (*Popular Culture in Early Modern Europe* [London: Temple Smith, 1978], p. 186 [and chapter 7 for "The World of Carnival"]).

73. Francis Bacon, "The Advancement of Learning [Books I–II]," in idem, *The Philosophical Works,* vol. 3, pp. 253–491, on p. 373; see also Shapin, "Descartes and the Doctor" (chapter 15 in this volume).

74. Francis Bacon, "Of Regiment of Health," in idem, *The Essayes or Counsels, Civill and Morall,* ed. Michael Kiernan (Cambridge, MA: Harvard University Press, 1985; orig. publ. 1625), pp. 100–102, on p. 101. Bacon's was a creative reading of what Celsus actually had to say on the matter in *De medicina,* while the editor of this edition of the *Essayes* more circumspectly judges that "the notion of the 'benigne Extreme' is Bacon's emphasis" (ibid., p. 237).

75. Bacon, "History of Life and Death," pp. 261–262.

76. William Rawley, "The Life of the Right Honourable Francis Bacon," in *The Works of Francis Bacon,* ed. James Spedding, Robert Leslie Ellis, and Douglas Denon Heath, 15 vols. (Boston: Brown and Taggard, 1860–1864; art. comp. 1670), vol. 1, pp. 3–18, on pp. 16–17.

77. Quoted in *The Works of Francis Bacon,* vol. 14, p. 567.

78. *The Works of Francis Bacon,* vol. 14, pp. 261, 277, 295. Like Robert Burton (quoted above), Bacon was here explicitly criticizing the pedantic dietary precision of Luigi Cornaro's influential *De vita sobria* (1558).

79. Bacon, "History of Life and Death," p. 304. This respecification of dietary moderation is briefly noted in Shapin, "The Philosopher and the Chicken" (chapter 11 in this volume). This is as far as Celsus went on the matter: "It is well . . . to attend at times a banquet, at times to hold aloof; to eat more than sufficient at one time, at another no more; to take food twice rather than once a day, and always as much as one wants provided one digests it" (Aulus Cornelius Celsus, *De medicina,* trans. W. G. Spencer, 3 vols. [London: Heinemann; Cambridge, MA: Harvard University Press, 1960], vol. 1, p. 43). But slightly later (ibid., p. 49) Celsus wrote that

"Coming to food, a surfeit is never of service, [and] excessive abstinence is often unserviceable."

80. Rawley, "Life of Bacon," p. 17 (for "grosser humours"); Bacon to Sir Humphrey May (May [1623]), in *The Works of Francis Bacon,* vol. 14, p. 515: "You may perhaps think me partial to Potycaries, that have been ever puddering in physic all my life." See also Lisa Jardine, *Ingenious Pursuits: Building the Scientific Revolution* (New York: Nan A. Talese, Doubleday, 1999), pp. 294–295. For details of Bacon's self-medication, and also what he took from apothecaries, see "The Letters and the Life, Vol. IV," in *The Works of Francis Bacon,* vol. 14, pp. 28–29, 53–54, 78–80 (the *Memoriæ valetudinis,* recording Bacon's struggles with stone, gout, and melancholy); vol. 8, pp. 200, 209, 328, 335; vol. 14, pp. 10, 335, 398–399, 431, 514–515, 566–567 (his recipe for the rhubarb purgative and how he used it). On the mend from an illness, Bacon wrote to Buckingham (29 August 1623): "I thank God, I am prettily recovered; for I have lain at two wards, the one against my disease, the other against my physicians, who are strange creatures" (*The Works of Francis Bacon,* vol. 14, p. 431).

81. Michel Eyquem de Montaigne, "Of Experience," in *The Complete Essays of Montaigne,* trans. Donald M. Frame (Stanford, CA: Stanford University Press, 1965; orig. publ. 1580, 1588), pp. 815–857, on p. 830.

82. Thomas Browne, *Pseudodoxia Epidemica: Or, Enquiries into Very Many Received Tenents, And Commonly Presumed Truths* (London, 1650), p. 229.

83. Laurent Joubert, *Popular Errors,* trans. Gregory David de Rocher (Tuscaloosa: University of Alabama Press, 1989; orig. publ. 1579), p. 247; idem, *The Second Part of Popular Errors,* p. 263.

84. John Aubrey, "Thomas Hobbes," in *Aubrey's Brief Lives,* ed. Oliver Lawson Dick (Ann Arbor: University of Michigan Press, 1975), pp. 147–159, on p. 155.

85. James Mackenzie, *The History of Health, and the Art of Preserving It,* 3rd ed. (Edinburgh, 1760; orig. publ. 1758), pp. 125–126. Later on (p. 135n.), Mackenzie made it clear that he was specifically criticizing Bacon's commendation of excess, and also the legitimacy of recruiting Celsus as an approving authority. See also George Cheyne, *An Essay of Health and Long Life* (London, 1724), pp. 47–48. Others attributed the medical authority of such dangerous counsel to Avicenna.

86. Bacon, "Of Regiment of Health," p. 101. A historian of political thought perceptively discerns in this aphorism the "summation" of Bacon's moral philosophy: Ian Box, "Bacon's Moral Philosophy," in *The Cambridge Companion to Bacon,* ed. Markku Peltonen (Cambridge: Cambridge University Press, 1996), pp. 260–282, on p. 278.

87. See, esp., Julian Martin, *Francis Bacon, the State, and the Reform of Natural Philosophy* (Cambridge: Cambridge University Press, 1992).

88. La Rochefoucauld, *Maxims,* p. 155.

89. Rawley, "Life of Bacon," p. 17; see also Shapin, "The Philosopher and the Chicken" (chapter 11 in this volume).

90. Celsus, *De medicina,* vol. 1, p. 43; see also Mark Grant, *Dieting for an Emperor: A Translation of Books 1 and 4 of Orabasius' 'Medical Compilations' with an Introduction and Commentary* (Leiden: Brill, 1997), p. 12.

91. For example, Joubert, *The Second Part of the Popular Errors,* p. 263 (the healthy man "while he is feeling well, belongs to himself and does not have to follow any rule or diet nor consult a physician"); John Arbuthnot, *An Essay Concerning the Nature of Aliments, and the Choice of Them, according to the Different Constitutions of Human Bodies . . .,* 4th ed. (London: J. and R. Tonson, 1756; orig. publ. 1731), pp. 178–179 ("a healthy Man, under his own Government, ought not to tie himself to strict Rules, nor to abstain from any Sort of Food in common Use"); Mackenzie, *The History of Health,* p. 135 ("A man who is sound and strong should ty himself down to no particular rule of diet, nor imagine that he stands in need of a physician"); Benito Jerónimo Feijóo y Montenegro, *Rules for Preserving Health, particularly with regard to Studious Persons,* trans. anon. from the Spanish (London, [1800?]), pp. 79–80, 85; see also Mikkeli, *Hygiene,* pp. 93–94, 102, 106, 108, 125, 130–131, 141. For the Rule in a popular medical text by a nonprofessional, see Elyot, *Castel of Helthe,* p. 45ʳ ("[A]s Cornelius Celsus saith, A man that is hole and well at ease, and is at his lybertie, ought not to bynde him selfe to rules, or nede a phisition"); also Leonard Lessius, *Hygiasticon: Or, A Treatise of the Means of Health and Long Life,* in *A Treatise of Health and Long Life, with the Sure Means of Attaining It, in Two Books. The First by Leonard Lessius, The Second by Lewis Cornaro, a Noble Venetian: Translated into English, by Timothy Smith, Apothecary* (London, 1743; orig. publ. ca. 1600), pp. 23–24 ("For [generally speaking] any Sort of Food that is common to one suits agreeably enough with hale Constitutions").

92. King James, *Basilicon Doron,* pp. 125, 127; and, following the king, Cleland, *The Instruction of a Young Noble-man,* p. 212: "I thinke it best to accustome your selfe unto the Countrie where you are."

93. De Courtin, *Rules of Civility,* pp. 128–130.

94. Peacham, *The Complete Gentleman,* p. 151; see also Lingard, *Letter of Advice,* p. 41.

95. Locke, *Some Thoughts Concerning Education,* pp. 11–12.

96. Gailhard, *The Compleat Gentleman,* part 1, pp. 85–86; Edward Panton, *Speculum Juventutis: or, a True Mirror . . .* (London, 1671), p. 188.

97. Montaigne, "Of Experience," p. 821.

98. See Shapin, "Descartes the Doctor" (chapter 15 in this volume).

99. Montaigne, "Of Experience," pp. 827, 830, 832. See also Margaret Brunyate, "Montaigne and Medicine," in *Montaigne and His Age,* ed. Keith Cameron (Exeter: University of Exeter Press, 1981), pp. 27–38.

100. Montaigne, "Of Experience," p. 827.

101. Montaigne, "Of Experience," p. 833.

102. Montaigne, "Of Experience," p. 832.

103. For a summary of the links between free-action and gentlemanly identity in early modern English culture, see Steven Shapin, *A Social History of Truth: Civility and Science in Seventeenth-Century England* (Chicago: University of Chicago Press, 1994), esp. chapters 2–3.

104. Montaigne, "Of Experience," pp. 830–831. Cf. Charron, *Of Wisdome* (1612), p. 197: "It is one of the vanities & follies of man, to prescribe lawes and rules that exceed the use and capacitie of men, as some Philosophers and Doctors have done. They propose strange and elevated formes or images of life, or at least-wise so difficult and austere, that the practice of them is impossible at least for a long time, yea the attempt is dangerous to manie."

105. Montaigne, "Of Experience," p. 835.

106. Joubert, *The Second Part of the Popular Errors,* p. 262.

107. Lessius, *Hygiasticon,* pp. 2–3. Lessius was not himself a physician, teaching philosophy and divinity at the Jesuits' college at Louvain.

108. Cheyne, *Essay of Health and Long Life,* p. 4; also Shapin, "Trusting George Cheyne" (chapter 13 in this volume).

109. On these points, see the classic papers by N. D. Jewson, "Medical Knowledge and the Patronage System in Eighteenth-Century England," *Sociology* 8 (1974): 369–385, and "The Disappearance of the Sick-Man from Medical Cosmology, 1770–1870," *Sociology* 10 (1976): 225–244.

110. For example, Charles E. Rosenberg, "Banishing Risk: Continuity and Change in the Moral Management of Disease," in *Morality and Health,* ed. Brandt and Rozin, pp. 35–51.

111. Steven Shapin, "Expertise, Common Sense, and the Atkins Diet, in *Public Science in Liberal Democracy,* ed. Peter W. B. Phillips (Toronto: University of Toronto Press, 2007), pp. 174–193.

112. Julia Child, quoted in Calvin Tomkins, "Table Talk," *New Yorker* (8 November 1999), p. 32.

Chapter 13. Trusting George Cheyne

1. There were, of course, non-instrumental reasons one might engage a physician, for instance, secular counsel and comfort in times of distress. I use the masculine pronoun here to refer to early modern physicians because of the overwhelming facts of the historical matter, and, later, when references are made to early modern common sayings and proverbs that used the masculine form to designate "people in general."

2. The surgeon was in a different position: in Antiquity expertise in, say, cutting for the stone was narrowly distributed and understood to be so. For the notion of a

coordinating common medical culture, see, notably, Charles E. Rosenberg, "The Therapeutic Revolution: Medicine, Meaning, and Social Change in Nineteenth-Century America," *Perspectives in Biology and Medicine* 20 (1977): 485–506; W. F. Bynum, "Health, Disease and Medical Care," in *The Ferment of Knowledge: Studies in the Historiography of Eighteenth-Century Science* ed. G. S. Rousseau and Roy Porter (Cambridge: Cambridge University Press, 1980), pp. 211–253, esp. pp. 228–230; and such exemplars of "the patient's point of view" history as Dorothy Porter and Roy Porter, *Patient's Progress: Doctors and Doctoring in Eighteenth-Century England* (Stanford, CA: Stanford University Press, 1989).

3. Descartes's version ran this way: "So, as Tiberius Caesar said (or Cato, I think), no one who has reached the age of thirty should need a doctor, since at that age he is quite able to know himself through experience what is good or bad for him, and so be his own doctor" (John Cottingham, ed. and trans., *Descartes' Conversations with Burman* [Oxford: Clarendon Press, 1976], p. 51). The source is in fact Suetonius's life of Tiberius. In 1645, Descartes told the Earl of Newcastle that "I share the opinion of Tiberius, who was inclined to think that everyone over thirty had enough experience of what was harmful or beneficial to be his own doctor. Indeed it seems to me that anybody who has any intelligence, and who is willing to pay a little attention to his health, can better observe what is beneficial to it than the most learned doctors": quoted in Shapin, "Descartes the Doctor" (chapter 15 in this volume).

4. Michel Eyquem de Montaigne, "Of Experience," in *The Complete Essays of Montaigne,* trans. Donald M. Frame (Stanford, CA: Stanford University Press, 1965; orig. publ. 1580, 1588), pp. 815–857.

5. George Cheyne, *An Essay of Health and Long Life* (London, 1724), p. 1 ("It is a common Saying, That every Man past Forty is either a *Fool* or a *Physician*"); idem, *Natural Method of Cureing the Diseases of the Body, and the Disorders of the Mind depending on the Body* (London, 1742), p. 47 ("These . . . *general Laws* . . . will best serve the End of Health, especially when every one, after a certain Age, becomes his own *Physician*"); cf. Anon., *A Letter to George Cheyne, M.D. F.R.S. Shewing the Danger of laying down General Rules to Those Who are Not Acquainted with the Animal Oeconomy, &c. . . . Occasion'd by his Essay on Health and Long Life* (London, 1724), p. 6: "It is a common Saying, as you observe, that a Man past Forty, is either a Fool, or a Physician; and if it means any thing, it only implys, that by so long experience, a Man that is not a Fool, will know what things have best agreed with him; and his Reason will direct him to continue the Use of them, till some alteration in his Constitution makes them hurtful."

6. In its original version, the Rule went: "A man in health, who is both vigorous and his own master, should be under no obligatory rules, and have no need, either for a medical attendant, or for a rubber and anointer. His kind of life should afford him variety" (Aulus Cornelius Celsus, *De medicina,* trans. W. G. Spencer, 3 vols. [Cambridge, MA: Harvard University Press, 1960], vol. 1, p. 43).

7. For the Rule in a popular medical text by a nonprofessional, see Thomas Elyot, *The Castel of Helthe* (London, 1541), p. 45^r: "As Cornelius Celsus saith, A man that is hole and well at ease, and is at his lybertie, ought not to bynde him selfe to rules, or nede a phisition." See also Shapin, "How to Eat Like a Gentleman" (chapter 12 in this volume).

8. For example, Laurent Joubert, *The Second Part of the Popular Errors,* trans. Gregory David de Rocher (Tuscaloosa: University of Alabama Press, 1995; orig. publ. 1587), p. 263 (the healthy man "while he is feeling well, belongs to himself and does not have to follow any rule or diet nor consult a physician"); John Arbuthnot, *An Essay Concerning the Nature of Aliments, and the Choice of Them, according to the Different Constitutions of Human Bodies . . . ,* 4th ed. (London, 1756; orig. publ. 1731), pp. 178–179 ("a healthy Man, under his own Government, ought not to tie himself to strict Rules, nor to abstain from any Sort of Food in common Use"); James Mackenzie, *The History of Health, and the Art of Preserving It,* 3rd ed. (Edinburgh, 1760; orig. publ. 1758), p. 135 ("A man who is sound and strong should ty himself down to no particular rule of diet, nor imagine that he stands in need of a physician").

9. Rosenberg, "The Therapeutic Revolution"; also John Harley Warner, *The Therapeutic Perspective: Medical Practice, Knowledge, and Identity in America, 1820–1885* (Princeton, NJ: Princeton University Press, 1997; orig. publ. 1986).

10. For dietetics and the care of the self, see Michel Foucault, *The History of Sexuality,* trans. Robert Hurley, 3 vols. (New York: Vintage Books, 1988–1990), vol. 2, pp. 97–139; see also vol. 3, pp. 140–141. Note that "dietetics" classically included the management of *all* the non-naturals, and thus that advice on food and drink was only one part of dietetics. Here I use the term in the classical sense, occasionally departing into a more modern restricted usage when the context makes that clear.

11. Cheyne, *Health and Long Life,* p. 5.

12. This is systematically argued in Shapin, "How to Eat Like a Gentleman" (chapter 12 in this volume); see also Keith Thomas, "Health and Morality in Early Modern England," in *Morality and Health,* ed. Allan M. Brandt and Paul Rozin (New York: Routledge, 1997), pp. 15–34; Margaret Pelling, "Food, Status and Knowledge: Attitudes to Diet in Early Modern England," in idem, *The Common Lot: Sickness, Medical Occupations and the Urban Poor in Early Modern England* (London: Longman, 1998), pp. 38–62.

13. For ancient natural philosophy and dietetics, see esp. Owsei Temkin, *Galenism: Rise and Decline of a Medical Philosophy* (Ithaca, NY: Cornell University Press, 1973), and Raymond Klibansky, Erwin Panofsky, and Fritz Saxl, *Saturn and Melancholy: Studies in the History of Natural Philosophy, Religion, and Art* (London: Thomas Nelson, 1964), chapter 1. See also Ludwig Edelstein, "The Dietetics of Antiquity," in idem, *Ancient Medicine: Selected Papers of Ludwig Edelstein,* ed. Owsei Temkin and C. Lilian

Temkin (Baltimore: Johns Hopkins Press, 1967; art. orig. publ. 1931), pp. 303–316; Owsei Temkin, *Hippocrates in a World of Pagans and Christians* (Baltimore: Johns Hopkins University Press, 1991); and Henry E. Sigerist, "Galen's *Hygiene*," in idem, *Landmarks in the History of Hygiene* (London: Oxford University Press, 1956), pp. 1–19. For a valuable recent survey of early modern dietetics, see Heikki Mikkeli, *Hygiene in the Early Modern Medical Tradition,* Annals of the Finnish Academy of Sciences and Letters, Humaniora, no. 305 (Helsinki: Academia Scientiarum Fennica, 1999).

14. The first two proverbs are cited as examples of popular error by Laurent Joubert (*Popular Errors,* trans. Gregory David de Rocher [Tuscaloosa: University of Alabama Press, 1989; orig. publ. 1579], p. 247); the next two are proverbs common in early modern England; the last is found in Sir John Harington's *The English Mans Doctor. Or the Schoole of Salerne* (London, 1607), embodying both formal medical wisdom *and* robust lay common sense. See also Morris Palmer Tilley, *A Dictionary of the Proverbs in England in the Sixteenth and Seventeenth Centuries* (Ann Arbor: University of Michigan Press, 1950). Some other texts in the "popular medical errors" tradition include Gaspard Bachot, *Erreurs populaires touchant la medecine et regime de santé* (Lyons, 1626); Jacques Primerose, *De vulgi erroribus in medicina* (Amsterdam, 1639); Luc d'Iharce, *Erreurs populaires sur la médecine* (Paris, 1783); and Balthasar-Anthelme Richerand, *Des erreurs populaires relatives à la médecine* (Paris, 1810).

15. Joubert, *Second Part of Popular Errors,* pp. 117–118.

16. See Theodore M. Browne, "The College of Physicians and the Acceptance of Iatro-Mechanism in England, 1665–1695," *Bulletin of the History of Medicine* 44 (1970): 12–30; idem, "Physiology and the Mechanical Philosophy in Mid-Seventeenth-Century England," *Bulletin of the History of Medicine* 51 (1977): 25–54; Harold J. Cook, "Physicians and the New Philosophy: Henry Stubbe and the Virtuosi-Physicians," in *The Medical Revolution of the Seventeenth Century,* ed. Roger French and Andrew Wear (Cambridge: Cambridge University Press, 1989), pp. 246–271; Arnold Thackray, *Atoms and Powers: An Essay on Newtonian Matter-Theory and the Development of Chemistry* (Cambridge, MA: Harvard University Press, 1970), pp. 49–51; Robert E. Schofield, *Mechanism and Materialism: British Natural Philosophy in an Age of Reason* (Princeton, NJ: Princeton University Press, 1970), chapter 3; Anita Guerrini, "James Keill, George Cheyne, and Newtonian Physiology, 1690–1740," *Journal of the History of Biology* 18 (1985): 247–266; idem, "The Tory Newtonians: Gregory, Pitcairne and Their Circle," *Journal of British Studies* 25 (1986): 288–311; idem, "Archibald Pitcairne and Newtonian Medicine," *Medical History* 31 (1987): 70–83; idem, "Isaac Newton, George Cheyne and the 'Principia Medicinae,'" in *The Medical Revolution of the Seventeenth Century,* pp. 222–245; idem, "The Varieties of Mechanical Medicine: Borelli, Malpighi, Bellini, Pitcairne," in *Marcello Malpighi: Anatomist and Physician,* ed. D. Bertoloni Meli, *Nuncius* 27

(1997): 111–128; Andrew Cunningham, "Sydenham versus Newton: The Edinburgh Fever Dispute of the 1690s between Andrew Brown and Archibald Pitcairne," in *Theories of Fever from Antiquity to the Enlightenment,* ed. W. F. Bynum and Vivian Nutton, Suppl. to *Medical History* (London: Wellcome Institute, 1981), pp. 71–98; G. A. Lindeboom, *Descartes and Medicine* (Amsterdam: Rodopi, 1979).

17. For example, Francis Bacon, "The Advancement of Learning [Books I–II]," in idem, *The Philosophical Works of Francis Bacon,* ed. James Spedding, Robert Leslie Ellis, and Douglas Denon Heath, 5 vols. (London: Longman and Co., 1857–1858), vol. 3, pp. 253–491.

18. Shapin, "Descartes the Doctor" (chapter 15 in this volume).

19. Archibald Pitcairne, *The Whole Works of Dr Archibald Pitcairn,* trans. George Sewell and J. T. Desaguliers, 2nd ed. (London, 1715), p. 19, quoted in Anita Guerrini, *Obesity and Depression in the Enlightenment: The Life and Times of George Cheyne* (Norman: University of Oklahoma Press, 2000), p. 40; see also p. 66.

20. Richard Mead, *A Mechanical Account of Poisons in Several Essays* (London, 1702), unpaginated preface, quoted in Guerrini, *Obesity and Depression,* p. 67.

21. John Woodward, *The Art of Getting into Practice in Physick, Here at Present in London* (London, 1722), p. 10.

22. George Cheyne, *A New Theory of Continual Fevers* (London, 1701; 2nd ed. with additions London, 1702).

23. George Cheyne, *The English Malady: Or, A Treatise of Nervous Diseases of All Kinds, as Spleen, Vapours, Lowness of Spirits, Hypochondriacal, and Hysterical Distempers, &c.* (London, 1733), p. 325.

24. Theodore M. Brown, "Cheyne, George," in *Dictionary of Scientific Biography,* vol. 3, pp. 244–245.

25. Eric T. Carlson, "Introduction" to Cheyne, *The English Malady,* facsimile edition (Delmar, NY: Scholars' Facsimiles and Reprints, 1976), pp. v–xii, on p. vii.

26. Indeed, a few colleagues even feared that Cheyne's turn to dietetics, by identifying general rules of health, would undermine that medical authority that flowed uniquely from practitioners' arduously acquired knowledge of particularities: Guerrini, *Obesity and Depression,* pp. 129–130.

27. George Cheyne, *An Essay on Regimen* (London, 1740), pp. xiv–xv: "I think [it] is true, just and *philosophical* [that] while Youth and tolerable Health continues, none ought to *alter* the common temperat Diet of the middling Rank of those among whom he lives, for a particular or *artificial* one, without a particular *Call,* and the best Advice"; see also p. xxxvi.

28. Cheyne, *Health and Long Life,* p. 36. "Seeds" in Cheyne's usage seem to have designated grains and grain-derived foods, from sago and rice puddings to porridge and bread.

29. For critical responses, see, for example, Edward Strother, *An Essay on Sickness and Health; . . . in which Dr. Cheyne's Mistaken Opinions in His Late Essay, are*

occasionally taken Notice of, 2nd ed. (London, 1725), esp. pp. 28, 209–222; Anon., *A Letter to George Cheyne,* esp. pp. 11–13; and Guerrini, *Obesity and Depression,* pp. 128–131.

30. Cheyne, *Essay on Regimen,* pp. xiii–xiv, xliii.

31. *The Letters of Dr. George Cheyne to the Countess of Huntingdon,* ed. Charles F. Mullett (San Marino, CA: Huntington Library, 1940) (hereafter *LH*); *The Letters of Doctor George Cheyne to Samuel Richardson (1733–1743),* ed. Charles F. Mullett, The University of Missouri Studies, vol. 18, no. 1 (Columbia: University of Missouri Press, 1943) (hereafter *LR*). These letters are at least as important for writing the history of the English book-trade as for the history of medical practice. The former set contains about forty letters written from 1730 to 1739; the latter about eighty. The reciprocal letters to Cheyne do not survive, though much of their content can be inferred from Cheyne's end of the correspondence. The letters also permit an assessment of aspects of Nick Jewson's important thesis about the relationship between the social standing of patients and the physician's position vis-à-vis disease individuality: N. D. Jewson, "Medical Knowledge and the Patronage System in Eighteenth-Century England," *Sociology* 8 (1974): 369–385; idem, "The Disappearance of the Sick-Man from Medical Cosmology, 1770–1870," *Sociology* 10 (1976): 225–244; see also Roy Porter, "Lay Medical Knowledge in the Eighteenth Century: The Evidence of the *Gentleman's Magazine,*" *Medical History* 29 (1985): 138–168; Malcolm Nicolson, "The Metastatic Theory of Pathogenesis and the Professional Interests of the Eighteenth-Century Physician," *Medical History* 32 (1988): 277–300. In these cases, while the correspondence is strongly marked by recognition of relative social standing, both the printer and the Countess are granted the personal individuality of their diseases. Appearing after this essay was published is Wayne Wild, *Medicine-by-Post: The Changing Voice of Illness in Eighteenth-Century British Consultation Letters and Literature* (Amsterdam: Rodopi, 2006).

32. Cheyne, *Essay on Regimen,* p. iii.

33. Cheyne, *Essay on Regimen,* p. iv.

34. Cheyne, *Essay on Regimen;* see also *LR,* 2 February 1742, pp. 82–83.

35. Cheyne, *Health and Long Life,* p. 220.

36. Cheyne, *Health and Long Life,* p. 222.

37. Cheyne, *English Malady,* p. 6. See also idem, *Health and Long Life,* pp. 172–175; idem, *The English Malady,* p. 5. Cheyne tended to avoid the practical management of "acute," "epidemical," or "contagious" diseases—for example, fevers, plague, and smallpox—noting that his dietary regime might not preserve those who embraced it from such distempers: Cheyne, *Essay on Regimen,* p. xv. But his dietetic books repeated his earlier causal accounts of fevers, and Cheyne's suggested remedies bore a close resemblance to his dietary regime for chronic conditions: Cheyne, *Natural Method,* pp. 96–111.

38. Cheyne, *Health and Long Life,* p. 224.

39. Cheyne, "The Case of the Author," in *The English Malady,* pp. 222–251; and see Anita Guerrini, "Case History as Spiritual Autobiography: George Cheyne's 'Case of the Author,'" *Eighteenth-Century Life* 19 (May 1995): 8–27.

40. Cheyne, *Essay on Regimen,* p. x.

41. Cheyne, *Health and Long Life,* p. 225.

42. On this crucial, but little noticed, feature of seventeenth- and eighteenth-century micromechanism, see Alan Gabbey, "The Mechanical Philosophy and Its Problems: Mechanical Explanations, Impenetrability, and Perpetual Motion," in *Change and Progress in Modern Science,* ed. Joseph C. Pitt (Dordrecht: D. Reidel, 1985), pp. 9–84.

43. Cheyne, *Essay on Regimen,* p. xxiii.

44. Cheyne, *Essay on Regimen,* p. xxv.

45. Cheyne, *Essay on Regimen,* p. xxiv.

46. Cheyne, *Essay on Regimen,* p. lviii.

47. Cheyne, *Natural Method,* pp. 57–58.

48. Cheyne, *Essay on Regimen,* p. lxii.

49. Cheyne, *Essay on Regimen,* p. 117. As a spa physician, Cheyne, like many others of his sort, advertised special expertise in knowing the constituents and effects of mineral waters from different sources, especially those of Bath, Bristol, Cheltenham, Clifton, Islington, Pyrmont, Spa (or "Spaw"), and Tunbridge; see Cheyne, *Observations Concerning the Nature and Due Method of Treating the Gout . . . Together With an Account of the Nature and Qualities of the Bath Waters* (London, 1720); also *The Medical History of Waters and Spas,* ed. Roy Porter, *Medical History,* 1990, suppl. no. 10; Katharine Park, "Natural Particulars: Medical Epistemology, Practice, and the Literature of Healing Springs," in *Natural Particulars: Nature and the Disciplines in Renaissance Europe,* ed. Anthony Grafton and Nancy Siraisi (Cambridge, MA: MIT Press, 1999), pp. 347–367 (for fourteenth- and fifteenth-century thinking).

50. Cheyne, *Essay on Regimen,* p. 56.

51. Cheyne, *Essay on Regimen,* p. 56.

52. Cheyne, *Essay on Regimen,* p. 56.

53. Cheyne, *Essay on Regimen,* p. 57.

54. Cheyne, *Essay on Regimen,* p. 60.

55. Cheyne, *Essay on Regimen,* p. xxvi.

56. Cheyne, *Essay on Regimen,* p. 266.

57. Cheyne, *Essay on Regimen,* p. 266.

58. Cheyne, *Essay on Regimen,* pp. 266–267; see also ibid., pp. 109–111; Cheyne, *Natural Method,* pp. 119–126, 150; *LH,* 28 August 1734, p. 43 (for practical directions).

59. Cheyne, *Natural Method,* p. 191.

60. Cheyne, *Natural Method,* pp. 191–192.

61. Cheyne, *Natural Method*, pp. 193, 195.

62. Cheyne, *Natural Method*, p. 195. And see Cheyne, *Health and Long Life*, pp. 178–180 (where Cheyne identified scurvy as "a kind of *Catholick* Distemper here in Britain," owing to climate and dietary customs).

63. Cheyne's medical practice is treated in Guerrini, *Obesity and Depression*, chapter 5. Beside the Richardson and Huntingdon cases, Guerrini also uses letters Cheyne wrote to Hans Sloane about his medical management of Catherine Walpole in the early 1720s.

64. *LR*, 10 February 1738, p. 36 ("I have a sincere Regard for you and am convinced that you are a Man of Probity and Worth beyond what I have met among Tradesmen"); on Cheyne instructing Richardson how to write a better novel than *Pamela*, ibid., 24 August 1741, pp. 67–70. Cheyne took no fees from Richardson for his medical counsel, but did extract at least part of the equivalent in printing and publishing services: ibid., 9 August 1735, p. 32.

65. Nevertheless, if he was so inclined, Cheyne was quick to reject patients' identification of the condition from which they were suffering; see, for example, his dissent from Lady Huntingdon's report that she was then afflicted with the "stone, gravel, or [a] hurt bowel": *LH*, 19 November 1733, p. 29; see also ibid., 18 February 1734, p. 37.

66. "We call the Hyp every Distemper attended with Lowness of Spirits. Whether it be Flatulence from Indigestion, Wind Cholic, Head-Pains, or an universal relaxed State of the Nerves, with Numbness, Weakness, Startings, Tremblings, etc., so that the Hyp is only a Short Expression for any Kind of nervous Disorder with whatever Symptoms (which are various nay infinite) or from whatever Cause": *LR*, 5 September 1742, p. 108.

67. *LH*, 12 August 1732, p. 9.

68. Cheyne, *Health and Long Life*, p. xvi.

69. *LH*, 19 July 1732, p. 4; 12 August 1732, p. 8.

70. For specimen complicated prescriptions, see *LR*, 9 August 1735, p. 33; *LH*, 19 November 1733, p. 29. For disapproval of Tunbridge water, see *LR*, 24 October 1741, p. 72. See also Guerrini, *Obesity and Depression*, p. 132. Like Lady Huntingdon, Richardson disliked Bath, though Cheyne tried hard to get both of them to take the waters there.

71. *LH*, 19 November 1733, p. 26; 29 December 1733, p. 31; quotation from *LR*, 20 April 1740, pp. 59–60. Walking was all right too, even to the point of "lassitude": *LR*, 12 February 1741, p. 66.

72. *LR*, 2 April 1742, p. 90

73. *LR*, 30 June 1742, p. 101.

74. *LH*, 19 November 1733, p. 29. See also *LH*, 29 December 1733, pp. 31–32; *LR*, 24 August 1741, p. 69; and see the editor's introduction to *LR*, pp. 18–19 (for Richardson's enthusiastic use of the chamber-horse: he never learned to ride a real horse).

75. See James C. Whorton, *Inner Hygiene: Constipation and the Pursuit of Health in Modern Society* (Oxford: Oxford University Press, 2000), for the nineteenth- and twentieth-century history of costiveness.

76. *LR,* 10 January 1742, p. 81.

77. *LR,* 16 July 1739, p. 54.

78. *LR,* 12 September 1739, p. 57.

79. *LR,* 3 February 1739, p. 47.

80. *LR,* 10 January 1742, p. 80.

81. *LR,* 22 June 1738, p. 38.

82. *LR,* 19 April 1742, p. 91.

83. *LR,* 26 April 1742, p. 92; see also ibid., 19 April 1742, p. 91, and 2 May 1742, p. 95. Evidently, Richardson accepted some of this advice over a long period of time, as he became widely known for his vegetarianism.

84. *LH,* [ca. 1732], p. 16.

85. *LH,* 28 August 1732, p. 10; see also ibid., [ca. 1732], p. 15.

86. *LH,* 19 November 1733, p. 28.

87. *LH,* November 1733, p. 29.

88. *LH,* 7 January 1734, p. 33.

89. *LH,* 19 October 1734, p. 46.

90. *LH,* 6 September 1735, p. 51.

91. *LH,* 20 August 1737, p. 59.

92. *LR,* 20 June 1739, p. 52.

93. *LR,* 12 December 1741, p. 75.

94. *LR,* 26 April 1742, p. 92.

95. *LR,* 7 December 1741, p. 74

96. *LR,* 23 December 1741, p. 76.

97. See my treatment of Francis Bacon's respecification of dietary moderation in Shapin, "How to Eat Like a Gentleman" (chapter 12 in this volume).

98. *LH,* 3 August 1734, p. 41

99. *LR,* 7 December 1741, p. 74.

100. *LR,* undated [ca. April 1740], p. 61.

101. *LR,* 23 December 1741, p. 77.

102. *LR,* 10 January 1742, p. 79.

103. *LR,* 12 September 1739, p. 57. Proverbs and aphorisms were used to similar effect in Cheyne's published work: "Diseases are always to be cured by their *Contraries*" (Hippocratic); "He that would be soon *well,* must be long *sick*"; "He that is *old* when he is young . . . will be young when he grows old in Years" (*Essay on Regimen,* pp. lx, lxii). But, as scholars of orality have pointed out, the natural home of proverbs is the oral rather than the literate domain: see Shapin, "Proverbial Economies" (chapter 14 in this volume).

104. *LH,* 3 November 1735, pp. 52–53. This medical casuistry—the vivid and

dramatic retelling of successfully treated cases—was also a notable feature of Cheyne's published work, particularly *English Malady,* pp. 177–256. My interest here is in the different moral texture and force such case histories had when they figured in the context of intimate relations.

105. *LH,* 15 April 1734, p. 39.

106. *LH,* 18 February 1734, p. 38.

107. *LH,* 19 November 1733, pp. 26–27.

108. *LR,* 30 December 1741, pp. 78–79. The testimonial was enclosed with the letter of 2 November 1742, pp. 115–117. See also ibid., 9 March 1742, p. 86.

109. *LH,* 29 December 1733, p. 30.

110. *LR,* 9 March 1742, p. 88.

111. *LR,* 27 March 1738, p. 48.

112. *LR,* 7 December 1741, p. 74; 30 December 1741, p. 78.

113. *LH,* 12 May 1736, p. 57.

114. *LH,* 12 August 1732, p. 8.

115. *LH,* 28 August 1732, p. 10.

116. *LH,* 28 August 1734, p. 44.

117. *LR,* 2 February 1742, p. 83; see also ibid., 14 July 1742, pp. 103–104: "Continue your Diet for God's Sake, your Life's Sake, and your Family's Sake."

118. Given the diffuseness of eighteenth-century vocabularies for designating chronic illnesses, that was not a difficult matter, though physicians still had to make moral and practical judgments that such shared experiences should be a basis for interaction.

119. *LR,* 4 June 1739, p. 50.

120. *LR,* 6 June 1740, p. 62.

121. *LR,* July 1742, pp. 102–103. See also Guerrini, *Obesity and Depression,* pp. 103, 169.

122. *LR,* 23 December 1741, p. 76.

123. *LR,* 23 December 1741, pp. 76–77.

124. *LH,* 18 February 1734, p. 38.

125. *LH,* 25 February 1737, p. 58.

126. The religious framing of much of Cheyne's dietary counsel is pervasive, and his close involvement with Lady Huntingdon's Methodist circle is thoroughly documented in Guerrini, *Obesity and Depression,* chapters 5–6.

127. *LH,* 6 September 1735, p. 49.

128. *LR,* 21 December 1734, p. 31; 31 January 1736, p. 33; 29 November [1739], p. 44; 7 November 1740, p. 62.

129. *LH,* 20 September 1734, p. 45; 4 September 1733, p. 24.

130. *LH,* 4 September 1733, p. 23.

131. *LH,* 9 August 1735, p. 48.

132. *LR,* 12 September 1739, p. 57.

133. *LR*, 2 May 1742, p. 95.

134. *LR*, 2 May 1742, p. 94; 17 May 1742, p. 98; 10 January 1742, p. 81.

135. *LR*, 9 March 1742, p. 86; see also 17 May 1742, p. 98 ("What Interest can I have in being thus bigoted?").

136. *LH*, 19 November 1733, p. 29.

137. *LH*, 6 September 1735, p. 49; see also ibid., 14 April 1736, p. 55.

138. *LR*, 6 June 1740, p. 61.

139. Such consultations were not exceptional in either case. Cheyne knew that both of his patients were receiving advice from other practitioners and even from friends and family. Whether or not this bothered him depended upon the precise circumstances of the consultations.

140. *LR*, [1742?], p. 96.

141. *LH*, 2 February 1742, p. 82; see also ibid., 19 November 1733, p. 29; ibid., 3 November 1735, p. 54.

142. *LH*, 19 May 1739, p. 61.

143. For example, *LR*, 13 May 1739, p. 49; 12 September 1739, p. 57: "pray read and consider my Essay on Regimen"; 23 December 1741, p. 77 "I wish you would look into the Essay on Regimen"; 30 December 1741, p. 79 "I wish you would only read the Cases in my Book of the English Malady for your Amusement."

144. *LR*, 2 February 1742, p. 82.

145. "Prudential expertise" is preferred here to the apparently more straightforward "experience" because of the latter's referential richness and because prudence carries with it the wanted sense of orientation to judgment and action: wisdom or experience applied to practical action.

146. To be sure, there might be other attributions involved in the identity of the empiric in the early modern period, such as a penchant for medical "specifics" and attendant "quackery"; see here Roy Porter, *Health for Sale: Quackery in England 1650–1850* (Manchester: Manchester University Press, 1989).

147. For literature on the importance of fashionability for the eighteenth-century British medical career, see, for example, Christopher J. Lawrence, "Ornate Physicians and Learned Artisans: Edinburgh Medical Men, 1726–1776," in *William Hunter and the Eighteenth-Century Medical World,* ed. W. F. Bynum and Roy Porter (Cambridge: Cambridge University Press, 1985), pp. 153–176; Roy Porter, "William Hunter: A Surgeon and a Gentleman," in ibid., pp. 7–34.

148. Roy Porter rightly emphasized the extent of eighteenth-century lay medical literacy, but he offered no evidence that such interest extended to micromechanical theorizing. The "theoretical matrix" for lay medical thinking, Porter wrote, "was the ingrained Hippocratic leanings of the Enlightenment gentleman": "Lay Medical Knowledge," p. 151. Guerrini similarly says (*Obesity and Depression*, p. 132) that in Cheyne's time "physicians and patients shared the same explanatory model of

disease," but on the evidence presented in this article, such sharing did not commonly include micromechanical explanatory structures.

149. For the informational richness and moral consequence of the face-to-face domain, see the classic work of Erving Goffman, *The Presentation of Self in the Everyday Life* (London: Allen Lane, 1959); also Steven Shapin, *A Social History of Truth: Civility and Science in Seventeenth-Century England* (Chicago: University of Chicago Press, 1994), esp. chapters 3 and 6.

Chapter 14. Proverbial Economies

1. The defining exercise can in principle go the other way round—common people can take a view of, and condemn, learned fools—but the learned have, until recently, controlled the presses.

2. See, among many examples, Galileo Galilei, "Letter to Madam Christina of Lorraine, Grand Duchess of Tuscany, Concerning the Use of Biblical Quotations in Matters of Science [1615]," in *Discoveries and Opinions of Galileo,* ed. and trans. Stillman Drake (Garden City, NY: Doubleday Anchor, 1957), pp. 196, 200; William Gilbert, *De Magnete,* trans. P. Fleury Mottelay (New York: Dover, 1958; orig. publ. 1600), p. 318; John Wilkins, *The Mathematical and Philosophical Works of the Right Rev. John Wilkins,* 2 vols. in 1 (London: Frank Cass & Co., 1970; facsimile reproduction of 1802 edition; tract orig. publ. 1638), vol. 1, p. 11; Isaac Newton, *The Mathematical Papers of Isaac Newton,* vol. 6, *1684–1691,* ed. D. T. Whiteside (Cambridge: Cambridge University Press, 1974), p. 192.

3. Ralph Waldo Emerson, "Nature," in Emerson, *The Complete Essays . . . ,* ed. Brooks Anderson (New York: The Modern Library, 1940; art. orig. publ. 1836), pp. 1–42, on p. 18: "The memorable words of history and the proverbs of nations consist usually of a natural fact, selected as a picture or parable of a moral truth."

4. Thomas Browne, *Pseudodoxia Epidemica: Or, Enquiries into Very Many Received Tenets, And Commonly Presumed Truths* (London, 1650; orig. publ. 1646), pp. 4–6.

5. Alfred Sidgwick, *Fallacies: A View of Logic from the Practical Side* (London: Kegan Paul, Trench, 1883), p. 266.

6. See here Barry Barnes and John Law, "Whatever Should Be Done with Indexical Expressions?" *Theory and Society* 3 (1976): 223–237.

7. This tendency is documented and criticized in Daniel M. Wegner and Robin R. Vallacher, "Common-Sense Psychology," in *Social Cognition: Perspectives on Everyday Understanding,* ed. Joseph P. Forgas (London: Academic Press, 1981), pp. 225–246.

8. For entry into the secondary literature on scholarly attitudes to proverbs as folkish or learned items, see, for example, Natalie Zemon Davis's superb "Proverbial Wisdom and Popular Errors," in idem, *Society and Culture in Early Modern France*

(Stanford, CA: Stanford University Press, 1975), pp. 227–267; also excellent work by James Obelkevich, "Proverbs and Social History," in *The Social History of Language,* ed. Peter Burke and Roy Porter (Cambridge: Cambridge University Press, 1987), pp. 43–72; Mary Thomas Crane, *Framing Authority: Sayings, Self, and Society in Sixteenth-Century England* (Princeton, NJ: Princeton University Press, 1993), esp. pp. 50–51, 206 (n23); Ann Moss, *Printed Commonplace-Books and the Structuring of Renaissance Thought* (Oxford: Clarendon Press, 1996); Kevin Sharpe, *Reading Revolutions: The Politics of Reading in Early Modern England* (New Haven, CT: Yale University Press, 2000), for example, pp. 101–105, 193–198, 320–322; and Adam Fox, *Oral and Literate Culture in England 1500–1700* (Oxford: Clarendon Press, 2000), chapter 2.

9. It would be very valuable to have a systematic "history of ideas" survey of the temporally changing, and synchronically varying, references and evaluations of "common sense": the *sensus communis* (the sense that mediates the five special senses); the shared sensibilities of Common Sense Philosophy; what everybody ought to know; what everybody knows if they have not received special instruction; the intellectual basis for prudence (or merely for prudence); what the common people only think they know; how everybody ought to think; how the common people think they think; and so on.

10. The indispensable source here is Archer Taylor, *The Proverb* (Cambridge, MA: Harvard University Press, 1931). Of course, in lay society proverbs are probably most commonly defined through ostention, by giving examples and trusting the inquirer to see the pertinent family resemblances. For a few other definitional exercises, see George B. Milner, "What is a Proverb?" *New Society,* no. 332 (6 February 1969): 199–202; Wolfgang Mieder and Alan Dundes, eds., *The Wisdom of Many: Essays on the Proverb* (Madison: University of Wisconsin Press, 1994), pp. vii–xiii; W. Carew Hazlitt, *English Proverbs and Proverbial Phrases* (London: Reeves & Turner, 1907), pp. vii–xxx; Richard Chenevix Trench, *Proverbs and Their Lessons* (London: George Routledge & Sons, 1905), pp. 1–25; Nigel Barley, "'The Proverb' and Related Problems of Genre Definition," *Proverbium* 23 (1974): 880–884; Neal R. Norrick, *How Proverbs Mean: Semantic Studies in English Proverbs* (Berlin: Mouton, 1985), pp. 65–79.

11. Taylor, *The Proverb,* p. 3.

12. Thomas Fuller, *The Worthies of England,* ed. John Freeman (London: Allen & Unwin, 1952; orig. publ. 1662), p. 5.

13. Aristotle, *Rhetoric,* 1413a 15. All quotations from Aristotle in this essay are from *The Complete Works of Aristotle,* 2 vols., ed. Jonathan Barnes (Princeton, NJ: Princeton University Press, 1984). See also Jan Fredrick Kindstrand, "The Greek Concept of Proverbs," *Eranos* 76 (1978): 71–85, on pp. 78–80. For metaphor in the definition of proverbs, see, for example, Peter Seitel, "Proverbs: A Social Use of Metaphor," *Genre* 2 (1969): 143–161.

14. Beatrice Silverman-Weinrich, "Towards a Structural Analysis of Yiddish Proverbs," in *The Wisdom of Many,* ed. Mieder and Dundes, pp. 65–85, on pp. 78, 80. Silverman-Weinrich acknowledges that no single semantic, phonic, or rhetorical marker is found in all proverbs, but she insists that most proverbs—properly so-called— have at least one such marker.

15. It is fair to say that the extent of our culture's present-day experience with the ways of farmyard animals has greatly diminished, and hence that this metaphorical base is being hollowed out. One has only to turn the pages of a dictionary of proverbs to see how many now need explication just because we no longer have the stock of familiar experience, or of common usage, to make sense of them. And, on these grounds alone, it is very likely that many proverbs current several generations ago have now passed out of use, and that many others will soon do so. But it is hard to understand how it can be seriously maintained that late modernity does not use and make proverbs, and the concluding section of this essay will retrieve a sample arising from present-day technical subcultures. For a claim that "The proverb is a language form which has largely passed from usage in contemporary American culture," see William Albig, "Proverbs and Social Control," *Sociology and Social Research* 15 (1931): 527–535. Albig reckoned that proverb-use was an inverse index of the extent of social differentiation, conflict, and change—so to speak, a marker of modernity.

16. For proverbs and orality, see Walter J. Ong, S.J., *Rhetoric, Romance, and Technology: Studies in the Interaction of Expression and Culture* (Ithaca, NY: Cornell University Press, 1971), esp. pp. 29–31, 78–81, 286–287; idem, *Orality and Literacy: The Technologizing of the Word* (London: Routledge, 1988), pp. 16ff., 34–35; also Obelkevich, "Proverbs and Social History," and Fox, *Oral and Literate Culture.*

17. Barbara Herrnstein Smith, *On the Margins of Discourse: The Relation of Literature to Language* (Chicago: University of Chicago Press, 1978), p. 69; Obelkevich ("Proverbs and Social History," p. 62) describes a shift in polite fashion during the nineteenth century from the anonymous proverb to the specifically authored quotation and aphorism: "To cite a quotation," Obelkevich says, "is to identify with the genius of the author and to lift oneself above the common herd."

18. Obelkevich, "Proverbs and Social History," p. 65.

19. Early anthropology is a rich source for characterizations of proverbs as expressions of the "primitive mentality" and its faulty modes of reasoning; see, for example, Edward B. Tylor, *Primitive Culture,* 2 vols. (London: John Murray, 1920; orig. publ. 1870), vol. 1, pp. 83–90 (for proverbs as "mines of historical knowledge" and as "survivals" of past culture, and the view that the age of proverb-making was definitively past); for more recent anthropological evaluations, see Christopher R. Hallpike, *The Foundations of Primitive Thought* (Oxford: Clarendon Press, 1979), pp. 110–112, and Jack R. Goody, *The Domestication of the Savage Mind* (Cambridge: Cambridge University Press, 1977), pp. 125–126.

20. The list originated with the seventeenth-century English churchman and antiquary Thomas Fuller (*Worthies of England,* p. 5).

21. Erving Goffman, *Frame Analysis: An Essay on the Organization of Experience* (New York: Harper & Row, 1974), though proverbial utterances do not usually lead to the bewilderment and annoyance Goffman ascribed to the frame-breaking passages he noted.

22. Thomas A. Sebeok, "The Structure and Content of Cheremis Charms," in *Language in Culture and Society: A Reader in Linguistics and Anthropology,* ed. Dell Hymes (New York: Harper & Row, 1964), pp. 356–371. Some of proverbs' authority may possibly flow from these linguistic signs linking them to other forms of utterance—religious, legal, and magical—known to be authoritative. Treating aphorisms rather than proverbs, Murray Davis has usefully drawn attention to the capacity of short-generic formulae to *interest:* Murray S. Davis, "Aphorisms and Cliches: The Generation and Dissipation of Conceptual Charisma," *Annual Review of Sociology* 25 (1999): 245–269.

23. Iona Opie and Peter Opie, *The Lore and Language of Schoolchildren* (Oxford: Clarendon Press, 1959).

24. Harvey Sacks, "On Proverbs," in idem, *Lectures on Conversation,* ed. Gail Jefferson (Oxford: Blackwell, 1992), vol. 1, pp. 104–112, on pp. 109, 111. This is a combination of several lectures given in 1964–1965. Note, however, that proverbs can be purposefully adapted to achieve twists on the original message and, when this is understood to be done, the effect can be totally different from that of a mistake. Take the present-day statistician's proverb, "The n justifies the mean"; the economist's expansion, "There ain't no such thing as a free lunch, but there is a cheap one"; the professor's, "You can lead a boy to college but you can't make him think"; Corporal Klinger's, "The nose is the window of the soul"; and Mae West's, "A hard man is good to find." In each case, the authority of the modified form piggy-backs on that of the undeformed original.

25. Latour himself, however, is notably unimpressed with the proverb, using it as an example of a "soft fact," something that may indeed spread stably in its expression over time and space, but whose reference is easily adapted and transformed by its users: Bruno Latour, *Science in Action: How to Follow Scientists and Engineers through Society* (Cambridge, MA: Harvard University Press, 1987), pp. 207–209.

26. For example, Heda Jason, "Proverbs in Society: The Problem of Meaning and Function," *Proverbium* 17 (1971): 617–623, on p. 618.

27. George Herzog, *Jabo Proverbs from Liberia: Maxims in the Life of a Native Tribe* (London: Oxford University Press, 1936), p. 2; Ruth Finnegan, *Oral Literature in Africa* (Oxford: Oxford University Press, 1970), p. 419.

28. Kenneth Burke, "Literature as Equipment for Living," in idem, *The Philosophy of Literary Form,* 3rd ed. (Berkeley: University of California Press, 1973; orig. publ. 1941), pp. 293–304, on pp. 296–297; see also Paul D. Goodwin and Joseph

W. Wenzel, "Proverbs and Practical Reasoning: A Study in Socio-Logic," *Quarterly Journal of Speech* 65 (1979): 289–302, on p. 290.

29. Sacks, "On Proverbs," p. 107.

30. I will, however, note that translation from one experiential domain to another is not a necessary feature of proverb use and meaning, even where proverbs are metaphorical. See also Seitel, "Proverbs," p. 145.

31. For example, Aristotle, *Nicomachean Ethics,* 1112a 17–1113a 14: "Deliberation is concerned with things that happen in a certain way for the most part, but in which the event is obscure, and with things in which it is indeterminate. We call in others to aid us in deliberation on important questions, distrusting ourselves as not being equal to deciding."

32. Stephen Toulmin, *Return to Reason* (Cambridge, MA: Harvard University Press, 2001).

33. Herzog's study of a West African proverbial economy (*Jabo Proverbs*, p. 2) makes a large claim for the place of proverbs in everyday generalization: "In the realm of conversation and verbal expression proverbs furnish almost exclusively the means by which generalizations are made explicit." It would be interesting to have an ethnographic study of how generalization actually happens in ordinary modern Western social life and in its technical subcultures.

34. This claim opposes an evidently common view that translation *must* be involved: see, for example, Geofirey M. White, "Proverbs and Cultural Models: An American Psychology of Problem Solving," in *Cultural Models in Language and Thought,* ed. Dorothy Holland and Naomi Quinn (Cambridge: Cambridge University Press, 1987), pp. 151–172, on pp. 152–155.

35. We now say "The exception proves the rule" mainly to express a view that rules just do have exceptions, while the historical sense was that apparent counter-instances *tested* a rule which, if legitimate, should have *no* exceptions. That sense of "proving" survives among bread-makers: proving the dough, letting it rise once to test whether the yeast is active. See Taylor, *The Proverb*, p. 78, and Linda and Roger Flavell, *Dictionary of Proverbs and Their Origins* (London: Kyle Cathie, 1993), p. 91.

36. Susan Kemper, "Comprehension and the Interpretation of Proverbs," *Journal of Psycholinguistic Research* 10 (1981): 179–198.

37. That's an interesting proverb because there is radical contemporary divergence about what it means when applied to human conduct. Some reckon that it is an injunction to keep active, commending the virtues of busy-ness (keep on the move and you won't get mentally fusty). Others use the proverb to convey quite opposite sentiments (if you don't stay in one situation long enough, you won't acquire those desirable softening and stabilizing qualities signified by the moss coating stones in a quiet stream). And there are still other interpretations and uses: see Barbara Kirshenblatt-Gimblett, "Toward a Theory of Proverb Meaning," *Proverbium* 22 (1973): 821–827.

38. Failure to "get the point" of metaphorical proverbs, or a tendency to take them literally, is treated as a pathological sign. In a vast literature on this subject, see, for example, Donald R. Gorham, "A Proverbs Test for Differentiating Schizophrenics from Normals," *Journal of Consulting Psychology* 20 (1956): 435–440, and Wolfgang Mieder, "The Use of Proverbs in Psychological Testing," *Journal of the Folklore Institute* 15 (1978): 45–55.

39. Clifford Geertz characterizes common sense as representing "the simple nature of the case": "An air of 'of-courseness,' a sense of 'it figures' is cast over things," or at least of certain things; "They are depicted as inherent in the situation, intrinsic aspects of reality, the way things go": Geertz, "Common Sense as a Cultural System," in idem, *Local Knowledge: Further Essays in Interpretive Anthropology* (New York: Basic Books, 1983), pp. 73–93, on p. 85; also idem, *Islam Observed* (Chicago: University of Chicago Press, 1968), pp. 90–94.

40. Two British epidemiologists have recently subjected to skeptical statistical inquiry the health, economic, and cognitive benefits proverbially associated with being "early to bed and early to rise": "Our results suggest that . . . the time of going to and getting up matters little." The Christmas issue of the medical journal in which the paper was published is, however, well known for its attempts at medical levity: Catharine Gale and Christopher Martyn, "Larks and Owls and Health, Wealth, and Wisdom," *British Medical Journal* 137 (19–26 December 1998): 1675–1677, on p. 1677; cf. Wolfgang Meder, "Early to Bed and Early to Rise: From Proverb to Benjamin Franklin and Back," in idem, *Proverbs are Never Out of Season: Popular Wisdom in the Modern Age* (New York: Oxford University Press, 1993), pp. 98–134.

41. Sacks, "On Proverbs," pp. 105, 109. Chaim Perelman and Lucie Olbrechts-Tyteca argue a somewhat weaker case with respect to "maxims": "a maxim can always be rejected . . . the agreement it calls forth is never compulsory, but so great is its force, so great is the presumption of agreement attaching to it, that one must have weighty reasons for rejecting it": idem, *The New Rhetoric: A Treatise on Argumentation,* trans. John Wilkinson and Purcell Weaver (Notre Dame, IN: University of Notre Dame Press, 1969), p. 165.

42. Latour, *Science in Action,* pp. 206–207. Latour imagines the proverbial utterance accompanying and justifying the doling out of a maternal apple. In my experience, however, mothers tend to say this proverb as part of a general injunction to sensible eating, or, sometimes specifically, to consuming fresh fruit and vegetables— in which case, the son would in fact be hard put to cite expert studies contradicting the mother's counsel. Later, I warn against, so to speak, comparing proverbial apples with scientific oranges: it will be useful to consider how real, as opposed to ideal, scientific practice itself may incorporate features of a proverbial economy.

43. Sacks, "On Proverbs," pp. 105, 109; Aristotle, *Rhetoric,* 1395a 12–13; see also Quintilian, *The Institutio Oratoria,* trans. Harold E. Butler, 4 vols. (Cambridge, MA: Harvard University Press, 1960), vol. 2, pp. 11, 295 (book V, chapter 11,

4142): "Generally received sayings . . . would not have acquired immortality had they not carried conviction of their truth to all mankind."

44. Take, for instance, this concluding sentence of a measuredly skeptical survey of management consultancy: "As the Chinese proverb has it, in a good wind even turkeys can fly; but management consultancies may be headed for the doldrums": *The Economist* (22 March 1997), p. 20. (It may be part of the intended pawky humor that there were no turkeys in China until modern times.)

45. Sacks, "On Proverbs," p. 105.

46. Burke, *Philosophy of Literary Form*, pp. 301–302; see also Taylor, *The Proverb*, p. 82; Perelman and Olbrechts-Tyteca, *The New Rhetoric*, p. 166. Sounding here more like a cognitive scientist than a literary critic, Herrnstein Smith argues (*Margins of Discourse*, p. 70) that *selective* mechanisms underwrite the broad range of situations to which a proverb can be "appropriately affirmed": "By a sort of natural selection, those proverbs that survive are literally the *fittest;* that is, they fit the widest variety of circumstances or adapt most readily to emergent environments."

47. For example, E. E. Evans-Pritchard, "Meaning in Zande Proverbs," *Man* 62 (1962): 4–7.

48. Sacks, "On Proverbs," p. 110; also Herrnstein Smith, *Margins of Discourse,* p. 72.

49. Gilat Hasan-Rokem, "The Pragmatics of Proverbs: How the Proverb Gets Its Meaning," in *Exceptional Language and Linguistics,* ed. Loraine K. Obler and Lise Menn (New York: Academic Press, 1982), pp. 169–173.

50. See, for example, Dell H. Hymes, "The Ethnography of Speaking," in *Anthropology and Human Behavior,* ed. Thomas Gladwin and William C. Sturtevant (Washington, DC: Anthropological Society of Washington, 1962), pp. 13–53 (for programmatic statements); E. Ojo Arewa and Alan Dundes, "Proverbs and the Ethnography of Speaking Folklore," in *The Ethnography of Communication,* ed. John J. Gumperz and Dell Hymes, Special Issue, *American Anthropologist* 66, no. 6, part 2 (December 1964): 70–85; and Seitel, "Proverbs."

51. Raymond Firth, "Proverbs in Native Life, with Special Reference to Those of the Maori," *Folk-lore* 37 (1926): 134–153, 245–270, on p. 134.

52. Indeed, sometimes proverbs are intentionally used as code—so that the translation from their metaphorical base is *not* transparent to certain auditors, or so that other aspects of their linguistic specialness restrict access to their meaning. So a mother may say to her husband—*devant les enfants*—"Little pitchers have big ears," confident (at least for a while) that the children will not understand that parental discretion is being counseled in a sensitive matter.

53. Aristotle, *Rhetoric,* 1295a 3–6. For proverb-speaking in relation to age-and-experience, see Goodwin and Wenzel, "Proverbs and Practical Reasoning," pp. 291–292; Finnegan, *Oral Literature in Africa,* p. 417; and William P. Murphy, "Oral Literature," *Annual Review of Anthropology* 7 (1978): 113–136, on pp. 128–129.

54. Herzog, *Jabo Proverbs,* p. 2. See, in this connection, the nauseatingly pack-aged, but famously popular, collections of proverbs and aphorisms marketed in America as the *Chicken Soup for the Soul* series.

55. Jesus Ben-Sira, *Ecclesiasticus or The Wisdom of Jesus Son of Sirach,* ed. John G. Snaith (Cambridge: Cambridge University Press, 1974; orig. comp. ca. 190 BCE), p. 103.

56. Miguel de Cervantes Saavedra, *The Adventures of Don Quixote,* trans. J. M. Cohen (Harmondsworth: Penguin, 1950; orig. publ. 1604–1614), p. 742; also pp. 510, 591, 655, 688, 695, 741–744. Nevertheless, Sancho knows enough to up-braid those whose proverb-citing is even more undisciplined than his own (ibid., p. 500). See also Fox, *Oral and Literate Culture,* pp. 165–167.

57. Seitel, "Proverbs," pp. 147–148, 155–158.

58. Aristotle, *Rhetoric,* 1358a 12–13.

59. Murray Davis's very funny essay on how sociological theories become "inter-esting" stresses the necessity that they visibly set themselves against what is suppos-edly taken-for-granted, and, in passing, he identifies the taken-for-granted with the proverbial: "A new theory will be noticed only when it denies an old truth (proverb, platitude, maxim, adage, saying, common-place, etc.)": Murray S. Davis, "That's Interesting: Towards a Phenomenology of Sociology and a Sociology of Phenome-nology," *Philosophy of the Social Sciences* 1 (1971): 309–344. Here I mean to show that proverbs can be "interesting" in just the same way that their theoretical nega-tion is "interesting."

60. For celery, see Christopher R. Henke, "Making a Place for Science: The Field Trial," *Social Studies of Science* 30 (2000): 483–511, on p. 501.

61. Tim B. Rogers, "Proverbs as Psychological Theories . . . Or Is It the Other Way Around?" *Canadian Psychology* 31 (1990): 195–207.

62. This is precisely the intention of E. D. Hirsch, Jr.'s repellent *A First Diction-ary of Cultural Literacy: What Our Children Need to Know* (Boston: Houghton Mifflin, 1991), the very first section of which (pp. 1–5) is a compilation of proverbs, accompanied by short interpretative homilies.

63. Compare, however, "A black hen may lay a white egg"; "A wise man com-monly has foolish children"; and "From the thornbush comes the rose." The corpus of proverbial wisdom does indeed recognize "regression to the mean."

64. For surprising scientific confirmation of Murphy's Law, see Robert A. J. Mat-thews, "The Science of Murphy's Law," *Scientific American* 276, no. 4 (April 1997): 88–91. Note also the recently overheard "The tendency of an event to occur varies inversely with one's preparation for it." For oral citation of Sod's Law in the scien-tific training of Ph.D. students, see Sara Delamont and Paul Atkinson, "Doctoring Uncertainty: Mastering Craft Knowledge," *Social Studies of Science* 31 (2001): 87–107, on pp. 97, 99.

65. See, for example, Goodwin and Wenzel, "Proverbs and Practical Reason-

ing," p. 301, and, for the civic setting of proverbial probabilism, see Steven Shapin, *A Social History of Truth: Civility and Science in Seventeenth-Century England* (Chicago: University of Chicago Press, 1994), chapter 3.

66. Ecclesiastes 3:1.

67. See Steven Shapin, "Rarely Pure and Never Simple," *Configurations* 7 (1999): 1–14.

68. Taylor, *The Proverb,* pp. 20–21, 168–169. On proverbs as a "subversive kind of wisdom" among the common people, see Obelkevich, "Proverbs and Social History," p. 49.

69. Geertz, "Common Sense," *Local Knowledge,* p. 89. Geertz rightly picked out (p. 91) the "anti-expert, if not anti-intellectual" tone of many proverbial expressions.

70. Taylor, *The Proverb,* pp. 4, 9. For example, "If Candlemas day [2 February] be sunny and bright, winter will have another flight; if Candlemas day be cloudy with rain, winter is gone and won't come again." This can hardly have any predictive value in a range of settings where such sayings continue to circulate, including Punxsutawney, Pennsylvania (where the official American groundhog lives), and would only seem to have any chance of being correct in Mediterranean conditions. On the other hand, lots of proverbs about diet and regimen smack of *dernier cri* medical thinking: see Shapin, "How to Eat Like a Gentleman" (chapter 12 in this volume). The notion of "phatic communion"—linguistic acts whose purpose is just the establishment and maintenance of social bonds—belongs to Bronislaw Malinowski: see his "The Problem of Meaning in Primitive Languages," in *The Meaning of Meaning,* ed. C. K. Ogden and I. A. Richards, 8th ed. (New York: Harcourt, Brace, 1956), pp. 296–336, esp. pp. 315–316.

71. For example, John Heywood, *A Dialogue of Herbs* (London, 1546); Nicholas Breton, *A Crossing of Proverbs* (London, 1616); and Michael Drayton's sonnet "To Proverbe [ca. 1602]," in *Minor Poems of Michael Drayton,* ed. Cyril Brett (Oxford: Clarendon Press, 1907), p. 45. For general introduction to the topic of crossed proverbs, see Archer Taylor, "Proverb," in *Dictionary of Folklore, Mythology and Legend,* ed. Maria Leach, 2 vols. (New York: Funk & Wagnalls, 1949–1950), vol. 2, pp. 902–906, on p. 903.

72. Wegner and Vallacher, "Common Sense Psychology," p. 231; cf. Jon Elster, *Alchemies of the Mind: Rationality and the Emotions* (Cambridge: Cambridge University Press, 1999), esp. pp. 10–13. See in this connection Merton's famous discussions of "norms and counternorms"—some of which take proverbial forms—in Robert K. Merton and Elinor Barber, "Sociological Ambivalence," in *Sociological Theory, Values, and Sociocultural Change,* ed. Edward A. Tiryaluan (Glencoe, IL: The Free Press, 1963), pp. 91–120; also Robert S. Lynd, *Knowledge for What? The Place of Social Science in American Culture* (Princeton, NJ: Princeton University Press, 1940), chapter 3.

73. Goody, *Domestication of the Savage Mind,* pp. 125–126; on this point, see also Sacks, "On Proverbs," p. 105. In this connection, Goody is preferred to Geertz, who invokes contradictory proverbs to illustrate what he calls the "immethodical-ness" of common sense. When Geertz says of proverbs ("Common Sense," p. 90) that "it is not their interconsistency that recommends them but indeed virtually the opposite," he is implicitly retaining the conception of proverbs as an inspectable set of propositions—whether consistent or inconsistent—that Goody rightly challenges.

74. See, for example, H. M. Collins, "The TEA Set: Tacit Knowledge and Scientific Networks," *Science Studies* 4 (1974): 165–186; idem, "Tacit Knowledge, Trust and the Q of Sapphire," *Social Studies of Science* 31 (2001): 71–85.

75. It would be pedantic to provide references for all of these, and I assume readers will have their own samples readily at hand. Almost needless to say, what's wanted in this connection is a much more systematic study of proverbial genres in expert practices. This part of the chapter is intended to stimulate interest in the subject, and it is offered as the result of merely casual—though long-standing—inquiry on my part. A crude compilation of some social scientific and natural scientific proverbs is Alexander E. Chamberlain, "Proverbs in the Making: Some Scientific Commonplaces," *Journal of American Folk-lore* 17 (1904): 161–170, 268–278. For important brief remarks on the "maxims" of science, see Michael Polanyi, *Personal Knowledge: Towards a Post-Critical Philosophy* (Chicago: University of Chicago Press, 1958), pp. 30–31, 50, 54, 88, 90, 125, 153–158, 162, 170, 311–312. For the biochemist's advice, see Delamont and Atkinson, "Doctoring Uncertainty," p. 95; for the machinists' maxim, see David E. Noble, "Social Choice in Machine Design: The Case of Automatically Controlled Machine Tools," in *Case Studies on the Labor Process,* ed. Andrew Zimbalist (New York: Monthly Review Press, 1979), pp. 18–50, on p. 44; for rules of thumb in technicians' practice, circulating as "as stories or snippets of advice," see Stephen R. Barley and Beth A. Bechky, "In the Backrooms of Science: The Work of Technicians in Science Labs," *Work and Occupations* 21 (1994): 85–126, on pp. 109–110.

76. Sir William Osler, *Aphorisms: From His Bedside Teachings and Writings,* ed. William Bennett Bean (Springfield, IL: Charles C. Thomas, 1961), pp. 103, 105, 137, 141.

77. William S. Reveno, *711 Medical Maxims: Diagnostic and Therapeutic Aids for the Physician,* 3 vols. (Westport, CT: Technomic Publishing Co., 1951–1976); also Fielding H. Garrison, "Medical Proverbs, Aphorisms, and Epigrams," *Bulletin of the New York Academy of Medicine* 4 (1928): 979–1005.

78. Harold Bursztajn and Robert M. Hamm, "Medical Maxims: Two Views of Science," *Yale Journal of Biology and Medicine* 52 (1979): 483–486, on p. 485; see also Harold Bursztajn et al., *Medical Choices, Medical Chances: How Patients, Families, and Physicians Can Cope with Uncertainty* (New York: Delacorte Press,

1981), esp. pp. 69–70, 204; Paul Atkinson, *Medical Talk and Medical Work: The Liturgy of the Clinic* (London: Sage, 1995), pp. 140–147; and Arthur S. Elstein, with Linda Allal et al., *Medical Problem Solving: An Analysis of Clinical Reasoning* (Cambridge, MA: Harvard University Press, 1978), esp. chapter 10.

79. Henry E. Ledgard, *Programming Proverbs* (Rochelle Park, NJ: Hayden Book Co., 1975), p. 3. Jerry Ravetz, who elsewhere strongly defended a view of science as "craftwork," nevertheless regarded the use of "aphorisms" as a sign of scientific "immaturity," to be eventually replaced by more certain "universal laws": Jerome R. Ravetz, *Scientific Knowledge and Its Social Problems* (New Brunswick, NJ: Transaction Publishers, 1996; orig. publ. 1971), pp. 375–376.

80. G[eorge] Polya, *How to Solve It: A New Aspect of Mathematical Method*, 2nd ed. (Princeton, NJ: Princeton University Press, 1973; orig. publ. 1945), pp. 3, 221–225; for the invocation of folkish proverbs in surgical problem-solving, see Trevor J. Pinch, H. M. Collins, and Larry Carbone, "Inside Knowledge: Second Order Measures of Skill," *The Sociological Review* 44 (1996): 163–186, on p. 163; and, in the training of biochemists, Sara Delamont, Paul Atkinson, and Odette Parry, *The Doctoral Experience: Success and Failure in Graduate School* (London: Falmer Press, 2000), p. 60.

81. Notably, Amos Tversky and Daniel Kahneman, "Judgment under Uncertainty: Heuristics and Biases," *Science* 185 (27 September 1974): 1124–1131, on pp. 1124, 1131.

82. To be fair, Tversky and Kahneman themselves stipulated ("Judgment under Uncertainty," p. 1130) that "The reliance on heuristics and the prevalence of biases are not restricted to laymen. Experienced researchers are also prone to the same biases—when they think intuitively."

83. Gerd Gigerenzer, Peter M. Todd, and the ABC Research Group, *Simple Heuristics That Make Us Smart* (New York: Oxford University Press, 1999).

84. Gigerenzer, Todd, and the ABC Research Group, *Simple Heuristics*, pp. 7–10 (for the model of "unbounded rationality"); p. 9 (for philosophers' practical rationality); pp. 25–26 (for remarks on the history of heuristics); 363 (for proverbs; emphasis added). See also Gerd Gigerenzer, *Adaptive Thinking: Rationality in the Real World* (Oxford: Oxford University Press, 2000), pp. 15–25 (for criticisms of cognitive psychological tendencies to conceive of everyday heuristics as biases); pp. 290–291 (for the Gambler's Fallacy); Robert Nozick, *The Nature of Rationality* (Princeton, NJ: Princeton University Press, 1993), pp. 163–172 (for sympathetic views of "Philosophical Heuristics"); William C. Wimsatt, "Heuristics and the Study of Human Behavior," in *Metatheory in Social Science: Pluralisms and Subjectivities*, ed. Donald W. Fiske and Richard A. Shweder (Chicago: University of Chicago Press, 1986), pp. 292–314; Michael E. Gorman, *Simulating Science: Heuristics, Mental Models, and Technoscientific Thinking* (Bloomington: Indiana University Press,

1992), esp. pp. 193, 210–213; and Herbert Simon and Allen Newell, *Human Problem Solving* (Englewood Cliffs, NJ: Prentice-Hall, 1972). The specific objection to Tversky and Kahneman on the Gambler's Fallacy is my own, not Gigerenzer's, though I take it to be broadly compatible with his point of view.

85. Geertz, "Common Sense," pp. 89–91.

86. Galileo Galilei, *Dialogue Concerning Two Chief World Systems,* trans. Stillman Drake, 2nd ed. (Berkeley: University of California Press, 1967; orig. publ. 1632), p. 203.

87. Following Alfred Schutz, Harold Garfinkel set up a contrast between "the rational properties of scientific and common sense activities," identifying the special "maxims of conduct" that distinguish the two modes. Yet he later felt obliged to warn against the domains such a distinction legitimately marks out: "To avoid misunderstanding I want to stress that the concern here is with the attitude of scientific *theorizing.* The attitude that informs the activities of actual scientific inquiry is another matter entirely": Harold Garfinkel, "The Rational Properties of Scientific and Common Sense Activities," in idem, *Studies in Ethnomethodology* (Englewood Cliffs, NJ: Prentice-Hall, 1967), chapter 8, on pp. 269–270, 272 (note).

88. Thomas Henry Huxley, "On the Educational Value of the Natural History Sciences," in idem, *Collected Essays,* vol. 3, *Science and Education: Essays* (New York: D. Appleton, 1900; orig. publ. 1854), pp. 38–65, on p. 45. For experimentally based claims that scientists are not notably better at formal logic than other sorts of people, see Michael J. Mahoney, "Psychology of the Scientist: An Evaluative Review," *Social Studies of Science* 9 (1979): 349–375, and Michael J. Mahoney and B. G. DeMonbreun, "Psychology of the Scientist: An Analysis of Problem-Solving Bias," *Cognitive Therapy and Research* 1 (1977): 229–238.

89. Albert Einstein, *Ideas and Opinions* (New York: Crown Publishers, 1954), p. 319.

90. Max Planck, *Scientific Autobiography and Other Papers,* trans. Frank Gaynor (New York: Philosophical Library, 1949), p. 88.

91. J. Robert Oppenheimer, "The Scientific Foundations for World Order," in Ernest Llewellyn Woodward et al., *Foundations for World Order* (Denver, CO: University of Denver Press, 1949), pp. 35–51, on p. 51; James B. Conant, *Science and Common Sense* (New Haven, CT: Yale University Press, 1951), p. 32; C. H. Waddington, *The Scientific Attitude* (West Drayton, Middlesex: Penguin, rev. ed., 1948), p. 117. For a famously splenetic attempt at putting Huxley, Einstein, Planck, Oppenheimer, Conant, Waddington (and the sociologists of science) in their place, see Lewis Wolpert, *The Unnatural Nature of Science* (Cambridge, MA: Harvard University Press, 1992), for example, p. 11: "I would almost contend that if something fits in with common sense it almost certainly isn't science." Also see Shapin, "How to Be Antiscientific" (chapter 3 in this volume).

Chapter 15. Descartes the Doctor

1. For a better manner of living in the world as a legitimate test of philosophical knowledge, see for example, Alexander Nehamas, *The Art of Living: Socratic Reflections from Plato to Foucault* (Berkeley: University of California Press, 1998); Pierre Hadot, *Philosophy as a Way of Life: Spiritual Exercises from Socrates to Foucault,* trans. Michael Chase, ed. Arnold Davidson (Oxford: Blackwell, 1995); and John Cottingham, *Philosophy and the Good Life: Reason and the Passions in Greek, Cartesian and Psychoanalytic Ethics* (Cambridge: Cambridge University Press, 1998).

2. Deirdre N. McCloskey, *If You're So Smart: The Narrative of Economic Expertise* (Chicago: University of Chicago Press, 1990).

3. All three of these "test-questions" should be reasonably familiar to modern historians of ideas: the first and second through the work of such modern "virtue theorists" as Martha Nussbaum and Alasdair MacIntyre; the third through historical studies of science-technology relations and of the rhetoric used to legitimate the place of science in capitalist societies.

4. Of course, other conceptions of philosophy, for example, Epicurean and Stoic, did not run this rule over knowledge: the purpose of philosophy was to console and to reconcile people to their inevitable fate.

5. See Owsei Temkin, *Hippocrates in the World of Pagans and Christians* (Baltimore: Johns Hopkins University Press, 1991), pp. 8–9; idem, *The Double Face of Janus and Other Essays in the History of Medicine* (Baltimore: Johns Hopkins University Press, 1977), pp. 187–188.

6. There is no special epistemological problem associated with recognizing therapeutic "success" in premodern medicine. From within a modernist and realist perspective, it is not difficult to identify premodern practices that "worked" and plausible reasons that might save cherished intellectual principles when interventions "did not work": see, for example, Erwin H. Ackerknecht, *Medicine and Ethnology: Selected Essays,* ed. H. H. Walser and H. M. Koelbing (Baltimore: Johns Hopkins University Press, 1971), pp. 120–134. When that realism is supplemented by a symbolic and cultural framework, "working" is recognized as dramatically visible—as when purgatives, emetics, and carminatives visibly, and sometimes spectacularly, "worked" on the body: see Charles E. Rosenberg, "The Therapeutic Revolution: Medicine, Meaning, and Social Change in Nineteenth-Century America," *Perspectives in Biology and Medicine* 20 (1977): 485–506 (reprinted in idem, *Explaining Epidemics and Other Studies in the History of Medicine* [Cambridge: Cambridge University Press, 1992), pp. 9–31]; and, for sociological sensibilities toward knowledge and its efficacy, see Michael J. Mulkay, "Knowledge and Utility: Implications for the Sociology of Knowledge," *Social Studies of Science* 9 (1977): 63–80.

7. For an entry into the large secondary literature on medicine and natural phi-

losophy in the Scientific Revolution, see the fine essay by H. J. Cook, "The New Philosophy and Medicine in Seventeenth-Century England," in *Reappraisals of the Scientific Revolution,* ed. David C. Lindberg and Robert S. Westman (Cambridge: Cambridge University Press, 1990), pp. 397–436; also idem, "Physicians and the New Philosophy: Henry Stubbe and the Virtuosi-Physicians," in *The Medical Revolution of the Seventeenth Century,* ed. Robert French and Andrew Wear (Cambridge: Cambridge University Press, 1989), pp. 246–271; Theodore M. Brown, "The College of Physicians and the Acceptance of Iatro-Mechanism in England, 1665–1695," *Bulletin of the History of Medicine* 44 (1970): 12–30; idem, "Physiology and the Mechanical Philosophy in Mid-Seventeenth-Century England," *Bulletin of the History of Medicine* 51 (1977): 25–54.

8. The "Epistle Dedicatory" to Hobbes's "De corpore," in *The English Works of Thomas Hobbes,* ed. Sir W. Molesworth, 11 vols. (London: Bohn, 1839–1845), vol. 1, referred (p. viii) to "the science of *man's body*" as "the most profitable part of natural science," but specified that this science was "first discovered" by Harvey. That science, Hobbes continued (p. xi), has been advanced in our times "by the wit and industry of physicians, the only true natural philosophers, especially of our most learned men of the College of Physicians in London." Hobbes had very little to say about medicine and it is not included in his map of the branches of philosophy in chapter 9 of *Leviathan,* ed. C. B. Macpherson (Harmondsworth: Penguin, 1968 [1651]), p. 149. For the significance of this omission, see Tom Sorell, "Hobbes's Scheme of the Sciences," in idem, ed., *The Cambridge Companion to Hobbes* (Cambridge: Cambridge University Press, 1996), pp. 45–61, on pp. 52–54. For scattered remarks on Hobbes's diet and regimen of health, see Arnold A. Rogow, *Thomas Hobbes: Radical in the Service of Reaction* (New York: W. W. Norton, 1986), pp. 224–226.

9. Of course, as in Antiquity, while those early moderns able to afford their services did employ professional physicians, the practice of diagnosing and treating oneself was common.

10. Francis Bacon, "The Advancement of Learning [Books I–II]," in *The Philosophical Works of Francis Bacon,* ed. James Spedding, Robert Leslie Ellis, and Douglas Denon Heath, 5 vols. (London: Longman, 1857–1858), vol. 3, pp. 253–491, on pp. 367, 373.

11. Francis Bacon, "Of the Dignity and Advancement of Learning, Book IV," in idem, *Philosophical Works,* pp. iv, 372–404, 390; idem, "The History of Life and Death," ibid., pp. v, 213–335. See also Graham Rees's introductory essay to his edition of Bacon's *De vijs mortis*: Bacon, *Philosophical Studies c. 1611–c. 1619,* ed. Graham Rees (Oxford: Oxford University Press, 1996), pp. xvii–cx. For documentation of seventeenth-century English medical concern with the prolongation of life, see Charles Webster, *The Great Instauration: Science, Medicine and Reform 1626–1660* (London: Duckworth, 1975), pp. 246–323.

12. Robert Boyle, "Usefulness of Experimental Natural Philosophy," in idem,

The Works of the Honourable Robert Boyle, ed. Thomas Birch, 2nd ed., 6 vols. (London: Rivington, 1772), vol. 2, pp. 1–246, on p. 66; see also pp. 185–186: "it is scarce to be expected, that till men have a better knowledge of the principles of natural philosophy . . . it is hard to arrive at a more comprehensive theory of the various possible causes of diseases, and of the contrivance and uses of the parts of the body, the method which supposes this knowledge should be other than in many things defective, and in some erroneous." See also ibid., p. 199, quoting Celsus: "The contemplation of nature, though it maketh not a physician, yet it fits him to learn physick." Boyle went on: "A deeper insight into nature may enable men to apply the physiological discoveries made by it (though some more immediately, and some less directly) to the advancement and improvement of physick."

13. Boyle's therapeutic activism sat alongside considerable stress on the curative power of nature: see Barbara Beigun Kaplan, *"Divulging of Useful Truths in Physick": The Medical Agenda of Robert Boyle* (Baltimore: Johns Hopkins University Press, 1993), esp. chapter 5. What Boyle called the body's "strainers" (for example, the liver, spleen, and kidneys) can alter a medicine's corpuscular texture or recombine it with other corpuscles, thus admitting it to certain bodily sites and not others. Or the specific medicine might itself "restore the strainers to their right tone and texture" (Boyle, "Usefulness of Experimental Natural Philosophy," p. 192). Boyle was skeptical of Paracelsian and Helmontian talk of a "universal medicine" (ibid., pp. 196–197).

14. Boyle, "Usefulness of Experimental Natural Philosophy," pp. 65, 112.

15. René Descartes, *Discourse on the Method,* published in *The Philosophical Writings of Descartes* (hereafter *PWD*), trans. John Cottingham, Robert Stoothoff, and Dugald Murdoch, 3 vols. (Cambridge: Cambridge University Press, 1985–1991), vol. 1, pp. 111–151, on pp. 142–143. Note also that Descartes claimed that medicine had the capacity to make "men in general wiser and more skilful" since "the mind depends so much on the temperament and disposition of the bodily organs," so that one could imagine a virtuous cycle in which the practitioners of reformed medicine would become more clever and, hence, capable of making even more discoveries, which would in turn make them cleverer still. See also Geneviève Rodis-Lewis, *Descartes: His Life and Thought,* trans. Jane Marie Todd (Ithaca, NY: Cornell University Press, 1998), pp. 127–128; and Temkin, *Hippocrates,* p. 13 (for the ancient view that "human intelligence could be changed by diet"). And see Steven Shapin, "Feeding, Feeling, Thinking," in *Gefühle zeigen: Manifestationesformen emotionaler prozesse,* ed. Johannes Fehr and Gerd Folkers (Zürich: Chronos Verlag, 2009), pp. 445–466.

16. Descartes, *Discourse,* pp. 143, 151. Descartes's medical promises, and especially his views about the extension of human life, are briefly sketched in Gerald J. Gruman, *A History of Ideas about the Prolongation of Life* (New York: Arno Press, 1977 [1966]), pp. 77–80.

17. Descartes, *Discourse*, p. 113; Stephen Gaukroger, *Descartes: An Intellectual Biography* (Oxford: Clarendon Press, 1995), pp. 20, 64 (generally endorsing the idea that Descartes formally studied medicine), and Rodis-Lewis, *Descartes*, pp. vii, ix, 1–2, 18–19; and idem, "Descartes' Life and the Development of His Philosophy," in *The Cambridge Companion to Descartes*, ed. John Cottingham (Cambridge: Cambridge University Press, 1992), pp. 21–57, on pp. 28–29 (judging it unlikely that Descartes did study medicine and disputing traditional early datings of his medical interests).

18. Letter from Descartes to Newcastle, October 1645, in *PWD*, vol. 3, p. 275 (emphasis added); see also Richard B. Carter, *Descartes' Medical Philosophy: The Organic Solution to the Mind-Body Problem* (Baltimore: Johns Hopkins University Press, 1983), p. 31.

19. Adrien Baillet, *The Life of Monsieur Des Cartes . . . Translated from the French by S. R.* (London, 1693 [1691]), pp. 79–80.

20. Letter from Descartes to Mersenne, January 1630, in *PWD*, vol. 3, p. 17; Baillet, *Monsieur Des Cartes*, p. 81; G. A. Lindeboom, *Descartes and Medicine* (Amsterdam: Rodopi, 1979), p. 43.

21. Letter from Descartes to Mersenne, 15 April 1630, in *PWD*, vol. 3, p. 21.

22. Letter from Descartes to Mersenne, 20 February 1639, in *PWD*, vol. 3, p. 134. On Descartes's dissections, see also Descartes to Mersenne, early June 1637, in ibid., p. 59; Baillet, *Monsieur Des Cartes*, p. 80; T. S. Hall, "Foreword," in Descartes, *Treatise of Man*, ed. and trans. T. S. Hall (Cambridge, MA: Harvard University Press, 1972), pp. xii–xiii.

23. Descartes, *Description of the Human Body*, in *PWD*, vol. 1, pp. 313–324, on p. 314.

24. John Cottingham, ed. and trans., *Descartes' Conversation with Burman* (Oxford: Clarendon Press, 1976), p. 50. Descartes refused to be drawn on the question of whether "man was immortal before the Fall." Although both the fact of and the explanation for Edenic immortality were much debated by Renaissance and early modern physicians, Descartes's judgment was that this was "not a question for the philosopher, but must be left to the theologians." Frans Burman was a twenty-year-old student who interviewed Descartes in April 1648 at the philosopher's home in Egmond. See also Lindeboom, *Descartes and Medicine*, pp. 96–97. The "Preface" to *The Passions of the Soul* noted that God had undoubtedly provided people "with all things necessary . . . to be preserved in perfect health to an extreme old age": what was lacking was the knowledge of what these necessary things were: "Preface" to Descartes, *The Passions of the Soul*, ed. and trans. Stephen Voss (Indianapolis: Hackett, 1989 [1649]), p. 7.

25. Descartes, *Principles of Philosophy* (extracts), in *PWD*, vol. 1, pp. 179–291, on p. 186. As Richard Carter puts it, for Descartes "physics deals with the general body out of which particular bodies are formed; medicine deals with particular,

mortal bodies that are alive and that have souls united to them": Carter, *Descartes' Medical Philosophy*, p. 31. Descartes's well-known stipulation—both in the *Principles* and in the *Treatise of Man*—that he was giving an account not of real but of imaginary human bodies did not, in his estimation here, diminish its significance for medical practice. The characteristics he attributed to these imaginary bodies were supposed to be "such as to correspond accurately with all the phenomena of nature." And this "will indeed be sufficient for application in ordinary life, since medicine and mechanics . . . are directed only towards items that can be perceived with the senses" (Descartes, *Principles*, p. 289 [part IV, sec. 204]). That is to say, philosophically informed medical interventions might work even if the philosophically posited underlying causal structures were not the real ones.

26. "Preface" to Descartes, *Passions*, p. 7.

27. Roger French, "Harvey in Holland: Circulation and the Calvinists," in *The Medical Revolution*, pp. 46–86, 53–54; also Baillet, *Monsieur Des Cartes*, pp. 121–122, 127–128.

28. Descartes, *Description*, p. 319.

29. And also how Descartes distinguished his cardiac physiology from Harvey's. See also Descartes, *Passions*, pp. 21–22; idem, *Description*, part 2. Also Thomas S. Hall, "The Physiology of Descartes," in Descartes, *Treatise*, pp. xxvi–xxxiii; idem, *History of General Physiology*, 2 vols. (Chicago: University of Chicago Press, 1975), vol. 1, pp. 250–264; Gary Hatfield, "Descartes' Physiology and Its Relation to His Psychology," in *The Cambridge Companion*, pp. 335–370; Gaukroger, *Descartes*, pp. 269–276. For the physiological writings of Descartes's major medical disciple, the Utrecht professor of theoretical medicine Henricus Regius, see Theo Verbeek, *Descartes and the Dutch: Early Reactions to Cartesian Philosophy, 1637-1650* (Carbondale: Southern Illinois University Press, 1992), chapter 2.

30. French, "Harvey in Holland," pp. 50–51; Thomas S. Hall, "First French Edition: Synopsis of Contents," in Descartes, *Treatise*, p. xxxvi.

31. Descartes, *Description*, pp. 316, 319.

32. The most important site of such discussion is Descartes's late writings on the passions of the soul.

33. Letter from Descartes to Newcastle, October 1645, in *PWD*, vol. 3, pp. 275–276.

34. Descartes, *Conversations*, p. 51. The source is in fact Suetonius's life of Tiberius. See also Baillet, *Monsieur Des Cartes*, p. 260 (who put Descartes's age at nineteen or twenty when he came to that opinion); Lindeboom, *Descartes and Medicine*, pp. 94–95; Rodis-Lewis, *Descartes*, p. 19.

35. Letter from Descartes to Princess Elizabeth, May or June 1645, in *PWD*, vol. 3, p. 251. His mother's disease was caused, he said, by "distress" (*déplaisirs*). See also Descartes to Mersenne, 30 July 1640, in ibid., p. 148 (for the maternal role in inheritance), and Baillet, *Monsieur Des Cartes*, pp. 3–5, 260.

36. Baillet, *Monsieur Des Cartes,* pp. 14, 66; see also Peter Dear, "A Mechanical Microcosm: Bodily Passions, Good Manners, and Cartesian Mechanism," in *Science Incarnate: Historical Embodiments of Natural Knowledge,* ed. Christopher Lawrence and Steven Shapin (Chicago: University of Chicago Press, 1998), pp. 51–82, on pp. 54–55.

37. Letter from Descartes to Jean-Louis Guez de Balzac, 5 April 1631, in *PWD,* vol. 3, p. 30; Descartes to Elizabeth, 1 September 1645, in ibid., p. 263; Baillet, *Monsieur Des Cartes,* p. 259 (who said that Descartes commonly spent as much as twelve hours a day in bed).

38. Letter from Descartes to Princess Elizabeth, 28 June 1643, in *PWD,* vol. 3, p. 227; also Baillet, *Monsieur Des Cartes,* pp. 35–36: high speculations "threw his mind into such violent Agitations . . . He wearied it out to that degree that his brain took fire, and he falls into a spice of enthusiasm." For the contemporary charge that Descartes was an "enthusiast," see Michael Heyd, *'Be Sober and Reasonable': The Critique of Enthusiasm in the Seventeenth and Eighteenth Centuries* (Leiden: E. J. Brill, 1995), chapter 4, esp. pp. 116–117. For Henry More, Anne Conway, and the dietetic management of philosophical heat, see Shapin, "The Philosopher and the Chicken" (chapter 11 in this volume); and, for treatment of pertinent aspects of the relationship between the passions and enthusiasm, see Adrian Johns, "The Physiology of Reading and the Anatomy of Enthusiasm," in *Religio Medici: Medicine and Religion in Seventeenth-Century England,* ed. Ole Peter Grell and Andrew Cunningham (Aldershot: Scholar Press, 1996), pp. 136–170.

39. Letter from Descartes to Constantijn Huygens, 5 October 1637, in *Oeuvres de Descartes,* ed. Charles Adam and Paul Tannery, 11 vols. (Paris: J. Vrin, 1964–1976), vol. 2, pp. 434–435; also Baillet, *Monsieur Des Cartes,* pp. 122, 258.

40. Letter from Descartes to Huygens, 4 December 1637, in *PWD,* vol. 3, p. 76; see also Gaukroger, *Descartes,* pp. 332–333.

41. Letter from Descartes to Mersenne, 9 January 1637, in *PWD,* vol. 3, p. 131; also Baillet, *Monsieur Des Cartes,* pp. 23, 260.

42. Letter from Descartes to Huygens, 6 June 1639, in *PWD,* vol. 3, p. 136.

43. Letter from Huygens to Descartes, 23 November 1637, in *Oeuvres,* vol. 1, p. 463; cf. Huygens to Descartes, 8 September 1637, in ibid., pp. 396–397; also Lindeboom, *Descartes and Medicine,* pp. 44, 96.

44. Reported in Pierre Des Maizeaux's *Life of St. Evremond* (1728) and quoted in Descartes, *Oeuvres,* vol. 9, p. 671; see also Gruman, *Prolongation of Life,* p. 79.

45. See Descartes, *Oeuvres,* vol. 9, pp. 670–671; Adrien Baillet, *La vie de Monsieur Des-Cartes,* 2 vols. (Paris, 1691), vol. 2, pp. 449–454.

46. Baillet, *Monsieur Des Cartes,* pp. 252–253; idem, *La vie,* vol. 2, pp. 414–423.

47. Reported in Baillet, *La vie,* vol. 2, pp. 452–453; also Descartes, *Oeuvres,* vol. 2, p. 671; and see Lindeboom, *Descartes and Medicine,* p. 95. Note that a disease—such as what we would now call pneumonia—that was precipitated by an alteration

in regimen, or by exposure to extreme conditions, would then have been regarded as a "violent" or "external" cause of death. So Baillet said (*Monsieur Des Cartes,* p. 252) that Descartes's last illness was partly caused by "the disorder of [his] regular way of living." The general idea was that one's natural life span was the length of time one would live without the violent causes that might shorten it artificially. The relationship recognized by the early moderns between disease and old age was, however, contested; some saw old age as a disease, others as a state that made one susceptible to disease. See, for representative discussion, Laurent Joubert, *Popular Errors,* trans. Gregory David de Rocher (Tuscaloosa: University of Alabama Press, 1989 [1579]), pp. 41–43; also M. D. Grmek, *On Ageing and Old Age: Basic Problems and Historic Aspects of Gerontology and Geriatrics,* Monographiae Biologicae, vol. 5, no. 2 (The Hague: Kluwer, 1958), esp. pp. 5–10.

48. Hector-Pierre Chanut to Princess Elizabeth, 19 February 1650, in Descartes, *Oeuvres,* vol. 5, p. 471; Samuel Sorbière to Pierre Petit, 20 February 1657, in ibid., p. 485.

49. *Extra ordinarisse Posttijdinghe* (10 April 1650); quoted and translated in Lindeboom, *Descartes and Medicine,* p. 94.

50. Lindeboom, *Descartes and Medicine,* p. 94; Dear, "A Mechanical Microcosm," pp. 61–62. For the circumstances of, and stories about, Descartes's death, see esp. Gaukroger, *Descartes,* pp. 415–417, and various sources assembled in Descartes, *Oeuvres,* vol. 5, pp. 470–500.

51. Baillet, *Monsieur Des Cartes,* p. 260. Descartes rarely approved of phlebotomy—which he accounted "extream dangerous to most People"—though, as noted, he did tolerate moderate bleeding in certain limited circumstances.

52. Letter from Descartes to Mersenne, 23 November 1646, in *PWD,* vol. 3, p. 301; Lindeboom, *Descartes and Medicine,* p. 44. Around the same time, in a case of nosebleed, Descartes warned against the use of wine, vinegar, mustard, and saffron. A small amount of blood might be let, but care had to be taken to take blood from the foot on the same side of the body as the bleeding nostril: Descartes to [Boswell?], [1646?], in Descartes, *Oeuvres,* vol. 4, pp. 694–700, on pp. 698–699; Lindeboom, *Descartes and Medicine,* pp. 43–44.

53. Letter from Jacqueline Pascal to her sister Gilberte Périer, 25 September 1647, in *Lettres, Opuscules et Mémoires de Madame Perier et de Jacqueline, sœurs de* Pascal, et *de Marguerite Perier, sa niece,* ed. M. P. Faugère (Paris: Auguste Vaton, 1845), pp. 309–312. Pascal suffered badly from dyspepsia and migraine, and, from around the time he met Descartes, probably from a kind of motor neuropathy. For an account of this momentous meeting—at which Descartes (as he later claimed) suggested the Puy-de-Dôme experiment—see Rodis-Lewis, *Descartes,* pp. 178–181; also E. T. Bell, *Men of Mathematics* (New York: Simon and Schuster, 1937), p. 80.

54. Letter from Descartes to Princess Elizabeth, December 1646, in *PWD,* vol. 3, pp. 304–305; Baillet, *Monsieur Des Cartes,* p. 260.

55. Baillet, *La vie*, vol. 2, p. 452. For notes on materia medica by Descartes, see "Remedia, et vires medicamentorum," in *Oeuvres*, vol. 11, pp. 641–644; also idem, "Excerpta anatomica," in ibid., pp. 543–634, on p. 606.

56. Letter from Descartes [to Alphonse Pollot], mid-January 1641, in *PWD*, vol. 3, p. 168.

57. Letters from Descartes to Princess Elizabeth, May or June 1645, in *PWD*, vol. 3, p. 250; 22 July 1645, in ibid., p. 255; October or November 1646, in ibid., p. 298. See in this connection L. W. B. Brockliss, "The Development of the Spa in Seventeenth-Century France," in *The Medical History of Waters and Spas, Medical History*, ed. Roy Porter, Supplement No. 10 (1990), pp. 23–47.

58. It should go almost without saying that choice in such matters was available only to those classes that had the resources to choose. Medical advice on dietetics and regimen was therefore geared to the élite. On this point, see Ludwig Edelstein, "The Dietetics of Antiquity," in idem, *Ancient Medicine: Selected Papers of Ludwig Edelstein*, ed. Owsei Temkin and C. Lilian Temkin (Baltimore: Johns Hopkins University Press, 1967 [1931]), pp. 303–316, on pp. 305–306; A. Emch-Dériaz, "The Non-Naturals Made Easy," in *The Popularization of Medicine 1650–1850*, ed. Roy Porter (London: Routledge, 1992), pp. 134–159, esp. pp. 135–136; and William Coleman, "Health and Hygiene in the *Encyclopédie*: a Medical Doctrine for the Bourgeoisie," *Journal of the History of Medicine* 29 (1974): 399–421, on pp. 399, 401. The usual list of early modern non-naturals included: ambient air, diet (in the strict sense of food and drink), sleeping and waking, exercise and rest, retentions and evacuations (including sexual release), and the passions of the mind. For debates over the Galenic sources of the doctrine and phrase, see L. J. Rather, "The 'Six Things Non-Natural': A Note on the Origins and Fate of a Doctrine and a Phrase," *Clio Medica* 3 (1968): 337–347; Saul Jarcho, "Galen's Six Non-Naturals: A Bibliographic Note and Translation," *Bulletin of the History of Medicine* 44 (1970): 372–377; P. Niebyl, "The Nonnaturals," *Bulletin of the History of Medicine* 45 (1971): 486–492; Jerome J. Bylebyl, "Galen on the Non-Natural Causes of Variation in the Pulse," *Bulletin of the History of Medicine* 45 (1971): 482–485. I deal extensively with early modern polite dietetics in Shapin, "How to Eat Like a Gentleman" (chapter 12 in this volume).

59. Descartes, *Conversations*, p. 51.

60. Baillet, *Monsieur Des Cartes*, p. 259. This was just the kind of thing that was commonly said of the dietetics of the early modern gentleman-philosopher. Indeed, it closely parallels Gilbert Burnet's funeral sermon preached over Robert Boyle's body in the year after Baillet wrote this account. It was a way of saying that the scholarly life had not "spoiled" the virtues and manners of the civic gentleman, and, of course, this kind of dietetics was a way for the gentleman-philosopher to present himself as unspoiled by scholarly moroseness, melancholy or asceticism: Shapin, "The Philosopher and the Chicken" and "How to Eat Like a Gentleman" (chapters 11 and 12 in this volume).

61. Baillet, *Monsieur Des Cartes,* pp. 259–260.

62. Descartes, *Conversations,* p. 50; for his liking of vegetables from his own garden, see Baillet, *La vie,* vol. 2, p. 450.

63. Baillet, *Monsieur Des Cartes,* p. 260.

64. Baillet, *Monsieur Des Cartes,* p. 260.

65. Letter from Descartes to Princess Elizabeth, 8 July 1644, in *PWD,* vol. 3, p. 237.

66. Letter from Descartes to Princess Elizabeth, May or June 1645, in *PWD,* vol. 3, pp. 249–251.

67. Both ancient (Galenic) and early modern medical thought did in fact accept the possibility that habit might gradually change innate temperament; see, for example, Joubert, *Popular Errors,* p. 43, and [Thomas Tryon], *The Way of Health, Long Life and Happiness, or, a Discourse of Temperance,* (London, 1683), p. 19. The author of the article on the non-naturals in the *Encyclopédie* agreed: see Emch-Dériaz, "The Non-Naturals Made Easy," pp. 138–139. So the proverbial "habit is a second nature" can be understood to express the view that habit can *give you* another nature, not just the notion that habit is almost as strong as innate endowment. For early modern Stoicism and the management of the passions, see Dear, "A Mechanical Microcosm," pp. 68–72.

68. Letter from Descartes to Princess Elizabeth, 6 October 1645, in *PWD,* vol. 3, p. 270; also Descartes, *Passions,* pp. 338 339; and, for the most philosophically sensitive treatment of Descartes on the passions, Susan James, *Passion and Action: The Emotions in Seventeenth-Century Philosophy* (Oxford: Clarendon Press, 1997), pp. 92–100. See also idem, "Reason, the Passions, and the Good Life," in *The Cambridge History of Seventeenth-*Century *Philosophy,* ed. Daniel Garber and Michael Ayers, 2 vols. (Cambridge: Cambridge University Press, 1998), vol. 2, pp. 1358–1396; idem, "Explaining the Passions: Passions, Desires, and the Explanation of Action," in *The Soft Underbelly of Reason: The Passions in the Seventeenth Century,* ed. Stephen Gaukroger (London: Routledge, 1998), pp. 17–33.

69. Dear, "A Mechanical Microcosm," pp. 68–69.

70. Gaukroger (*Descartes,* p. 388) interestingly suggests a shift occurring around the time of his correspondence with Princess Elizabeth "from a somatopsychic account, in which the influence of bodily dispositions on the state of the soul is stressed, to a psychosomatic account in which are stressed the effects of the soul on bodily dispositions."

71. James, *Passion and Action,* pp. 106–108 (emphasis added); Amélie O. Rorty, "Descartes on Thinking With the Body," in *Cambridge Companion to Descartes,* led. Cottingham, pp. 371–392; and Cottingham, *Philosophy and the Good Life,* pp. 87–96.

72. Descartes, *Treatise,* pp. 5–10, 17–19.

73. Descartes, *Treatise,* p. 70 ("when the blood that goes into the heart is more

pure and subtle and is kindled more easily than usual, this arranges the little nerve that is there in the manner that is required to cause the sensation of *joy*"), and p. 111 (for dry air); letter from Descartes to Princess Elizabeth, May or June 1645, in *PWD*, vol. 3, p. 250; Descartes, *Principles*, pp. 280–281.

74. Descartes, *Treatise*, pp. 108–112.

75. Michel Eyquem de Montaigne, *The Complete Essays of Montaigne*, trans. Donald M. Frame (Stanford, CA: Stanford University Press, 1965 [1580–1588]), esp. "Of Experience," pp. 815–857. See, for example, the similarity between Montaigne's view of the appetites in sickness and views Descartes expressed in his conversation with Burman: "Both in health and in sickness I have readily let myself follow my urgent appetites. I give great authority to my desires and inclinations" ("Of Experience," p. 832). See also M. Brunyate, "Montaigne and Medicine," in *Montaigne and His Age*, ed. Keith Cameron (Exeter: University of Exeter Press, 1981), pp. 27–38, and, for general remarks on the conservatism of medical practice against the background of change in medical theory, see John Henry, "Doctors and Healers: Popular Culture and the Medical Profession," in *Science, Culture and Popular Belief in Renaissance Europe*, ed. Stephen Pumfrey, Paolo L. Rossi, and Maurice Slawinski (Manchester: Manchester University Press, 1991), pp. 191–221, esp. pp. 211–212: "the mechanical philosophy made no significant impact on the practice of medicine."

76. See, for example, Henry E. Sigerist, *Landmarks in the History of Hygiene* (London: Oxford University Press, 1956), chapter 2. The most influential English version in this period was Sir John Harington's *The English Mans Doctor. Or the Schoole of Salerne* (London, 1607). For the proverbial element in early modern medical culture, see also Henry, "Doctors and Healers," pp. 198–201, and Shapin, "How to Eat Like a Gentleman" and "Proverbial Economies" (chapters 12 and 14 in this volume).

77. For the original, see Aulus Cornelius Celsus, *De medicina*, trans. W. G. Spencer, 3 vols. (Cambridge, MA: Harvard University Press, 1960), vol. 1, p. 43 (book 1, p. i); also p. 57. Celsus probably lived in Tiberius's reign, the same emperor whose opinion that every man should be his own physician Descartes so much liked; cf. Plutarch: "A diet which is very exact and precisely according to rule puts one's body both in fear and danger" (Plutarch, "Rules for the Preservation of Health," in idem, *Plutarch's Lives and Miscellanies*, ed. A. H. Clough and W. W. Goodwin, 5 vols. [New York: The Colonial Company, 1905], vol. 1, pp. 251–279, on p. 263). For many early modern endorsements of the Rule of Celsus, see Shapin, "How to Eat Like a Gentleman" and "Trusting George Cheyne" (chapters 12 and 13 in this volume).

78. Paul Slack, "Mirrors of Health and Treasures of Poor Men: The Use of the Vernacular Medical Literature of Tudor England," in *Health, Medicine and Mortality in the Sixteenth Century*, ed. Charles Webster (Cambridge: Cambridge University Press, 1979), pp. 237–273, on p. 268.

79. Archer Taylor, *The Proverb* (Berlin: P. Lang, 1985 [1931]), pp. 121–129; Dear, "A Mechanical Microcosm"; Baillet, *Monsieur Des Cartes,* pp. 18–19, 59 (for Descartes's refusal to be "a Slave" to passions, and how he overcame a youthful addiction to gambling).

80. Descartes, *Discourse,* p. 151.

81. Descartes, *Principles,* p. 189.

82. Letter from Descartes to Mersenne, 20 February 1639, in *PWD,* vol. 3, p. 135.

83. Letter from Descartes to Mersenne, 9 January 1639, in *PWD,* vol. 3, p. 131. In these connections, providence and the "grace of God" were repeatedly invoked, and Descartes fell in with traditional Renaissance and early modern medical sensibilities about the complicated relationships between, on the one hand, using one's best efforts to maintain health and extend life and, on the other, acknowledging the limits to these efforts set by God's will and plan: see on this point Joubert, *Popular Errors,* pp. 41–43.

84. Letter from Descartes to Chanut, 15 June 1646, in *PWD,* vol. 3, p. 289; cf. Baillet, *Monsieur Des Cartes,* p. 53.

85. Cf. Gaukroger's contention (*Descartes,* p. 388) that the sentiments expressed to Chanut represent an intellectual "shift" brought about by the reflections on which Descartes was engaged *c.* 1645–1646 about "the nature of the substantial union of mind and body." The preface to *The Passions of the Soul* (1649) was, however, a quite typical Cartesian expression of medical optimism.

86. This is Descartes's own example: letter from Descartes to Hyperaspistes, August 1641, in *PWD,* vol. 3, pp. 189–190. Roger French ("Harvey in Holland," pp. 78–79) puts the point at issue well: "As a *scientia,* medicine is 'assent to the conclusion of a demonstrative syllogism' [here French is quoting Vopiscus-Fortunatus Plemp, Dutch medical professor and friend of Descartes], but its individuality and autonomy lie in the fact that it is also an art. In the art of medicine, the writ of the philosophers does not run. In the art of medicine, there are no rules for certain knowledge: as Galen says there is no sure way of telling a nephritic from a colic pain, nor how a medicine acts by its 'whole substance.' As Celsus says, in the art of medicine there are no sure precepts, as in other natural sciences and there are many possible conclusions (and, by implication, the accumulated experiential knowledge of the centuries is important)."

87. Descartes, *Discourse,* p. 122.

88. See Shapin, "Trusting George Cheyne" (chapter 13 in this volume).

89. Owsei Temkin, *Galenism: Rise and Decline of a Medical Philosophy* (Ithaca, NY: Cornell University Press, 1973), pp. 39–40.

90. Slack, "Mirrors of Health," pp. 271–272; also Alan Macfarlane, *The Family Life of Ralph Josselin, a Seventeenth-Century Clergyman: An Essay in Historical Anthropology* (New York: Norton, 1970), esp. pp. 173–176; Keith Thomas, "Health

and Morality in Early Modern England," in *Morality and Health*, ed. Allan M. Brandt and Paul Rozin (New York, 1997), pp. 15–34, esp. 20–24; Shapin "How to Eat Like a Gentleman" (chapter 12 in this volume).

91. For a brilliant survey of this general tendency in seventeenth-century mechanical philosophy, see Alan Gabbey, "The Mechanical Philosophy and Its Problems: Mechanical Explanations, Impenetrability, and Perpetual Motion," in *Change and Progress in Modern Science,* ed. J. C. Pitt (Dordrecht: D. Reidel, 1985), pp. 9–84, esp. pp. 10–12. T. S. Hall shows that these kinds of apparently revolutionary, but practically insubstantial, transformations are also at the heart of Descartes's physiology. Descartes claimed to reject Galen while recasting many Galenic concepts in the vocabulary of Cartesian physics: see Hall's notes to the text of Descartes's *Treatise,* for example, nn. 13, 21, 35, 85; see also Hatfield, "Descartes' Physiology," pp. 341–344; and, for overall continuities between Cartesian thought and tradition, Etienne Gilson, *Etudes sur le rôle de la pensée médiévale dans la formation du système cartésien,* 4th ed. (Paris: J. Vrin, 1975), and Dennis Des Chene, *Physiologia: Natural Philosophy in Late Aristotelian and Cartesian Thought* (Ithaca, NY: Cornell University Press, 1996).

92. For attempts at genuinely historical explanation of mechanical intelligibility, see Dear, "A Mechanical Microcosm."

Chapter 16. Science and the Modern World

1. Andrew Dickson White, *The Warfare of Science* (New York: D. Appleton, 1876), and then developed as idem, *A History of the Warfare of Science with Theology in Christendom,* 2 vols. (New York: D. Appleton, 1896). White was following in the tradition of John William Draper, whose *History of the Conflict between Religion and Science* (New York: D. Appleton, 1874) similarly announced the inevitable triumph of science over religion.

2. Max Weber, "Science as a Vocation," in *From Max Weber: Essays in Sociology,* ed. and trans. H. H. Gerth and C. Wright Mills (London: Routledge, 1991; art. orig. publ. 1919, from a speech in 1917), pp. 129–156, on p. 142.

3. Thorstein Veblen, "The Place of Science in Modern Civilization," *American Journal of Sociology* 11 (1906): 585–609, on pp. 585–588.

4. A. N. Whitehead, *Science and the Modern World: Lowell Lectures* (London: The Scientific Book Club, 1946; orig. publ. 1925), p. 2.

5. George Sarton, *The Study of the History of Science* (Cambridge, MA: Harvard University Press, 1936), p. 5.

6. Charles Coulston Gillispie, *The Edge of Objectivity: An Essay in the History of Scientific Ideas* (Princeton, NJ: Princeton University Press, 1960), p. 9.

7. George Sarton, "The History of Science," in idem, *The Life of Science: Essays in the History of Civilization* (New York: Henry Schuman, 1948), pp. 29–58, on p. 55.

8. Herbert Butterfield, *The Origins of Modern Science, 1300–1800,* rev. ed. (New York: The Free Press, 1957; orig. publ. 1949), pp. 7–8.

9. A. C. Crombie, *Medieval and Early Modern Science,* 2 vols., rev. 2nd ed. (Garden City, NY: Doubleday Anchor Books, 1959; orig. publ. 1952), vol. 1, p. 7.

10. Gillispie, *Edge of Objectivity,* p. 8.

11. Richard S. Westfall, "The Scientific Revolution," *History of Science Society Newsletter* 15, no. 3 (July 1986): http://www.clas.ufl.edu/users/rhatch/pages/03-Sci-Rev/SCI-REV-Home/05-RSW-Sci-Rev.htm [accessed 18 July 2006].

12. See, for example, David A. Hollinger, "Money and Academic Freedom a Half-Century after McCarthyism: Universities and the Force Fields of Capital," in *Unfettered Expression: Freedom in American Intellectual Life,* ed. Peggie J. Hollingsworth (Ann Arbor: University of Michigan Press, 2000), pp. 161–184.

13. Michael Specter, "Political Science: The Bush Administration's War on the Laboratory," *New Yorker* (13 March 2006), pp. 58–69, on p. 61.

14. Derek J. deSolla Price, *Little Science, Big Science* (New York: Columbia University Press, 1968; orig. publ. 1963), p. 19.

15. http://www.cbsnews.com/stories/2002/04/29/opinion/polls/main507515 .shtml [accessed 12 February 2006].

16. National Science Foundation, Division of Science Resources Statistics, *Survey of Public Attitudes toward and Understanding of Science and Technology, 2001:* http://www.nsf.gov/statistics/seind04/c7/fig07–06.htm; European Commission, Eurobarometer 55.2, *Europeans, Science and Technology,* December 2001: http://europa .eu.int/comm/public_opinion/archives/eb/ebs_154_en.pdf [both accessed 12 February 2006].

17. Jon D. Miller, Eugenie C. Scott, and Shinji Okamoto, "Public Acceptance of Evolution," *Science* 313, no. 5788 (11 August 2006): 765–766.

18. http://www.pollingreport.com/science.htm; http://www.unl.edu/rhames/courses/current/creation/evol-poll.htm [both accessed 18 July 2006].

19. J. H. Leuba, *The Belief in God and Immortality: A Psychological, Anthropological and Statistical Survey* (Boston: Sherman, French & Co., 1916).

20. Tim Radford, "'Science Cannot Provide All the Answers': Why Do So Many Scientists Believe in God?" *The Guardian* (4 September 2003): http://www.guardian .co.uk/life/feature/story/0,13026,1034872,00.html [accessed 15 February 2006]; Edward J. Larson and Larry Witham, "Scientists Are Still Keeping the Faith," *Nature* 386 (3 April 1997): 435–436.

21. Edward J. Larson and Larry Witham, "Leading Scientists Still Reject God," *Nature* 394 (23 July 1998): 313.

22. http://www.religioustolerance.org/ev_publi.htm (These figures are from a poll conducted in November 1991; accessed 19 August 2009)

23. Andrew Kohut and Bruce Stokes, *America Against the World: How We Are Different and Why We Are Disliked* (New York: Times Books/Henry Holt, 2006), p.

61: "In 2004, by a 52 percent to 34 percent margin, Americans said it was more important to conduct such research, which might result in new cures for human diseases, than to avoid destroying the potential life of embryos. Two years earlier, only a plurality of Americans supported stem-cell research (43 percent in favor to 38 percent against)."

24. http://pewglobal.org/commentary/display.php?AnalysisID=66 [accessed 26 July 2006].

25. Ronald Dworkin, "Three Questions for America," *The New York Review of Books* 53, no. 14 (21 September 2006), pp. 24–30, on p. 24. We can set aside without comment, as an instance of a lawyer's scientific naivete, the fact that much radioactivity is indeed "harmless."

26. See, for example, Frank M. Turner, "Rainfall, Plagues, and the Prince of Wales: A Chapter in the Conflict of Science and Religion," *Journal of British Studies* 13 (1974): 46–65.

27. Jim Holt, "Madness About a Method: How Did Science Become So Contentious and Politicized?" *The New York Times Magazine* (11 December 2005), pp. 25, 28.

28. Stephen B. Withey, "Public Opinion about Science and Scientists," *Public Opinion Quarterly* 23 (1959): 382–388. Etzioni and Nunn argue convincingly that the public mind makes little, if any, distinction between science and technology: Amitai Etzioni and Clyde Nunn, "The Public Appreciation of Science in Contemporary America," *Daedalus* 103, no. 3 (Summer 1974): 191–205.

29. Georgine M. Pion and Mark W. Lipsey, "Public Attitudes toward Science and Technology: What Have the Surveys Told Us?" *Public Opinion Quarterly* 45 (1981): 303–316, on p. 304 (Table 1). The National Opinion Research Center (NORC) has compiled time-series data on public confidence in various institutions. The data show a decline in confidence in science from the early 1960s to the late 1970s, but this follows a drop in confidence for *all* major public institutions, and the decline for science was notably *less* than it was for others: ibid., p. 307.

30. Figures quoted in Holt, "Madness About a Method," p. 25, from a NORC survey conducted between August 2004 and January 2005.

31. Kohut and Stokes, *America Against the World,* pp. 60, 86.

32. Specter, "Political Science," p. 61.

33. See, for example, Shapin, "How to Be Antiscientific" (chapter 3 in this volume, pp. 36–37).

34. http://teacher.pas.rochester.edu/phy_labs/AppendixE/AppendixE.html; http://physics.ucr.edu/~wudka/Physics7/Notes_www/node5.html [accessed 18 July 2006].

35. For example, http://www.sciencemag.org/feature/data/scope/keystone1/ [accessed 18 July 2006].

36. "1. Define the question; 2. Gather information and resources; 3. Form hypothesis; 4. Perform experiment and collect data; 5. Analyze data; 6. Interpret data

and draw conclusions that serve as a starting point for new hypotheses; 7. Publish results.": http://en.wikipedia.org/wiki/Scientific_method [accessed 18 July 2006].

37. http://www2.selu.edu/Academics/Education/EDF600/Mod3/sld001.htm [accessed 18 July 2006].

38. Thomas Henry Huxley, "On the Educational Value of the Natural History Sciences," in idem, *Collected Essays,* vol. 3, *Science and Education: Essays* (New York: D. Appleton, 1900; art. orig. publ. 1854), pp. 38–65, on p. 45; Peter B. Medawar, *The Art of the Soluble* (London: Methuen, 1967), p. 132.

39. For example, John Dupré, *The Disorder of Things: Metaphysical Foundations of the Disunity of Science* (Cambridge, MA: Harvard University Press, 1993); Alexander Rosenberg, *Instrumental Biology, or the Disunity of Science* (Chicago: University of Chicago Press, 1994); Nancy Cartwright, *The Dappled World: A Study of the Boundaries of Science* (Cambridge: Cambridge University Press, 1999); see also Peter Galison and David J. Stump, eds., *The Disunity of Science: Boundaries, Contexts, and Power* (Stanford, CA: Stanford University Press, 1996).

40. Edward Teller, "Back to the Laboratories," *Bulletin of the Atomic Scientists* 6, no. 3 (March 1950): 71–72.

41. See, for example, J. Robert Oppenheimer, "Communication and Comprehension of Scientific Knowledge," in Melvin Calvin et al., *The Scientific Endeavor: Centennial Celebration of the National Academy of Sciences* (New York: The Rockefeller University Press, 1965), pp. 271–279, on p. 272. On the normalization of the scientist, see Steven Shapin, *The Scientific Life: A Moral History of a Late Modern Vocation* (Chicago: University of Chicago Press, 2008), chapter 3.

42. Stephen Jay Gould, "Nonoverlapping Magisteria—Evolution versus Creationism," *Natural History* 106, no. 2 (March 1997): 16–25.

43. For example, Ralph E. Lapp, *The New Priesthood: The Scientific Elite and the Uses of Power* (New York: Harper & Row, 1965); Spencer Klaw, *The New Brahmins: Scientific Life in America* (New York: William Morrow, 1968).

44. Quoted in Henry S. Hall, "Scientists and Politicians," in *The Sociology of Science,* ed. Bernard Barber and Walter Hirsch (New York: Free Press, 1962; art. orig. publ. in *Bulletin of the Atomic Scientists* [February 1956]), pp. 269–287, on pp. 270–271.

45. See, for example, Peter Dear, *Discipline and Experience: The Mathematical Way in the Scientific Revolution* (Chicago: University of Chicago Press, 1995); Robert S. Westman, "The Astronomer's Role in the Sixteenth Century: A Preliminary Study," *History of Science* 18 (1980): 105–147.

46. The material in this and the next several paragraphs is included in Shapin, *The Scientific Life,* chapters 2–3.

47. Henry A. Rowland, "The Highest Aim of the Physicist, Presidential Address Delivered at the Second Meeting of the Society, on October 28, 1899," *Bulletin of the American Physical Society* 1 (1899): 4–16, on p. 13; also in *Science* n.s. 10, no.

258 (8 December 1899): 825–833. For a pertinent Hopkins context to Rowland's remarks, see Maryann Feldman and Pierre Desrochers, "Truth for Its Own Sake: Academic Culture and Technology Transfer at Johns Hopkins University," *Minerva* 42 (2004): 105–126, esp. pp. 117–118.

48. Albert Einstein, "Scientific Truth [1929]," in idem, *Ideas and Opinions* (New York: Crown Publishers, 1954), pp. 261–262; for Einstein's early operationalism, influenced by Mach, see Gerald Holton, "Mach, Einstein, and the Search for Reality," in *The Twentieth-Century Sciences: Studies in the Biography of Ideas,* ed. Gerald Holton (New York: W. W. Norton, 1972), pp. 344–381.

49. C. P. Snow, "Address by Charles P. Snow [to Annual Meeting of American Association for the Advancement of Science]," *Science* n.s. 133, no. 3448 (27 January 1961): 256–259, on p. 257.

50. Pion and Lipsey, "Public Attitudes to Science and Technology," pp. 308 (Table IV) and 309.

INDEX

abstinence, 242, 244, 248, 302; heroic, 245–247; of women, 246–247. *See also* asceticism

academic science, 212–213, 229; as distinguished from industrial science, 229–233; versus the humanities, 379

Addison, Joseph, 168

Administrative Science Quarterly (*ASQ*), 214, 231

Aesop, 38–39

AIDS vaccines, 24

air-pump, Boyle's experiments involving, 89–90; drawing of, 93, 98–99; language used to describe, 106–108; material technology of, 92–94; and matters of fact, 113–114; replication of, 96–97; satire of, 172

alchemists, 96, 105, 114

Alexander the Great, 159, 241

aliments: Cheyne's advice regarding, 297–298. *See also* dietetics

Allen, Woody, 1, 2

Allestree, Richard, 163

American Cyanamid, 224

Ames test, 24

Anaxagoras, 139

Anthony, St., 242–243

antiscience, 41–46. *See also* metascience

Apollo, 291

Apollonius, 249

Arendt, Hannah, 125, 140

argon (Ar), 23, 25

Aristotle, 49, 124, 151, 159, 251, 328, 331, 357

Arrowsmith (Lewis), 255

Arundel House, 66, 73

asceticism, dietary, 235; and its Christian tropes, 242–247; and its classical tropes, 240–242; dangers of, 272; Jewish traditions of, 244; and spiritual knowledge, 242–243. *See also* dietetics

Ascham, Roger, 146, 155

Ashmolean Museum, 63

AT&T, 221

Atkinson, Paul, 344

Aubrey, John, 69, 161, 182, 187, 196, 201, 260, 275

Augustine, St., 244

authority: Descartes as, 372–373; of physicians, 292, 304–310; of scientific knowledge, 26–28, 386–391

Bacon, Francis, 17, 48, 49, 63, 103, 134, 155, 277, 351; *Advancement of Learning*, 159–160; and dietary moderation, 249, 270–271; on medical dietetics, 273–274; on medical practice, 276, 293, 353–354; reforms in learning proposed by, 129–130; on truth, 207–208

Bacon, Roger, 126

Baillet, Adrien, 363, 364

Barber, Bernard, 213, 214

Barnes, Barry, 28–29, 231

barometers, 22–23

Bauman, Zygmunt, 26, 28

Bechler, Zev, 134–135

Bell Labs, 222, 225

Bennett, J. A., 193, 201

Berlin, Isaiah, 54

Bernard, Claude, 9

Biagioli, Mario, 152
bloodletting. *See* phlebotomy
Bloor, David, 18, 19
Boas, Marie. *See* Hall, Marie Boas
Bohr, Niels, 33
Borelli, Giovanni Alfonso, 293
Boulliau, Ismael, 81
Box, Steven, 231
Boyle, Robert, 17, 22–23, 49, 52, 84, 177,
    180, 239, 293; accessibility of, 69–71,
    72–73, 188; air-pump experiments of,
    89–90, 92–94, 97, 98–99, 172; and
    anomalous suspension, 86; *Certain
    Physiological Essays,* 96; as Christian
    gentleman, 163–164, 165, 197–199,
    211; *The Christian Virtuoso,* 197–199;
    Continuation of New Experiments, 97,
    98–99; on credibility, 62; critics of, 108;
    eulogy for, 164–165, 251; experimental
    work sites of, 60, 64–65, 68, 69–70;
    experimental work questioned, 81–82;
    on fact as aspect of scientific knowl-
    edge, 102–103, 108–109; fragile health
    of, 251; *History of Colours,* 96, 97;
    Hobbes as critic of, 105, 108, 110–112;
    and Robert Hooke, 67, 78, 182, 184,
    192–193, 190; importance of solitude
    to, 131–132, 136; literary technology
    of, 91–97, 115–116; medical views
    of, 354; modesty of, 101–103; *New
    Experiments Physico-Mechanical,* 89,
    96, 97, 98, 99, 108, 112; *A Proëmial
    Essay,* 108; and the Royal Society, 162–
    163; *The Sceptical Chymist,* 105, 109,
    110; *Seraphick Love,* 132; technicians
    and assistants of, 77–78; and visual
    representations as text, 98–99; writings
    of, 96–103, 108–110, 115–116
Brannigan, Augustine, 20
Brathwait, Richard, 129, 163, 269
Brewster, David, 134
bricolage, 143
Bridgman, Percy, 37
Brillat-Savarin, Jean Anthelme, 235
Britaine, William de, 269
Brougham, Henry, 9
Brouncker, William, 84, 188, 190, 194
Brown, Peter, 242–243
Browne, Thomas, 141, 163, 275, 316
Bryskett, Lodowick, 155, 269

Bryson, Anna, 264
Burghley, Lord, 267, 268
Burke, Edmund, 55
Burke, Kenneth, 324, 330
Burman, Frans, 358, 363, 364
Burnet, Gilbert, 69, 70, 132, 164–165,
    251, 262
Burns, Robert, 142
Burton, Robert, 128, 152, 155, 250, 272
Butler, Samuel, 128, 166; satires of, 156,
    171–172, 176
Butterfield, Herbert, 378
Bynum, Caroline, 247

caffeine: risks of, 24–25
Carlyle, Thomas, 180
Carneades: as mouthpiece for Boyle, 105,
    109–110
Carothers, Wallace, 220–221
Carr, E. H., 13
Carrier Corporation, 230
Cartwright, Nancy, 7
Carty, John J., 221
Castiglione, Baldassare, 147, 148, 153,
    265, 269
Catherine of Siena, St., 246
Cato the Censor, 278
Cavendish, Henry, 239, 254
Cecil, William, 150
celibacy, 344. *See also* asceticism
Celsus, Rule of, 273, 274–275, 277, 286,
    290–291, 369
Chamberlain, Lesley, 1–2
Chamberlayne, Edward, 207
Chargaff, Erwin, 33, 44
Charleton, Walter, 251
Charron, Peter, 271
Chesterfield, Lord, 168–169, 294
Cheyne, George, 253, 270, 271; common-
    sense advice of, 304–305; dietary advice
    of, 282–283, 294–296, 302–303, 304,
    312; drugs prescribed by, 301; exercise
    prescribed by, 297, 301; Lady Hunting-
    don as patient of, 294, 295, 300, 301–
    304, 305, 306, 307, 308–309, 310;
    medical expertise of, 290, 296–300,
    304–310; obesity of, 307–308; patients
    of, 294, 295, 300–304; patients' trust
    in, 304–310; Richardson as patient of,
    300–303, 304, 305, 306, 307, 308, 309,

fact: as aspect of scientific knowledge, 90–91; Boyle on, 102–103, 108–109; debate over, 108–113; as distinguished from artifact, 113; as distinguished from theory, 107–108; as established by multiple witnesses, 91, 94–98; establishment of, 91–92, 113–116

Faraday, Michael, 136

fasting: in the Christian tradition, 246–247. *See also* abstinence

Feinstein, Alvan, 24

Feisenberger, H. A., 205

Feuerbach, Ludwig, 235, 242

Feyerabend, Paul, 386

Ficino, Marsilio, 250

Fieser, Louis, 224

Finch, John, 252

Firth, Raymond, 331

Fisch, Harold, 132

Fisk, James, 222

Flamsteed, John, 177

Fleck, Ludwik, 115–116

Fleischmann, Martin, 21

Florio, John, 279

Fodor, Jerry, 7

Foucault, Michel, 261

Fox, George, 120

Francis of Assisi, St., 246

Franklin, Benjamin, 51

French, Roger, 357

Fuller, Thomas, 321

Gailhard, Jean, 263, 266

Gainsford, Thomas, 262, 269

Galen, 248, 261

Galenic medicine, 261, 269–270, 279, 283, 352, 367, 372

Galenic non-naturals. *See* non-naturals

Galileo, 152, 287–288, 349

Galison, Peter, 7

Gambler's Fallacy, 348

Gassendi, Pierre, 103, 293

Gay, John, 294

Geertz, Clifford, 348

Gellner, Ernest, 375

General Electric (GE), 220, 224, 227

General Mills, 226

General Motors, 223

gentility. *See* gentlemen

gentlemen: conduct expected of, 153; and

dietary moderation, 262–265, 284–286; as distinguished from scholars, 128–129; and medical care, 283–284; and new scientific practices, 169–178; as participants in experimental pursuits, 79–83; proper education of, 146–150, 157–160, 178–181; as scholars, 164–169. *See also* Christian virtuoso; courtesy books

Gibbon, Edward, 50, 121

Giddens, Anthony, 30

Gigerenzer, Gerd, 347, 348

Gillispie, Charles, 379

Glanvill, Joseph, 79, 82, 161, 162, 163, 164

gluttony, 264, 291

God, belief in, 381–382. *See also* religion

Goddard, Jonathan, 187, 190

Goffman, Erving, 137, 322

Golden Mean, dietary, 265–266, 273, 269, 270, 273–275, 285, 291–292, 304

Gooding, David, 136

Goody, Jack, 340

Gouldner, Alvin, 214–215

governmental research facilities, 215

Gray, Thomas, 120

Greek philosophers. *See* classical philosophers

Gresham, Thomas, 66

Gresham College, 66, 67; Hooke at, 183–184, 185–187

Grew, Nehemiah, 187

Gross, Paul, 43

Guazzo, Stefano, 129, 146–147, 148, 153, 265

Guerrini, Anita, 294

Gwyn, Nell, 65

Haak, Theodore, 187, 188, 190, 200

Habermas, Jürgen, 380

Hacking, Ian, 90

Hagstrom, Warren, 213, 214

Hall, A. Rupert, 6

Hall, Marie Boas, 6, 71

Hall, T. S., 357

Halley, Edmond, 237

Hartlib, Samuel, 64, 186

Harvey, Gabriel, 148

Harvey, William, 353

Haskins, Charles Homer, 4

Helmont, Francis Mercury van, 65
Henshaw, Thomas, 64, 188
Herbert, Lord, 157
Herzig, Rebecca, 10
heuristics, 319–320, 346–348
Hevelius, Johannes, 194, 202
Hill, Abraham, 187, 188
Hirsch, Walter, 213
historicity: of science, 5, 6–8; of Scripture, 6
Hobbes, Thomas, 27, 74, 104, 177, 275–276, 352–353; Boyle's view of, 105; as critic of Boyle, 105, 108, 110–112; on Scholastic universities, 158
Holder, William, 188
Hollinger, David, 12
Holmes, Richard, 4
Hooke, Grace, 202, 203
Hooke, Robert, 177; and Boyle's air-pump, 92, 95; as Boyle's assistant, 78, 182, 192–193; as curator of experiments for the Royal Society, 182, 185, 186, 190, 191–192, 193–194, 209; a day in the life of, 183–185; diary of, 188, 190–191, 205–206; experimental performances of, 85–87; experimental testimony of, 209–211; experimental work sites of, 60, 67–68, 77, 83–84; friends and acquaintances of, 187–188, 190–191; household of, 187; identity of, 182–183, 208–209, 211; and Newton, 134–135, 195; and Oldenburg, 191, 193, 194–195, 200, 202; patents and inventions of, 200–202; perceived failures of, 209–211; private life of, 203–204; private world of, 185–190; published works of, 192, 204; religious perspective of, 204–206; as servant to other practitioners of science, 191–195; and technicians, 195–196; as tradesman in deportment, 202–203; wealth of, 200; work habits of, 83–84
Hoskins, John, 187, 188, 191
Houghton, Walter, 174
Hounshell, David A., 220, 224
Howard, Henry, 66
Howard, Thomas, 66
human body: Aristotelian view of, 352; Cheyne's conception of, 296–300; and the emotions, 364–367; fluids of, 296, 297, 367; micromechanical account of, 289, 293–300, 312. *See also* dietetics; Galenic medicine; medical expertise
humanism, 260
humanist writers, 146–147, 159; dietetics of, 250
Hume, David, 50, 256, 317
Humphrey, Lawrence, 146
Hunt, Henry, 196, 200, 209
Huntingdon, Lord, 310
Huntingdon, Selina, Countess of: as Cheyne's patient, 294, 295, 301–304
Huxley, T. H., 9, 36, 349, 350, 386
Huygens, Christiaan, 81, 86, 93, 113, 201, 202, 209
Huygens, Constantijn, 359, 360

iatromathematics, 293, 312
iatromechanism, 293, 312; Cheyne's conception of, 296–300
Idea of Science, 386–387
Industrial Research Institute, 218
industrial scientists, 212–233; autonomy of, 220–224; in the corporate environment, 215–233; as distinguished from academic scientists, 229–233; internal conflict experienced by, 213–215; journals focusing on, 217–218; publication by, 227–228; scientific values of, 212–220, 228–229
intellectual property, 213; in the corporate environment, 226; secrecy of, 228
Intelligent Design, 384
intercontinental ballistic missiles, 23–24

Jacob, J. R., 132
James VI, King (later James I of England), 262, 264; dietary advice of, 266–267, 277–278
James, Susan, 366–367
James, William, 317
Jasanoff, Sheila, 26
Jefferson, Thomas, 225
Jerome, St., 126
Jesus, asceticism of, 242
Jewett, Frank, 221
John the Baptist, 120
Johnson, Samuel, 121, 168, 175, 178, 294, 328

Jones, Richard, 96
Joubert, Laurent, 267, 275, 281, 282, 292

Kafka, Franz, 255
Kahneman, Daniel, 347, 348
Kant, Immanuel, 1
Kaplan, Norman, 231
Kearney, Hugh, 148
Keats, John, 120
Keegan, John, 4
Kettering, Charles "Boss," 223, 226
Keynes, Geoffrey, 191
King, William, 172, 175
knowledge: conceptions of, 257–258; vulgar, 315–316. See also scientific knowledge
Kornhauser, William, 214
Koyré, Alexandre, 6
Kuhn, Thomas, 6, 7, 36, 39, 84, 133, 215, 342, 386

laboratories, 63. See also experimental work sites
Lakatos, Imre, 386
Langmuir, Irving, 220
La Rochefoucauld, duc de, 265, 276
Latimer, Hugh, 148
Latour, Bruno, 23, 113, 324, 328
Lawrence Livermore National Laboratory, 227
Lazarsfeld, Paul, 232–233
Lear, King, 17–18
learning: distempers of, 159–160. See also education; pedantry; Scholasticism
Leermakers, John, 222
Leeuwenhoek, Antoni van, 53, 172
Leibniz, Gottfried, 50, 158, 203
Lessius, Leonard, 282
Letters Recommendatory, 71
Levine, Joseph, 173–174
Levitt, Norman, 43
Lewis, Sinclair, 255
Lewontin, Richard C., 33, 44
Life of Brian, 2
Lindeboom, G. A., 360
Lingard, Richard, 166, 268–269
Linus, Franciscus, 105, 108, 110–11
Lister, Martin, 177, 210
literary technology, 91, 113; as employed

by Boyle, 91–103, 115–116; function of, 114–115
Little, Arthur D., 224
Livermore. See Lawrence Livermore National Laboratory
Locke, John, 49, 52, 61, 278; on learning, 169–170, 263, 270
Lodwick, Francis, 188
logic, study of, 157
longevity, 297–298; Descartes' reflections on, 370–371; as goal of medicine, 354
LSD: risks of, 24–25
Luhmann, Niklas, 30
Lynch, Michael, 20

Macarius, Abba, 243
MacKenzie, Donald, 23
Mackenzie, George, 163
Mackenzie, James, 276
Magalotti, Lorenzo, 70, 73, 75–76, 85
Maimonides, Moses, 244
Malcolm, Norman, 238
Marcson, Simon, 214
marijuana: risks of, 24–25
material technology, 91, 92, 113
mathematical knowledge, 133, 389
Mather, Cotton, 165
Mayow, John, 195
McGuire, J. E., 7
Mead, Richard, 294
Medawar, Peter, 33, 38, 42
medical culture: and diet, 271–276, 283–286; moral aspect of, 260–262; and Rule of Celsus, 273, 274, 277; skepticism regarding, 276–283, 286, 351–354, 361–362
medical expertise: Cheyne's claims to, 290, 296–300; and common sense, 289, 304–305; Descartes' views of, 354–373; in early modern England, 289–290; laypeople's familiarity with, 290–291; and micromechanism, 293–296; relevant, 313–314; trust as basis for, 304–310, 312; uncertainty as aspect of, 371–372
medical texts, 259. See also Cheyne, George
Mees, Kenneth, 222, 223, 224–226, 228–229
mercury, 299, 301

Mersenne, Marin, 355–356, 359, 370, 371

Merton, Robert K., 12, 51–52, 144, 145, 213, 214, 218

metascience, 32–46; claims of, 32–34, 45–46. *See also* science; scientific method; scientists

metonymy, 22–23

Milton, John, 140, 156, 157

mineral waters, 362

Minneapolis-Honeywell, 221, 225

mock-heroic poems, 171–172

moderation, dietary: Cheyne as advocate of, 302–303; and gentlemanly health, 245, 246–247, 262–265; as medically beneficial, 271–276; as practiced by Descartes, 363–364; as virtue, 261–262, 265–271. *See also* dietetics; temperance

modernity: and science, 375, 377–391

Mohammed, 119, 120, 121

Moliére, 373

Molyneux, Thomas, 203

monasticism, 126

Monconys, Balthasar de, 74, 75

Montaigne, 128, 140, 155, 208, 248–249, 275, 278–279, 286, 290, 317, 372; "Of Experience," 279–281

Montesquieu, 50

Montmor, Henri-Louis Habert de, 81

Monty Python, 2

Moore, G. E., 387–388

Moore, Jonas, 187, 188, 191

moral virtue: medical aspect of, 260–262. *See also* virtue

Moran, Bruce, 152

Moray, Robert, 193, 203

More, Henry, 81, 108, 110, 239, 251–253, 254

More, L. T., 254

More, Thomas, 146

Moses, 119

Murphy, Edward A., 336

Muses, 240

Namier, Lewis, 9

Napoleon, 278

Nash, Beau, 294

National Cash Register, 223

National Endowment for the Humanities, 379–380

National Institutes of Health, 379–380

National Science Foundation, 379–380

Naturalistic Fallacy, 3, 11, 388

Nazi Germany, 51–52

New Historicists, 7

new scientific practices: charged with pedantry, 169–178; criticism of, 174–175; legitimacy of in seventeenth-century England, 144–145; satires of, 171–173; solitude as essential for, 130–132; value of, 169–170

Newman, Cardinal, 180

Newton, Isaac, 1, 8, 10, 122, 177, 195; dietary habits of, 237, 239, 253–254, 256–257; and Hooke, 134–135, 195, 202; importance of solitude to, 122, 133–134, 180; rational principles proposed by, 293

Newtonian principles, 293

Nicolson, Marjorie Hope, 170

Nietzsche, Friedrich, 258; diet of, 1–2

nobility: and virtue, 149. *See also* gentlemen

Noble, David, 223, 344

non-naturals, 269–270, 279, 283, 291, 367

nuclear weapons, 227

nutrition. *See* aliments; dietetics

Office of Naval Research (ONR), 215

Oldenburg, Henry, 65, 81–82, 104, 133, 135, 164; and Hooke, 191, 193, 194–195, 200, 202

Opie, Iona, 323

Opie, Peter, 323

Oppenheimer, J. Robert, 35, 233, 349–350

Organization Man, The (Whyte), 216–217, 226

Orth, Charles, 215–216

Osler, William, 345

Outram, Dorinda, 122–123

Owens-Illinois Glass Company, 221

Oxford Ashmolean Museum, 63

Papin, Denis, 78, 191, 195

Parmenides, 249

Pascal, Blaise, 22, 361

Pasteur, Louis, 23

patents and inventions, 213, 228

Paul (apostle), 119, 242

Peacham, Henry, 66, 147, 189, 202, 262, 264, 270, 272, 278
pedant, 162; attributes of, 154–155. *See also* scholar
pedantry, 154–155, 157–160, 164–169; and the new scientific practices, 169–178
Penn, William, 121
Pepys, Samuel, 260
Petley, Brian, 33
Petty, William, 64, 131, 190
philosophers: Christian views of, 125–127; classical views of, 124–125, 139–140, 151, 240–241; dietetics of, 251–254; ill health of, 251–254; and proverbs, 316–317. *See also* experimental philosophers
phlebotomy (bloodletting), 298–299, 300, 302, 362
physicians. *See* Cheyne, George; medical culture; medical expertise
Picot, Claude, 360
Pinch, Trevor, 23, 25, 26
Pitcairne, Archibald, 293
Planck, Max, 35, 36, 40, 349
Plato, 128, 240, 248, 249
Plato's Academy, 53
plenism, 106, 107
Plotinus, 242, 268
pneumatics. *See* air-pump
Polanyi, Michael, 224
Polya, George, 346
Pope, Alexander, 294
Pope, Walter, 187
Popper, Karl, 36, 386
Porphyry, 242
Porter, Roy, 264
Porter, Theodore, 29, 30
Power, Henry, 93, 186
pregnancy, 303
prejudice: and knowledge, 47, 52, 53–56; methodology as remedy for, 48–49
Priestley, Joseph, 51, 179
proverbs, 287; and common sense, 317, 318, 319, 348–350; and computer programming, 345–346; criteria for, 322–329; crossed, 339–341; definitions of, 319, 320–321; as deliberation, 325; flawed logic in, 338–341; as heuristic, 346–348; as inference instructors, 333–336; metaphor as used in, 315–316,

322–323, 325–327, 332; mnemonic robustness of, 323–324, 343–344; in oral culture, 321; and proverbial economies, 319, 329–333, 349–350; as reflective knowledge, 319, 333–336; reiteration of, 326–327; relating to health, 291, 292, 345; relativism of, 319, 336–338; as rule-like propositions, 329–330; scholarly study of, 330–331, 346–350; in scientific contexts, 343–346; truth of, 332–333, 338–341; as viewed by philosophers and scholars, 316–318; virtues of, 341–346
Pumfrey, Stephen, 210
Puritanism, 260
Pythagoras, 140, 241
Pythagorean asceticism, 241–242, 249

quackery, 284

Raleigh, Walter, 120
Rambler, 178
Ramesey, William, 163
RAND Corporation, 223
Ranelagh, Katherine, Lady, 65, 67, 68, 69; Hooke as guest and employee of, 189, 191
rationalism: as aspect of Descartes' dietetics and therapies of, 369–370, 372–373; and Descartes' pursuit of medical knowledge, 354–358
rational methods: as guard against prejudice, 48–49
Rattansi, P. M., 7
Ravetz, J. R., 7
RCA, 218, 227–228
Reich, Leonard, 228
relativism, 34–35, 43–44; of proverbs, 319, 336–338
religion, 2; poll of Americans' beliefs regarding, 381–383; science as, 2–3; science in conflict with, 381–384; as supportive of scientific practices, 163–164. *See also* Christianity; Christian thought; Christian virtuoso; courtesy literature
repertoires, 143
research management. *See* industrial scientists
respiration: Hooke's experimental work on, 210–211

rhetoric, 17; Boyle's use of, 99–103; of solitude, 138. *See also* literary technology
Richards, Graham, 27
Richardson, Samuel: as Cheyne's patient, 294, 295, 300–303
Roberval, Gilles, 81
Rorty, Richard, 317
Rosenberg, Alexander, 7
Rousseau, George, 264
Rousseau, Jean-Jacques, 121
Rowland, Henry, 389–390
Royal Society of London, 53; and access to experimental sites, 71–72, 73–74; as community of gentleman-scholars, 160–164; decorum at, 74–77; descriptions of, 74–75; and experimental knowledge, 114; experimental work sites of, 60, 62, 63, 65–66, 77; history of, 130–131; Hooke as curator of experiments for, 185, 186, 190, 191–192, 193–194, 209; and Newton's discoveries, 134–135; purpose of, 112; rituals of, 76; satires on, 170–173
Rule of St. Benedict, 246
rules of thumb, 322. *See also* common sense; proverbs

Sacks, Harvey, 323, 324, 328, 329, 330
Sarton, George, 3–4, 378
Schmitt, Charles, 7
scholar: attributes of, 151–154; in civil society, 127–129; definitions of, 150–151; as distinguished from the gentleman, 128–129; as gentleman, 164–169; melancholy as companion of, 128, 153–154, 251; as object of derision, 154–157; and public affairs, 129–130
Scholasticism: criticism of, 157–158
science: civic success of, 13–14; and common sense, 287–288, 349–350; conceptual unification of, 37, 40; in conflict with religion, 381–384; financial support for, 379–380; future of, 391; growth of, 380–381; heresies against authority of, 4–6; historicity of, 5, 6–8; internal criticisms of, 44–45; in the modern world, 375, 377–391; moral authority of, 3–4, 10–11; naturalism in study of, 15; as the new religion, 2–3; place of, 57; proverbs as applied in, 343–

346; and skepticism, 19; social dimension of, 43; as socially constructed, 34–35; sociology of, 51, 213–217, 229–233; specialized knowledge as basis for authority in, 26–27; trust as condition for, 43, 52–53, 202, 211; unity of, 7. *See also* industrial scientists; metascience
science, history of: aim of, 14; hagiographical conception of, 117; identity of, 7–8; lowering of tone in, 1–14; Sarton's view of, 3–4, 378; and the Scientific Revolution, 378–379
Science Wars, 32–34. *See also* metascience
scientific community: discourse within, 103–104; dispute and controversy within, 108–113; as egalitarianism ideal, 50–51; exclusions from, 53–54; language as used by, 114–116; linguistic boundaries of, 104–106; linguistic boundaries within, 106–108, 115; prejudice within, 47, 52, 53–56
scientific culture: support for, 179–180
scientific experimentation: condition of gentlemen involved in, 79–83; replication of, 96–97, 209; showing of, 85–87; trying of, 82, 83, 84–85. *See also* experimental work sites
scientific inquiry: nobility of, 172
scientific knowledge: access to, 94; authority of, 26–28, 386–391; and credibility, 18–31; British attitudes toward, 165–166; empiricist conceptions of, 60–62; foundations of, 90; man-made nature of, 115; mathematical, 133; as performance, 5; and prejudice, 52, 53–56; probabilistic conception of, 90–91; respecification of, 164, 178–181; and scientific method, 117; solitude as setting for, 122–124, 132–137, 138–141; threshold of, 60–62; and trust, 43, 52–53, 82, 202, 211; as truth, 5, 11, 20, 35, 206–208, 389–390; validation of, 89–91, 97–98, 114; and virtue, 47–48, 163–164; witnessing as aspect of, 60–61, 94–103. *See also* experimental knowledge
scientific method, 5, 7, 131, 287; and scientific knowledge, 117; varying views of, 32, 36–37, 42–44, 379, 385–386, 387
scientific research: autonomy of, 220–224,